PERSPECTIVES IN CARBONATE GEOLOGY

Other publications of the International Association of Sedimentologists

SPECIAL PUBLICATIONS

40 **Analogue and Numerical Modelling of Sedimentary Systems**
From Understanding to Prediction
Edited by P. de Boer, G. Postma, K. van der Zwan, P. Burgess and P. Kukla
2008, 336 pages, 172 illustrations

39 **Glacial Sedimentary Processes and Products**
Edited by M.J. Hambrey, P. Christoffersen, N.F. Glasser and B. Hubbard
2007, 416 pages, 181 illustrations

38 **Sedimentary Processes, Environments and Basins**
A Tribute to Peter Friend
Edited by G. Nichols, E. Williams and C. Paola
2007, 648 pages, 329 illustrations

37 **Continental Margin Sedimentation**
From Sediment Transport to Sequence Stratigraphy
Edited by C.A. Nittrouer, J.A. Austin, M.E. Field, J.H. Kravitz, J.P.M. Syvitski and P.L. Wiberg
2007, 549 pages, 178 illustrations

36 **Braided Rivers**
Process, Deposits, Ecology and Management
Edited by G.H. Sambrook Smith, J.L. Best, C.S. Bristow and G.E. Petts
2006, 390 pages, 197 illustrations

35 **Fluvial Sedimentology VII**
Edited by M.D. Blum, S.B. Marriott and S.F. Leclair
2005, 589 pages, 319 illustrations

34 **Clay Mineral Cements in Sandstones**
Edited by R.H. Worden and S. Morad
2003, 512 pages, 246 illustrations

33 **Precambrian Sedimentary Environments**
A Modern Approach to Ancient Depositional Systems
Edited by W. Altermann and P.L. Corcoran
2002, 464 pages, 194 illustrations

32 **Flood and Megaflood Processes and Deposits**
Recent and Ancient Examples
Edited by I.P. Martini, V.R. Baker and G. Garzón
2002, 320 pages, 281 illustrations

31 **Particulate Gravity Currents**
Edited by W.D. McCaffrey, B.C. Kneller and J. Peakall
2001, 320 pages, 222 illustrations

30 **Volcaniclastic Sedimentation in Lacustrine Settings**
Edited by J.D.L. White and N.R. Riggs
2001, 312 pages, 155 illustrations

29 **Quartz Cementation in Sandstones**
Edited by R.H. Worden and S. Morad
2000, 352 pages, 231 illustrations

28 **Fluvial Sedimentology VI**
Edited by N.D. Smith and J. Rogers
1999, 328 pages, 280 illustrations

27 **Palaeoweathering, Palaeosurfaces and Related Continental Deposits**
Edited by M. Thiry and R. Simon Coinçon
1999, 408 pages, 238 illustrations

26 **Carbonate Cementation in Sandstones**
Edited by S. Morad
1998, 576 pages, 297 illustrations

25 **Reefs and Carbonate Platforms in the Pacific and Indian Oceans**
Edited by G.F. Camoin and P.J. Davies
1998, 336 pages, 170 illustrations

24 **Tidal Signatures in Modern and Ancient Sediments**
Edited by B.W. Flemming and A. Bartholomä
1995, 368 pages, 259 illustrations

23 **Carbonate Mud-mounds**
Their Origin and Evolution
Edited by C.L.V. Monty, D.W.J. Bosence, P.H. Bridges and B.R. Pratt
1995, 543 pages, 330 illustrations

16 **Aeolian Sediments**
Ancient and Modern
Edited by K. Pye and N. Lancaster
1993, 175 pages, 116 illustrations

3 **The Seaward Margin of Belize Barrier and Atoll Reefs**
Edited by N.P. James and R.N. Ginsburg
1980, 203 pages, 110 illustrations

1 **Pelagic Sediments on Land and Under the Sea**
Edited by K.J. Hsu and H.C. Jenkyns
1975, 448 pages, 200 illustrations

REPRINT SERIES

4 **Sandstone Diagenesis: Recent and Ancient**
Edited by S.D. Burley and R.H. Worden
2003, 648 pages, 223 illustrations

3 **Deep-water Turbidite Systems**
Edited by D.A.V. Stow
1992, 479 pages, 278 illustrations

2 **Calcretes**
Edited by V.P. Wright and M.E. Tucker
1991, 360 pages, 190 illustrations

Special Publication Number 41 of the International Association of Sedimentologists

Perspectives in Carbonate Geology: a Tribute to the Career of Robert Nathan Ginsburg

EDITED BY

Peter K. Swart, Gregor P. Eberli and Judith A. McKenzie

SERIES EDITOR

Ian Jarvis
School of Geography, Geology and the Environment
Centre for Earth and Environmental Science Research
Kingston University, London
Penrhyn Road
Kingston upon Thames
KT1 2EE
UK

SERIES CO-EDITOR

Tom Stevens
Department of Geography
Royal Holloway, University of London
Egham, Surrey
TW20 0EX
UK

WILEY-BLACKWELL
A John Wiley & Sons, Ltd., Publication

This edition first published 2009, © 2009 by International Association of Sedimentologists

Blackwell Publishing was acquired by John Wiley & Sons in February 2007. Blackwell's publishing program has been merged with Wiley's global Scientific, Technical and Medical business to form Wiley-Blackwell.

Registered office
John Wiley & Sons Ltd, The Atrium, Southern Gate, Chichester, West Sussex, PO19 8SQ, UK

Editorial offices
9600 Garsington Road, Oxford, OX4 2DQ, UK
The Atrium, Southern Gate, Chichester, West Sussex, PO19 8SQ, UK
111 River Street, Hoboken, NJ 07030-5774, USA

For details of our global editorial offices, for customer services and for information about how to apply for permission to reuse the copyright material in this book please see our website at www.wiley.com/wiley-blackwell

Library of Congress Cataloguing-in-Publication Data

Perspectives in carbonate geology : a tribute to the career of Robert Nathan Ginsburg /
edited by Peter Swart, Gregor Eberli, Judith McKenzie.
p. cm. – (International association of sedimentologists special publications ; 41)
Includes bibliographical references and index.
ISBN 978-1-4051-9380-1 (hardback : alk. paper)
1. Ginsburg, Robert N. 2. Rocks, Carbonate. 3. Sediments (Geology) I. Swart, Peter K.
II. Ginsburg, Robert N. III. Eberli, Gregor Paul, 1956- IV. McKenzie, J.A. (Judith A.)

QE471.2.P46 2009
552'.58--dc22

2008042551

A catalogue record for this book is available from the British Library.

Set in 10/12 pt Melior by Newgen Imaging Systems (P) Ltd, Chennai, India
Printed and bound in Malaysia by Vivar Printing Sdn Bhd

1 2009

Contents

Preface

Robert Ginsburg in characteristic South Florida field gear.

Robert Nathan Ginsburg's career in carbonate sedimentology began in 1950 when he left the University of Chicago to become a research assistant at the University of Miami's Marine Laboratory, the precursor of the present Rosenstiel School of Marine and Atmospheric Science. Subsequently he moved, first to establish and lead a research and training programme on Quaternary carbonates for the Shell Development Company (1954–60), then to become Professor of Geology and Oceanography at The Johns Hopkins University (1960–65). In 1970 he was persuaded by Cesare Emiliani to come back to the University of Miami as Professor of Sedimentology. At that time he organized the T. Wayland Vaughan Laboratory for Comparative Sedimentology headquartered on ocean-facing Fisher Island at the entrance to the Port of Miami.

His first published paper appeared soon after his initial arrival in Miami, 'Intertidal Erosion on the Florida Keys' (1953). It was a harbinger of his future career as it questioned the prevailing chemical explanation for shoreline erosion by offering a biological alternative. In the following half century, with his associates, post-doctoral fellows and students he has authored or co-authored a series of seminal papers, books and reports on the links between contemporary and Holocene processes and products of carbonate deposition and their fossil counterparts. These publications have been on subjects ranging from the formation of dolomite (Shinn *et al.*, 1965), precipitation of cements in reefs (Ginsburg & James, 1976; James *et al.*, 1976), health of coral reefs (Ginsburg, 1997; Ginsburg *et al.*, 2001), sedimentation patterns on carbonate platforms (Beach & Ginsburg, 1980; Ginsburg, 2005), stromatolites (Logan *et al.*, 1964) and the history and development of carbonate platforms (Eberli & Ginsburg, 1987a,b; Ginsburg, 2001; Ginsburg *et al.*, 1991; Schlager & Ginsburg, 1981). Within this corpus of contributions are an editorial exhortation, 'So What', to develop the wider implications of our specific findings (Ginsburg, 1982), a paper emphasizing the feedback of sediments on their deposition 'Disobedient Sediments' (Ginsburg, 2005) and what is termed the Ginsburg Model of autocyclic accumulation of shoaling-upward successions (Ginsburg, 1971).

These works, combined with the several issues of two informal periodicals he founded and edited (*Sedimenta* and *Geological Milestones*), together with countless field trips he has led and lecture tours in North America, Europe, North Africa and Australia, have had a most significant influence world-wide on research, teaching and the petroleum geology of carbonate deposits. A measure of this impact is the award of Fellowship in the American Association for the Advancement of Science and the Geological Society of America, the Twenhofel Medal of the Society for Sedimentary Geology, the Sorby Medal of the International Association of Sedimentology and honorary membership in four professional societies.

The 22 papers in this volume *Perspectives in Carbonate Geology: a Tribute to the Career of Robert Nathan Ginsburg* were mainly presented at a special symposium at the 2005 meeting of the Geological Society of America meeting at Salt Lake City to mark Robert's 80th birthday. Of the 60 authors on the 22 papers, 20 are either former or current associates, students, postdoctoral associates, or students of former students and associates.

A young Robert Ginsburg (centre) at the Shell Research Laboratory in Coral Gables.

Robert Ginsburg at the time of his selection as AAPG distinguished lecturer, circa 1987.

The fact that the majority of the papers in this publication are connected with modern carbonate sediments is appropriate as Robert pioneered the concept of comparative sedimentology, that is using the modern to compare to and relate to and understand the ancient. These studies are concerned with Robert's areas of passion, coral reefs and sea-level (Hubbard), submarine cementation and formation of beach rock (Shinn), surface sediments on Great Bahama Bank and other platforms (James, Gischler, Swart and Reijmer), origin of ooids (Kindler), coastal sediments (Vlaswinkel), formation of stromatolites (Andres), impact of storms on sediments (Reeder), and the formation of dolomite (Mónica Sánchez-Román). Two of these studies in particular draw on some of Robert's and his colleagues seminal papers. For example, Mónica Sánchez-Román demonstrates the formation of dolomite under the influence of aerobic bacteria, a mode of formation which could explain the dolomite forming on the tidal flats of Andros Island, Bahamas described in the classic 1965 paper (Shinn *et al.*, 1965). The second set of papers (John Reijmer and Peter Swart) follows up work carried out by the Shell group in the mid-1950s and not published until over a decade later on the distribution of sedimentary facies on Great Bahama Bank (Traverse & Ginsburg, 1966). In two papers in this volume the issue of the sediment distribution and their isotopic and mineralogical compositions are revisited, making use of navigational and computer methods not available to the workers in the mid-1950s. It is also gratifying to see that many of Robert's early colleagues (Shinn, James, Gischler and Kindler) have contributed papers in this section. Gene Shinn, a research technician with Robert at Shell Development Laboratory in Coral Gables in the early 1960s, revisits one of Robert's favourite subjects, syndepositional cementation, in his tale of Beach Rock formation and a subject discussed in Robert's earlier papers. Noel James, a former postdoctoral research assistant at the Fisher Island Station in the early 1970s, discusses epiphyte production of carbonate sediment in cool-water carbonates, once again an area Robert has extensively investigated in subtropical carbonates (Nelsen & Ginsburg, 1986). Finally in this section there are papers by the 'grandchildren' of Robert Ginsburg, Miriam Andres and Brigitte Vlaswinkel. Miriam, who recently completed a postdoctoral position with Pam Reid at the University of Miami, describes modern stromatolites from the Bahamas, while Brigitte, a recent graduate student from Miami, presents a paper on the effect of sea-level rise and sedimentation on the coast of the Everglades. These are all areas that were pioneered in studies by Robert.

Application of the study of modern environments to ancient sediments is the theme of six papers which study rocks ranging in age from the Pleistocene to the Proterozoic. The first of these by Albert Hine draws on work published by Ginsburg, that showed the presence of siliciclastic sediments derived from the north as underpinning the Pleistocene of South Florida (Chung & Ginsburg, 1985; Warzeski *et al.*, 1996). Several papers in this section (Strasser, Forkner and Pope) draw on the lessons learned in the Bahamas (Ginsburg *et al.*, 1977) and use what is apparently one of the most cited abstracts in geology (Ginsburg, 1971) describing the landward movement of mud on carbonate platforms. These include the paper by André Strasser and Stéphanie Védrine dealing with the Jurassic, that by Robert Forkner, Linda Hinnov, Robert Goldhammer and Laurie Hardie with the Triassic, and by Mike Pope, Steve Holland and Mark Patkowsky with the Ordovician. The contribution by Mike Grammer (a former Ginsburg student) and Audrey Ritter describes phylloid algae similar to that in a previous Ginsburg collaborative study (James *et al.*, 1988) and the paper by Linda Kah, Julie Bartley and Alice Stagner deals with Proterozoic stromatolites. All these papers take the lessons learned from the modern and apply them to the ancient; comparative sedimentology in action and application.

Finally there are five papers dealing with processes common to all time periods. The first of these, by Harold Wanless, deals with layering, a subject often discussed in Robert's writings (Ginsburg *et al.*, 1977). Wolfgang Schlager, a colleague of Robert's at the University of Miami in the 1970s and 1980s, together with Georg Warrlich deals with changing sedimentation during sea-level oscillations, a topic previously investigated by Robert in the Bahamas (Ginsburg, 2001). The third paper by a former student (Don McNeill) and postdoctoral associate (Gregor Eberli), addresses fracturing in carbonate rocks collected during the Bahamas Drilling Project, an ambitious drilling operation conceived and organized by Robert in the late 1980s (Ginsburg, 2001). In this expedition two deep holes were drilled from a jack-up barge on the western margin of the Bahamas. The last two papers in this section, one by Bernard Riegl and Sam Purkis and one by Carl Drummond and Liliah Marlow, present modelling aspects of the development of facies on carbonate platforms supporting some of Robert Ginsburg's original work (Ginsburg, 1971, 1974; Shinn *et al.*, 1969).

So that is our (Swart, Eberli and McKenzie) and the authors of this volume tribute to Robert Nathan Ginsburg on his 80th birthday. I am sure that the countless students, young and old, who have been influenced by Robert will share in the following sentiments. Robert's enthusiasm has shaped our lives and we hope that this compilation in some way pays tribute to his career, although that would be a difficult task.

The editors would like to thank all those who participated in the completion of this volume including the reviewers listed below in alphabetical order: Tom Aigner, Miriam Andres, Christian Betzler, Steve Boss, Stephen Burns, Carl Drummond, Christophe Dupraz, Paul Enos, Elrick Ettensohn, Eberhardt Gischler, Kate Gilis, Mike Grammer, Laura Guertin, Mitch Harris, Paul Hearty, Stephen Hesselbo, Al Hine, Linda Hinnov, Hans Hoffman, Adrian Immenhauser, Xavier Janssen, Chris Kendall, Pascal Kindler, David Mallinson, Mitch Malone, Don McNeill, Florian Mauser, Lucian Montaggione, Hank Mullins, David Osleger, William Patterson, Werner Piller, John Reijmer, Harry Roberts, Eugene Rankey, Robert Riding, Bernard Riegl, Elias Samankassou, Art Saller, Diethard Sanders, Wolfgang Schlager, Toni Simo, Taury Smith, George Stanley, André Strasser and Ralf Weger.

In particular we would like to thank Greta Mackenzie, Tom Stevens, Ian Jarvis and Stella Bignold who dealt with the final copy editing issues.

Peter K. Swart, Gregor Eberli, and Judith McKenzie
December 2007

REFERENCES

Beach, D.K. and **Ginsburg, R.N.** (1980) Facies succession of Pliocene–Pleistocene carbonates, Northwestern Great Bahama Bank. *Am. Assoc. Petrol. Geol. Bull.*, 64, 1634–1642.

Chung, G.S. and **Ginsburg, R.N.** (1985) Siliciclastic incursion in Southern Florida and development of Florida

Reefs during Late Cenozoic. *Am. Assoc. Petrol. Geol. Bull.*, **69**, 244.

Eberli, G.P. and **Ginsburg, R.N.** (1987a) Segmentation and coalescence of Cenozoic carbonate platforms, Northwestern Great Bahama Bank. *Geology*, **15**, 75–79.

Eberli, G.P. and **Ginsburg, R.N.** (1987b) Segmentation and coalescence of Cenozoic carbonate platforms, Northwestern Great Bahama Bank – Reply. *Geology*, **15**, 1082.

Ginsburg, R.N. (1953) Intertidal erosion on the Florida Keys. *Bull. Marine Sci. Gulf Caribbean*, **3**, 55–69.

Ginsburg, R.N. (1971) Landward movement of carbonate mud – New model for regressive cycles in carbonates. *Am. Assoc. Petrol. Geol. Bull.*, **55**, 340.

Ginsburg, R.N. (1974) Introduction to comparative sedimentology of carbonates. *Am. Assoc. Petrol. Geol. Bull.*, **58**, 781–786.

Ginsburg, R.N. (1982) Seeking answers: Suggestions for students. *Journal of Sedimentary Petrology*, **52**, 351–352.

Ginsburg, R.N. (1997) Gauging the health of the world's coral reefs – Monitoring vs. mapping. *Geotimes*, **42**, 404.

Ginsburg, R.N. (Ed.) (2001) *Subsurface geology of a prograding carbonate platform margin, Great Bahama Bank: Results of the Bahamas Drilling Project. SEPM Special Publication*, **70**, Tulsa.

Ginsburg, R.N. (2005) Disobedient sediments can feedback on their transportation, deposition and geomorphology. *Sedimentary Geol.*, **175**, 9–18.

Ginsburg, R.N. and **James, N.P.** (1976) Submarine botryoidal aragonite in Holocene reef limestone, Belize. *Geology*, **4**, 431–436.

Ginsburg, R.N., **Gischler, E.** and **Kiene, W.E.** (2001) Partial mortality of massive reef-building corals: an index of patch reef condition, Florida Reef Tract. *Bull. Marine Sci.*, **69**, 1149–1173.

Ginsburg, R.N., **Hardie, L.A.**, **Bricker, O.P.**, **Garrett, P.** and **Wanless, H.R.** (1977) Exposure Index: a quantitative approach to defining position within the tidal zone. In: *Sedimentation on the Modern Carbonate Tidal Flats of Northwest Andros Island, Bahamas* (Ed. L.A. Hardie), Vol. **22**, pp. 7–11. The Johns Hopkins University Studies in Geology.

Ginsburg, R.N., **Harris, P.M.**, **Eberli, G.P.** and **Swart, P.K.** (1991) The growth-potential of a bypass margin, Great Bahama Bank. *J. Sedimentary Petrol.*, **61**, 976–987.

James, N.P., **Ginsburg, R.N.**, **Marszalek, D.S.** and **Choquette, P.W.** (1976) Facies and fabric specificity of early subsea cements in shallow Belize (British Honduras) reefs. *J. Sedimentary Petrol.*, **46**, 523–544.

James, N.P., **Wray, J.L.** and **Ginsburg, R.N.** (1988) Calcification of encrusting aragonitic algae (Peyssonnellaceae) – Implications for the origin of Late Paleozoic reefs and cements. *J. Sedimentary Petrol.*, **58**, 291–303.

Logan, B.W., **Rezak, R.** and **Ginsburg, R.N.** (1964) Classification and environmental significance of algal stromatolites. *J. Geol.*, **72**, 68–83.

Nelsen, J.E. and **Ginsburg, R.N.** (1986) Calcium carbonate production by epibionts on Thalassia in Florida Bay. *J. Sedimentary Petrol.*, **56**, 622–628.

Schlager, W. and **Ginsburg, R.N.** (1981) Bahama carbonate platforms – the deep and the past. *Marine Geol.*, **44**, 1–24.

Shinn, E.A., **Ginsburg, R.N.** and **Lloyd, R.M.** (1965) Recent supratidal dolomite from Andros Island, Bahamas. In: *Dolomitization and Limestone Diagenesis* (Eds L.C. Pray and R.C. Murray), Vol. **13**, pp. 112–124. *SEPM Special Publication*, Tulsa.

Shinn, E.A., **Lloyd, R.M.** and **Ginsburg, R.N.** (1969) Anatomy of a modern carbonate tidal-flat, Andros Island, Bahamas. *J. Sedimentary Petrol.*, **39**, 1202–1228.

Traverse, A. and **Ginsburg, R.N.** (1966) Palynology of the surface sediments of Great Bahama Bank, as related to water movement and sedimentation. *Marine Geol.*, **4**, 417–459.

Warzeski, E.R., **Cunningham, K.J.**, **Ginsburg, R.N.**, **Anderson, J.B.** and **Ding, Z.D.** (1996) A Neogene mixed siliciclastic and carbonate foundation for the Quaternary carbonate shelf, Florida Keys. *J. Sedimentary Res.*, **66**, 788–800.

Dedication to Robert N. Ginsburg

With this volume, his friends wish to honour Robert N. Ginsburg, one of the founders of carbonate sedimentology and a driving force in sedimentary geology for more than half a century.

Bob Ginsburg started his university education right after World War II – at a time when Georges de Buffon's advice 'collect facts, from those arises the thought' was a popular motto in the natural sciences. It was an understandable attitude under the circumstances. During the war, technological development had advanced much faster than scientific observation, and there were many powerful tools – such as isotope chemistry – ready to be applied in the natural sciences. During his education at the University of Chicago, Ginsburg certainly came to appreciate the potential of the new techniques but he also realized that collecting data is much more productive when guided by focused questions. He became a master of linking sharp perception with asking the right questions about what has been observed. 'What? Why? So what?' These questions became his compass to both discovery and synthesis. They already guided his early work on carbonate sediments of South Florida as well as the studies on modern and ancient stromatolites. Many subsequent contributions on the modern carbonate sediments of Florida, the Bahamas and the Caribbean demonstrate Ginsburg's mastery of linking process to product by careful observation and asking fundamental questions.

Two other qualities contribute to Ginsburg's deep impact in geology. He is a superb organizer and he is magically attracted by unknown territory. In this spirit, he launched the studies of the reef walls and the subsurface of the Bahamas. In typical Ginsburgian fashion, the subsurface work started as a nickel-and-dime operation using cores from water wells; it culminated with two deep wells drilled by a jack-up rig. The results, in turn, led to ODP drilling in the Florida Straits. The sum total of the Bahamian research drilling is a platform-to-basin transect that sets a new standard for the study of carbonate platforms around the world.

Ginsburg has always been able to mobilize people and set things in motion. Initiatives such as the Global Sedimentary Geology Program or the International Year of the Reef found world-wide response. However, he prefers to work in small, family-size groups where people know each other well and the formal structure is kept to a minimum. One such group was the Shell Research Laboratory at Coral Gables, 1954–65; another one developed in 1965–70 in the Department of Geology at Johns Hopkins University when Ginsburg held the Chair of Geology and oceanography; the most enduring group was the Fisher Island Laboratory of the Rosenstiel School of Marine and Atmospheric Science, University of Miami, 1971–90. Each group earned international fame in carbonate field research and in every instance Bob Ginsburg was the decisive force that brought them to life.

Finally, a word on Bob Ginsburg's style and attitude in scientific matters. Spreading the word by heated scientific debate and dogmatic declaration never was his style. He is at his best in the informal ambience of field seminars and short courses where he can both teach and learn by observing and questioning Mother Nature. And Nature, he keeps telling us, should not only be studied but also cared about. I vividly remember a scene that encapsulates his style and attitude. It happened in Heidelberg, Germany, 1996. Ginsburg, silver-haired but in youthful enthusiasm, was delivering the evening lecture on coral reefs to introduce the International Year of the Reef that he had just initiated. He outlined the scientific challenges of the complex reef systems, but he also called for empathy with the reefs, likening them to cities under the sea that needed our attention and protection. I could not help but imagine quite different scholarly disputes that happened in this very room centuries ago: on the galleries above our heads still rose the pulpits of the opponents in theological disputes of the past when scholars were facing each other across the hall in often war-like confrontations. I was happy to find myself back in our time, with teachers and friends such as Bob Ginsburg, who embodies this special combination of qualities so well captured in French: 'grand géologue, grand savant, grand monsieur'!

Wolfgang Schlager

Int. Assoc. Sedimentol. Spec. Publ. (2009) **41**, 1–18

Depth-related and species-related patterns of Holocene reef accretion in the Caribbean and western Atlantic: a critical assessment of existing models

DENNIS K. HUBBARD

Department of Geology, Oberlin College, Oberlin, OH 44074, USA (E-mail: dennis.hubbard@oberlin.edu)

ABSTRACT

Reef-accretion rate was measured in 151 core intervals from 12 Caribbean and western Atlantic locations. Palaeowater depth for each core interval was determined by comparing its position to the curve of Lighty *et al.* (1982), based on calendar years before 1950. While the majority of the data points fell within the upper 15 m of the water column, no clear depth-related pattern of reef accretion emerged. This is in sharp contrast to the widely held assumption that reef accretion will decrease exponentially with water depth at rates approximately an order of magnitude below the corresponding rates of coral growth at the same depths. Similarly, reef-accretion rates from facies dominated by branching *Acropora palmata* (3.83 m kyr^{-1}) versus those associated with massive corals (3.07 m kyr^{-1}) were not significantly different ($\alpha = 0.05$), owing to high variance across all water depths. Reef accretion showed a tendency to increase at higher rates of sea-level rise, but that relationship was also non-significant.

It is proposed that the known depth-related decrease in carbonate production may be offset by a parallel drop in bioerosion. While available data generally support this hypothesis, quantitative verification must await careful measurements of both biological degradation and transport along a depth gradient. Nevertheless, bioerosion appears to not only play an important role in creating reef fabric, but to perhaps affect patterns of reef accretion as much as initial calcification. Regardless of the cause, the patterns revealed by this study fly in the face of the assumptions that underlie our most widely accepted Holocene reef models. Clearly new ones are needed that emphasize the varying contribution of biological material to what is largely a process of physical aggradation – in short reef corals grow, coral reefs accrete.

Keywords Reef accretion, Caribbean, Holocene, bioerosion, sea level, *Acropora.*

INTRODUCTION

What emerges from the myriad definitions of reefs is that they are resistant structures that stand above their surroundings, thus exerting some degree of influence over local circulation. Reefs can be built by organisms as large as modern corals or Cretaceous rudists, or as small as Precambrian microbes. They might emerge as fringing and barrier reefs or form submerged features along shelf margins (Macintyre *et al.*, 1981; Hubbard *et al.*, 1997, 2005), or even in very deep water well below the photic zone (Fosså *et al.*, 2002; Reed *et al.*, 2005). Disagreements focus on how resistant these features need to be and what is responsible for that rigidity. Throughout most of the twentieth century, descriptions of modern reefs emphasized the constructive role of corals, corallines and other organisms that secrete calcium carbonate and 'build reefs'. Mention is made of physical damage and bioerosion by grazers and infauna (e.g. Ginsburg, 1958; Scoffin, 1992) but the role of construction was until recently perceived as overwhelmingly dominant. As a result, discussion of reefs throughout the later twentieth century focused on reef 'framework' as the architect of this rigidity (Lowenstam, 1950; Newell *et al.*, 1953) and the importance of 'large, colonial or gregarious, intergrown skeletal organisms in general growth position' in creating it (Fagerstrom, 1987).

More recently, the interiors of some Caribbean coral reefs have been likened to 'garbage piles' that comprise as much sediment and toppled corals as in-place or interlocking organisms (Hubbard *et al.*, 1990). These ideas challenge the primacy of biological production in coral-reef assembly. It is ironic that as early as 1888, Johannes Walther (translated in Ginsburg *et al.,* 1994) noted a dominance of detrital material in Quaternary reefs exposed along the Sinai Penninsula. This idea was apparently ignored in favour of more-popular 'framework' models derived from the discussions of Lowenstam and Newell, despite an apparent recantation by the latter in Newell 1971. As we swim over a modern reef, there is little disagreement over its status. Less readily apparent is the fact that the interiors of many (or possibly most?) of these structures contain less than 30% recognizable coral, with only a portion of that in place (Conand *et al.*, 1997; Hubbard *et al.*, 1998). The remainder is sand and rubble caused by ubiquitous bioerosion plus void space. In as much as this composition probably cannot by itself create the rigidity that is the hallmark of 'true reefs', the roles of encrustation and cementation must rise to equal those of initial calcification and subsequent bioerosion.

How are coral reefs built?

The symbiotic relationship between many corals and their endolithic zooxanthellae, leads to calcification being strongly dependant on light intensity and character ('photosynthetically active radiation' occurs at the red end of the spectrum). This relationship and the maximum depth for corals have been generally understood since the nineteenth century (Quoy & Gaimard, 1825). Light and, therefore, photosynthesis decrease exponentially with depth. Accordingly, the ratio between light intensity (I) at any depth and the light-saturating intensity (I_k: the amount of light that will result in maximum photosynthesis) for a particular coral species also drops. Chalker (1981) proposed that photosynthesis by scleractineans can be approximated by this ratio and the hyperbolic tangent function. Bosscher (1992) showed that such relationships could be used to predict the growth rate of corals with increasing depth (Fig. 1).

It has been generally assumed that reef accretion is a biological process that is dominated by the growth of corals that are largely in place or at least have not been moved far from where they grew. While it has been long understood that post-mortem degradation of corals is commonplace, it has been largely assumed that much of the detritus from bioerosion will stay within the reef, and that the depth-related pattern of coral growth will be reflected in trends in reef accretion along the same gradient. Schlager (1981) summarized available data for reef aggradation relative to sea-level rise and coral growth (Fig. 2). While accretion was an order of magnitude slower in his model, it generally mimicked the rapid decrease of coral growth with depth. Citing anecdotal data from other studies, it was argued that most reefs in shallow water have been capable of building at rates greater than the maximum rate of glacio-eustatic sea-level rise during the Holocene (*c.* 7 m kyr^{-1}). Based on this, it was suggested that drowned reefs (or platforms) in the ancient record represent a 'paradox' that can be resolved only by invoking extreme and short-lived conditions associated with either rapid (and probably tectonically induced) sea-level rise or degraded environmental conditions (e.g. 'inimical bank waters' or larger-scale marine 'crisis events').

These concepts represent the foundation upon which most current models of Holocene reef development are based (Adey, 1978; Schlager, 1981; Macintyre, 1988). Linear-extension rates of corals in shallow water can exceed 10 cm yr^{-1}, but coral growth generally ranges from around 10 mm yr^{-1} in shallow water to less than 1 mm yr^{-1} at depth (for an excellent compilation, see Dullo, 2005). Reefs are thought to build at a rate roughly an order of magnitude slower (Schlager, 1981; Bosscher, 1992) but still following the general depth-related pattern of coral growth (Fig. 3). From this, it may be deduced that areas dominated by branching coral will build faster than those inhabited by slower-growing massive species (Adey & Burke, 1976), while reefs in shallow water will build faster than their deeper counterparts (Schlager, 1981; Bosscher, 1992).

Despite the overwhelming acceptance of these axioms, they have never been systematically tested. This paper attempts to quantify depth- and species-related patterns of Caribbean and western Atlantic reef accretion using a preliminary survey of information from the literature as well as unpublished core data. It attempts to convince the reader that the relationship between coral growth and reef accretion is not as has been generally assumed. More specifically, it is proposed that the pattern that emerges from systematic analyses of existing core data is controlled no more by coral growth than by the myriad physical and biological processes that come afterwards. If this is the case, then it will have significant bearing on existing models of coral-reef accretion.

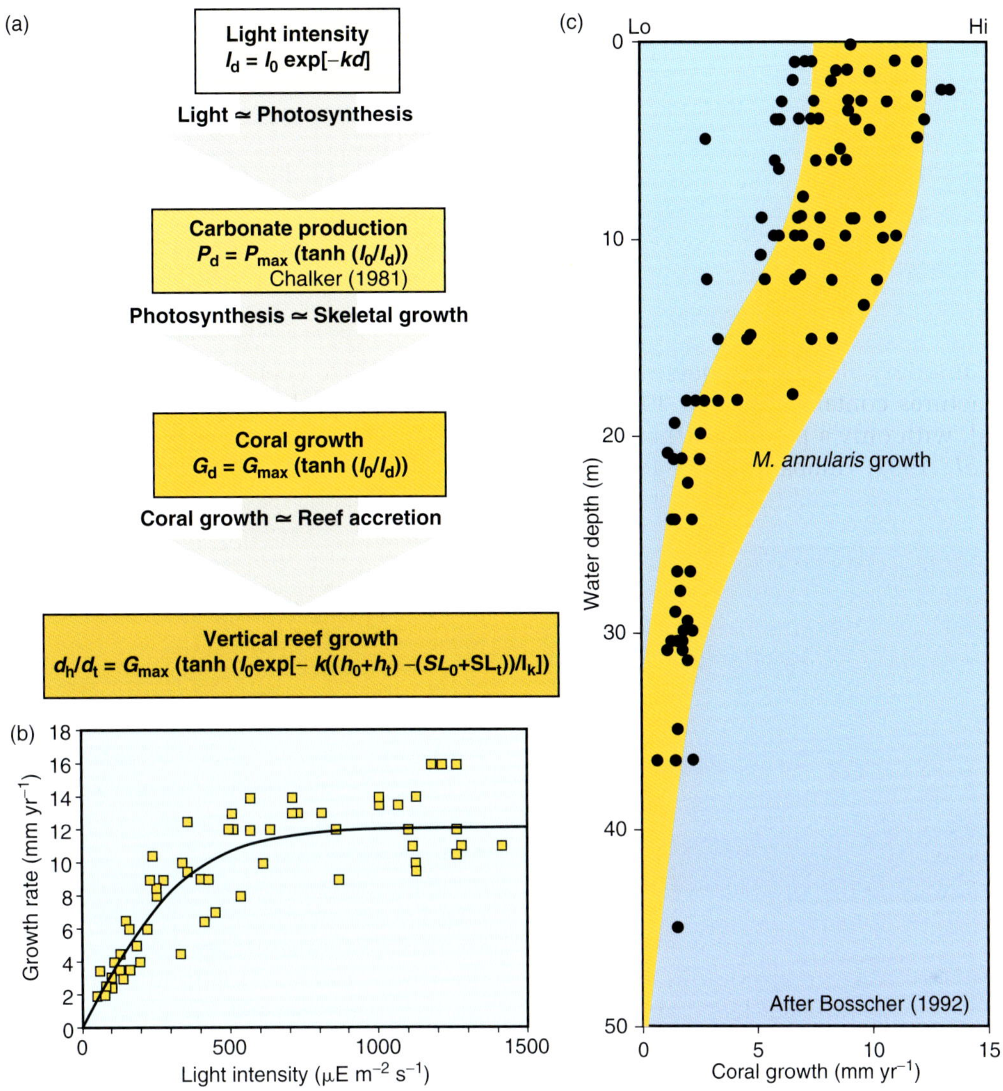

Fig. 1. (a) Summary of a quantitative model to predict coral growth and reef accretion with depth. (b) Growth rate for the *Montastraea annularis* complex versus light intensity (measured in microEinsteins). The best-fit line mimics the hyperbolic tangent function, as predicted by Chalker (1981). (c) The relationship between coral growth and water depth. Circles are actual growth rates determined by X-radiography of modern colonies. The shaded area is the predicted range of coral-growth rates based on A and B. It has been generally assumed that while reef accretion will proceed at rates an order of magnitude slower than coral growth, a similar depth-related trend (i.e. exponentially decreasing) is present. All after Bosscher (1992).

METHODS

Reef accretion is generally defined as vertical aggradation over time. This is in part an artefact of the vertical orientation of cores from which measurements are made. However, it also stems from the realization that coral growth is strongly controlled by light from the surface. While reefs may build horizontally, this is generally held to be the lateral expression of largely upward extension. In this study, accretion rates were calculated using all data that could be recovered from the literature as well as unpublished information in the author's archives. In the literature, each sample was reported by species, depth below present sea level and radiometric age. The analyses discussed below compare accretion rates for reef intervals dominated by *Acropora palmata* (Lamarck) versus massive species across a wide palaeodepth range. Because *A. palmata* is easily distinguished from massive species in cores, no effort was made to check the identifications reported in the

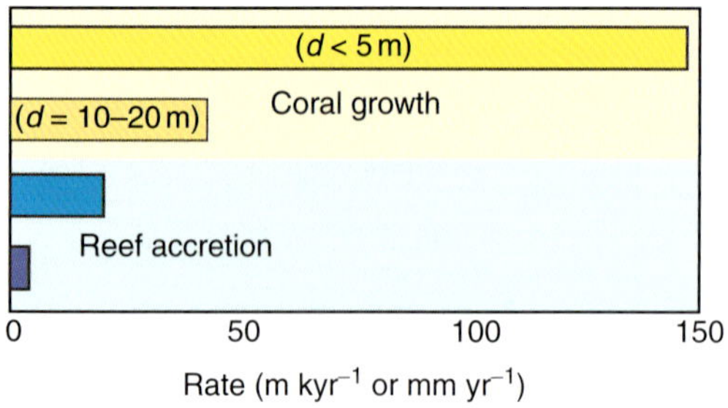

Fig. 2. Reef accretion versus sea-level rise and coral growth (d=water depth). In this scenario by Schlager (1981), shallow reefs (depth less than 5 m) are expected to build an order of magnitude faster than their deeper-water counterparts. This is explained largely by the dominance of faster-growing branching corals in shallow water.

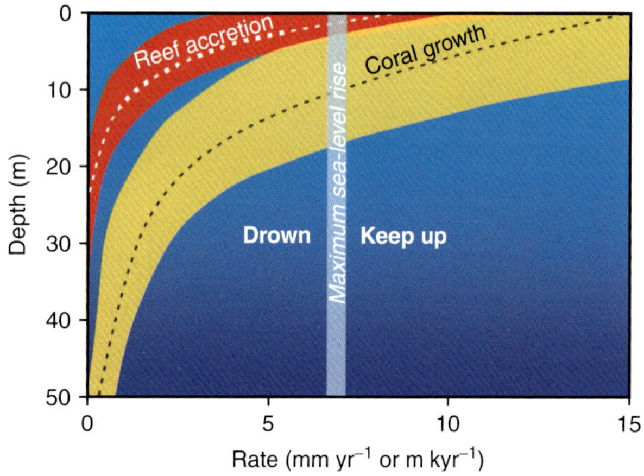

Fig. 3. Summary of presumed depth-related patterns of reef accretion and coral growth. While coral growth (yellow) can exceed 10 cm yr^{-1} in some shallow-water species, it generally drops from a maximum of *c.* 10–20 mm yr^{-1} in shallow water to less than 1 mm yr^{-1} at depth. Based on rates quoted in the previous literature, reef accretion (red) should decrease following a similar pattern, but at rates an order of magnitude slower. According to Schlager's (1981) 'Drowning Paradox', many shallow-water reefs in the past have built up faster than the highest rates of glacio-eustatic sea-level rise (~7 m kyr^{-1}: light blue bar). Those reefs should have been able to keep up with rising sea level, while their slower-building cohorts drowned.

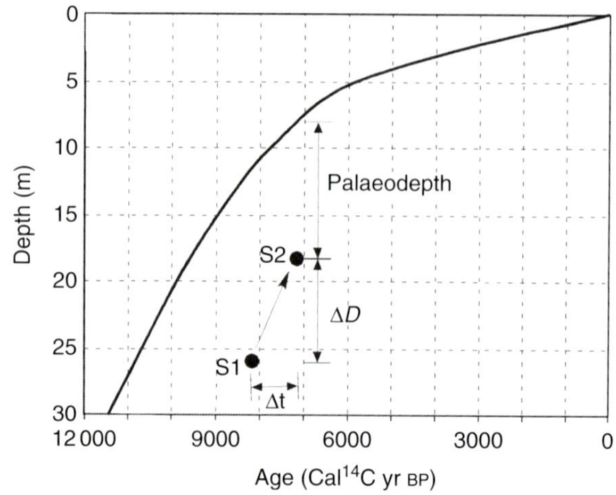

Fig. 4. Illustration of measurements used in this paper. Palaeodepth (D_p) is computed as the difference between the position of a recovered sample and sea level at the time it was deposited. Time of deposition is based mostly on calibrated radiocarbon measurements. Accretion was calculated as $\Delta D/\Delta t$.

literature. The precision of the depth information varied among studies, but errors are most likely random and are not thought to impact the general patterns shown below. Reported radiocarbon ages were calibrated by Beta Analytic, Inc. using the INTCAL-98 data set (Stuiver *et al.*, 1998) and methods similar to those of Talma & Vogel (1993). This transformation takes into account metabolic variations between sample types (i.e. coral vs. shell, wood, etc.), global/local reservoir effects, and biological fractionation for marine carbonates using a presumed isotopic value of $\delta^{13}C \sim 0$, unless otherwise specified. The procedure is similar to that used by the freeware program CALIB, except that it operates on smoothed data by visually scanning for short-term perturbations in the calibration curve (D. Hood, personal communication). U/Th ages were used as reported. Errors in age were typically 100 years or less, and are assumed to be randomly distributed about the reported dates.

Palaeowater depth

Water depth at the time of deposition was calculated as the vertical difference between the present depth of a coral sample and the height of sea level at the time corresponding to its radiometric age (Fig. 4). It is assumed that the recovered coral was either in place or was alive close to the time when it was deposited. Hindcast sea level is based on the corrected Lighty *et al.* (1982) sea-level curve as presented in Hubbard *et al.* (2005). This curve is virtually identical to the curve of Toscano & Macintyre (2003), which used many of the same samples and a nearly identical correction algorithm (i.e. CALIB vs. the proprietary Beta calibration). The rate of accretion was calculated using the formula:

$$A = (\Delta D/\Delta t)$$

where A is the accretion rate (in metres per thousand years), ΔD is the vertical difference between two core samples (in metres) and Δt is the time difference (in thousands of years) between the two samples.

Choosing the right sea-level curve

Recent discussions have challenged the Lighty *et al.* (1982) curve and its subsequent iterations. The objections generally relate to either the validity of the samples used for the curve (Blanchon, 2005) or the occurrence of numerous coral and peat samples, largely from Belize, that sit above it (Gischler, 2006). Figure 5 summarizes the available *A. palmata* data from the Caribbean. It includes all the coral data from Toscano & Macintyre (2003) plus the Belize and Campeche Bank samples of Gischler & Hudson (2004) and Blanchon & Perry (2004), respectively; additional, unpublished data from the Virgin Islands and Puerto Rico are also plotted. The sea-level curve that was used in this paper (from Hubbard *et al.*, 2005) is shown in black. It is virtually identical to the curve of Toscano & Macintyre (2003) (orange), because both are little more than isotopically based corrections of the Lighty *et al.* (1982) curve using either the Beta Analytic algorithm or the similar CALIB routine. A. Conrad Neumann's peat curve from Bermuda (unpublished personal communication), similarly corrected and allowing for differences between coral and peat (*c.* 400 years) is shown in green. Except for minor variations that are probably related to smoothing versus straight-line interpolation, the curves are again identical. The curve proposed by Blanchon (red) displays a series of step-like changes, reminiscent of the 'melt-water pulses' of Fairbanks (1989). Unlike the events proposed for Barbados, no separate isotopic evidence for sudden melt-water releases have been provided for any of the variations from a smooth curve. Coral samples well above his curve during all three proposed steps argue against their existence. In addition, the inflections in the curve are in part based on an interpretation of mangrove peat that is in fundamental conflict with the arguments of Gischler (2006), who argues that the elevated *A. palmata* and peat samples from Belize (plus Jamaica) support a fundamentally elevated Caribbean sea-level curve (blue line).

Verifying that a particular sample is in place has been the fundamental 'Achilles heel' of coral-based sea-level curves. Likewise, discriminating between reef material and allochthonous debris

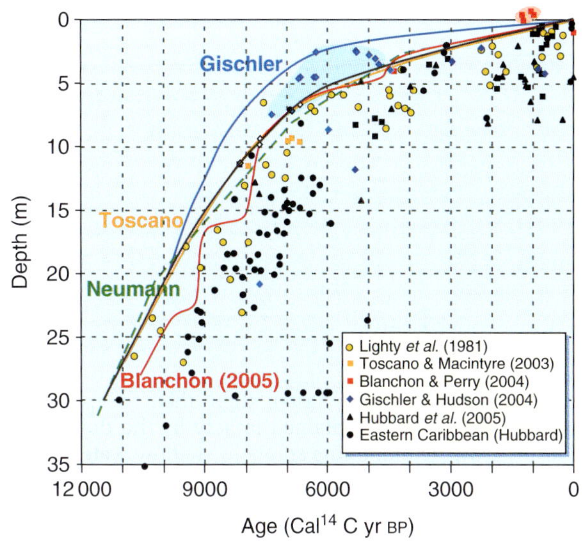

Fig. 5. Variability in proposed Caribbean sea-level curves. The data points represent the radiometric age and vertical position of individual *Acropora palmata* samples from cores throughout the region. The curve used in this paper (black: after Hubbard *et al.*, 2005) is based on both *A. palmata* and peat. This curve and the one by Toscano & Macintyre (2003; orange) are little more than isotopically corrected versions of the one originally proposed in Lighty *et al.* (1982; not shown). Neumann's (unpubl. pers. comm.) curve for Bermuda (dashed green) is based on mangrove peats, and is in general agreement. The stepped curve of Blanchon (2006; red) was not considered because (a) it is inconsistent with the entirety of the data, and (b) no separate isotopic evidence has been provided to support these drastic deviations from a smoothed trend. The curve of Gischler (2006; blue) sits above the calibrated Lighty curve with a maximum deviation of *c.* 3 m. The Toscano & Macintyre curve is based on the assumption that the curve should be placed at the bottom of what they interpreted as intertidal to slightly supratidal peats. Gischler's curve (Gischler & Hudson, 2004) straddles the top of the same peat–coral package.

thrown up onto supratidal storm ridges further complicates the matter. A detailed discussion of these problems and a solid defence of one particular curve over another is beyond the scope of this paper. However, it is worth noting that all the curves follow the same general pattern, and an approximate average of all the curves falls somewhere close to the black line in Fig. 5. The maximum difference between the three more central curves and either of the other two amounts to 3 m. The absolute depth of one particular sample might vary by up to that much, but it seems unlikely that the overwhelming patterns described below could be fundamentally changed. Nevertheless, the general impacts of choosing one curve over another will be discussed later.

Accretion rate

Reef-accretion rates have typically been reported for whole cores or reefs (for an excellent summary, see Dullo, 2005). In this paper, accretion rates from individual intervals (i.e. between two adjacent samples) within cores were chosen over those for whole cores or facies, because palaeowater depth could be more accurately assigned to a shorter time step. Intervals that contained largely mixed coral types or were dominated by detrital deposition were not included. Samples that plotted above sea level or resulted in negative accretion rates (i.e. the lower sample was younger than the upper one) were not used. Finally, rates calculated from closely spaced samples with age differences less than the reported analytical errors were also discarded. Of the 11 discarded accretion calculations, all were faster than 10 m kyr^{-1}. All but three were from palaeowater depths greater than 5 m. Futhermore, these shorter-term rapid rates were invariably contained within longer intervals that reflected much lower net accretion rates.

Separate analyses of reef accretion were run for (A) all corals regardless of species, and (B) a comparison of *A. palmata* versus massive-coral facies. The dominant corals within an interval were determined by examining core logs. The 11 data points from abandoned *A. palmata* reefs off Barbados (Fairbanks, 1989; Bard *et al.*, 1990) were used for the comparison between accretion rates in massive and branching facies only. They were eliminated from the depth-related comparison because no independent verification exists for sea level at the time these deeper reefs were active. All the Belize corals above the curve used in this paper (Fig. 5) occurred within mixed-coral intervals that violated the acceptance criteria. Thus, all the data used from the cores of Gischler & Hudson (2004) fell below the curve.

RESULTS

A total of 151 core intervals met the criteria described above. Of these, 79 were from branching-coral facies, compared with 72 from massive-coral intervals. Samples were recovered from 11 Caribbean and Atlantic locations (and from multiple reefs at many of these), ranging from central Florida in the north to Panama and Belize in the southwest and Antigua/Barbados in the easternmost Caribbean (Fig. 6; Table 1). Present-day environments ranged from emergent barrier and fringing reefs to submerged but still active shelf-edge reefs (e.g. Lang Bank and SW Puerto Rico) and abandoned reefs that are now below the depth where accretion can occur (e.g. Barbados). The data are tabulated in Appendix 1.

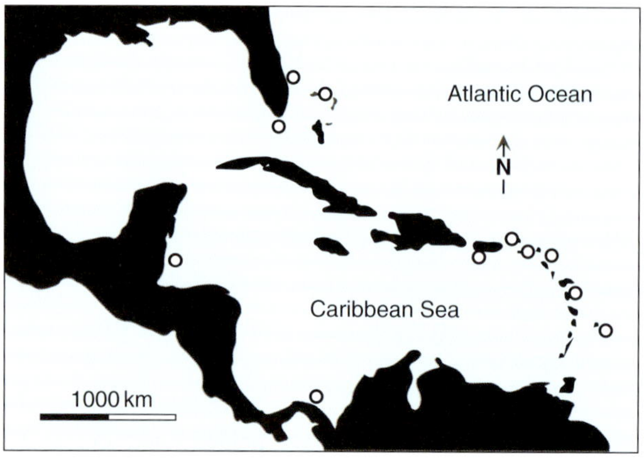

Fig. 6. Map showing the locations of the sites from which core data were derived (open circles).

Table 1. Sources of core data used in this study.

Florida	Lighty *et al.* (1982), Toscano & Lundberg (1998). Toscano & Macintyre (2003)
Bahamas	Lighty *et al.* (1982)
Martinique	Lighty *et al.* (1982)
Puerto Rico	Hubbard *et al.* (1997, unpublished data), Lighty *et al.* (1982)
Antigua	Macintyre *et al.* (1985)
St. Croix	Adey *et al.* (1977), Burke *et al.* (1989), Hubbard *et al.* (1990), Hubbard *et al.* (2005), Hubbard (unpublished data), Lighty *et al.* (1982), Macintyre & Adey (1990)
Barbados	Fairbanks (1989)
Panama	Macintyre & Glynn (1976)
Belize	Gischler & Hudson (2004)

The majority of the data points fell within the upper 15 m of the water column (Fig. 7). As expected, the fastest accretion rates were derived from core intervals in water <10 m deep at the time of deposition. However, accelerated accretion was evenly split between branching- and massive-coral intervals, and any depth-related coral affinities were weak above 15 m. *A. palmata* clearly thrived at depths much greater than is generally accepted as its maximum, commonly deeper than 10 m and occasionally down to 20 m. More important, they were capable of significant accretion well below the 5-m depth limit proposed by Lighty *et al.* (1982) and Macintyre (1988), among others.

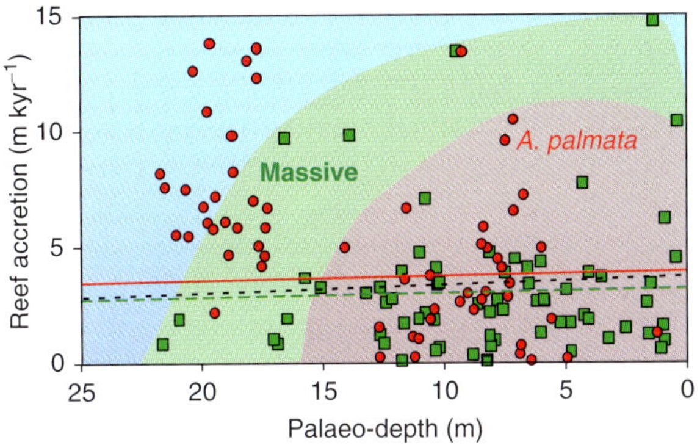

Fig. 7. Reef accretion versus palaeowater depth for branching (red dots) versus massive corals (green squares). Approximate limits for branching and massive facies (shaded) reflect a general overlap in the depth ranges for the two facies, but with a slight tendency for branching *Acropora palmata* to inhabit shallower water than massive species. The linear regression lines for branching (red solid: $y = 0.02x + 3.98 - R^2 = 0.001$), massive (green dashed: $y = 0.02x + 3.20$; $R^2 = 0.0001$) and all corals combined (black dotted: $y = 0.03x + 3.52$; $R^2 = 0.003$) show no strong depth-related relationship and R^2 values are very low, reflecting the large variance across all depth ranges (Fig. 8). The difference between the regression lines for branching (solid) and massive corals (dashed) is not statistically significant.

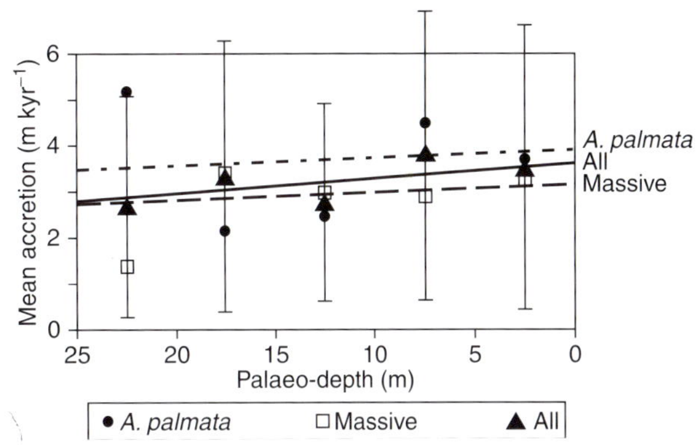

Fig. 8. Reef accretion versus palaeowater depth in 5-m increments. The depth-related trend is weak (lines shown here are based on linear regression) regardless of coral type, and standard deviations are larger than the difference between mean accretion rates at either depth extreme. The anomalously high accretion rate for *A. palmata* between 20 and 25 m is based on only one core interval.

No strong depth-related pattern emerged either by facies type or for all corals combined. The highest R^2 value for linear regression was only 0.003. Exponential and polynomial best-fit lines yielded similarly poor results. The difference between means for shallow- and deep-water accretion (<1.0 m kyr^{-1}) was lower than standard deviations within 5-m depth intervals (Fig. 8). The mean accretion rate for *A. palmata* intervals was faster than that for massive-coral facies (3.83 vs. 3.07 m kyr^{-1}), but this difference was not significant at $\alpha = 0.05$ using a Student's *t*-test, owing to these large variances. Additional data could raise the significance levels for this comparison, but the absolute difference would remain small, and certainly not approach the relationship that has been presumed for the two coral types.

Accretion rate was weakly and positively correlated with the rate of sea-level rise, but the R^2 value for the relationship was only 0.015. Only 16 of the measured intervals accreted at rates greater than 7 m kyr^{-1}, the upper threshold for Holocene sea-level rise proposed by Schlager (1981); these were equally distributed between 'deep' and 'shallow' reefs. In all but five cases, the intervals of unusually high accretion were small and were contained within a larger interval over which mean accretion rate was below that threshold. Of those five intervals, only one was associated with a vertical section of significant thickness (*A. palmata* facies at $d = -1.4$ m).

DISCUSSION

The higher concentration of measured intervals in palaeowater depths less than 15 m is consistent with the presumption that light intensity exerts a primary control on coral growth, and by extension, reef accretion. Also, the tendency for massive-coral intervals to extend into deeper water than branching-coral sections (compare shaded areas in Fig. 7) mimics modern reef zonation. However, neither the species- nor depth-related differences in average accretion rate that are the foundations of most modern-reef models clearly emerge from the data.

Schlager (1981) proposed that shallow reefs (set at $d < 5$ m) accrete between 1 and 20 m kyr^{-1}, compared to less than 2 m kyr^{-1} at depths between 10 and 20 m (Fig. 2). The depth-related model of Bosscher (1992; Fig. 1) was built on these assumptions, as are most discussions of reef accretion versus sea-level rise, both past (Macintyre, 1988) and future (Graus & Macintyre, 1999). Figure 9 compares these presumptions to the results of this study. The maximum 'shallow-water' accretion rate (16.6 m kyr^{-1}) was higher than that for 'deep reefs' (9.73 m kyr^{-1}). However, the majority of the accretion rates for the Caribbean cores analysed

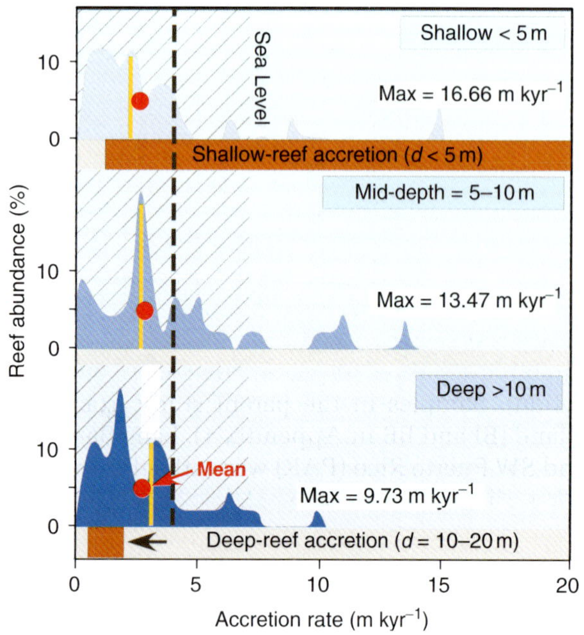

Fig. 9. Depth-related patterns of reef-accretion rates from this study (blue curves), compared with rates suggested by Schlager (1981; brown bars) for reefs in water depths less than 5 m and greater than 10 m (d=depth). Means and medians for each distribution are also shown. The range of Holocene sea level proposed by Schlager (up to 7 m kyr^{-1}) is shown by the diagonal pattern. Maximum accretion rate decreased with depth (16.66 m kyr^{-1} shallower than 5 m; 9.73 m kyr^{-1} deeper than 10 m). However, the primary mode of reef accretion remained below 4 m kyr^{-1} (dashed black line) for all depths. Mean (large red dots) and median accretion rates (yellow lines) for the three depth ranges are nearly identical for all three depth ranges.

for this study fell below 4 m kyr^{-1}, regardless of palaeowater depth. Moreover, mean accretion rates for the three depth intervals were virtually identical, as were the medians and the positions of the primary modes.

If these patterns are representative of what is actually happening on a larger scale, then we are faced with a number of questions. What is responsible for this surprising and seemingly counter-intuitive finding? How do we explain a system in which branching corals grow faster than their massive counterparts, but where a depth-related decrease in accretion is clearly absent? Can reefs in water depths greater than 10 m actually keep pace with those in shallow water where light intensity and coral growth are at least an order of magnitude higher? And, if so, how? Finally, what are the implications for reefs keeping up or being left behind by the sea-level rise in the distant past, or the near future?

It comes as no surprise that reef-accretion rates are significantly below the rate of coral growth. Coral cover is rarely 100% on any reef. It has been long understood that bioerosion and physio-chemical damage occur on reefs (Grant, 1826), and can easily reduce 70% of the carbonate produced in a reef to sediment (Stearn & Scoffin, 1977; Hubbard *et al.*, 1990; Conand *et al.*, 1997). Much of this is exported from the reef, especially by storms (Woodley *et al.*, 1981; Hubbard, 1992), leaving only a portion of the original carbonate fixed by coral, corallines and other calcifiers within the reef edifice. What is surprising is the apparent disconnection between the depth-related patterns of coral growth and reef accretion that has emerged from these analyses. The presumption that the dominance of light-mediated calcification will remain as the fundamental underlying control of reef accretion has been widely held. The results of this synthesis indicate otherwise.

Possible sources of error

As in any budgeting exercise, an unexpected result begs for consideration of two possibilities: (1) the measurements are incorrect or somehow inappropriate, or (2) some factor may have been omitted from consideration. There are numerous sources of inaccuracy in the data used for the calculations. Precisely determining the position of a sample within a core can be difficult, and surely some of these reported values are less than totally accurate. However, these are undoubtedly evenly distributed around the true values and would logically cancel out one another. Also, palaeowater depth is based on a sea-level curve that is by necessity generalized, and represents a regional average. As discussed above, honest disagreements persist over the details of the Caribbean sea-level curve over the past 12,000 years. Some of this stems from different interpretations of the same data. Toscano & Macintyre (2003) chose to place their curve above the majority of the *A. palmata* and below the mangrove peats from Florida, Jamaica and Belize. Gischler (2006) argued that the curve should be placed over both. One might reasonably argue that there is some validity in both arguments and that the truth lies somewhere in between. If this is the case, then the maximum variation from the curve used here would be in the order of 1.5 m. The curve used in this paper (from Hubbard *et al.*, 2005) is based on both mangrove peat (Florida and Bermuda) and *A. palmata*. In that case, the vertical variation of the peat, which was short-rooted, was small and agreed with the line skirted over the top of the coral package (with the exception of the 11 samples from Belize and the four young samples from Campeche Bank that plot above present sea level). If this agreement in any way validates the curve, then the possibility that the difference seen in the Belize samples represents local tectonic differences cannot be ruled out. The point in all of this is that the actual depth values assigned to any individual coral based on using the 'wrong' sea-level curve might be in error by 1–3 m. It will be left to the reader to decide if this sort of difference could be used to explain the lack of variability in reef accretion seen in this compilation over a depth range of some 22 m.

Another source of possible error is related to the assumption that recovered corals were either in growth position or ultimately came to rest close to the place where they lived. This lay at the core of the argument between Toscano & Macintyre (2003) and Blanchon (2005) over whether there are 'good' and 'bad' corals, and which should and should not be used to reconstruct sea level. The Faustian bargain of Lighty *et al.* (1982) was to lay a curve over the top of the 'coral envelope', assuming that the occasional upward deviants represented rubble tossed up on beaches or storm ridges. However, another reliable independent data source (e.g. peat) is required to support or refute this approach.

The identification of basal coral attachments strongly argues for a colony being in place. Convex upward coral-growth bands similarly lend credibility to a colony in growth position, albeit with less confidence. Unfortunately, the likelihood of encountering either in a core with any regularity is small. Coral-growth rates consistent with the hindcast depth can be useful (Hubbard *et al.*, 1986), but this will not resolve issues related to *A. palmata*. It has been proposed that the differences between polyp structure on the top and bottom of *A. palmata* branches can be used to divine orientation, and that 'right-side-up' branches reflect colonies in growth position. However, measurement of over 500 broken and in-place branches from the same localities reveals no statistical difference (D.K. Hubbard, unpublished data).

It is necessary to consider *in situ* colonies – those that are either in growth position or have not moved significant distances from where they grew. While great strides have been made in identifying taphonomic signatures associated with different styles of coral deposition (Perry, 1999, 2000, 2001; Blanchon & Perry, 2004), this is not yet a panacea. When available, the greatest assurance can be derived from either direct observation in submarine outcrop/excavation or reconstructing topography using multiple cores along transects. The data from northern Florida (in Lighty *et al.*, 1982; Toscano & Macintyre, 2003) were from excavations where the in-place nature of the sampled corals could be directly observed. The samples from Hess Ship Channel (Adey *et al.*, 1977) similarly benefited from direct observation (W. Adey, personal communication). Samples in the parent study from Buck Island (BI and BB in Appendix 1), Lang Bank (LB) and SW Puerto Rico (PAR) were all collected from cores aligned in transects where time lines could be used to reconstruct palaeotopography. In all these cases, the samples were recovered from either the apex of the topographic surface or along a broad horizontal bank. The result is that there was no shallower site from which to derive the samples. Also, these features were of sufficiently limited lateral extent that any disrupted sample had only limited space to be moved from where it grew. Similar reconstructions can be made from the well-constrained transects at Galeta Point, Panama (Macintyre & Glynn, 1976). Even if it is assumed that all the remaining studies suffered from dramatic downward transport of samples, the patterns reflected in the curves of Lighty *et al.* (1982), Toscano & Macintyre (2003), and Hubbard *et al.* (2005) remain.

This is not to say that any of these curves are a perfect representations of Holocene sea-level rise. Certainly, the package of mangrove and coral data above the curve that fueled the debate between Toscano & Macintyre (2003) and Gischler (2006) cannot be ignored. However, the author has not examined any of these samples and is, therefore, not in a position to comment (for example, are the peats short-rooted? What are the species, and what is their relationship to sea level? What is the condition of the corals and what evidence is there that might be used to reconstruct palaeotopography?). Nevertheless, the bulk of the data available support the curve used in this paper as a realistic approximation of sea level over the past 12,000 years. Even if this position were abandoned, the maximum spread among all the curves is less than 3 m, a value that is not likely to reconcile the trends shown in Fig. 7 with present models based on an order-of-magnitude decrease in reef accretion between the sea surface and a depth of 20–30 m. It is worth noting that using the same method and the uppermost Belize curve as a datum, Gischler found that reef accretion increased significantly with depth, and intervals dominated by massive corals accreted faster than those dominated by branching acroporids (E. Gischler, personal communication, 2006).

A final possibility relates to the way in which the data were collected. Existing models are more often based on generalized patterns seen throughout a core or along an entire reef over its lifetime. In contrast, the measurements reported here are from discrete intervals within single cores. Furthermore, the data do not include intervals with a mixture of branching and massive corals or a predominance of detritus. This approach was chosen to (1) allow for a more accurate characterization of average palaeowater depth across a single accretionary interval, and (2) address the presumption that branching-coral reefs in shallow water will accrete much faster than their massive counterparts at depth. What effect might this have on the data, and would it be sufficient to explain the difference between the expected depth- and species-related patterns and those seen here?

Reefs do not accrete evenly, either spatially or temporally. Using each interval between dated samples adds variability and undoubtedly contributed to the high variance seen in Fig. 8. However, when the patterns from a particular reef system were examined (Fig. 10a), wholesale accretion rates between coral types and with increasing depth remained similar, both to one another and to the average rates for smaller intervals reported above. For example, *A. palmata* reefs from the shelf edge off both St. Croix and Puerto Rico accreted at rates similar to those from massive-coral reefs in deeper water, at the same place and time. Similarly, branching- and massive-coral reefs near Buck Island (St. Croix) accreted at similar rates, regardless of the dominant coral type. Whole cores showed the same pattern; cores from the same time frame within a single reef yielded accretion rates similar to each other and to the averages reported above for smaller intervals regardless of depth or facies (Fig. 10b).

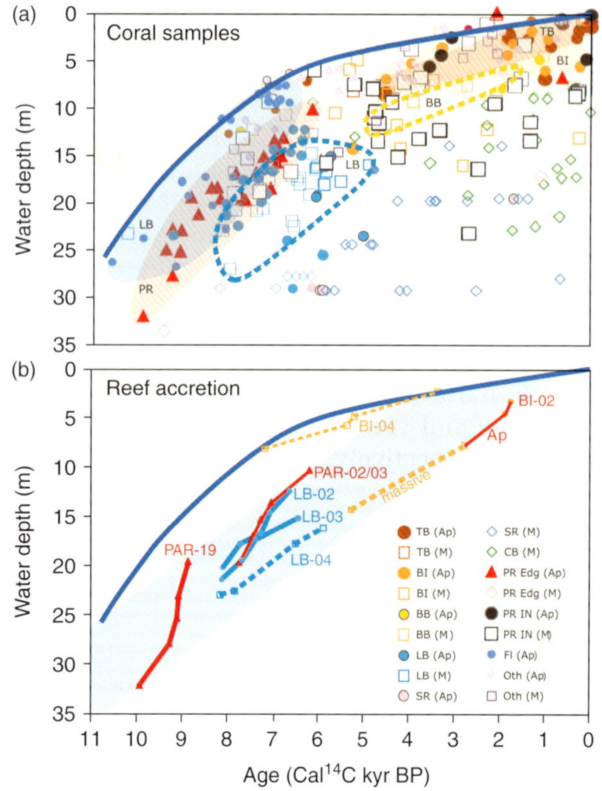

Fig. 10. (a) Ages versus water depths of all samples examined for this synopsis. The coloured envelopes delineate samples grouped by species for six reefs on St. Croix and Puerto Rico. The general trend of each envelope is interpreted as the overall rate of reef building by that species in a single reef. Earlier reefs accreted at the same rate regardless of species (filled = *Acropora palmata*; open = massive) or water depth. Similar parity existed in later reefs, except they built at slower rates. (b) Reef-accretion pathways for selected cores examined for this study. Note that, with the exception of the massive-coral interval constrained by accommodation space (BI-04) and one deep reef (PAR-19), accretion rates were similar. *Acropora palmata* intervals are shown by solid lines; massive intervals are dashed. LB = Lang Bank (E St. Croix, US Virgin Islands); BI = Buck Island (N St. Croix), PAR = La Parguera (SW Puerto Rico). The shaded blue band corresponds to the mean accretion rate for this study (3.45 m kyr^{-1}).

The possible role of bioerosion

References to biological destruction of substrate date at least to Grant's 1825 description of temperate species of *Cliona*, and were more formally recognized when Neumann (1966) introduced the term 'bioerosion'. More recently, Scoffin *et al.* (1980) concluded that, 'the internal structure of the reef is controlled not only by the nature of the dominant builders but also by the type and rates of bioerosion'. Chazottes *et al.* (1995) proposed that the greatest substrate loss off Morea was due to grazing fish and urchins (2.33 kg m^{-2} yr^{-1}), followed by macroborers (e.g. molluscs, worms: 0.20 kg m^{-2} yr^{-1}) and microborers (algae, fungi: 0.009: 2.33 kg m^{-2} y^{-1}). A similar hierarchy has been proposed for the Great Barrier Reef (Kiene &

Hutchings, 1994; Tribollet *et al.*, 2002), and the Caribbean (Stearn & Scoffin, 1977).

A quantitative assessment of bioerosion rates across the full depth range of modern reefs is not available. However, anecdotal information is available that may allow at least a qualitative characterization of depth-related patterns. In shallow water off Barbados, Scoffin *et al.* (1980) noted increased bioerosion in shallower water, owing to a greater abundance of *Diadema antillarum Philippi* closer to shore. Steneck (1994) reported *Diadema* populations of 16 individuals per square metre at a depth of 3 m in Jamaica, compared with 2 individuals/m^2 and <1 individual/m^2 at depths of 10 m and 20 m, respectively. Kiene & Hutchings (1994) measured grazing rates of 0.30–1.96 kg m^{-2} yr^{-1} in shallow to mid-range depths off Lizard Island, Australia compared with only 0.08–0.29 kg m^{-2} yr^{-1} in deeper water on the forereef; a similar but variable trend was reported for total bioerosion. Off Bonaire, Steneck & McClanahan (2004) measured a 75% decrease in bite density and size on naturally occurring flora between shallow water and 30 m depth.

Data for infaunal boring are less clear but, on balance, infer a depth-related decrease in substrate destruction. By inference, the increased bioerosion by sponges within the wave-cut notches of Bermuda (Neumann, 1966) argues for greater sponge density close to sea level. Kiene & Hutchings (1994) measured higher rates of boring in shallower water, but the differences were not statistically significant. This is supported by data from Moore & Shedd (1977), who reported a consistent, depth-related decrease in sponge boring on three Jamaican reefs (8–80% decrease between 15 m and 27 m; 65% decrease between 27 m and 40 m). Vogel *et al.* (2000) reported a consistent decrease in microboring off Lee Stocking Island (Bahamas), from 0.2 kg m^{-2} yr^{-1} at 2 m to 0.1 kg m^{-2} yr^{-1} at 10 m and 0.01 kg m^{-2} yr at 30 m. Highsmith (1980) proposed that the abundance of boring bivalves is proportional to primary productivity, which should be higher in shallow water where light intensity is greater. By contrast, Macdonald & Perry (2003) noted an increase in bioerosion with depth within the lagoon of Discovery Bay, Jamaica (8–10% loss shallower than 16 m; 18% below 16 m). However, they also noted heavy nutrient inputs from surrounding development that, in this restricted environment, could exert a control that overwhelms depth-related patterns. Goreau & Hartman (1963) cited a higher density of sponge boring at depths of 30 m or more, but this was in very slow-growing plate corals. No actual bioerosion rates were reported, and this may have reflected low calcification rates rather than high infaunal excavation.

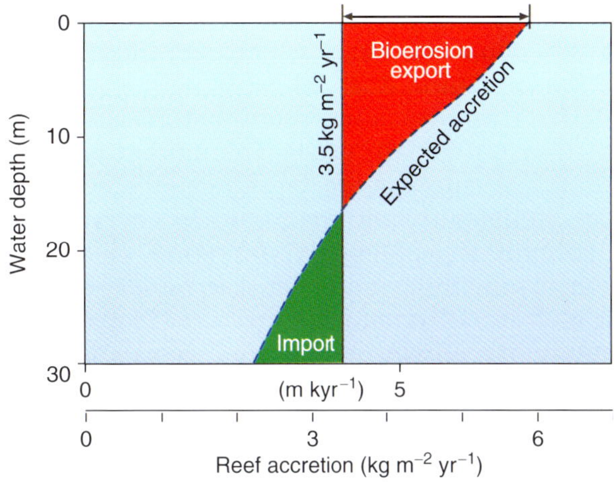

Fig. 11. Stylized model explaining depth-related differences between expected reef accretion based on prevailing models and the results of this study. The lower portion of the 'expected accretion' curve ($d > 10$ m) is based on the measured growth rates and species mix at Cane Bay Virgin Islands) and an assumed total-coral abundance of approximately 50%. Accretion rates in water shallower than 10 m were increased to allow for the greater accretion assumed to be associated with branching *Acropora palmata*. In this scenario, bioerosion decreases from a maximum rate of *c.* 2.5 kg m^{-2} yr^{-1} in shallow water to 0.7 kg m^{-2} yr^{-1} at a depth of 25 m, based on data from the literature cited in this paper. The remainder of the offsetting difference is related to sediment that is transported downslope but stays within the reef.

Figure 11 compares the depth-related trend in accretion from this study to a stylized carbonate-production curve based in part on data from Cane Bay in the US Virgin Islands (Hubbard *et al.*, 1990). In this scenario, increased bioerosion in shallow water (2.0–2.5 kg m^{-2} yr^{-1}) offsets higher calcification rates. With increasing depth, both calcification and bioerosion decrease on parallel tracks, resulting in the loss of a strong depth-related change in reef accretion. It must be kept in mind that bioerosion by itself cannot create the depicted result. Accretion will be affected only to the extent that the sediments produced by bioerosion are exported down the reef front. In this scenario, the apparent homogenization of depth-related accretion patterns also reflects material produced in shallow water being shifted into deeper environs but remaining within the reef.

This exercise makes a significant assumption in creating a stylized production curve from a single location and utilizing depth-related bioerosion data that are scant at best. While none of the values are inconsistent with those found in the literature, this cartoon is neither a quantitative treatment nor a proof that the proposed relationship actually exists. This must await more careful measurement of depth-related bioerosion and sediment-export patterns. It nevertheless highlights the potential importance of bioerosion as something more than a secondary control of reef accretion and architecture. It may ultimately prove to be as important as calcification by corals in determining the rates at which reefs are built.

The 'drowning paradox'?

Based on an assumption that 'many Holocene reefs can be shown to have outpaced even the fastest sea-level rise', Schlager (1981) proposed that reef drowning presents a paradox that can be resolved only by extreme events such as sudden and rapid sea-level rise or degraded oceanic conditions that severely compromise the accretionary capacity of a reef or platform. Based on the results of this study, reefs that accrete at rates in excess of $7\,m\,kyr^{-1}$ are in fact rare. Only one out of 151 intervals exceeded that threshold and was not part of a slower, long-term aggradational history. Given this, the drowning of a reef or platform seems much less surprising. Parts of reefs are capable of extraordinary 'sprints', but reef accretion is generally slower and sea level largely exerts the dominant control. At the average rate of reef accretion proposed by this study ($3–4\,m\,kyr^{-1}$; blue band in Fig. 10b), reefs would have lagged behind rising sea level prior to 7000–6000 years ago when the rate of sea-level rise slowed below this value. After this time, most Caribbean reefs were able to catch up in the absence of some environmental stress that severely compromised the ability of reef biota to make carbonate.

Significance to existing models

Two fundamental principles that underlie our existing Holocene coral-reef models are:

(1) reefs in shallow water (i.e. <5 m) build an order of magnitude faster than their deeper-water counterparts (i.e. >10–15 m);
(2) reefs dominated by branching corals accrete much faster than those primarily inhabited by slower-growing, massive species.

This study has shown that in the Caribbean and western Atlantic these relationships have not been as strong over the past 10–12,000 years, as has been assumed. It is not clear whether this pattern holds for other oceans or other time periods. Dullo (2005) has provided an excellent synthesis of coral growth and reef accretion in the Caribbean versus the Indo-Pacific, and has proposed that the early plateau in sea level *c.* 6000 years ago restricted reef-building and resulted in slower rates of reef accretion in the latter. A more-detailed look at shorter-term accretion within individual reefs (and cores) would provide an excellent comparison to this study and could serve to broaden our understanding of the role of sea-level rise and accommodation space in a pattern that seems considerably different than has been taught.

Whether or not something akin to the scenario depicted in Fig. 11 ultimately emerges as the correct explanation, the lack of a substantial decrease in reef accretion according to either depth or species cannot be ignored, and may have profound implications for models of coral-reef accretion over the past 10,000 years, and perhaps longer. At the very least, these findings reinforce the idea that reefs do not grow in the organic sense, rather they build or accrete as largely physical structures in which biological processes make varying contributions. Thus, although impossible, it would be fortuitous indeed if it were possible to simply strike the phrase 'reef growth' from the literature. While 'growth' in the strictest sense is not incorrect, the phrase 'reef growth' makes a strong inference of biological primacy that would not be of concern when discussing, for example, 'delta growth' or 'fault growth'.

In the absence of oceanographic stresses that limit calcification as a supplier of raw material for reef building, the rate of sea-level rise emerges as the primary determinant of whether or not a reef will "keep up" or "give up" (*sensu* Neumann & Macintyre, 1985). Prior to 7000 cal. yr BP, sea-level rose at a rate exceeding $5\,m\,kyr^{-1}$, faster than accretion in most of the core intervals measured in this study. After 6000 cal. yr BP, this rate dropped below $1\,m\,kyr^{-1}$, and virtually every reef shallower than 25 m was capable of keeping pace or exceeding on rising sea level. In some instances, accretion in reefs closer to sea level was probably constrained by a lack of accommodation space. While this undoubtedly skewed the pattern of shallow-water accretion rates reported in this study, its effect was not sufficient to return shallow-water accretion to a position of primacy.

It is interesting to note that a weak but positive correlation was found between the rates of sea-level rise and accretion. Again this was probably influenced by a lack of accommodation space for shallow-water reefs during intervals of slowed sea-level rise. However, the higher rates of accretion during periods of rapidly rising sea level were not found at mid-depths but rather in some of the deepest calculated palaeowater depths. Core PAR-19 (Fig. 10b) built at an average rate exceeding 9 m kyr^{-1}. Despite starting off in 15 m of water, a depth generally considered too deep to allow significant accretion by branching corals, this *A. palmata* reef was actually catching up with rapidly rising sea level when accretion suddenly stopped *c.* 9000 years ago; it was in water only 5 m deep by that time. Similarly, the submerged reefs off Barbados cored by Fairbanks (1989) accreted at rates higher than 6 m kyr^{-1} during perhaps the most rapid sea-level rise in the Holocene. All of this suggests reconsideration of existing Holocene reef models, which are based largely on strong depth- and species-dependant accretion.

CONCLUSIONS

The findings of this study appear to conflict with two of the most basic rules of Holocene coral-reef accretion. There are undoubted inaccuracies in the methods used to log the cores and the assumptions used in analysing the data from the literature. However, it is difficult to imagine any combination of factors that, if changed, could make the pattern seen in Figs 7–10 look like those in Figs 2 and 3. The next step is to revisit the studies from which the existing concepts were developed, and determine whether the reef-wide patterns that they examined were actually different than those shown here or whether the models derived from them were simply a very good explanation of what was known at the time, but now need to reconsider. Similarly, the results reported above should be used to resolve how such different conclusions can be reached and whether these should be used to formulate new models.

Based on the findings discussed above, the following conclusions are offered.

1. No strong depth-related decrease in reef-accretion rate was found in cores from reefs that built in water depths between 0 and 25 m.
2. The decrease in average accretion rate from 3.58 m kyr^{-1} in shallow water to 2.76 m kyr^{-1} between depths of 20 and 25 m was less than the standard deviations measured at any depth (2.20–3.13 m kyr^{-1}), and was not statistically significant.
3. Average accretion rates in reefs dominated by branching *A. palmata* (3.83 m kyr^{-1}) were not statistically different than those from intervals of massive species (3.07 m kyr^{-1}).
4. Bioerosion and export of detritus into deeper reef environs undoubtedly exert some control on this pattern. Whether they can make up the difference between previously assumed trends and those shown here awaits further study. Nevertheless, the role of bioerosion emerges as a factor that is quantitatively important in not only influencing the character of internal reef fabrics, but also affecting the rate at which reefs build as well.
5. Few reefs have built faster than the rate of sea-level rise in the early and mid-Holocene. Therefore, drowned reefs should be more expected than has been suggested by Schlager's (1981) 'drowning paradox'. This has an important bearing on understanding the relationship between reef accretion and rising sea level, both past and future.
6. Coral reefs are structures that are built as much by physical sedimentation as by calcification by corals, coralline algae and other large calcifiers. In short, reef corals grow; coral reefs accrete.

ACKNOWLEDGEMENTS

This paper is indirectly the result of nearly 30 years of core investigations that were supported by the National Science Foundation, the National Oceanic and Atmospheric Administration (National Undersea Research Program and Sea Grant), and the National Park Service. More recently, our understanding of Holocene and older fossil reefs in outcrop have been made possible through funding by the Petroleum Research Fund. I am indebted to an army of colleagues and students who have toiled with me to collect many of the samples upon which these conclusions are based, in particular, Ivan Gill, Randy Burke, Heinrich Zankl, Hank Tonnemacher and Wilson Ramirez. Their names are absent from this paper only to protect the innocent. I also want to acknowledge the support of colleagues at West Indies Laboratory who served as my mentors or provided much-needed support and valuable commentary throughout the years in which many of the cores in this study were recovered, especially Bob Dill, Bill and Betsy Gladfelter, Gray Multer and John Ogden. The manuscript also benefited from the comments of an anonymous reviewer, plus the review and thoughtful discussions/correspondence over the past year with Eberhard Gischler, with whom I am honoured to both agree and disagree.

Finally, individuals like Bob Ginsburg who asked the questions that have directed us for the three decades since our first holes were drilled with those crude, portable units ultimately set the stage for a new science that was to follow. The session in which this contribution was presented reflects the high esteem in which we hold, not only Bob, but all of those individuals who propose ideas that force us to abandon comfortable dogmas.

APPENDICES

Appendix 1. Data used for analyses discussed in this paper.

St. Croix (USVI)

Sample	SL [m]	Wtr D [m]	Accretion rate [m kyr^{-1}] *Acropora palmata*	Accretion rate [m kyr^{-1}] Massive	Site (source)
BI 01-02	−2.2	−1.7	2.3		Buck Island (Hubbard *et al.*, 2005)
BI 01-13	−2.8	−2.7	2.9		
BI 01-24	−3.4	−3.8	1.7		
BI 01-40	−6.8	−4.2		2.0	
BI 02-01	−1.0	−2.4	10.0		
BI 02-05	−1.2	−3.4	3.7		
BI 02-27	−2.0	−5.9		2.6	
BI 02-43	−3.2	−8.6		2.5	
BI 03-05	−0.1	−0.4	1.3		
BI 03-16	−0.8	−1.3	2.0		
BI 03-37	−3.3	−5.2		1.7	
BI 04-02	−2.4	0.0		1.4	
BI 04-14	−4.0	−0.9		6.3	
BI 04-21	−4.4	−1.6		1.3	
BI 05-24	−0.6	−5.6	1.9		
BI 05-40	−1.0	−6.2		2.7	
BI 07-07	−0.3	−7.6		2.3	
BI 07-20	−1.5	−10.8		7.1	
BB 01-02	−0.8	−4.3	3.8		BI Bar (Hubbard *et al.*, 2005)
BB 01-11	−1.3	−6.9	0.8		
BB 02-01	−1.2	−4.8		1.7	BI Bar (Macintyre & Adey, 1990)
BB 02-02	−2.0	−6.0		1.3	
BB 02-03	−2.8	−6.5		3.4	
CB 01-05	0.0	−10.4		0.6	Cane Bay (Hubbard *et al.*, 1990)
CB 01-11	−0.9	−10.2		0.7	
CB 02-01	−0.3	−15.1		3.3	
CB 02-07	−0.5	−16.9		0.9	
CB 03-01	−0.8	−8.2		0.1	
CB 04-03	−0.2	−7.9		1.0	
CB 04-09	−0.9	−8.1		0.7	
CB 04-16	−1.6	−8.1		1.2	
CB 04-23	−4.3	−8.8		0.4	
CB 05-15	−0.9	−12.5		0.8	
CB 05-23	−1.6	−12.7		1.3	
CB 06-02	−0.5	−20.9		1.9	
CB 06-07	−0.9	−21.6		0.9	
CB 07-02	−0.2	−17.0		1.1	
LBE 06-06	−3.8	−12.2		2.8	Lang Bank (Hubbard, unpublished data)
LBE 06-18	−4.5	−13.2		3.0	
LBE 06-30	−5.3	−14.1	5.0		
LBE 07-02	−4.1	−19.4	2.2		
LBE 07-18	−5.0	−20.5	5.5		
LBN 01-06	−5.5	−9.5		13.5	

Appendix 1. Continued.

St. Croix (USVI)

Sample	SL [m]	Wtr D [m]	Accretion rate [m kyr^{-1}] *Acropora palmata*	Massive	Site (source)
LBN 02-05	−6.5	−6.0	5.0		
LBN 02-18	−7.5	−7.1	10.5		
LBN 02-30	−8.0	−8.8	2.3		
LBN 02-41	−8.0	−11.3	1.2		
LBN 05-25	−5.5	−13.9		9.9	
LBN 05-53	−6.1	−16.6		9.7	
LBN 08-02	−5.8	−6.7	7.3		
LBN 08-22	−6.4	−8.5	5.1		
LBN 08-36	−7.7	−9.2	13.4		
LBN 09-03	−6.0	−7.7	4.2		Lang Bank (Hubbard, unpublished data)
LBS 03-03	−5.9	−9.3	2.6		
LBS 03-11	−8.5	−10.2		3.5	
LBS 03-15	−9.3	−10.5	3.8		
LBS 03-19	−11.0	−10.6	1.9		
LBS 04-04	−4.9	−11.2	0.3		
LBS 04-08	−5.0	−11.7		1.8	
LBS 04-14	−6.1	−11.7		4.0	
LBS 04-21	−8.2	−12.6		3.3	
LBS 04-34	−10.0	−12.7	1.6		
LBS 06-39	−6.4	−15.8		3.7	
S Lang Bk	−4.8	−10.3		3.4	Lang Bank (Adey *et al.*, 1977)
S Lang Bk	−6.4	−11.0		4.8	
S Lang Bk	−5.3	−11.1		1.9	
S Lang Bk	−10.0	−11.1	1.0		
Tague Bay	−1.0	0.0	2.9		Tague Reef (Burke *et al.*, 1989)
Tague Bay	−5.8	−1.4	4.4		
Tague Bay	−0.1	−1.5	6.0		
Tague Bay	−1.3	−1.6		2.6	
Tague Bay	0.0	−1.6	0.9		
Tague Bay	−2.5	−4.8		0.5	
Puerto Rico					
PAR 01-02	0.0	−5.0	0.2		Parguera (Hubbard *et al.*, 1997)
PAR 01-07	−2.8	−3.0	8.8		
PAR 01-11	−3.0	−5.0		3.2	
PAR 01-27	−3.9	−7.1		4.5	
PAR 01-39	−5.0	−10.8		2.2	
PAR 02-06	−7.0	−8.4	5.9		
PAR 02-13	−9.6	−10.4	2.3		
PAR 03-10	−7.1	−7.4	2.9		
PAR 03-12	−7.9	−7.5	9.6		
PAR 03-20	−8.9	−8.2	4.9		
PAR 05-03	0.0	−0.1	0.8		
PAR 05-09	−2.3	−0.5	8.5		
PAR 05-20	−2.5	−2.1	3.4		
PAR 05-29	−3.2	−4.3		7.7	
PAR 05-40	−3.8	−7.8		2.7	
PAR 05-53	−6.5	−10.4		1.9	
PAR 06-01	−0.4	−0.3	0.4		
PAR 06-11	−3.1	−0.9		1.0	Parguera (Hubbard *et al.*, 1997)
PAR 06-22	−3.9	−4.1		1.9	
PAR 07-05	−2.0	−1.4		14.7	
PAR 08-06	−1.0	−2.5	1.5		
PAR 08-18	−2.9	−4.0		3.9	
PAR 08-25B	−3.5	−5.9		2.7	
PAR 09-09	−0.2	−8.2		4.8	
PAR 09-11	−0.2	−8.5		2.7	
PAR 09-29	−1.0	−12.5		2.6	

(*Continued.*)

Appendix 1. Continued.

St. Croix (USVI)					
Sample	SL [m]	Wtr D [m]	Accretion rate [m kyr^{-1}] *Acropora palmata*	Massive	Site (source)
PAR 10-19	−1.0	−6.0		4.4	
PAR 10-25	−1.2	−6.6		4.1	
PAR 10-35	−1.4	−8.1		2.2	
PAR 10-63	−2.9	−10.3		4.2	
PAR 10-81	−3.4	−11.8		0.1	
PAR 11-11	−7.0	−9.0	3.0		
PAR 11-29	−11.5	−8.3	3.1		Parguera (Hubbard *et al.*, 1997)
PAR 12-01	0.0	−0.4	9.8		
PAR 12-17	−0.5	−6.4	0.1		
PAR 12-24	−4.3	−3.2		1.1	
PAR 17-09	−0.7	−16.6		2.0	
PAR 19-55	−13.7	−7.8	4.5		
PAR 19-87	−16.4	−11.5	6.7		
PAR A-02	−9.9	−7.3	3.4		
PAR A-14	−11.7	−6.9	0.4		
VH1-C1	−0.1	−1.1		0.6	Vieques (Lighty *et al.*, 1982)
Florida					
Sand Key	−7.1	−2.2	2.0		Toscano & Macintyre (2003)
Fla-SeqA-01	−10.4	−7.1	6.6		Lighty *et al.* (1982)
Fla-SeqB-01	−3.9	−12.7	0.3		
Fla-SeqB-02	−13.1	−4.5	4.4		
Fla-SeqC-01	−15.6	−2.5	7.1		
Fla-SeqD-01	−11.4	−11.6	3.6		
Panama					
H03C02 5/5	−2.0	0.2	1.4		Galeta (Macintyre & Glynn, 1976)
H03C06 2/2	−4.8	−1.2	1.3		
H03C7 2/3	−6.0	−0.8	9.3		
H06C01 3/3	−1.5	−0.1	2.0		
H10C04 1/7	−3.0	−2.5	2.3		
H11C01 4/4	−3.2	−0.6	3.2		
H11CC06 4/5	−4.6	−3.5		3.7	
Antigua					
N2-07	−1.6	−1.5		3.4	Nonesuch Bay (Macintyre *et al.*, 1985)
N2-38	−3.8	−7.5		3.9	
N4-04	−0.3	−1.0		1.6	
N5-05	−3.3	−0.5		4.5	
Belize					
BBR-4	−7.6	−0.4	3.1		Belize (Gischler & Hudson, 2004)
BBR-5	−7.0	−4.1		4.4	
BBR-6	−4.2	−1.9	3.2		
BBR-7	−4.2	−0.2	2.4		
BBR-8a	−6.0	−2.5	6.3		
BBR-8b	−5.5	−6.3		4.3	
BBR-9a	−2.5	−8.3		2.8	Belize (Gischler & Hudson, 2004)
BBR-9b	−5.2	−9.4		4.4	
Barbados					
RGF12-16-5	n/a	n/a	5.7		Deep Reefs (Fairbanks, 1989)
RGF12-21-10	n/a	n/a	0.5		
RGF12-6-7	n/a	n/a	4.1		
RGF12-9-5	n/a	n/a	9.5		
RGF7-12-2	n/a	n/a	10.8		
RGF7-16-2	n/a	n/a	8.1		
RGF7-4-2	n/a	n/a	1.0		
RGF7-5-5	n/a	n/a	7.4		
RGF9-21-11	n/a	n/a	4.6		
RGF9-32-4				2.8	
RGF9-8-2	n/a	n/a	11.1		

REFERENCES

Adey, W.H. (1978) Coral reef morphogenesis: a multidimensional model. *Science*, **202**, 831–837.

Adey, W.H. and **Burke, R.B.** (1976) Holocene bioherms (algal ridges and bank barrier reefs) of the eastern Caribbean. *Geol. Soc. Am. Bull.*, **87**, 95–109.

Adey, W.H., Macintyre, I.G., Stuckenrath, R. and **Dill, R.F.** (1977) Relict barrier reef system off St. Croix: its implications with respect to late Coenozoic coral reef development in the western Atlantic. *Proc. 3rd Intl. Coral Reef Symp.*, **2**, 15–21.

Bard, E., Hamelin, B., Fairbanks, R.G. and **Zindler, A.** (1990) Calibration of the ^{14}C timescale over the past 30,000 years using mass spectrometric U–Th ages from Barbados cores. *Nature*, **345**, 405–410.

Blanchon, P. (2005) Comments on 'Corrected western Atlantic sea-level curve for the last 11,000 years based on calibrated ^{14}C dates from *Acropora palmata* framework and intertidal mangrove peat'. *Coral Reefs*, **24**, 183–186.

Blanchon, P. and **Perry, C.T.** (2004) Taphonomic differentiation of *Acropora palmata* facies in cores from Campeche Bank Reefs, Gulf of Mexico. *Sedimentology*, **51**, 53–76.

Bosscher, H. (1992) *Growth Potential of Coral Reefs and Carbonate Platforms*. Proefschrift Vrije Universiteit Amsterdam, Utrecht, Netherlands, 157p.

Burke, R.B., Adey, W.H. and **Macintyre, I.G.** (1989) Overview of the Holocene history, architecture, and structural components of Tague Reef, St. Croix, U.S.V.I. In: *Terrestrial and Marine Geology of St. Croix, U.S. Virgin Islands* (Ed. D.K. Hubbard), *West Indies Laboratory Special Publ.*, **8**, 105–110.

Chalker, B.E. (1981) Simulating light-saturation curves for photosynthesis and calcification by reef-building corals. *Marine Biol.*, **63**, 135–141.

Chazottes, V., Le Champion-Alsumard, T. and **Peyrot-Clausade, M.** (1995) Bioerosion rates on coral reefs: interactions between macroborers, microborers and grazers (Morea, French Polynesia). *Paleogeogr., Paleoclimatol., Paleoecol.*, **13**, 189–198.

Conand, C., Chabanet, P., Cuet, P. and **Letourneur, Y.** (1997) The carbonate budget of a fringing reef in La Reunion Island (Indian Ocean): sea-urchin and fish bioerosion and net calcification. *Proc. 8th Int. Coral Reef Symp.*, **1**, 953–958.

Dullo, W.-C. (2005) Coral growth and reef growth: a brief review. *Coral Reefs*, **51**, 33–48.

Fagerstrom, A.J. (1987) *The Evolution of Reef Communities*. John Wiley & Sons, New York, 600 p.

Fairbanks, R.G. (1989) A 17,000 year glacio-eustatic sea level record: influence of glacial melting rates on the Younger Dryas event and deep-ocean circulation. *Nature*, **342**, 637–642.

Fosså, J.H., Mortensen, P.B. and **Fuevik, D.M.** (2002) The deep-water coral *Lophelia pertusa* in Norway waters: distribution and fishery impacts. *Hydrobiologia*, **47**, 1–12.

Ginsburg, R.N. (1958) Early diagenesis and lithification of shallow-water carbonate sediments in south Florida. *SEPM Spec. Publ.*, **5**, 80–100.

Ginsberg, R.N., Gischler, E. and **Schlager, W.** (1994) Johannes Walther on reefs. *Comparative Sedimentology Laboratory, University of Miami, Geological Milestones*, Vol. II, 141 p.

Gischler, E. (2006) Comments on 'Corrected western Atlantic sea-level curve for the last 11,000 years based on calibrated ^{14}C dates from *Acropora palmata* framework and intertidal mangrove peat'. *Coral Reefs*, **25**, 273–279.

Gischler, E. and **Hudson, J.H.** (2004) Holocene development of the Belize barrier reef. *Sedimentary Geol.*, **164**, 223–236.

Goreau, T.F. and **Hartman, W.D.** (1963) Control of coral reefs by boring sponges. In: *Mechanisms of Hard Tissue Destruction* (Ed. R.F. Sognnaes), *AAAS Publ.*, **75**, 25–54.

Grant, R.E. (1826) Notice of a new zoophyte (*Cliona celata* Gr.) from the Firth of Forth. Edinburgh. *New Philos. J.*, Apr–Oct, 78–81.

Graus, R.R. and **Macintyre, I.G.** (1999) Global warming and the future of Caribbean coral reefs. *Carbonates Evaporites*, **13**, 43–47.

Highsmith, R.C. (1980) Geographic patterns of coral bioerosion, a productivity hypothesisis. *J. Exp. Mar. Biol. Ecol.*, **46**, 177–196.

Hubbard, D.K. (1992) Hurricane-induced sediment transport in open-shelf tropical systems – an example from St. Croix, U.S. Virgin Islands. *J. Sedimentary Petrol.*, **62**, 946–960.

Hubbard, D.K., Burke, R.B. and **Gill, I.P.** (1986) Styles of reef accretion along a steep, shelf-edge reef, St. Croix, U.S. Virgin Islands. *J. Sedimentary Petrol.*, **56**, 848–861.

Hubbard, D.K., Burke, R.B. and **Gill, I.P.** (1998) Where's the reef: the role of framework in the Holocene. *Carbonates Evaporites*, **13**, 3–9.

Hubbard, D.K., Gill, I.P., Burke, R.B. and **Morelock, J.** (1997) Holocene reef backstepping – southwestern Puerto Rico Shelf. *Proc. 8th Int. Coral Reef Symp.*, **2**, 1779–1784.

Hubbard, D.K., Miller, A.I. and **Scaturo, D.** (1990) Production and cycling of calcium carbonate in a shelf-edge reef system (St. Croix, U.S. Virgin Islands): applications to the nature of reef systems in the fossil record. *J. Sedimentary Petrol.*, **60**, 335–360.

Hubbard, D.K., Zankl, H., van Heerden, I. and **Gill, I.P.** (2005), Holocene reef development along the northeastern St. Croix shelf, U.S. Virgin Islands. *J. Sedimentary Res.*, **75**, 97–113.

Kiene, W.E. and **Hutchings, P.A.** (1994) Bioerosion experiments at Lizard Island, Great Barrier Reef. *Coral Reefs*, **13**, 91–98.

Lighty, R.G., Macintyre, I.G. and **Stuckenrath, R.** (1982) *Acropora palmata* reef framework: a reliable indicator of sea level in the western Atlantic for the past 10,000 years. *Coral Reefs*, **1**, 125–130.

Lowenstam, H.A. (1950) Niagaran reefs of the Great Lakes area. *GSA Memoir*, **67**, 215–248.

Macdonald, I.A. and **Perry, C.T.** (2003) Biological degradation of coral framework in a turbid lagoon environment, Discovery Bay, Jamaica. *Coral Reefs*, **22**, 523–535.

Macintyre, I.G. (1988) Modern coral reefs of western Atlantic: new geologic perspective. *AAPG Bull.*, **72**, 1360–1369.

Macintyre, I.G. and **Adey, W.H.** (1990) Buck Island Bar, St. Croix, USVI: a reef that cannot catch up with sea level. *Atoll Res. Bull.*, **336**, 1–7.

Macintyre, I.G. and **Glynn, P.W.** (1976) Evolution of modern Caribbean fringing reef, Galeta Point, Panama. *AAPG Bull.*, **60**, 1054–1072.

Macintyre, I.G., Burke, R.B. and **Stuckenrath, R.** (1981) Core holes in the outer fore reef off Carrie Bow Cay, Belize: a key to the Holocene history of the Belizean barrier reef complex. *Proc. 4th Int. Coral Reef Symp.*, **1**, 567–574.

Macintyre, I.G., Multer, H.G., Zankl, H.L., Hubbard, D.K., Weiss, M.P. and **Stuckenrath, R.** (1985) Growth and depositional facies of a windward reef complex (Nonsuch Bay, Antigua, W.I.). *Proc. 5th Int. Coral Reef Symp.*, **6**, 605–610.

Moore, C.H. and **Shedd, W.W.** (1977) Effective rates of sponge bioerosion as a function of carbonate production. *Proc. 3rd Int. Coral Reef Symp.*, **2**, 499–505.

Neumann, A.C. (1966) Observations on coastal erosion in Bermuda and measurements of the boring rates of the sponge *Cliona lampa*. *Limnol. Oceanog.*, **11**, 92–108.

Neumann, A.C. and **Macintyre, I.** (1985) Reef response to sea-level rise: keep-up, catch-up, or give-up. *Proc. 5th Int. Coral Reef Symp.*, **3**, 105–110.

Newell, N.D., Rigby, J.K., Fischer, A.G., Whiteman, A.J. Hickox, J.E. and **Bradley, J.S.** (1953) *The Permian Reef Complex of the Guadalupe Mountains Region, Texas and New Mexico: a Study in Paleoecology*. W.H. Freeman, San Francisco, 236p.

Perry, C.T. (1999) Reef framework preservation in four contrasting modern reef environments, Discovery Bay, Jamaica. *J. Coastal Res.*, **15**, 796–812.

Perry, C.T. (2000) Macroboring of Pleistocene coral communities, Falmouth Formation, Jamaica, *Palaios*, **15**, 483–491.

Perry, C.T. (2001) Storm-induced coral rubble deposition: Pleistocene records of natural reef disturbance and community response. *Coral Reefs*, **20**, 171–183.

Quoy, J.R. and **Gaimard, J.P.** (1825) Memoire sur l'accroissment des polyps lithophytes consider e geologiquement. *Annales Sci. Naturelles*, **6**, 373–390.

Reed, J.K., Shepard, A.N., Koenig, C., Scanlon, K.M. and **Gilmore, R.G. Jr.** (2005) Mapping, habitat characterization, and fish surveys of the deep-water *Oculina* coral reef Marine Protected Area: a review of historical and current research. In: *Cold-Water Corals and Ecosystems* (Eds A. Freiwald and J.M. Roberts), Springer-Verlag, Berlin-Heidelberg, pp. 443–465.

Schlager, W. (1981) The paradox of drowned reefs and carbonate platforms. *GSA Bull.*, **92**, 197–211.

Scoffin, T.P. (1992) Taphonomy of coral reefs: a review. *Coral Reefs*, **11**, 57–77.

Scoffin, T.P., Stearn, C.W., Boucher, D., Frydll, P., Hawkins, C.M., Hunter, I.G. and **MacGeachy, J.K.** (1980) Calcium carbonate budget of a fringing reef on the west coast of Barbados. *Bull. Mar. Sci.*, **30**, 475–508.

Stearn, C.W. and **Scoffin, T.P.** (1977) Carbonate budget of a fringing reef, Barbados. *Proc. 3rd Intl. Coral Reef Symp.*, **2**, 471–476.

Steneck, R.S. (1994) Is herbivore loss more damaging to reefs than hurricanes? Case studies from two Caribbean Reef Systems (1978–1988). *Proc. Colloq. on Global Aspects of Coral Reefs, Miami*, 220–226.

Steneck, R.S. and **McClanahan, T.** (2004) *A Report on the Status of the Coral Reefs of Bonaire with Advice on the Establishment of Fish Protection Area.*, Univ. of Maine, Walpole, ME, 89 pp.

Stuiver, M., Reimer, P.J. and **Braziunas, T.F.** (1998) High-precision radiocarbon age calibration for terrestrial and marine samples. *Radiocarbon*, **40**, 1127–1151.

Talma, A. and **Vogel, J.** (1993) A simplified approach to calibrating C14 dates. *Radiocarbon*, **35**, 317–322.

Toscano, M.A. and **Lundberg, J.** (1998) Early Holocene sea-level record from submerged fossil reefs on the southeast Florida margin. *Geology*, **26**, 255–258.

Toscano, M.A. and **Macintyre, I.G.** (2003) Corrected western Atlantic sea-level curve for the last 11,000 years based on calibrated ^{14}C dates from *Acropora palmata* framework and intertidal mangrove peat. *Coral Reefs*, **22**, 257–270.

Tribollet, A., Decherf, G., Hutchings, P.A. and **Peyrot-Clausade, M.** (2002) Large-scale spatial variability in bioerosion of experimental coral substrates on the Great Barrier Reef (Australia): importance of microborers. *Coral Reefs*, **21**, 424–432.

Vogel, K., Gektidis, M., Golubic, S., Kiene, W.E. and **Radtke, G.** (2000) Experimental studies on microbial bioerosion at Lee Stocking Island: Bahamas and One Tree Island, Great Barrier Reef, Australia – implications for paleoecological reconstructions. *Lethaia*, **33**, 190–204.

Woodley, J.D., Chornesky, E.A., Clifford, P.A., Jackson, J.C.B., Kaufman L.S., Knowlton, N., Land, J.C., Pearson, M.P., Wulff, J.L., Curtis, A.S.G., Dallmeyer, M.D., Jupp, B.P., Koehl, M.A.R., Neigel, J. and **Sides, E.M.** (1981) Hurricane Allen's impact on Jamaican coral reefs. *Science*, **214**, 749–755.

Int. Assoc. Sedimentol. Spec. Publ. (2009) **41**, 19–28

The mystique of beachrock

EUGENE A. SHINN

University of South Florida Marine Science Center, 140 Seventh Avenue South, St. Petersburg, FL 33701, USA (E-mail: eshinn@marine.usf.edu)

ABSTRACT

Parallel rows of intertidal beachrock composed of large (up to 4 m), oblong, pillow-shape blocks submerged by rising Holocene sea level occur throughout the Caribbean. Such blocks have recently been discovered in 90 m of water off southwest Florida. The best known example is in 5–7 m of water off the northwest side of Bimini, Bahamas, where the feature is known as the Bimini Road. 'New-Age' alternative thinkers assert that the arrangement of the Bimini stones, and similar stones off Andros Island in the Bahamas, is actually man-made. The theory holds that ancient humans, principally citizens of the mythical city of Atlantis rearranged blocks of beachrock to form a harbour. This paper demonstrates that the stones and their arrangement are natural and suggests that some villages protected by ancient harbours in the Mediterranean may in fact have been developed on naturally occurring beachrock.

Keywords Atlantis, Bimini Road, Bahamas, Beachrock, Cayce Foundation, Keystone Vug.

INTRODUCTION

The purpose of this paper is to highlight opposing interpretations that currently exist between geologists and a growing number of 'New-Age' alternative thinkers/archaeologists. A general account of beachrock formation, especially beachrock submerged by rising Holocene sea level, is provided; however, the main intent is to discuss the various alternative origins for submerged beachrock. That beachrock forms in the intertidal zone has been known for almost two centuries, although the precise mechanism of origin has been debated for more than 50 years. For the most recent and detailed history of beachrock distribution and theories of origin, the reader is directed to Gischler (2007). A pivotal paper on beachrock at Loggerhead Key, Dry Tortugas in the Gulf of Mexico, was Robert N. Ginsburg's first paper in 1953.

ANTHROPOGENIC VERSUS NATURAL ORIGIN

Rows of large oblong and polygonal limestone blocks popularly called the 'Bimini Road', lie off the northwest coast of the Bahamian island of Bimini (Fig. 1). The underwater feature first attracted the attention of adventurers/divers in the mid-1960s. The mystique of these huge rocks, aligned in several rows in approximately 7 m of water (Fig. 2), has continued to resonate among alternative thinkers. Since their discovery, made possible after erosion of overlying sediment, the proposition that the stones are man-made and/or 'intelligently' placed by ancient cultures has proliferated and gained credibility. The anthropogenic-origin hypothesis initiated with Valentine (1969, 1976), Rebikoff (1972, 1979) and followed the suggestion of Cayce (1968). Berlitz (1969, 1984) and Zink (1978) related the origin of the Bimini Road to the legend of the lost continent of Atlantis. Zink's study was funded by the Cayce Foundation, a spirit-based group located in Virginia Beach, Virginia.

Little was published about the Bimini Road during the 1980s. However, after a decade-long hiatus of interest and activity, the anthropogenic-origin hypothesis was revived in a two-part pro/con article (Hearty & Donato, 1998) in which Donato argues for a human origin and Hearty argues for a natural origin. More recently, Menzies (2002) proposed that the stones were placed on the sea floor by ancient Chinese merchantmen who he maintains travelled to the Bahamas before Columbus. The stones were thought to be part of a system for hauling and repairing damaged ships (Menzies, 2002).

Although many scientists have examined the Bimini rocks and their peculiar alignment John

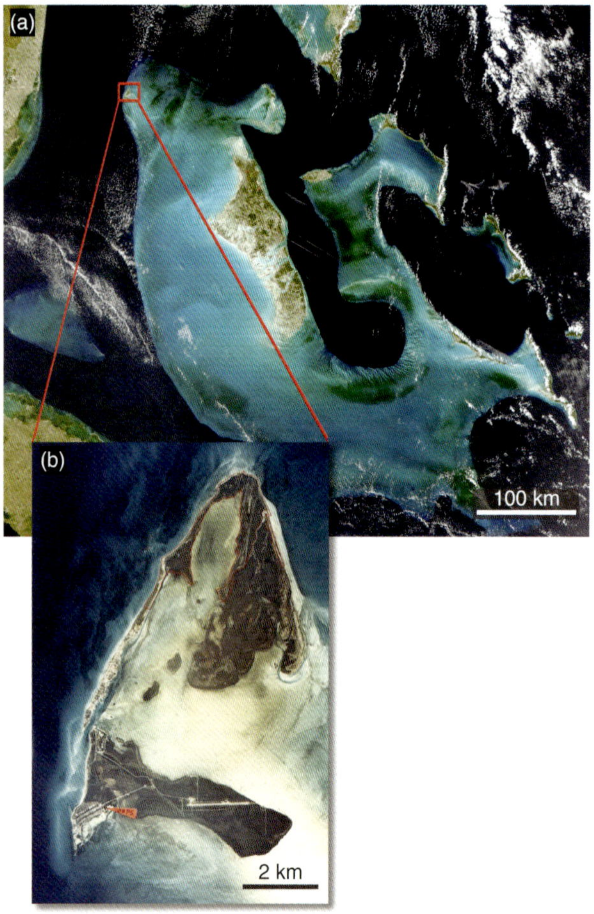

Fig. 1. (a) Satellite image of Great Bahama Bank and Andros Island in the Bahamas. North is at top. Florida is upper left corner. Inset (b) shows North and South Bimini. The underwater Bimini Road parallels the northwest coast of North Bimini. North Bimini and South Bimini are separated by a navigation channel that leads to Bimini Harbour. The harbour is partially protected by shallow water and a series of mangrove-populated sand spits to the east. The arrow indicates Port Royal.

Fig. 2. Underwater view of the submerged 'Bimini Road' beachrock stones.

Gifford conducted the first serious geological study in the 1970s. Relying on keen observation, hand specimens and some core samples, Gifford & Ball (1980) determined that the stones were naturally emplaced. In a study of supposedly anthropogenic objects from the vicinity Harrison (1971), also concluded the stones and their alignment were natural. Adventurer and book writer Peter Thompkins later initiated and financed a study during which the stones were extensively sampled with a newly developed coring device. A summary of the expedition is provided by Shinn (2004). The cores proved that the stones are composed of *in situ* natural beachrock that has been submerged by a combination of rising Holocene sea level and erosion of underlying sand (Shinn, 1978). Coring and examination showed that they rest directly on weathered Pleistocene limestone. The Pleistocene limestone is coated by a reddish-brown calcrete, indicating subaerial exposure preceding Holocene sedimentation and beachrock formation.

Later, McKusick & Shinn (1980) presented bulk 14C age data from the cores. The dates ranging from 3510 to 2745 years BP indicate the stones are much too young to be part of the mythological city/state of Atlantis, which was said to be a legend 7 ka when told to Plato 2 ka.[1] Randi (1981) also concluded that the stones are natural, as did geologist Paul Hearty (Hearty & Denato, 1998). Richards (1988), Cayce *et al.* (1988) and Denato (Hearty & Denato, 1998) had already challenged the earlier studies and at the time of writing, internet bloggers, including a sizeable group with a website and a journal called *Atlantis Rising* have challenged all science-based investigations. A psychologist (see Little (2006), recently joined the ranks of alternative believers. He claims in his website that the author (Shinn) is part of a government conspiracy to prevent people from learning the true origin of the stones.

Whereas the 1978 study by Shinn was based on more than a dozen *in situ* core borings and ^{14}C dates (McKusick & Shinn, 1980), human-origin theories of formation were, and still are, based almost entirely on the seemingly man-made pattern of the stones. These hypotheses state that beaches, and hence beachrock, cannot form such straight lines. However, proponents of the hypotheses, often point to the area of curved stones at their southern end, called the 'inverted J', as proof of their anthropogenic origin (Little, 2006). If ancient man did construct the so-called Bimini Road or harbour using beachrock or a man-made material, there should be abundant human artefacts at the site. Scoffin (1970) noted glass bottles being incorporated into contemporary beachrock at Bimini, as did Shinn (1978). However, to date no obvious human artefacts, such as tools, pottery, etc., have been found incorporated within the Bimini Road beachrock or otherwise conclusively linked with formation of the stones.

The lack of certifiable artefacts presents a problem for those who believe the stones are anthropogenic. The proponents generally describe the fabled city/state of Atlantis, following descriptions of Spence (1968), to be a highly technical society. An advanced technical society, similar to ours, should have produced many artefacts much as those of other ancient cultures. Certifiable artefacts that have been found in and near the site post-date the stones by as much as 3 kyr and consist mainly of shipwreck remnants such as ballast stones, cement barrels or other material discarded from ships. For example, Portland Cement was originally transported aboard ships in wooden barrels. When jettisoned overboard, the wood decays, leaving cement cylinders that can be confused with columns (Harrison, 1971). Finally, the alternative thinkers may be influenced by myths of the so-called Bermuda Triangle and the often repeated stories of buried pyramids and temples. The cement barrels misidentified as columns (Harrison, 1971) probably initiated the stories of submerged temples. These legends abound and appear to be proliferating with time.

How underwater 'roads' are created

To evaluate the Bimini Road more accurately, it is appropriate here to review the fundamentals of beachrock formation and configuration. For a thorough review of beachrock origins and theories, see Gischler (2007). His scholarly review includes observations by nineteenth-century scientists Chamisso (1821), Lyell (1832), Darwin (1842), Dana (1875) and Gardiner (1898). These were the earliest scientists to recognize beachrock formation. Later in the twentieth century, Vaughan (1914), Field (1919, 1920), Daly (1924) and Keunen (1933) described or noted the presence, and rapidity, of beachrock formation in various parts of the world. These twentieth-century researchers noted that beachrock is restricted to warm tropical seas, in essence the same latitudes to which coral reefs are restricted, the so-called Coral Seas.

Modern studies using petrography and carbonate chemistry began with Ginsburg (1953), Stoddart & Cann (1965), Dunham (1970, 1971) and Multer (1971). These workers, like those before, recognized that beachrock formation is restricted to the intertidal zone, but unlike previous studies, they elucidated diagnostic attributes such as the presence of internal laminations caused by grain-size variations, seaward-dipping layers and bubble-shape voids, i.e. keystone vugs (Dunham, 1970). Keystone vugs are features identical to those found in adjacent uncemented beach sands. As rising tides and waves cover the part of a sandy beach that dries during low tide, air trapped within the sand is confined and coalesces to form bubbles. Most such bubbles reach the surface, producing the circular pits so common to beachcombers. However, some air bubbles, however, remain over many tidal cycles. The bubbles tend to push grains upward, forming an arch. These arches function much like the keystone arches used in stone buildings for millennia. Some of these millimetre-size arches can remain locked in position long enough to be 'frozen' in place by precipitation of calcium-carbonate cement between the grains. This process results in the formation of near-spherical to elongate voids within the forming beachrock (Dunham, 1970).

Beachrock voids are similar to the more linear fenestral birds-eye structures that are usually restricted to fine-grained tidal-flat accumulations (Shinn, 1983). For a number of reasons, such voids rarely form in beaches composed of quartz sand. The lack of preserved voids in quartz-sand beach accumulations is probably due to a combination of mineralogy, shape of the grains, the apparent inability of carbonate cement to initiate precipitation on quartz grains and the colder water temperature often associated with quartz sand beaches. However, voids are very common in carbonate-sand beaches, even those composed of spherical ooid sands.

Many questions remain concerning beachrock genesis including (1) cement mineralogy (not all cements are aragonite; some are high-magnesium calcite), (2) proper identification of beachrock (sediment cemented in a freshwater lens beneath an island and later exposed in the intertidal or subtidal zone by rising sea level and erosion can be misidentified as beachrock), (3) rate of cementation, (4) influence of fresh water and temperature and (5) the role of microbes in the cementation process. These considerations and others are beyond the scope and purpose of this paper.

DISCUSSION

Arid versus humid climate

During the mid-1960s, the author examined many beaches and offshore sediments in the Persian and Arabian Gulf. The Persian Gulf climate is arid, and salinity is elevated (Purser & Seibold, 1973). Aridity and high salinity favour rapid calcium-carbonate precipitation, and with little rainfall the rate of freshwater cementation is retarded. Observations in the Arabian Gulf were biased toward areas of the southern windward shore (Trucial Coast), where sedimentation and shoreline accretion are most rapid. The author's observations in the more humid Florida and Bahamas platform environments were generally biased toward areas of erosion and locally to areas of complicated onlap sedimentation (Strasser & Davaud, 1986).

These contrasts in sedimentation style and diagenesis provided unique insights to the processes of beach formation as well as to both intertidal and submarine sediment diagenesis. For example, cementation in the Persian Gulf is so rapid that even sandy sediment a few centimetres below the intertidal surface can be rapidly lithified. The resulting rock often includes glass, pottery, nuts and bolts, and other human artfacts. Subtidal cementation of sandy sediment is rapid and has been shown to transform sediment into hard limestone in less than 30 years (Shinn, 1969). This rapidly forming submarine rock does not contain keystone vugs. In the intertidal beach sands that do contain keystone vugs, cementation proceeds even more rapidly. Tidal pumping in the intertidal beach provides a continual supply of supersaturated seawater, and during low tide, high evaporation and rapid degassing probably accelerate cementation (Fig. 3).

Ginsburg (1953), Evamy (1973) and many others have shown that the most common cementing agent is fibrous aragonite. In areas of rapid lateral beach accretion, cemented zones are continually buried beneath sand as fast as the biogenic and oolitic grains are created and deposited. In such areas, hidden from view, the continual formation of beachrock proceeds unnoticed. If the beach accretes rapidly beyond some unknown threshold, the sediment is taken out of contact with seawater and cementation is slowed or does not take place. When the accretion rate is retarded or periodically stopped, cementation is enhanced. The result of this fluctuating process was repeatedly observed by the author in excavations around the Qatar Peninsula in the mid-1960s, where construction workers mined sand with shovels. The mined areas revealed beachrock zones that were sufficiently lithified as to resist removal with picks and shovels. Such areas were generally left and because of the removal of soft sand around them, they could be observed in three dimensions. The rock areas formed distinct ridges parallel to the accreting beach.

Significantly, sandy sediment in the back-beach areas away from the active intertidal beaches was not being cemented either by aragonite or calcite. Lack of calcite cementation in back-beach areas is probably due to aridity and lack of a rainfall-source of fresh water. In contrast, back-beach areas in the humid Bahamas are quickly lithified (probably <1000 years) by freshwater diagenesis, namely by the precipitation of low-magnesium calcite (Halley & Harris, 1979). Thus, varying rates of sedimentation and absence of fresh water often control formation and distribution of beachrock. Numerous examples of the overriding effect of sedimentation rate, such as rapidly forming submarine sand spits, are provided in Shinn (1969, 1971).

Fishhook spits and 'J' shaped structures

Intertidal cementation in the Persian Gulf is especially rapid on the protected inner margins of curved, fishhook-shaped accreting sand spits. The rock forming on the sand spits is exposed to more than a metre of tidal fluctuation but is protected from waves that would disrupt cementation before it could occur. The formation of new spits is clearly documented in repeated photographs of an area on the northeast coast of Qatar (Shinn, 1973). Newly cemented sediment and increasingly more cementation occur in progressively older and older spits (Fig. 4). Microbial inducement cannot be ruled out, but low rainfall precludes freshwater involvement in the cementation process at this site. The large thick blocks such as those at Bimini and elsewhere in the Caribbean were not observed. Possibly the rapid rate of sedimentation does not allow sufficient subaerial exposure for thermal fracturing and creation of thick blocks. Beachrock that forms beneath accreting sand is not exposed to as much sun and weathering as rock that forms

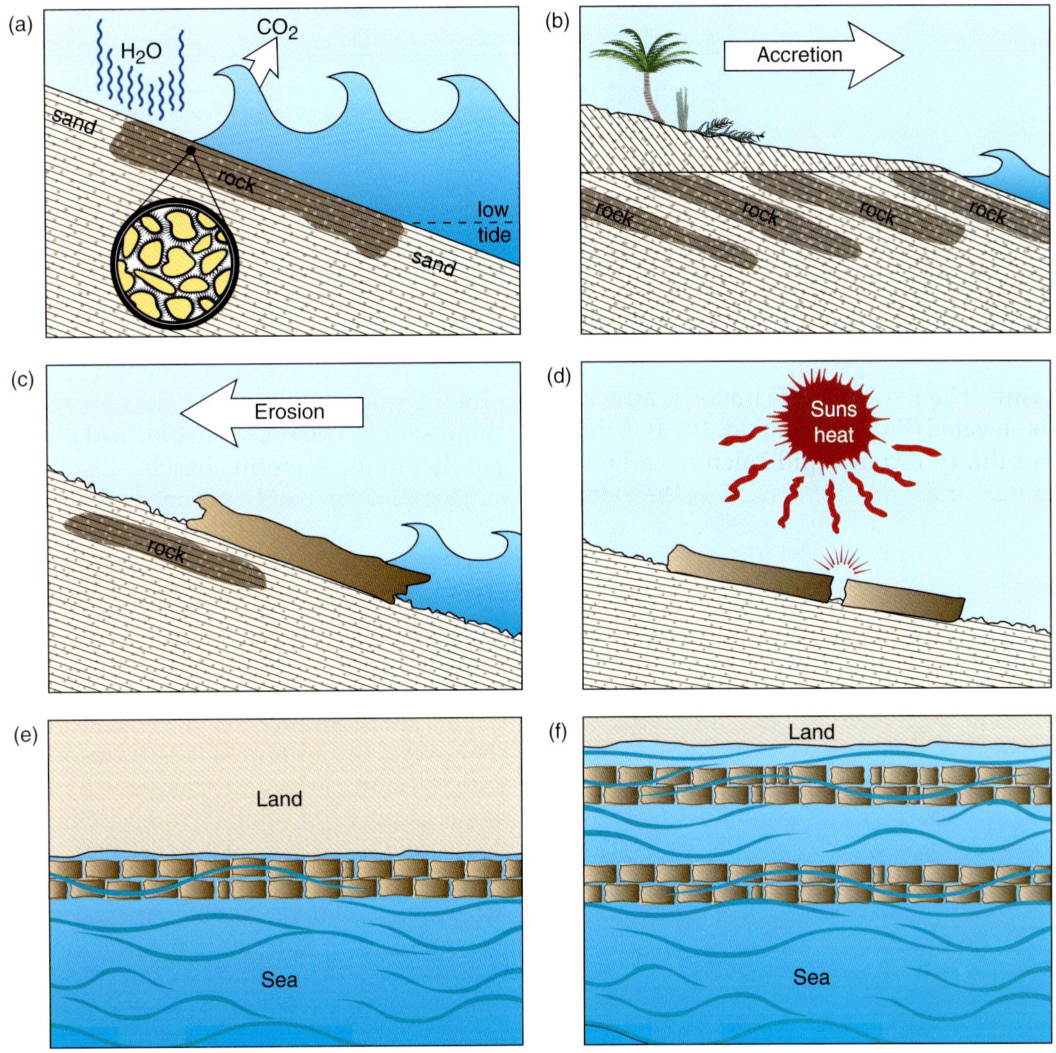

Fig. 3. Schematic drawings of beachrock formation that can lead to formation of a road-like feature. (a) Cross-section through a beach shows formation of beachrock beneath intertidal beach sands. Waves and tides serve to pump huge volumes of seawater through intertidal sands. Evaporation of seawater, which may aid in the cementation process, is enhanced by tidal fluctuation and waves. Circular inset shows microscope view of acicular aragonite growing on and fusing beach-sand grains together to form rock. (b) Beachrock forms beneath a seaward-prograding beach during intermittent beach accretion. Pulses of sedimentation cause progradation of beach sand beyond area of most active cementation. During this process, the forming beachrock lies hidden beneath soft sand. (c) Transgression of the sea over the beach occurs when supply of sand is halted or when sea level rises. Erosion of overlying sand exposes the previously formed beachrock while the intertidal (and submarine) process of aragonite cementation continues. Good examples of this process can be seen along a swimming beach at Bimini. Such quickly formed rock often contains broken bottles, pottery and other human artefacts. (d) Exposure to the sun and to constant wetting and drying causes the rock to crack into separate slabs in much the same way as a concrete road. The size of individual slabs is controlled by rock thickness. Uncemented sand underlying the rock also promotes cracking as the rock settles in much the same manner as ice on a frozen pond. Constant abrasion by moving beach sand rounds off corners to form pillow shapes. (e) Viewed from above, the straight row of broken rock slabs can resemble a road. (f) When sea level rises and erosion of underlying sand takes place, the stones are submerged to various depths (see Figs 5 and 6). Multiple parallel 'roads' may be produced in this manner. The 'road' may curve when beachrock forms on a hook-shape sand spit.

on a transgressive beach. In the Persian Gulf, both in the intertidal and subtidal zone, the rapidly forming rock is horizontal and thin. Although this rock cracks and buckles extensively, it tends to form polygonal shapes separated by expansion ridges (Shinn, 1969; Evamy, 1973; Assereto & Kendall, 1977).

The process of intertidal cementation on fish-hook sedimentary spits is relevant to the Bimini Road beachrock. Off Bimini, the area of curved

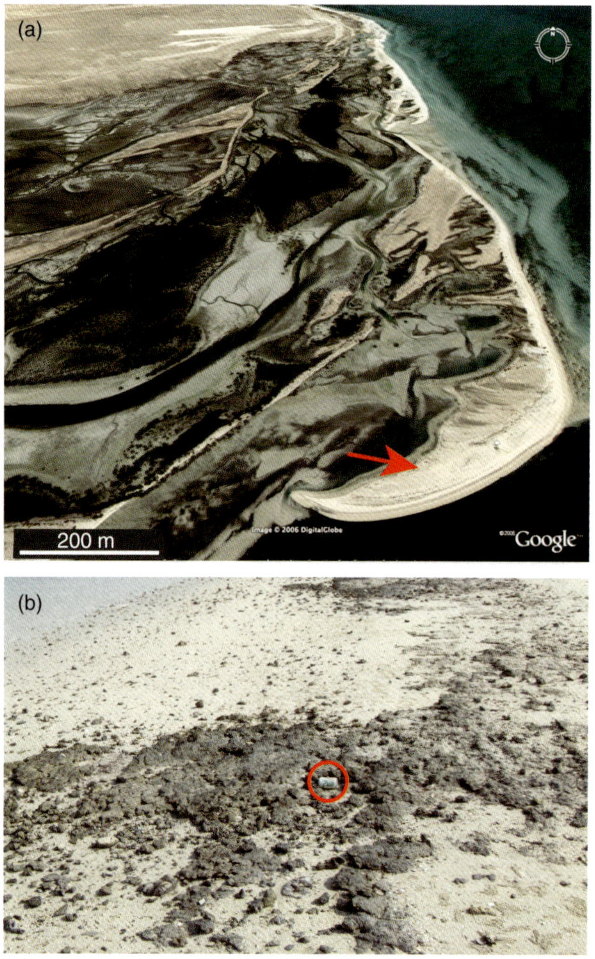

Fig. 4. (a) A recent Google Earth image (rotated to produce an oblique view) of accreting hook-shape sand spits on the northeast coast of Qatar in the Persian Gulf (Shinn, 1973). Arrow shows approximate location of forming beachrock shown in photograph (b) taken in 2005. Soft drink can (circled) provides scale. The spit indicated by arrow is the second spit to form since the original study by Shinn (1973) that was conducted in 1967. Beachrock occurs on all the spits and increases in hardness in the older spits.

stones, the so-called 'inverted J', is often cited as evidence of human construction. Observations of beachrock formation on curved spits in the Persian Gulf indicate that similar processes probably have produced the 'inverted J' off Bimini. Even at Bimini, a similar present-day 'J' probably would be evident at the entrance into Bimini Harbour were it not for human intervention. That area has been extensively modified by a seawall and a small boat channel.

Other areas of submerged beachrock

The 'road' of beachrock stones is not restricted to the Bimini area. Similar road-like alignments of oblong beachrock can be found throughout the Bahamas. A large area of identical pillow-shape rocks can be seen in 3–5 m of water off the east side of Andros Island near Nichols Town. Many of the rocks there have provided a foundation for reef initiation, but for the most part, the rocks are visible and easy to access by swimming off the beach. Little (2006) believes these stones, like those at Bimini, are anthropogenic. Other personal observations have been made in Abaco and the Exuma Islands in the Bahamas and off the northwest coast of Vieques Island, Puerto Rico.

Hurricanes Charley in 2004 and Katrina in 2005 exposed and undermined beachrock with identical shapes as those at Bimini on the east side of Loggerhead Key at Dry Tortugas (Fig. 5). These stones were in the intertidal zone when first observed by the author in the early 1970s. Because of undermining by storm waves, they are now in water approximately 1 m deep. Similar large beachrock slabs are present at nearby Hospital Key in water depths up to 3 m. These observations provide support for the conclusion of Shinn (1978) that the Bimini stones were submerged by both erosion of underlying sand and rising sea level. Based on the bulk ^{14}C ages of around 3 ka, they are too deep to be explained by sea-level rise alone.

Ginsburg (1953) conducted his research on beachrock that was mostly buried on the west side of Loggerhead Key. Recent erosion of overlying sand has exposed the rock and revealed a 10- to 15-cm diameter iron pipe extending several metres perpendicular to the beach. The pipe, now a part of the rock, is believed to have been a sewage pipe laid in sand for the Carnegie Institution Marine Laboratory that first opened in 1905 (Shinn & Jaap, 2005). Alfred G. Mayor, the first director of the laboratory, reported to R.A. Daly that sand deposited at Loggerhead Key by a hurricane in 1919 became beachrock in one year (Daly, 1924).

One of the more interesting recently discovered beachrock sites is at Pulley Ridge in 90 m of water on the west Florida shelf. Pulley Ridge is a drowned barrier island as detailed by recent multibeam surveys (Jarrett *et al.* 2005). Pulley Ridge is for the most part capped by a veneer of live coral but in places the underlying rock is exposed, revealing pillow-shape limestone blocks with dimensions identical to the stones off Bimini (Fig. 6). Pulley Ridge has a well-defined fishhook or J shape at its northern end (Fig. 7). Such J shapes are typical features of barrier

Fig. 5. Newly exposed submerged beachrock 'road' on east side of Loggerhead Key at Dry Tortugas. The 'Loggerhead Road' was exposed by Hurricane Katrina in 2005. Undermining of sand beneath the stones caused the rock to subside approximately 1 m. The blocks are 1–2 m long.

Fig. 6. A new beachrock road was recently discovered in 90 m of water at Pulley Ridge off the southwest coast of Florida (Jarrett *et al.*, 2005). The blocks are 1–2 m long. The discovery was made in a joint investigation by University of South Florida, US Geological Survey, and the National Oceanic and Atmospheric Administration.

islands near tidal inlets and occur on barrier islands all along the eastern USA, the west coast of Florida and the northern Gulf of Mexico.

Similar rows of carbonate beachrock to those at Pulley Ridge have been reported from the same water depth (90 m) in the northern Gulf of Mexico (unpublished Minerals Management Report). It is well documented that sea level has risen at least 100 m worldwide in the past 13–10 kyr world-wide (Fairbanks, 1989).

Speculation and future research possibilities

Alternative thinkers who believe the Bimini stones are anthropogenic features often cite historic harbours in the Mediterranean Sea as models for what they believe occurred at Bimini during prehistoric or 'Atlantian' times. The sand spits and offshore barrier islands that migrate and form harbours in the Persian Gulf may well indicate that some ancient harbours in the Mediterranean Sea developed geologically in a similar fashion. In the Persian Gulf or Arabian Gulf, one can observe that curved spits have provided natural sheltered harbours allowing villages to be established because of the shelter they provide. As the spits migrate laterally along the coast and the harbours fill in, villages adjacent to the shallowing end of the harbour are abandoned and rebuilt around the deepening newly forming harbour entrance.

Similar processes are likely to have occurred along the southern Mediterranean shore, where the climate is arid and seawater salinity is generally elevated. In areas where the harbours were especially important for commerce, it would be reasonable to build structures around previously formed natural beachrock-protected

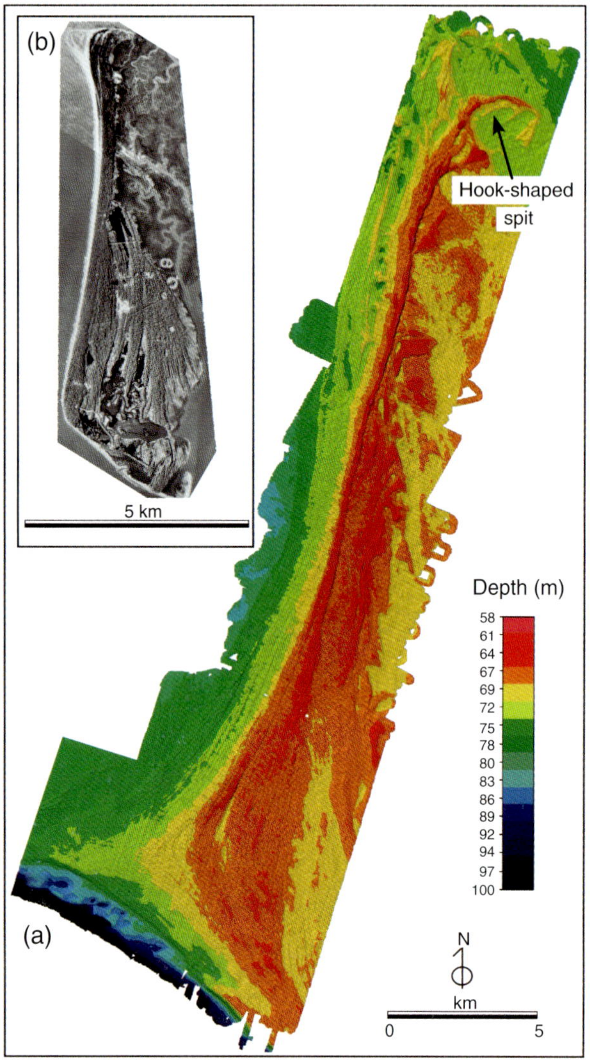

Fig. 7. (a) A multibeam map image of Pulley Ridge, compared with (b) an aerial image of Bull Island, South Carolina (image rotated for easy comparison). Hook-shape spit features typical of most barrier islands (Jarrett *et al.*, 2005).

harbours. 'J' shapes would be expected in such areas. Because of the abundant ancient cultures in the Mediterranean region, it would also be reasonable to expect an abundance of artefacts that might lead archaeologists, both conventional and alternative, to conclude that the harbours were entirely man-made. Ancient Mediterranean cultures may have simply taken advantage of and added to what nature had already provided. Therefore, the argument that the Bimini Road is anthropogenic because of similarities with stones around ancient Mediterranean harbours may not be valid.

ACKNOWLEDGEMENTS

The author thanks many people who made this research possible. None would have happened without the initial encouragement and support of Peter Tompkins. The early fieldwork at Bimini was made possible by Capt. Roy Gaensslen, Harold Hudson, Daniel Robbin and Lloyd Burkley. The draft manuscript was improved by the editing skills of Eberhard Gischler, and both Gray and Susan Multer. The final revision was reviewed and improved by Pascal Kindler, Paul Hearty and Barbara Lidz. The pioneering work on beachrock by R.N. Ginsburg at Dry Tortugas and his encouragement to convert the author from biology/technician to geologist is gratefully acknowledged. The author also thanks Marshall Payn of the Epigraphic Society for frequent discussions regarding antiquities. Finally, insistence of alternative thinkers/bloggers and others on the fringe provided additional stimulation to complete this paper. The author is especially indebted to Betsy Boynton for artwork and preparation of the figures.

ENDNOTE

1 The legend is told in two of Plato's works, *Timaeus* and *Critias*, written around 500 BC. For a more complete history of the legend, see Spence (1968). Lewis Spence was the author of five books on Atlantis of which the most complete, *The History of Atlantis*, was published in London in 1926. The cited book (Spence, 1968) is a republication of the original 1926 book. In 1932, Lewis Spence edited a journal, the *Atlantis Quarterly*. Clearly, Lewis Spence, a renowned writer and historian of legends and the occult, can be credited with stimulating popular interest in the legend as well as belief in the existence of Atlantis.

REFERENCES

Assereto, R.L.A.M. and **Kendall, C.G.St.C.** (1977) Nature, origin and classification of peritidal tepee structures and related breccias. *Sedimentology*, **24**, 153–210.

Berlitz, C. (1969) *The Mystery of Atlantis*. Grosset and Dunlap, New York, 149 pp.

Berlitz, C. (1984) *Atlantis: The Eighth Continent*. G.P. Putnam's Sons, New York, 160 pp.

Cayce, E.E. (1968) *Edgar Cayce on Atlantis*. Warner Books, New York, 175 pp.

Cayce, E.E., Cayce-Schwartzer, G. and **Richards, D.G.** (1988) *Mysteries of Atlantis Revisited*. Harper & Row, San Francisco, 201 pp.

Chamisso, A.V. (1821) Bemerkungen und Ansichten von dem Naturforscher der Expedition. In: *Entdeckungs-Reise* *in die Sud-See und nach der Berings-Strasse zur Erforschung einer nordostlichen Durchfahrt* (Ed. O.V. Kotzebue), Vol. **3**, Hofmann, Weimear, 240 pp.

Daly, R.A. (1924) *The Geology of American Samoa*, Vol. **340**. Carnegie Institution of Washington Publication, Washington, DC, pp. 93–143.

Dana, J.D. (1875) *Corals and Coral Islands*. Sampson Low, Marstonk Low and Searle, London, 338 pp.

Darwin, C.R. (1842) *Structure and Distribution of Coral Reefs*. Smith Elder, London, 214 pp.

Dunham, R.J. (1970) Keystone vugs in carbonate beach deposits (abstract). *Am. Assoc. Petrol. Geol. Bull.*, **54**, 845.

Dunham, R.J. (1971) Meniscus cement. In: *Carbonate Cements* (Ed. O.P. Bricker). Johns Hopkins University Studies in Geology, Vol. **19**, Johns Hopkins University, Baltimore, Maryland, USA, pp. 297–300.

Evamy, B.D. (1973) The precipitation of aragonite and its alteration to calcite on the Trucial Coast of the Persian Gulf. In: *The Persian Gulf – Holocene Carbonate Sedimentation and Diagenesis in a Shallow Epicontinental Sea* (Ed. B.H. Purser). Springer Verlag, Heidelberg, Berlin, pp. 329–341.

Fairbanks, R.G. (1989) A 17,000-year glacio-eustatic sea level record: Influence of glacial melting rates on the Younger Dryas event and deep ocean circulation. *Nature*, **342**, 637–642.

Field, R.M. (1919) *Investigations Regarding the Calcium Carbonate Oozes at Tortugas, and the Beachrock at Loggerhead Key: Carnegie Institution of Washington Yearbook*, Vol. **18**, Washington, DC, pp. 187–198.

Field, R.M. (1920) Origin of beachrock (coquina) at Loggerhead Key, Tortugas (abstract). *Geol. Soc. Am. Bull.*, **31**, 215.

Gardiner, J.S. (1898) The coral reefs of Funafuti, Rotuma and Fiji together with some notes on the structure and formation of coral reefs. *Cambridge Philos. Soc. Proc.*, **9**, 417–503.

Gifford, J.A. and **Ball, M.M.** (1980) Investigation of submerged beachrock deposits off Bimini, Bahamas. *Natl. Geogr. Soc. Res. Rep.*, **12**, 21–38.

Ginsburg, R.N. (1953) Beachrock in south Florida. *J. Sedimentary Petrol.*, **23**(2), 85–92.

Gischler, E. (2007) Beachrock and intertidal precipitates. In: *Geochemical Sediments and Landscapes* (Eds D.J. Nash and S.J. McLaren), Chapter **11**. Blackwell, Oxford, pp. 365–390.

Halley, R.B. and **Harris, P.M.** (1979) Fresh-water cementation of a 1,000 year-old oolite. *J. Sedimentary Petrol.*, **49**(3), 0969–0988.

Harrison, W. (1971) Atlantis undiscovered: Bimini, Bahamas. *Nature*, **230**, 287–289.

Hearty, P.J. and **Donato, W.M.** (1998) Atlantis at Bimini: Fantasy or fact. *Bahamas Handbook and Businessman's Annual*, Nassau, Bahamas, pp. 81–102.

Jarrett, B.D., Hine, A.C., Halley, R.B., Naar, D.F., Locker, S.D., Neumann, A.C., Twichell, D., Hu, C., Donahue, B.T., Jaap, W.C., Palandro, D. and **Ciembronowicz, K.** (2005) Strange bedfellows – a deep-water hermatypic coral reef superimposed on a drowned barrier island; southern Pulley Ridge, SW Florida platform margin. *Marine Geol.*, **214**, 295–307.

Keunen, P.H. (1933) Geology of coral reefs. *Snellius Expedition*, **5**, 1–125

Little, G. (2006) Underwater stone formation at Bimini, Bahamas reveals its secrets: Ancient maritime culture in Bahamas confirmed. *Alternate Perceptions Magazine*, Issue 100, http://www.mysteriousamerica.net/bimini-caysal200.html

Lyell, C. (1832) *Principles of Geology, Being an Attempt to Explain the Former Changes of the Earth's Surface, by Reference to Causes Now in Operation*. John Murray, London, Vol. **2**, 330 pp.

McKusick, M. and **Shinn, E.A.** (1980) Bahamian Atlantis reconsidered. *Nature*, **287**(4), 11–12.

Menzies, G. (2002) *1421*. Bantam Press, Uxbridge Road, Ealing, 408 p.

Multer, H.G. (1971) Holocene cementation of skeletal grains into beachrock, Dry Tortugas, Florida. In: *Carbonate Cements* (Ed. O.P. Bricker). Johns Hopkins University Studies in Geology, Vol. **19**, Johns Hopkins University, Baltimore, Maryland, USA, pp. 25–26.

Purser, B.H. and Seibold, E. (1973) The principal environmental factors influencing Holocene sedimentation and diagenesis in the Persian Gulf.I In: *The Persian Gulf – Holocene Carbonate Sedimentation and Diagenesis in a Shallow Epicontinental Sea* (Ed. B.H. Purser). Springer Verlag, Heidelberg, Berlin, pp. 1–10.

Randi, J. (1981) Atlantean Road: The Bimini Beach-Rock. *The Skeptical Inquirer*, **5**(3), 42–43.

Rebikoff, D. (1972) Precision underwater photomosaic techniques for archaeological mapping: Interim experiment on the Bimini 'cyclopean' complex. *Int. J. Nautl. Archaeol. Underwater Expl.*, **1**, 184–186.

Rebikoff, D. (1979) Underwater archaeology: Photogrammetry of artifacts near Bimini. *Explorers J.*, 122–125.

Richards, D.G. (1988) Archaeological anomalies in the Bahamas. *J. Sci. Expl.* **2**(2), 181–201.

Scoffin, T.P. (1970) A conglomeratic beachrock in Bimini, Bahamas. *J. Sedimentary Petrol.*, **40**(2), 756–758.

Shinn, E.A. (1969) Submarine lithification of Holocene carbonate sediments in the Persian Gulf. *Sedimentology*, **12**, 109–144.

Shinn, E.A. (1971) Submarine cementation in the Persian Gulf. In: *Atlas of Carbonate Cements* (Ed. O. Bricker). John Hopkins University Press, Baltimore, Maryland, pp. 63–65.

Shinn, E.A. (1973) Coastal accretion in an area of longshore transport, Persian Gulf. In: *Persian Gulf Sedimentation* (Ed. B.H. Purser). Springer Verlag, New York, pp. 179–191.

Shinn, E.A. (1978) Atlantis: Bimini hoax. *Sea Frontiers*, **24**, 130–141.

Shinn, E.A. (1983) Tidal flat environment. In: *Carbonate Depositional Environments* (Ed. P.A. Scholle). American Association of Petroleum Geologists Memoir, **33**, 171–210. American Association of Petroleum Geologists, Tulsa, Oklahoma, USA.

Shinn, E.A. (2004) A geologist's adventures with Bimini beachrock and Atlantis true believers. *Skeptical Inquirer*, **28**(1), 38–44.

Shinn, E.A. and **Jaap, W.C.** (2005) *Field Guide to the Major Organisms and Processes Building Reefs and Islands of the Dry Tortugas: The Carnegie Dry Tortugas Laboratory Centennial Celebration (1905–2005)*. October 13–15, 2005, Key West, Florida. U.S. Geological Survey Open-File Report 2005-1357, 43 pp.

Spence, L. (1968) *The History of Atlantis*. University Books Inc. Library of Congress no. 67–26623, 238 pp.

Stoddart, D.R. and **Cann, J.R.** (1965) Nature and origin of beachrock. *J. Sedimentary Petrol.*, **35**, 243–273.

Strasser, A. and **Davaud, E.** (1986) Formation of Holocene limestone sequences by progradation, cementation, and erosion: two examples from the Bahamas. *J. Sedimentary Petrol.*, **56**, 422–428.

Valentine, J.M. (1969) Archaeological enigmas of Florida and the Western Bahamas. *Muse News, June (Miami Museum of Science)*, **1**, 26–29, 41–47.

Valentine, J.M. (1976) Underwater archeology in the Bahamas. *Explorers J.*, December, 176–182.

Vaughan, T.W. (1914) Building of the Marquesas and Tortugas atolls and a sketch of the geologic history of the Florida reef tract: Carnegie Institution of Washington Publication 182, *Papers of the Department of Marine Biology*, **5**, 55–67.

Zink, D. (1978) *The Stones of Atlantis*. Prentice-Hall, Englewood Cliffs, NJ.

Int. Assoc. Sedimentol. Spec. Publ. (2009) **41**, 29–46

A re-evaluation of facies on Great Bahama Bank I: new facies maps of western Great Bahama Bank

JOHN J.G. REIJMER*†1, PETER K. SWART[2], THORSTEN BAUCH*3, ROBERT OTTO[2], LARS REUNING*4, SVEN ROTH*5 and SUSANNE ZECHEL*6

**IFM-GEOMAR, Leibniz-Institut für Meereswissenschaften, Dienstgebäude Ostufer, Wischhofstr. 1-3, D-24148 Kiel, Germany*

†Centre de Sédimentologie-Paléontologie, Laboratoire (EA 4229) 'Géologie des Systèmes et des Réservoirs Carbonatés', Université de Provence (Aix-Marseille I), 3, place Victor Hugo, Case 67, F-13331 Marseille Cédex 3, France

ABSTRACT

A re-evaluation of the sediment distribution patterns on western Great Bahama Bank shows a facies distribution with two end-members. Coarse-grained sediments in the north, west and south of the bank surround a mud-dominated realm located on the western leeward side of Andros Island. This facies distribution is comparable to earlier maps, but shows considerably more detail and a complex distribution from grainstones to mud-rich wackestones. As in other carbonate platforms, sediment distribution appears to be influenced by (1) tidal currents, (2) prevailing wind direction, and (3) the interaction of the rate of Holocene sea-level rise with the pre-existing Pleistocene topography. The grain-size distribution very precisely reflects current-influenced and protected areas on the platform. The correlation between the distribution of pellets and the 63–125 μm grain-size fraction most probably reflects the predominantly biological origin of this grain size. Aragonite dominates the mineralogy on the platform, low-magnesium calcite and high-magnesium calcite occurs in higher quantities only in a few environments on the platform.

Keywords Facies distribution, grain-size, mineralogy, Great Bahama Bank, Holocene.

INTRODUCTION

The Great Bahama Bank (Fig. 1a) has served as an inspiration for geologists in understanding modern processes of carbonate sedimentation since the pioneering work of Field (1931) and Illing (1954). Most of our knowledge of the spatial distribution of sediments on Great Bahama Bank (GBB) is based on the work of Illing (1954), Ginsburg *et al.* (1958), Newell *et al.* (1959) and Purdy (1963a,b). Valuable as these early descriptions are, it is the aim of this study to re-evaluate the distribution of the surface sediments on GBB using samples collected from precisely located points (Fig. 1b; established using GPS) and to quantify not only the types of sediments present, but also the mineralogy, grain-size distribution and skeletal and non-skeletal content. This paper together with a companion paper (Swart *et al.*, 2009) takes a new look at the surface facies and geochemistry of sediments from Great Bahama Bank.

Various maps show sedimentary facies on GBB or in the Bahamas in general (Ginsburg *et al.*, 1958; Newell *et al.*, 1959; Purdy, 1963a,b; Ball, 1967; Enos, 1974). In all these studies sediments were classified using distinct facies types with comparable subdivisions. All maps show that skeletal sediments occur mainly on the margin of the bank whereas non-skeletal grains

[1]Present address: VU University, Faculty of Earth and Life Sciences (FALW), De Boelelaan 1085, 1081 HV Amsterdam, The Netherlands (Corresponding author: E-mail: John.Reijmer@falw.vu.nl).

[2]Marine Geology and Geophysics, Rosenstiel School of Marine and Atmospheric Science, 4600 Rickenbacker Causeway, Miami, FL 33149, USA.

[3]Present address: Schlumberger Oilfield Services D&M, Hamrasletta 15, 4056 Tananger, Norway.

[4]Present address: RWTH Aachen University, Wüllnerstr. 2, D-52062 Aachen, Germany.

[5]Present address: Nummerical Rocks, Stiklestadveien 1, 7041 Trondheim, Norway.

[6]Present address: TU Bergakademie Freiberg, Institut für Geologie und Paläontologie, Bernhardtvon-Cotta Straße 2, 09596 Freiberg, Germany.

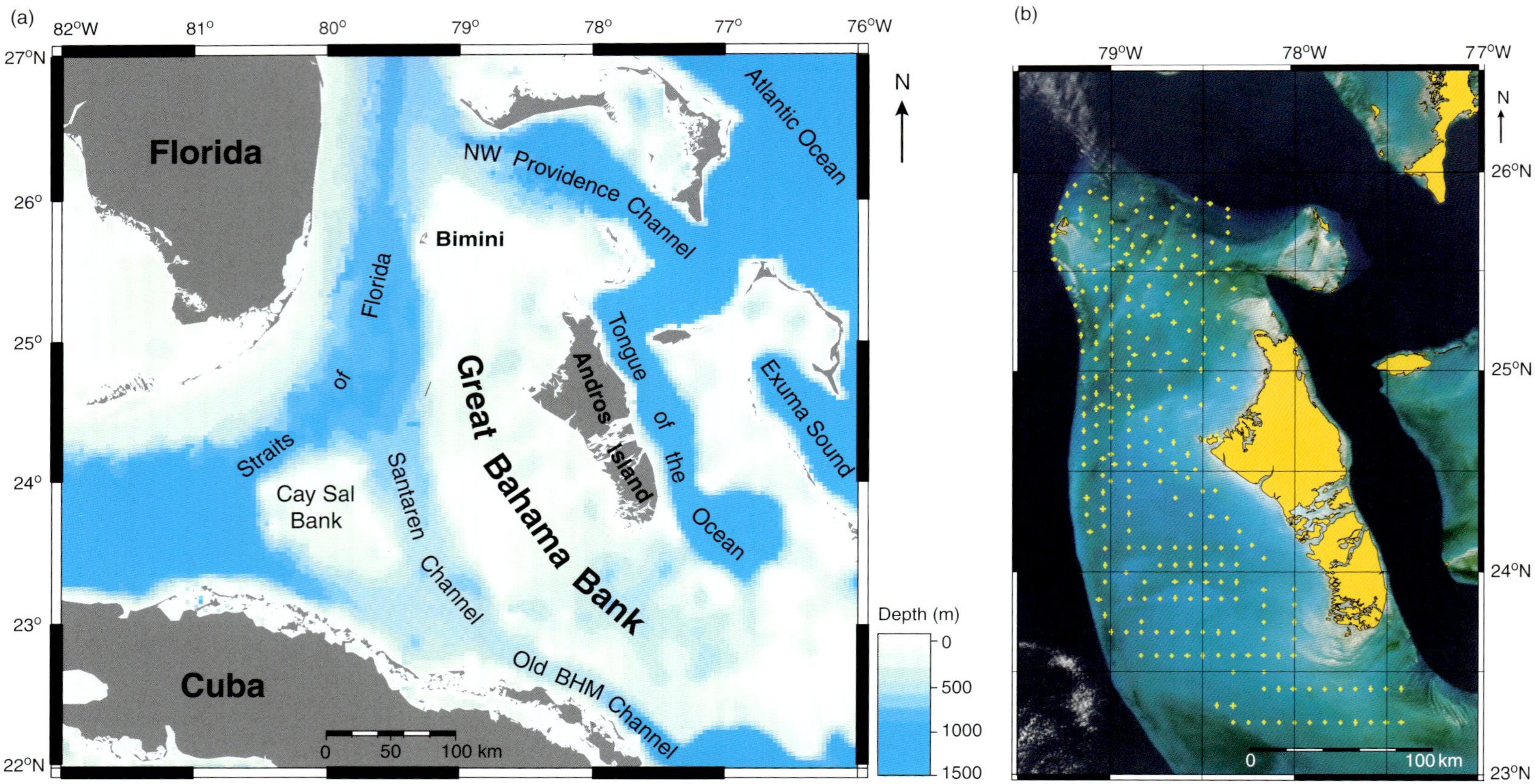

Fig. 1. (a) Location map of Great Bahama Bank with main topographic names. Old BHM Channel: Old Bahama Channel. (b) Location map samples (yellow crosses). Satellite image of Great Bahama Bank shows the shallow-water bank in light blue to green colours. Deep-water basins show up in dark blue colours. Morphology of the islands accentuated in yellow. Large island in the centre of the figure is Andros Island. Latitude and longitude indicated at the side. Kilometre bar for scale.

(mainly faecal pellets) dominate the interior of the platform. Very fine-grained sediments predominate in platform interior zones that are protected by topographic barriers such as marginal islands and submarine topography ('pelletoidal sands with lime mud matrix and lime muds', Ginsburg *et al.*, 1958; 'muddy sand and mud', Newell *et al.*, 1959; 'pellet-mud facies' and 'mud facies', Purdy, 1963a; 'pelletoidal packstones' and 'pelletoidal wackestones', Enos, 1974). Some differences, however, exist in the interpretation and, as seen on the maps, in the distribution of the facies types. For example, Ginsburg *et al.* (1958) and Enos (1974) show that the sediments in the platform interior mainly contain pellets, peloids or pelletoids. Newell *et al.* (1959) and Purdy (1963a,b), however, described sediments containing a larger percentage of grapestones. Enos (1974) also described the occurrence of grapestones without showing them on the map as a particular facies type. Enos (1974) was the only study in which the Dunham classification (1962) was used and thus showed the components to matrix ratio along the platform. All maps known from the literature were based on sampling along a series of profiles crossing the bank and additional datasets. All samples were taken without modern precision positioning systems and thus might contain significant uncertainties in the exact location of facies types.

The major question, however, concerns the processes that control the present facies distribution on western GBB, the Andros lobe. This paper attempts to determine the links between the sediment distribution and environmental parameters such as currents, tides, topography, evaporation (salinity), precipitation, water exchange with the surrounding ocean water masses and sediment export.

Currents and circulation

Most of the surface water masses of the Caribbean Sea, which eventually form the Florida Current and the Gulf Stream, enter in the Lesser Antilles through various passages (Wajsowicz, 2002). A smaller portion (approximately 20%) enters the Caribbean Sea through the Windward Passage between Cuba and Hispaniola (Schmitz & Richardson, 1991). The Florida Current is fed as well by water masses through the Old Bahama Channel and Santaren Channel and through the Northwest Providence Channel (Fig. 1a; Leaman *et al.*, 1995). Atkinson *et al.* (1995) and Leaman *et al.* (1995) showed that the average water transport through the Santaren Channel amounts to 1.9 Sv or 1.8 Sv, respectively. The transport through the Northwest Providence Channel reaches 1.2 Sv (Leaman *et al.*, 1995). On the western, leeward sides the current system on GBB is connected with the currents forming the sources of the Gulf Stream. On the windward sides a link exists with the North Atlantic Subtropical Gyre situated eastward in the Sargasso Sea. The GBB has a great lateral extent and shows a well-defined marginal escarpment with islands. Shallow waters cover most of GBB and consequently, the interaction with water masses from the open ocean is limited (Smith, 1940). Islands such as Andros Island deflect westward-moving currents generated by the trade winds to the north and south (Bathurst, 1975). As a result protected areas with limited water exchange are present on the western side of these islands (Bathurst, 1975). Water movement on GBB is mostly influenced by the wind and tides, but also by waves and storms (Smith, 1940; Cloud, 1962; Purdy, 1963a; Traverse & Ginsburg, 1966; Winland & Matthews, 1974; Gonzalez & Eberli, 1997).

Long-term net flow measurements over GBB show a very slow current velocity of 2 cm s^{-1} toward the north (002°N; Smith, 1995). This is thought to be the outcome of two almost opposing forces, a tide-induced flow toward the east-southeast and a wind- and density-driven northward flow (Smith, 1995). Tidal currents vary from near 0 to 4 km h^{-1} in the tidal channels (Newell & Rigby, 1957; Bathurst, 1975) and influence the entire platform to some degree.

Water temperature, wind and precipitation

The surface seawater temperature (SST) on GBB ranges from 18.5°C during winter to 28.5°C in summer (Cloud, 1962; Bathurst, 1975). During the summer, higher SSTs are reached locally. Mixing of bank and ocean waters is very low (Purdy, 1963a). Smith (1940) noted that the difference in SST between bank and surface ocean waters may be as large as 3.6°C. The lateral variation in the SST on the bank itself is very low, but on bank margins the slow mixing with ocean water is demonstrated by changes in SSTs (Purdy, 1963a).

The Bahama Platform is influenced by the trade winds blowing from east or southeast in the summer (March to August). In the winter months, the winds have a northeasterly direction.

Storms and hurricanes frequently affect the area and exhibit highly variable source direction, with winter storms generally coming from the northwest whereas the hurricane directions are highly variable. Only major storms affect sedimentation on the platform because of its great lateral extent and its protection by marginal shoals (rocky and sand shoals, and reefs) and islands (Tucker & Wright, 1990). The annual precipitation over the Bahama Platform amounts to 1355 mm (Carew & Mylroie, 1997), concentrated between May and November, with a maximum in September and October (Gebelein, 1974). The winter months (December–April) are relatively dry. The average monthly rainfall varies between 100 and 250 mm (Miller *et al.*, 1983).

Seasonal atmospheric circulation patterns strongly influence temperature and precipitation patterns on GBB. Circulation responds to changes in the position and strength of high- and low-pressure systems, the Intertropical Convergence Zone (ITCZ), the Subtropical Divergence Zone (STDZ), respectively, and the associated trade-wind belt. During May to November, the atmospheric highs and lows are less pronounced over the North Atlantic Ocean, and the ITCZ and the STDZ move northward. The trade winds weaken and high precipitation prevails over the northern Bahamas. The atmospheric regime migrates southward during the boreal winter (December–April) and the ITCZ shifts toward the Equator or slightly south of it (0–5°S). During this period the atmospheric high- and low-pressure systems strengthen and the easterly trade winds are enhanced. The result is a stronger influence of colder and dryer air masses. During this winter period, the evaporation minus precipitation (E − P) averages around 225–250 cm yr^{-1} while it only reaches about 50 cm yr^{-1} in the summer (Smith, 1995).

Salinity

In summer, salinity varies from 36‰ near the edge of the platform to 46‰ in the platform interior, protected environment west of Andros Island (Smith, 1940; Ginsburg *et al.*, 1958; Gebelein, 1974). Maximum winter salinity reaches 38‰ (Gebelein, 1974). Reasons for the persistence of increased salinity in the bank interior are the great lateral extension of the platform and the existence of marginal shoals and islands reducing the exchange of platform and open ocean waters (Tucker & Wright, 1990); tidal exchange with oceanic water is very small (Bathurst, 1975).

Carbonate mineralogy and grain size

Carbonate mineralogy on GBB is well documented and shows a dominance of aragonite on the platform top with minor high-magnesium calcite (HMC) and low-magnesium calcite (LMC) (Purdy, 1963a; Husseini & Matthews, 1972; Milliman, 1974; Mullins, 1986; Milliman *et al.*, 1993).

Ginsburg (1956) noted, 'Variations in the physical environment of carbonate depositional environments (bathymetry, areal geography, and hydrography) are reflected in the grain size and constituent particle composition of the sediments'. On the platform top grain size has been used as a relative indicator of current intensity (Purdy, 1963a). The *in situ* production of coarse-grained skeletal material and changes in the original particle size on the platform, e.g. grapestone formation and particle breakdown, interfere with this assumption (Purdy, 1963a).

METHODS

Grab samples

Approximately 300 samples were obtained using a Shipek sampler during four cruises aboard the RV Bellows between 2001 and 2004 (Fig. 1b). The design of this sampler prevents muds from escaping from the sample during retrieval from the sea bottom. It thus differs from the Van-Veen sampler used during earlier studies (Purdy, 1963a,b) in that it better retains the mud fraction.

Facies classification

Immediately after collection, samples were assigned to a modified Dunham scheme (Dunham, 1962). In order to complement the classification we added divisions between wackestone and mudstone (mud-rich wackestone; facies number 1.5), between packstone and wackestone (mud-rich packstone, facies number 2.5), and finally between packstone and grainstone (mud-lean packstone or poorly washed grainstone; facies number 3.5) (Fig. 2). Samples from all cruises were preserved in cold storage and after the 2004 cruise all samples were re-examined to ensure consistency between cruises.

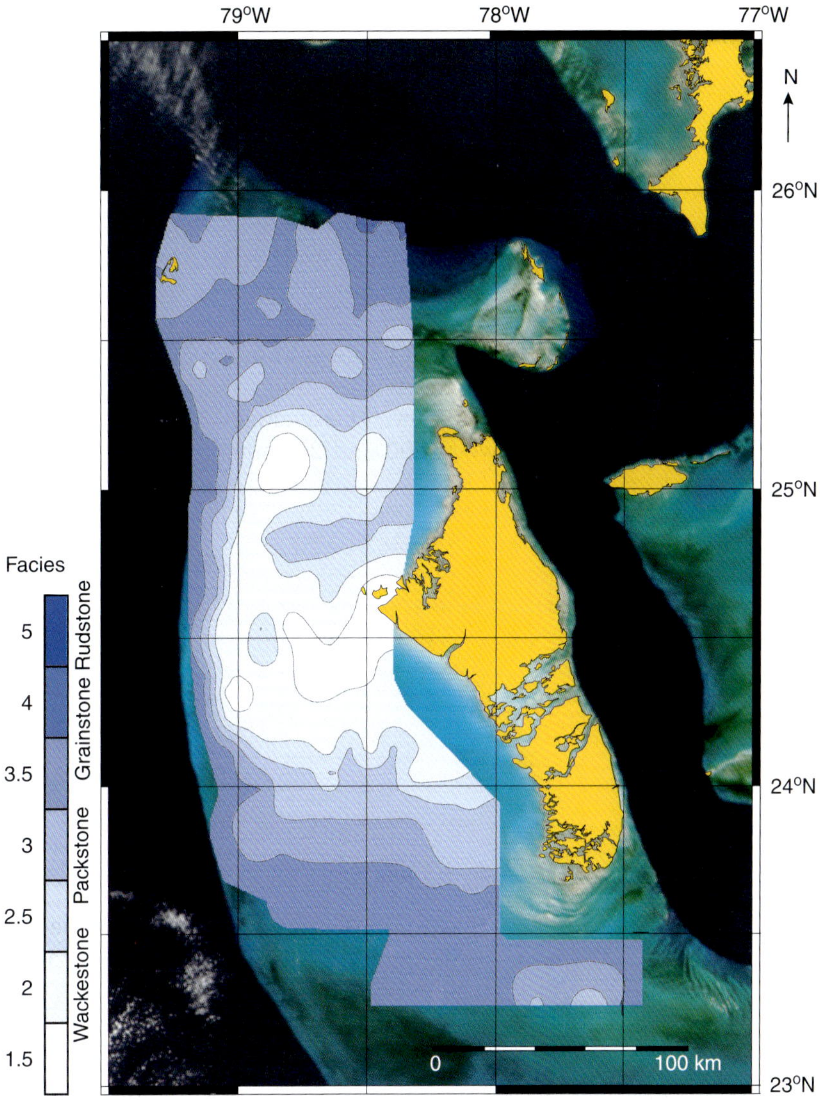

Fig. 2. Facies distribution along Great Bahama Bank. Mud-rich wackestone = 1.5; Wackestone = 2; Mud-rich packstone = 2.5; Packstone = 3; Mud-lean packstone or poorly washed grainstone = 3.5; Grainstone = 4; Rudstone = 5.

Grain-size variations

The samples were placed in 100 mL containers and oven dried at 40°C before sieve analysis. A small split of the dried samples was analysed by X-ray diffraction. The dried samples were wet sieved to separate the coarse (>63 µm) from the fine fraction (<63 µm). The fine fraction was left in the 5-L bottles to settle; water was then siphoned off; and the fine sediment was oven dried at 40°C. The coarse fraction was dried, weighed and split into five sub-fractions (very coarse sand to gravel >1000 µm; coarse sand, 500–1000 µm; medium sand, 250–500 µm; fine sand, 125–250 µm; very fine sand, 63–125 µm; after Wentworth, 1922) using a hand-held stainless-steel sieve set. Each sub-fraction was placed in a pre-weighed glass vial, which was weighed again, to determine the sub-fraction weight.

Mineralogy

Whole-rock samples were oven dried at 40°C and hand ground in a mortar to homogenize the sediment (Milliman, 1974). Additional samples representative of coarse and fine size fractions were similarly individually homogenized by grinding by hand in a mortar. The areas of the principal peaks of aragonite, HMC and LMC were measured with Scintag and Panalytical X-ray instruments using Cu-Kα radiation at RSMAS/University of Miami. The samples were scanned between 20° and 40° 2θ with a speed of 0.01° per second. The proportions of aragonite, HMC and LMC were quantified using a method outlined by Swart *et al.* (2002) assuming that the sample is composed entirely of aragonite, HMC, and LMC, which is generally the case for GBB samples.

The reproducibility of the method was determined through the replicate analysis of samples and is approximately ±2%.

Sediment composition

In order to identify individual facies groups, the coarse fraction (63 to >1000 μm) of each sample was analysed for its variation in skeletal (coral fragments, calcareous green algae, benthic foraminifera, gastropods, echinoderm spines and fragments, serpulids, ostracods, bryozoans, sponge spines and others) and non-skeletal components (peloids, ooids, grapestones, clasts) using a stereo binocular microscope (sieve samples) as well as a polarizing microscope (thin-sections). Only the distributions of pellets and ooids are described here.

For the identification of the skeletal and non-skeletal components, the following literature was used: Bolli *et al.* (1994), Haq & Boersma (1978), Illing (1954), Flügel (2004), Loeblich & Tappan (1978), and Scholle & Ulmer-Scholle (2003).

RESULTS

Facies

The facies distribution (Fig. 2) shows a dominance of mud-free to mud-lean sediments in the north, west and south of the northern part of GBB. A gradual transition exists between these sediments and mud-rich wackestones present near the west side of Andros Island. The facies pattern is more complex than that presented in earlier studies (Ginsburg *et al.*, 1958; Newell *et al.*, 1959; Purdy, 1963b; Ball, 1967; Enos, 1974). A somewhat concentric distribution of various facies types is striking. The muddier sediment types (mud-rich wackestone (facies 1.5) to packstone (facies 3)) clearly dominate the platform interior. These sediments form a lobe adjacent to Andros Island, but also form an elongate zone close to the western edge of the platform. The cleaner sediments (packstones to grainstones) form a concentric belt on the western edge of the platform extending to the northern and southern regions. The grainier tongues (facies 3 and 3.5) projecting from NW Andros Island in a WSW and WNW direction are also notable features.

Grain-size

The grain-size distribution on GBB shows a distinct trend from coarse sediments on the western edge of the platform towards finer and muddier sediments in the vicinity of Andros Island (Fig. 3a–d). The tidal flats with scrub mangroves on the western side of Andros Island further confirm the muddy character of the sediments in this inner platform environment. The coarsest mean grain-sizes are on the northern, western and southern margins of the bank. The mean grain-size map (Fig. 3a) precisely displays open and protected environments on the platform. Note especially the zones with relatively coarse-grained sediments associated with the ooid belts at the eastern edge of the platform, N and SSE of Andros Island, near the Tongue of the Ocean.

Mineralogy

Aragonite is by far the dominant mineral across the platform and varies between 77.7 and 100%, with a mean of 93.3% (Fig. 4a; Table 1). High- and low-magnesium calcite (HMC and LMC) content varies between 0 and 22.3%, and 0 and 3.9%, respectively, with a mean of 6.5% for HMC and 0.2% for LMC (Fig. 4b and c; Table 1). The analysis of the fine-fraction (<63 μm) of the 2004 dataset showed an appreciable increase in the mean values of HMC and LMC, with values ranging from 4.8% (bulk samples) to 10.6% (fine fraction) for HMC and from 0.5% (bulk) to 6.7% (fine fraction) for LMC (Table 1).

Facies variations

The distribution of pellets largely coincides with the mud-rich facies distribution; as they are more abundant in restricted areas of the platform (Fig. 5a). As might be expected, ooids preferentially occur on the edge of the platform and in the more open areas on the platform (south and north of Andros Island; Fig. 5b) and thus show a large overlap with the more grainy facies types on the platform. The largest quantities of *Halimeda* were present in deeper waters, on the northern edge of the platform. On the platform itself *Halimeda* sp. occurs in minor quantities only. Grapestones were found on the more open northern part of the platform and along the southern side of the investigated area (not shown). Part of the grapestones from the northern sector of the platform consist of non-indurated grains bound by soft algal mucus that thus differ from the indurated ones found in the southern part of the platform.

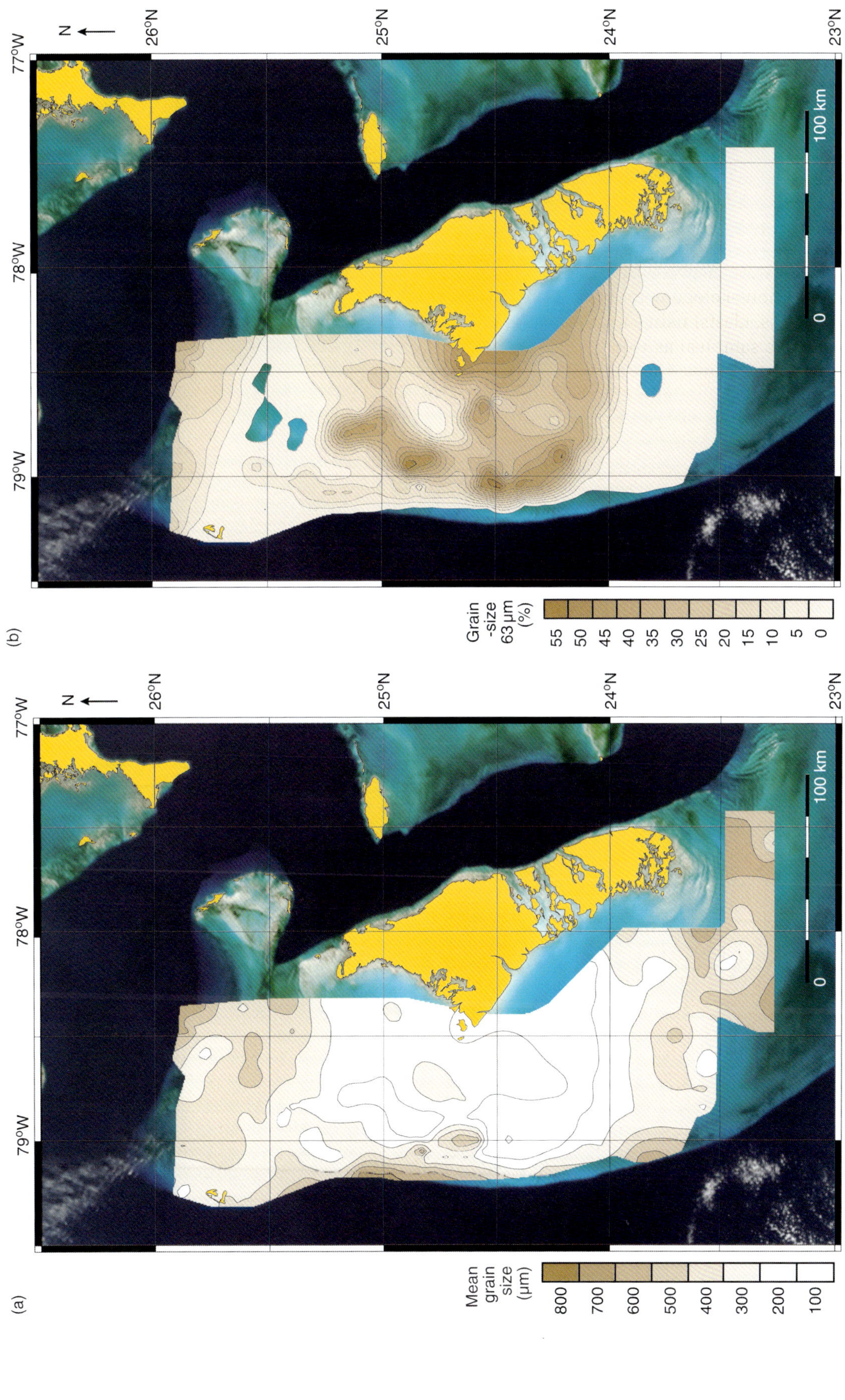

Fig. 3. Grain-size distribution across Great Bahama Bank. (a) Mean grain-size; (b) grain size <63 μm (% of total sample); (c) 63–125 μm (% of total sample); (d) grain size 500–1000 μm (% of total sample).

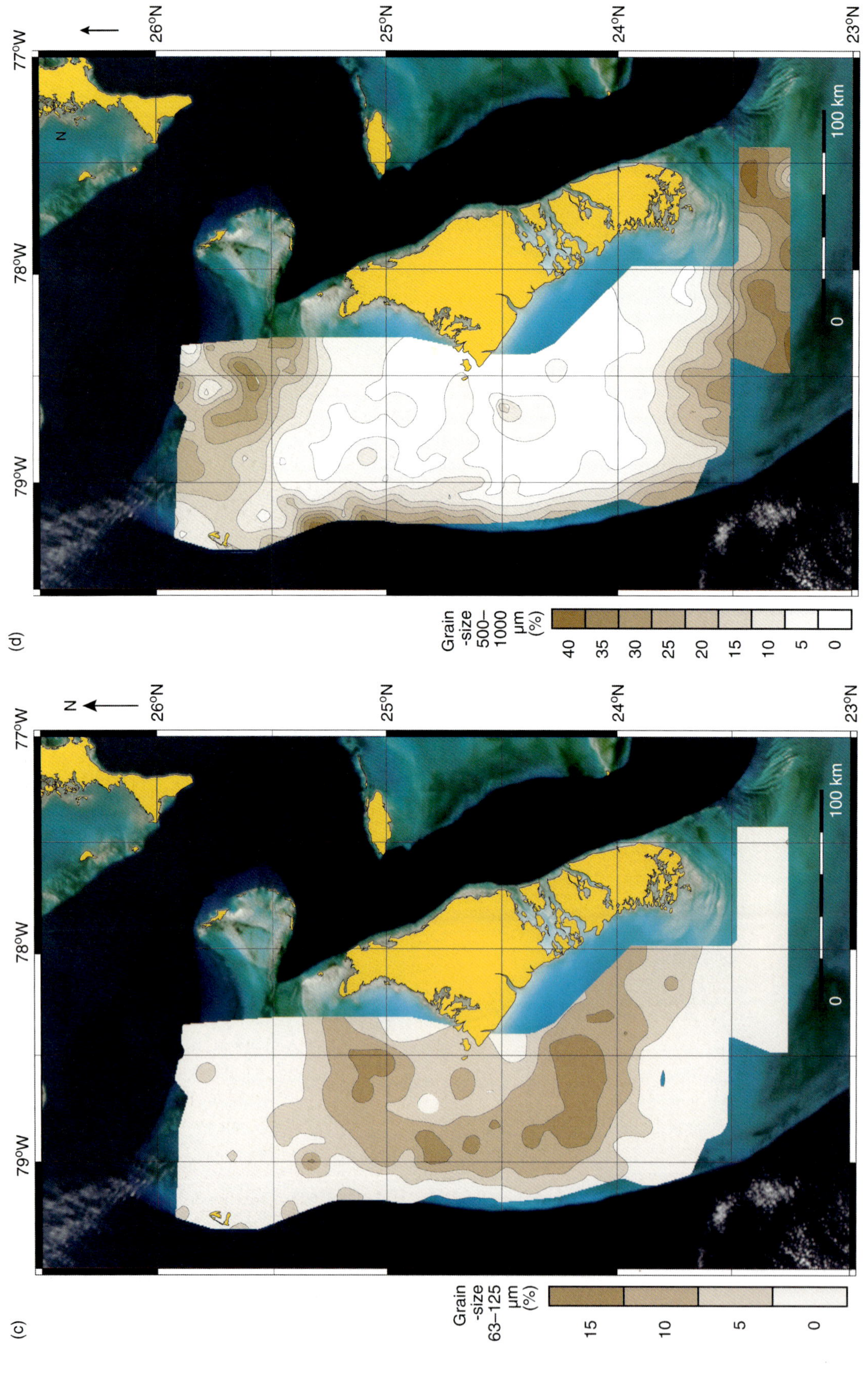

Fig. 3. Continued.

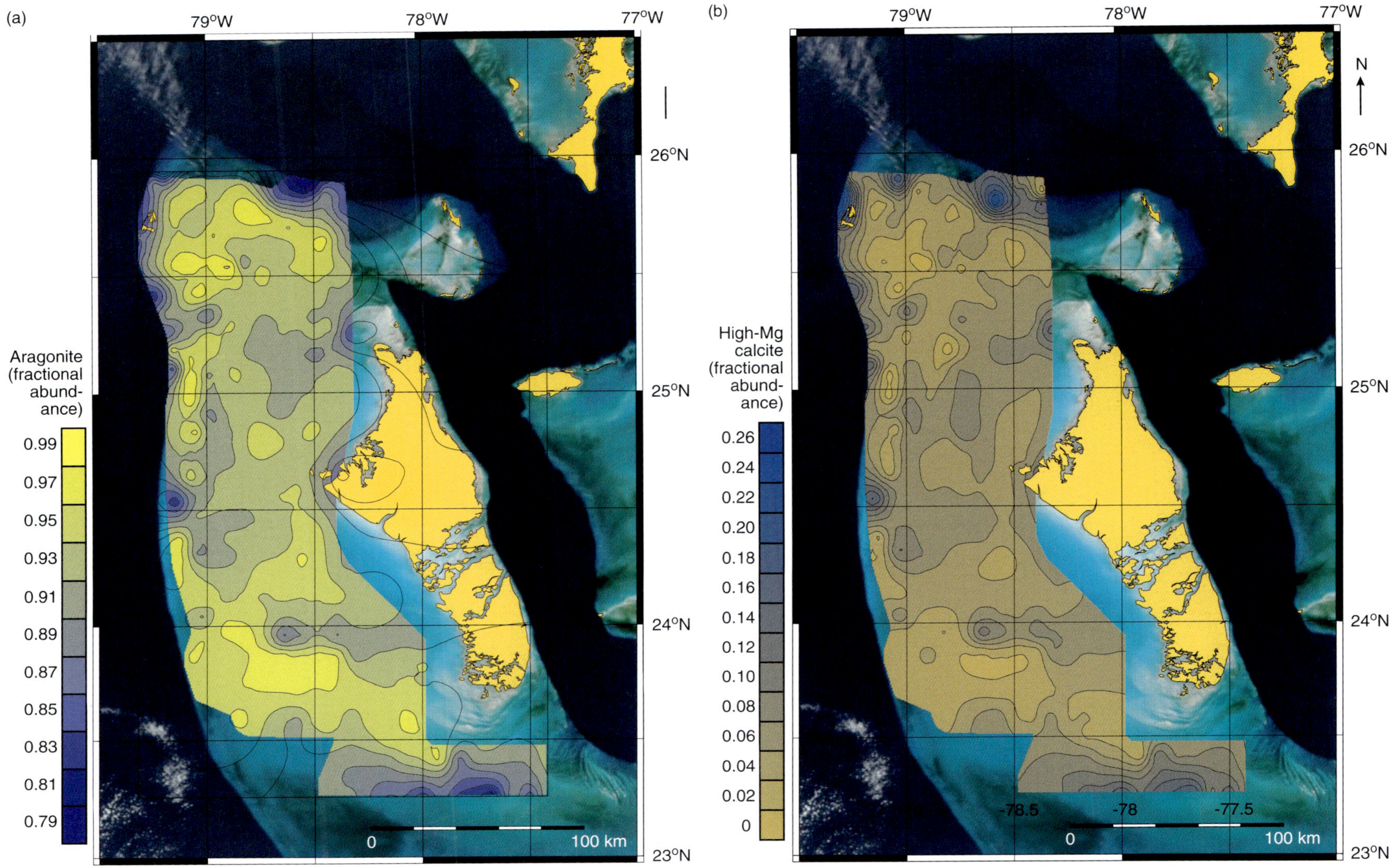

Fig. 4. Mineralogy distribution along Great Bahama Bank (bulk samples). (a) Fractional abundance aragonite (bulk samples); (b) fractional abundance HMC (bulk samples); (c) fractional abundance LMC (bulk samples). 1.0 represents 100%.

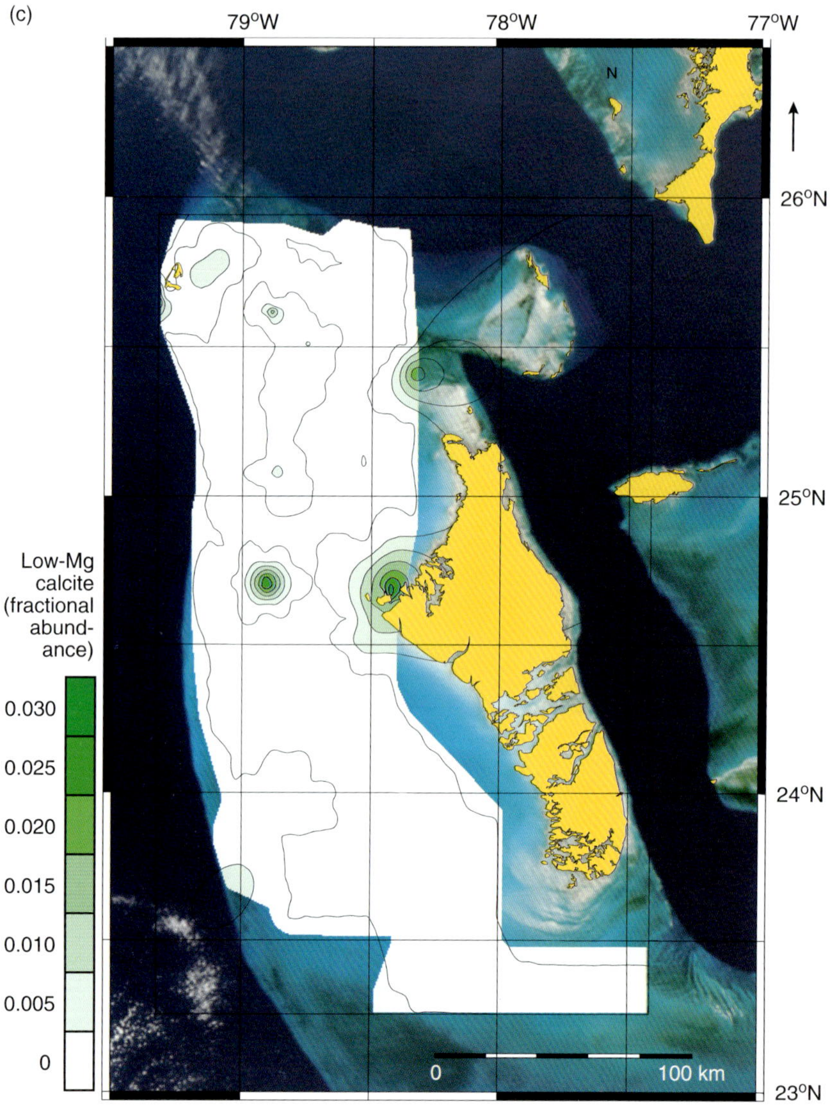

Fig. 4. Continued.

Table 1. Fractional abundance of carbonate mineralogies (% of total carbonate).

	Bulk total dataset			Bulk 2004 dataset			<63 µm 2004 dataset		
Mineralogy	Mean	Maximum	Minimum	Mean	Maximum	Minimum	Mean	Maximum	Minimum
Aragonite	93.3	100	77.7	94.8	100	79.9	82.7	94.5	36.5
HMC	6.5	22.3	0	4.8	19.5	0	10.6	20.3	2.8
LMC	0.2	3.9	0	0.5	1.6	0	6.7	47.2	0.4

DISCUSSION

Two major questions should be answered when discussing the sediment distribution along GBB: (i) What physical and biological processes control the present-day sediment distribution; and (ii) how does this distribution compare with facies distribution patterns observed on other carbonate platforms? To address these questions it is necessary to assess possible links between sediment distribution and physical processes such as currents, tides, platform topography, evaporation (salinity), water exchange with the surrounding ocean water masses, and sediment export.

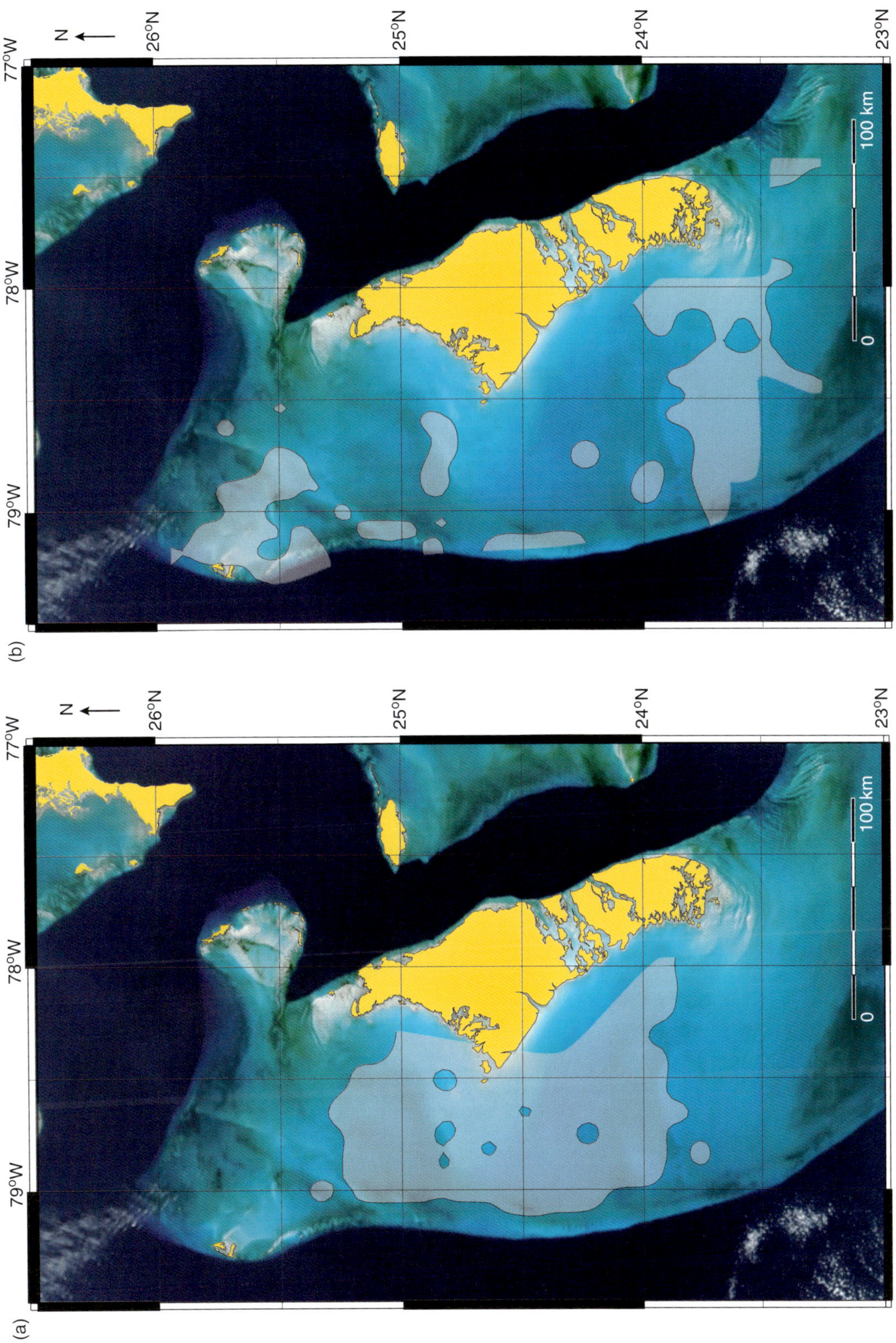

Fig. 5. Presence (shaded areas) on Great Bahama Bank of (a) pellets and (b) ooids.

Facies patterns

Purdy (1963b) hypothesized that the facies on GBB theoretically would be distributed in a series of bilaterally symmetrical bands parallel to the margin of the bank. However, the depositional topography of the Pleistocene substrate creates local current conditions that have resulted in the somewhat patchy facies distribution pattern observed (Purdy, 1963a,b). The early maps of Newell & Rigby (1957) and Newell *et al.* (1959), however, showed some concentric facies belts in the lee of Andros Island. All samples taken during these early studies were taken by hand or using a Van Veen grab-sampler. The Van Veen sampler has the disadvantage that it loses a proportion of the very fine fraction (<63 μm) during the retrieval process. This probably is the reason that differences in the mud content exist between the early facies maps and our data. The distribution pattern observed in the present study of GBB, however, shows a roughly concentric series of facies belts with an outer edge dominated by coarse-grained, mud-free to poorly washed grainstones, and an inner platform area with mud-rich wackestones forming a U-shaped band, situated close to Andros Island (Fig. 2).

Rankey & Morgan (2002) and Rankey (2002) suggested that the tidal flats west of Andros Island do not respond solely to the rate of sea-level rise, but also to the variation in circulation or storm intensity and direction. The tongue of less muddy sediment that extends from northwest Andros Island supports this scenario because it is clearly positioned at a location where the influx of waters from the east was observed on this side of the platform (Newell & Rigby, 1957; Harris, 1979, 1983). Another factor that seems to influence the distribution of mud-rich sediments is water depth and the related parameter, water energy. Mud-rich aragonitic sediments characterize the deeper areas while the shallower areas comprise coarser-grained sediments. As shown on the facies map (Fig. 2), Andros Island also plays an important role in the present-day sediment distribution as it protects the inner platform environments from the currents induced by the easterly trade winds and by tides. The facies change to grainstone dominance (facies 3.5–4) observed north and south of Andros Island precisely registers the influence of currents on the overall sediment distribution. The ooid-dominated tidal bars south of Tongue of the Ocean (Ginsburg, 2005) and north of Andros Island, on Joulters Cay (Newell & Rigby, 1957; Ball, 1967; Harris, 1979, 1983), also support the presence of strong currents that enter the shallow-water environments. The mud-free facies is positioned close to these facies belts, and thus confirms the influence of currents on the grain-size distribution in these parts of the platform. The grainstone belt (facies 4) crossing the platform south of Andros Island suggests that this zone acts as a conduit for currents of Tongue of the Ocean water. Another interesting feature is the agreement between the pellet distribution and the abundance of the 63–125 μm grain-size fraction showing the biological control on grain-size distribution (cf. Figs 3c and 5a).

Another process that influences the facies distribution is the tidal currents that freely enter the shallow-water platform. This is shown by the trend towards mud-free sediments in the southern part of the platform where no major barriers exist that would prevent the removal of the fine-grained material (Figs 2, 3a, 3b and 3d). The observed export of shallow-water muds by physical processes occurring during passage of winter cold fronts confirms this inference (Wilson & Roberts, 1992, 1995). The precise correlation of variations in sediment accumulation on the western slope of GBB with the scaling of atmospheric and oceanographic processes underlines the subtle character of the processes influencing sediment export from the inner platform areas (Roth & Reijmer, 2004, 2005).

Topography and sediment distribution

Another important question is whether the platform topography inherited from the last glacial cycle influenced the facies distribution pattern found today. The present-day distribution shows a link between water depth on the platform and the type of sediment present. Mud-rich facies (facies 1.5) shows a clear relation to deeper areas (Fig. 2). Coarse-grained, mud-free facies occur either close to the edge of the platform, or on elevated areas within the interior of the platform (e.g. southeast of Bimini; Figs 1 and 3a–d), or within the more open parts of the platform, in the southern part of the study area. Newell *et al.* (1959) described two broad shoals running across the Andros lobe, (i) the Bimini axis, which extends from Bimini in the west to the northern tip of Andros Island, and (ii) the Billy Island axis running westward from the middle promontory of Andros Island. The map of mean grain-size reflects these topographic features (Fig. 3a). The more open character of the southern part of the platform is also clearly shown by coarser, cleaner sediments on these maps (Fig. 3a–d). Tidal currents entering the platform in combination with wave action most likely caused this distribution.

Comparisons with other platform lagoons

Other carbonate platform lagoons such as those from Mayotte in the Indian Ocean (Zinke *et al.*, 2001, 2003a), show a facies distribution pattern that is clearly influenced by the interplay of the antecedent topography and the rate of the Holocene sea-level rise (Zinke *et al.*, 2003b, 2005). The rate with which the former topography was flooded determined the development of a current pattern within the lagoon and thus steered the facies distribution. This current pattern also depended on the continuity of the barrier surrounding the lagoon before the platform interior was reflooded after the last glacial. Other present-day facies variations were related to either the vicinity of the barrier reef or the reefs surrounding the inner lagoon of the island or the input of terrigenous material. The link between facies variations and the distance to the edge of the platform also can be observed on the Bahamian platform. The same holds for the influence of the antecedent topography (see discussion below).

In a discussion of the sediment distribution of three isolated carbonate platforms seaward of the Belize barrier reef system (Glovers Reef, Lighthouse Reef and Turneffe Islands) Gischler (1994) and Gischler & Lomando (1999) suggested that variations in antecedent topography and exposure to waves and currents were the dominant factors influencing the sediment distribution of these three relatively small-scale carbonate platforms. In analogy with the scenario developed for the facies distribution of the inner lagoon of Mayotte (Zinke *et al.*, 2001, 2003a), Gischler (2003) showed that the timing of reflooding of the Pleistocene bedrock in combination with the rate of sea-level rise determined the sedimentation patterns in the lagoons of the Belize atolls. The early flooded atolls (Glovers Reef and Lighthouse Reef) developed an open-circulation pattern, while the Turneffe Islands, flooded in a relatively late stage, at present has restricted circulation. The latter might also be related to its present rather low-energy position in the lee of the other platforms. Purdy & Gischler (2003) also noted that the distribution and overall character of the Holocene facies in the lagoons of Belize is strongly influenced by processes such as siliciclastic input related to the uplift of the hinterland, and by antecedent topography of structural origin influenced by karst processes during glacial lowstands (Purdy, 1974; Purdy *et al.*, 2003).

The data available on the antecedent pre-Holocene topography of GBB (Boss & Rasmussen, 1995; Boss, 1996) also suggest that the observed facies patterns are to some extent linked to the antecedent topography. The correlation between the timing of the last sea-level rise flooding the shallow-water areas of the platform and the onset of sediment production, without any significant lag time, suggests that the antecedent pre-Holocene topography acted as sediment production and export site immediately after reflooding (Roth & Reijmer, 2004).

The facies distribution shows the interaction of the prevailing winds and currents with the pre-existing topography. At present considerable accommodation space is unfilled, which probably results from a combination of a pre-existing topography, flat-topped platform with no distinct barrier, and a rate of sea-level rise that resulted in the establishment of a current system that removes sediment, soon after the platform was flooded. This early export of platform sediments towards the surrounding slopes and basins started at 7.2 ka, shortly after reflooding of the shallow-water areas, as discussed by Wilber *et al.* (1990), McNeill *et al.* (1998) and Roth & Reijmer (2004, 2005).

Mineralogy

Appreciable differences exist between the carbonate mineralogy of the bulk sediments (Fig. 4) and the <63 μm grain-size fraction (not shown). The percentage of HMC and LMC increases in the <63 μm fraction when compared with the bulk sediments (Table 1). These differences are most likely related to the export of the fine aragonite mud whitings produced in the water column (Shinn *et al.*, 1989; Robbins & Blackwelder, 1992; MacIntyre & Read, 1992, 1995; Milliman *et al.*, 1993; Thompson *et al.*, 1997; Yates & Robbins, 1998, 1999). This production of aragonite fines in the water column makes them very sensitive to transport during tides, storms and passing winter fronts (McCave, 1972; Neumann & Land, 1975; Wilson & Roberts, 1992, 1995; Wilber *et al.*, 1993). The precise recording of climatic variations on the slopes of GBB most likely reflects this process (Roth & Reijmer, 2005).

Values for HMC are highest along the northern edge of the platform (Fig. 4b). Slightly increased values occur in the mud-dominated facies belt in the lee of Andros Island. The cause of this latter increase remains uncertain, but could have a biological (e.g. pellet formation) or a diagenetic origin.

Two small zones with slightly elevated LMC values occur on the bank (Fig. 4c). One near the western tip of Andros, which is probably related to the input of Pleistocene eroded material from the island. The second zone relates to an inlet on the eastern side of the platform in which the ooid shoals occur also, evidencing strong currents that enter the platform from the Tongue of the Ocean.

Facies patterns and isotope signals

Mud-dominated, aragonite-rich sediments (facies types 1.5–4.5) form the majority of the sediments in the interior of GBB (Figs. 2, 3a, 4a). These facies belts are surrounded by skeletal and non-skeletal aragonite sediments on the northern and western outer edges of the platform and in the southern parts of the study area (Fig. 2). The predominantly aragonite sediments on GBB have similar $\delta^{13}C$ and $\delta^{18}O$ compositions (Swart *et al.*, 2009). A comparison between the pellet distribution and the distribution of the mud-rich sediments (cf. Figs 2, 3c and 5a) shows that muds and related sediment particles are important sediment components on the platform. The muds and their derivates seem to dominate the isotope distribution observed on GBB (Swart *et al.*, 2009).

Comparison between new facies map and literature data

The facies maps that are widely used in the literature are based on the data sets presented by Illing (1954), Ginsburg *et al.* (1958; Fig. 6a), Newell & Rigby (1957), Newell *et al.* (1959), Cloud (1962), Purdy (1963a) and Traverse & Ginsburg (1966; Fig. 6a). The most widely used facies distribution maps of Purdy (1963b; Fig. 6b) and Enos (1974; Fig. 6c) make use of these data sets.

The facies, grain-size and mineralogy maps presented in the present study show a more diverse distribution of all sediment parameters along the platform. The gross facies distribution with a skeletal-dominated, coarse-grained platform edge and a mud-dominated interior in the lee of Andros Island shows up on all maps, however (Figs 2 and 6a–c). Another feature that shows up on all maps is the sharp transition from mud-dominated sediments (facies type 1.5) to grainstone facies (facies type 4) on the western boundary of the platform (Figs 2 and 6a–c). A detailed comparison, however, is difficult, because the majority of the maps use a classification scheme partly based on the occurrence of single grain types and carbonate muds. A significant difference is that the dominant grapestone occurrence in the northern part of the platform as shown by Newell *et al.* (1959) and Purdy (1963b) could not be confirmed by this study (cf. Fig. 2 with 6b). On the Enos (1974) map this northern grapestone-dominated facies belt was classified as a peloid-dominated grainstone facies, which agrees more with our results (cf. Fig. 2 with Fig. 6c). Despite the fact that the Enos (1974) map also used the Dunham classification scheme (1962), large differences in facies distribution exist between that map and this study. Further detailed studies on the composition of the different grain-size fractions in combination with a detailed petrographic analysis are needed to quantify the skeletal and non-skeletal component distribution on the platform.

CONCLUSIONS

The facies, grain-size and mineralogy distribution patterns on GBB show a current-dominated sedimentation pattern with a clear distinction between the outer edge and the inner platform. Inner-platform facies distribution relates to the present-day bathymetry, which might represent in part an inherited Pleistocene topography. Seismic studies are needed to verify this.

Grain-size differences reflect the bathymetry and show a more coarse-grained outer platform and a mud-dominated interior. The grainstone distribution north and south of Andros Island registers the presence of strong currents that enter and cross the shallow-water environment. The correlation between the distribution of the 63–125 µm grain-size fraction and the distribution of pellets suggests a predominantly biological origin of this grain size.

The mineralogy of the sediments shows a clear dominance of aragonite mixed with HMC and minor LMC. Concentration of both calcite types in the fine fraction suggests the preferential transport of aragonite-rich muds from the platform through various processes.

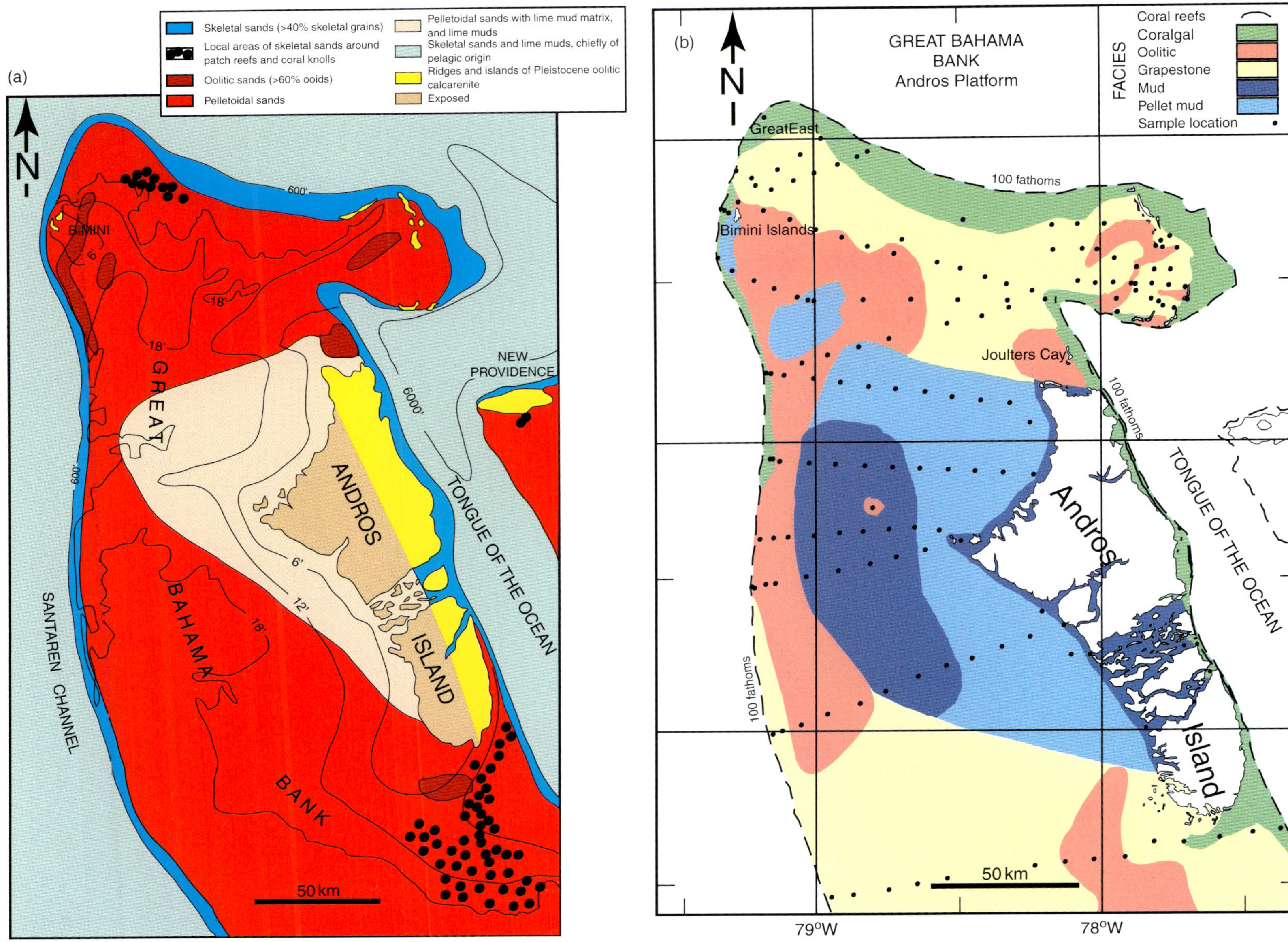

Fig. 6. (a) Facies map (modified) after Ginsburg *et al.* (1958) and Traverse & Ginsburg (1966); (b) facies map (modified) after Purdy (1963b); (c) facies map (modified) after Enos (1974).

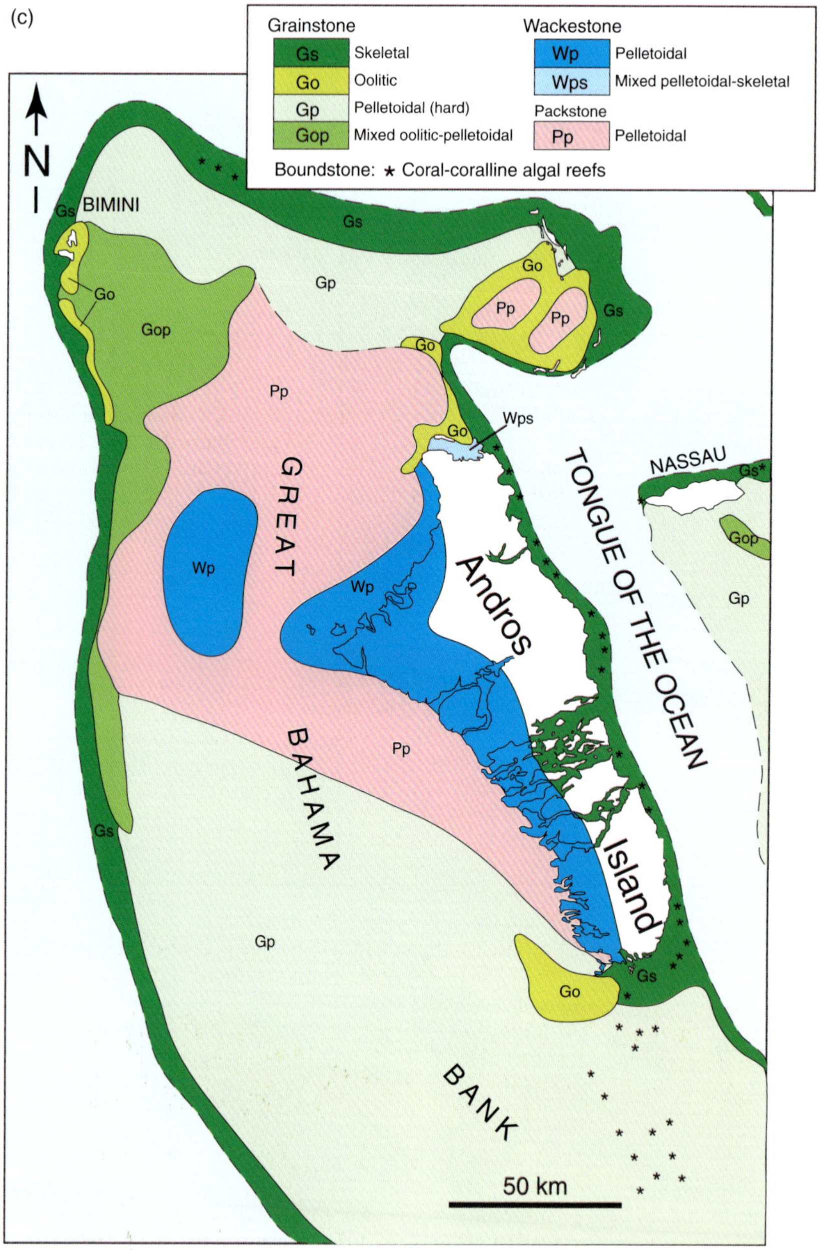

Fig. 6. Continued.

ACKNOWLEDGEMENTS

The authors thank the crew of the RV *Bellows* for their assistance in gathering this data set throughout four cruises between 2001 and 2004. We also thank Jill Falk, Hendrik Lantzsch and Astrid Ryba (Kiel) and Amel Saied and Corey Schroeder (Miami) for their help preparing the samples. This project was funded in part by the German Science Foundation (Projects Re1051/8 and 9) and Leibniz funds of W.-Chr. Dullo (IFM-GEOMAR, Kiel). Additional support for the use of the RV *Bellows* was provided by Florida Institute of Oceanography and Comparative Sedimentology Laboratory. We thank Paul (Mitch) Harris, Pascal Kindler and Paul Enos for their constructive reviews of the manuscript.

REFERENCES

Atkinson, L.P., Berger, T., Hamilton, P., Waddell, E., Leaman, K. and **Lee, T.N.** (1995) Current meter observations in the Old Bahama Channel. *J. Geophys. Res.*, **100**, 8555–8560.

Ball, M.M. (1967) Carbonate sand bodies of Florida and the Bahamas. *J. Sed. Petrol.*, **37**(2), 556–591.

Bathurst, R.G.C. (1975) *Carbonate Sediments and their Diagenesis. Developments in Sedimentology*, Vol. 12. Elsevier, Amsterdam, 439 pp.

Bolli, H.M., Beckmann, J.-P. and **Saunders, J.B.** (1994) *Benthic Foraminiferal Biostratigraphy of the South Caribbean Region.* Cambridge University Press, Cambridge, 408 pp.

Boss, S.K. (1996) Digital shaded relief image of a carbonate platform (northern Great Bahama Bank): scenery seen and unseen. *Geology*, **24**, 985–988.

Boss, S.K. and **Rasmussen, K.A.** (1995) Misuse of Fischer plots as sea-level curves. *Geology*, **23**, 221–224.

Carew, J.L. and Mylroie, J.E. (1997) Geology of the Bahamas. In: *Geology and Hydrogeology of Carbonate Islands* (Eds H.L. Vacher and T.M. Quinn), pp. 91–139. Elsevier Science, New York, USA.

Cloud, P.E. Jr. (1962) Environment of calcium carbonate deposition west of Andros Island. *U.S. Geol. Surv. Profess. Papers*, **350**, 1–138.

Dunham, R.J. (1962) Classification of carbonate rocks according to depositional texture. In: *Classification of Carbonate Rocks – A Symposium* (Ed. W.E. Ham), *AAPG Memoir* **1**, pp. 108–121. American Association of Petroleum Geologists, Tulsa, Oklahoma, USA.

Enos, P. (1974) *Surface Sediment Facies of the Florida-Bahamas Plateau Map Series MC-5 no.*, 4 edn. Geological Society of America, Boulder, Colorado, USA.

Field, R.M. (1931) Geology of the Bahamas. *GSA Bull.*, **42**, 759–784.

Flügel, E. (2004) *Microfacies of Carbonate Rocks.* Springer, Berlin, Germany, 976 pp.

Gebelein, C.D. (1974) Guidebook for modern Bahamian platform environments. *Geol. Soc. Am. Ann. Mtg Field Guide*, 93 pp. Geological Society of America, Boulder, Colorado, USA.

Ginsburg, R.N. (1956) Environmental relationships of grain size and constituent particles in some South Florida carbonate sediments. *AAPG Bull.*, **40**, 2384–2427.

Ginsburg, R.N. (2005) Disobedient sediments can feedback on their transportation, deposition and geomorphology. *Sed. Geol.*, **175**, 9–18.

Ginsburg, R.N., Lloyd, R.M., McCallum, J.S., Stockman, K.W. and **Moody, R.A.** (1958) Surface sediments of Great Bahama Bank, Shell Development Company, Houston, Texas, USA.

Gischler, E. (1994) Sedimentation on three Caribbean atolls: Glover Reef, Lighthouse Reef and Turneffe Islands, Belize. *Facies*, **31**, 243–254.

Gischler, E. (2003) Holocene lagoonal development in the isolated carbonate platforms off Belize. *Sed. Geol.*, **159**, 113–132.

Gischler, E. and **Lomando, A.J.** (1999) Recent sedimentary facies of isolated carbonate platforms, Belize-Yucatan system, Central Ameri *J. Sed. Res.*, **69**, 747–763.

Gonzalez, R. and **Eberli, G.P.** (1997) Sediment transport and bedforms in a carbonate tidal inlet; Lee Stocking Island, Exumas, Bahamas. *Sedimentology*, **44**, 1015–1030.

Haq, B.U. and **Boersma, A.** (1978) *Introduction to Marine Micropaleontology.* Elsevier, New York, 376 pp.

Harris, P.M. (1979) Facies anatomy and diagenesis of a Bahamian ooid shoal. *Sedimenta*, **7**, 175.

Harris, P.M. (1983) The Joulters ooid shoal, Great Bahama Bank. In: *Coated Grains* (Ed. T.M. Peryt), pp. 132–141. Springer, Berlin, Heidelberg, Germany.

Husseini, S.I. and **Matthews, R.K.** (1972) Distribution of high-magnesium calcite in lime muds of the Great Bahama Bank; diagenetic implications. *J. Sed. Petrol.*, **42**(1), 179–182.

Illing, L.V. (1954) Bahaman calcareous sands. *AAPB Bull.*, **38**, 1–95.

Leaman, K.D., Vertes, P.S., Atkinson, L.P., Lee, T.N., Hamilton, P. and **Waddell, E.** (1995) Transport, potential vorticity, and current/temperature structure across Northwest Providence and Santaren Channels and the Florida Current off Cay Sal Bank. *J. Geophys. Res.*, **100**, 8561–8569.

Loeblich, A.R. Jr. and **Tappan, H.** (1978) Sarcodina chiefly 'Thecamoebians' and Foraminiferida. In: *Treatise on Invertebrate Paleontology*, Part C, Protista 2 (Ed. R.C. Moore). Geological Society of America and University of Kansas Press, Vol. **1/2**. Geological Society of America, Boulder, Colorado, USA.

MacIntyre, I.G. and **Reid, R.P.** (1992) Comment on the origin of aragonite needle mud: a picture is worth a thousand words. *J. Sed. Petrol.*, **62**(6), 1095–1097.

MacIntyre, I.G. and **Reid, R.P.** (1995) Crystal alteration in a living calcareous alga (Halimeda): implications for studies in skeletal diagenesis. *J. Sed. Res.*, **A65**, 1143–1153.

McCave, I.N. (1972) Transport and escape of fine-grained sediment from shelf areas. In: *Shelf Sediment Transport* (Eds D.J.P. Swift, D.B. Duane and O.H. Pilkey), pp. 225–248. Van Nostrand Reinhold, New York, USA.

McNeill, D., Grammer, G.M. and **Williams, S.C.** (1998) A 5 My chronology of carbonate platform margin aggradation, southwestern Little Bahama Bank, Bahamas. *J. Sed. Res.*, **68**, 603–614.

Miller, A., Thompson, J.C., Peterson, R.E. and **Haragan, D.R.** (1983) *Elements of Meteorology*, 4th ed. Columbus (Merill Publ. Co.), USA.

Milliman, J.D. (1974) *Marine Carbonates.* Springer Verlag, Berlin, Heidelberg, New York, 375 pp.

Milliman, J.D., Freile, D., Steinen, R. and **Wilber, R.J.** (1993) Great Bahama Bank aragonitic muds: mostly inorganically precipitated, mostly exported. *J. Sed. Petrol*, **63**(4), 589–595.

Mullins, H.T. (Ed.) (1986) *Carbonate Depositional Environments, Modern and Ancient. Part 4: Periplatform Carbonates* (Eds J.E. Warme and K.W. Shanley), *Quarterly*, **81**, 1–63. Colorado School of Mines Press, Golden (Colorado), USA.

Neumann, A.C. and **Land, L.S.** (1975) Lime mud deposition and calcareous algae in the Bight of Abaco, Bahamas: a budget. *J. Sed. Petr.*, **45**, 763–786.

Newell, N.D. and **Rigby, J.K.** (1957) Geological studies on the Great Bahama Bank. In: *Regional Aspects of Carbonate Deposition: A Symposium with Discussion* (Eds R.J. Le Blanc and J.G. Breeding), SEPM Spec. Publ., **5**, pp. 15–72. Society of Economic Paleontologists and Mineralogists, Tulsa, Oklahoma, USA.

Newell, N.D., Imbrie, J., Purdy, E.G. and **Thurber, D.L.** (1959) Organism communities and bottom facies, Great Bahama Bank. *Bull. Am. Museum Nat. Hist.*, **117**, 177–228.

Purdy, E.G. (1963a) Recent calcium carbonate facies of the Great Bahama Bank. 1. Petrography and reaction groups. *J. Geol.*, **71**, 334–355.

Purdy, E.G. (1963b) Recent calcium carbonate facies of the Great Bahama Bank. 2. Sedimentary facies. *J. Geol.*, **71**, 472–497.

Purdy, E.G. (1974) Karst-determined facies patterns in British Honduras: Holocene carbonate sedimentation model. *AAPG Bull.*, **58**, 825–855.

Purdy, E.G. and **Gischler, E.** (2003) The Belize margin revisited: 1. Holocene marine facies. *Int. J. Earth Sci. (Geol. Rundsch.)*, **92**, 532–551.

Purdy, E.G., Gischler, E. and **Lomando, A.J.** (2003) The Belize margin revisited. 2. Origin of Holocene antecedent topography. *Int. J. Earth Sci. (Geol. Rundsch.)*, **92**, 552–572.

Rankey, E.C. (2002) Spatial patterns of sediment accumulation on a Holocene carbonate tidal flat, Northwest Andros Island, Bahamas. *J. Sed. Res.*, **72**, 591–601.

Rankey, E.C. and **Morgan, J.** (2002) Quantified rates of geomorphic change on a modern carbonate tidal flat, Bahamas. *Geology*, **30**, 583–586.

Robbins, L.L. and **Blackwelder, P.L.** (1992) Biochemical and ultrastructural evidence for the origin of whitings: a biologically induced calcium carbonate precipitation mechanism. *Geology*, **20**, 464–468.

Roth, S. and **Reijmer, J.J.G.** (2004) Holocene Atlantic climate variations deduced from carbonate periplatform sediments (leeward margin, Great Bahama Bank). *Paleoceanography*, **19**, PA1003, doi:10.1029/2003PA000885.

Roth, S. and **Reijmer, J.J.G.** (2005) Holocene millennial to centennial carbonate cyclicity recorded in slope sediments of the Great Bahama Bank and its climatic implications. *Sedimentology*, **52**, 161–181.

Schmitz, W.J. and **Richardson, P.L.** (1991) On the sources of the Florida Current. *Deep-Sea Res.*, **38**, 379–401.

Scholle, P.A. and **Ulmer-Scholle, D.S.** (2003) *A Color Guide to the Petrography of Carbonate Rocks*, AAPG Memoir, **77**. American Association of Petroleum Geologists, Tulsa, Oklahoma, USA.

Shinn, E.A., Steinen, R.P., Lidz, B.H. and **Swart, P.K.** (1989) Whitings, a sedimentologic dilemma. *J. Sed. Petrol.*, **59**, 147–161.

Smith, C.L. (1940) The Great Bahama Bank. *J. Mar. Res.*, **3**, 147–189.

Smith, N.P. (1995) On long-term net flow over Great Bahama Bank. *J. Phys. Oceanogr.*, **25**, 679–684.

Swart, P.K. James, N.P., Mallinson, D., Malone, M.J., Matsuda, H. and **Simo, T.** (2002) Data report: Carbonate mineralogy of sites drilled during Leg 182, In: *Proceedings of the Ocean Drilling Program, Scientific Results* (Eds A.C. Hine, D.A. Feary and M.J. Malone), **182** (CD Rom). ODP, College Station, Texas, USA.

Swart, P.K., Reijmer, J.J.G. and **Otto, R.** (2009) A re-evaluation of facies on Great Bahama Bank II: variations in the $\delta^{13}C$, $\delta^{18}O$, and mineralogy of surface sediments. In: *Perspectives in Sedimentary Geology: A Tribute to the Career of Robert Nathan Ginsburg* (Eds P.K. Swart, G.P. Eberli and J.A. McKenzie). Blackwell, Oxford. *Spec. Publ. Int. Assoc. Sedimentol.*, **41**, 47–60.

Thompson, J.B., Schultze-Lam, S., Beveridge, T.J. and **Des Marais, D.J.** (1997) Whiting events: biogenic origin due to the photosynthetic activity of cyanobacterial picoplankton. *Am. Soc. Limnol. Oceanogr.*, **42**(1), 133–141.

Traverse, A. and **Ginsburg, R.N.** (1966) Palynology of the surface sediments of Great Bahama Bank, as related to water movement and sedimentation. *Mar. Geol.*, **4**, 417–459.

Tucker, M.E. and **Wright, V.P.** (1990) *Carbonate Sedimentology*. Blackwell Scientific Publications, Oxford, UK, 482 pp.

Wajsowicz, R.C. (2002) A modified Sverdrup Model of the Atlantic and Caribbean circulation. *J. Phys. Oceanogr.*, **32**, 973–993.

Wenthworth, C.K. (1922) A scale of grade and class terms for clastic sediments. *J. Geol.*, **30**, 377–392.

Wilber, R.J., Milliman, J.D. and **Halley, R.B.** (1990) Accumulation of bank-top sediment on the western slope of Great Bahama Bank: rapid progradation of a carbonate megabank. *Geology*, **18**, 970–974.

Wilber, R.J., Whitehead, J., Halley, R.B. and **Milliman, J.D.** (1993) Carbonate periplatform sedimentation by density flows: a mechanism for rapid off-bank and vertical transport of shallow-water fines: Comment and reply. *Geology*, **21**, 667–669.

Wilson, P.A. and **Roberts, H.H.** (1992) Carbonate-periplatform sedimentation by density flows: a mechanism for rapid off-bank and vertical transport of shallow-water fines. *Geology*, **20**, 713–716.

Wilson, P.A. and **Roberts, H.H.** (1995) Density cascading: off-shelf sediment transport, evidence and implications, Bahama Banks. *J. Sed. Res.*, **A65**, 45–56.

Winland, H.D. and **Matthews, R.K.** (1974) Origin and significance of grapestone, Bahama Islands. *J. Sed. Petrol.*, **44**, 3, 921–927.

Yates, K.K. and **Robbins, L.L.** (1998) Production of carbonate sediments by a unicellular green alga. *Am. Mineral.* **83**, 1503– 1509.

Yates, K.K. and **Robbins, L.L.** (1999) Radioisotopic tracer studies of inorganic carbon and Ca in microbially derived $CaCO_3$. *Geochim. Cosmochim. Acta*, **63**, 129–136.

Zinke, J., Reijmer, J.J.G. and **Thomassin, B.A.** (2001) Seismic architecture and sediment distribution within the Holocene barrier reef-lagoon complex of Mayotte (Comoro archipelago, SW Indian Ocean). *Palaeogeo., Palaeoclim., Palaeoecol.*, **175**, 343–368.

Zinke, J., Reijmer, J.J.G. and **Thomassin, B.A.** (2003a) Systems tracts sedimentology in the lagoon of Mayotte associated with the Holocene transgression. *Sed. Geol.*, **160**, 57–79.

Zinke, J., Reijmer, J.J.G., Thomassin, B.A., Dullo, W.-C., Grootes, P.M. and **Erlenkeuser, H.** (2003b) Postglacial flooding history of Mayotte Lagoon (Comoro Archipelago, southwest Indian Ocean). *Mar. Geol.*, **194**, 181–196.

Zinke, J., Reijmer, J.J.G., Taviani, M., Dullo, W.-C. and **Thomassin, B.A.** (2005) Facies and faunal assemblage changes in response to the Holocene transgression in the lagoon of Mayotte (Comoro Archipelago, SW Indian Ocean). *Facies*, **50**, 391–408.

Int. Assoc. Sedimentol. Spec. Publ. (2009) **41**, 47–59

A re-evaluation of facies on Great Bahama Bank II: variations in the $\delta^{13}C$, $\delta^{18}O$ and mineralogy of surface sediments

PETER K. SWART*, JOHN J.G. REIJMER†‡1 and ROBERT OTTO*

**Marine Geology and Geophysics, Rosenstiel School of Marine and Atmospheric Science, 4600 Rickenbacker Causeway, Miami, FL 33149, USA (E-mail: pswart@rsmas.miami.edu)*

†IFM-GEOMAR, Leibniz-Institut für Meereswissenschaften, Dienstgebäude Ostufer, Wischhofstr. 1-3, D-24148 Kiel, Germany

‡Centre de Sédimentologie-Paléontologie, FRE CNRS 2761 "Géologie des Systèmes Carbonatés", Université de Provence (Aix-Marseille I), 3, place Victor Hugo, Case 67, F13331 Marseille Cédex 3, France

ABSTRACT

In order to investigate the spatial distribution of $\delta^{13}C$ and $\delta^{18}O$ of modern carbonate sediments on Great Bahama Bank, ~290 surface samples were collected from a grid of stations approximately 10 km apart between 2001 and 2004. These samples were classified using a modified Dunham scheme, physically separated into six size fractions and subsequently analysed for their mineralogy (aragonite, low-Mg calcite and high-Mg calcite) and $\delta^{13}C$ and $\delta^{18}O$ values. A striking feature of these data is the relatively positive $\delta^{13}C$ values of all the samples. Based on measurements of $\delta^{13}C$ and $\delta^{18}O$ of the dissolved inorganic carbon and the water, most of the sediments can be considered to be in C and O isotopic equilibrium with the ambient waters. The high $\delta^{13}C$ values are suggested to arise from isotopic enrichment of the dissolved inorganic carbon pool by photosynthesis of seagrasses, benthic algae and cyanobacteria on the platform and through the fractionation of HCO_3^- during the precipitation of calcium carbonate. Sediments that are not in C and O isotopic equilibrium are dominated more by skeletal material. The data showed an absence of significant spatial variation in $\delta^{13}C$ of the sediments on the Great Bahama Bank and no clear spatial patterns relative to the margin of the platform. The $\delta^{18}O$ of the sediment showed more variation, with the interior sediments being isotopically enriched relative to the platform margin. The absence of significant variations in the $\delta^{13}C$ in the modern surface sediments of Great Bahama Bank irrespective of facies type suggests that in the case of Great Bahama Bank, downcore variations in $\delta^{13}C$ cannot be related to changes in facies.

Keywords Stable isotope, shallow-water carbonates, Bahamas, grain-size, carbonate platforms, salinity.

INTRODUCTION

The descriptions of Great Bahama Bank (GBB) and its surface sediments (Illing, 1954; Newell *et al.*, 1959; Purdy, 1963a, 1963b; Traverse & Ginsburg, 1966; Enos, 1974) made between 1950 and 1970 are widely used for modern biological and geological studies on GBB. In addition, GBB is often applied as an analogue for the study of ancient carbonate platforms. This paper together with a companion study (Reijmer *et al.*, 2009) examines spatial variations in the facies and geochemistry of surface sediments on GBB. Although there have been several papers that provided some general geochemical characteristics on the sediments on GBB (Lowenstam & Epstein, 1957; Shinn *et al.*, 1989), there have been no studies which have described spatial variation in the inorganic and organic mineralogy of the sediments on the same scale as the original sedimentary facies were investigated. Such data are important as changes in the $\delta^{13}C$ of platform and periplatform carbonates are increasingly being used for stratigraphic purposes (Vahrenkamp, 1996; Saltzman *et al.*, 2004) and for providing information on the global CO_2 cycle in periods prior to the

[1]Present address: Vrije Universiteit, Faculty of Earth and Life Sciences (FALW), De Boelelaan 1085, 1081 HV Amsterdam, The Netherlands.

existence of pelagic records (Berner, 1987; Kump & Arthur, 1999). It is well documented that global changes in the $\delta^{13}C$ of platform carbonates can be correlated between various strata during certain periods of the geological record (Föllmi *et al.,* 1994; Vahrenkamp, 1996; Valladares *et al.,* 1996; Weissert *et al.,* 1998; Kump & Arthur, 1999; Veizer *et al.,* 1999; Immenhauser *et al.,* 2002; Saltzman, 2002a,b; Saltzman *et al.,* 2004; Saltzman & Young, 2005) as well as related to the pelagic record of $\delta^{13}C$ change (Gale, 1993; Weissert *et al.,* 1998). Such variations are normally interpreted as reflecting changes in the partitioning of the global carbon cycle, with elevated values indicating the increased burial of organic carbon (Berner *et al.,* 1983; Kump & Arthur, 1999). However, it is well established that changes in $\delta^{13}C$ can also reflect diagenesis (Allan & Matthews, 1982; Immenhauser *et al.,* 2003) as well as variations in the relative proportion of periplatform and pelagic material (Swart & Eberli, 2005; Reuning *et al.,* 2006). In this paper the origin of the $\delta^{13}C$ and $\delta^{18}O$ and mineralogy of ~290 samples collected from Great Bahama Bank on a regular grid spacing of ~10 km between 2001 and 2004 are reported.

SAMPLES

Samples were collected using a Shipek sampler during four cruises aboard the RV *Bellows* between 2001 and 2004 (Fig. 1). The Shipek sampler differs from the Van-Veen sampler used during earlier studies (Purdy, 1963a,b) in that it is better at retaining the mud material. In total 291 bulk samples were analysed. Of these approximately 120 were separated into size fractions (>1000, 500–1000, 250–500, 125–250, 125–63 and

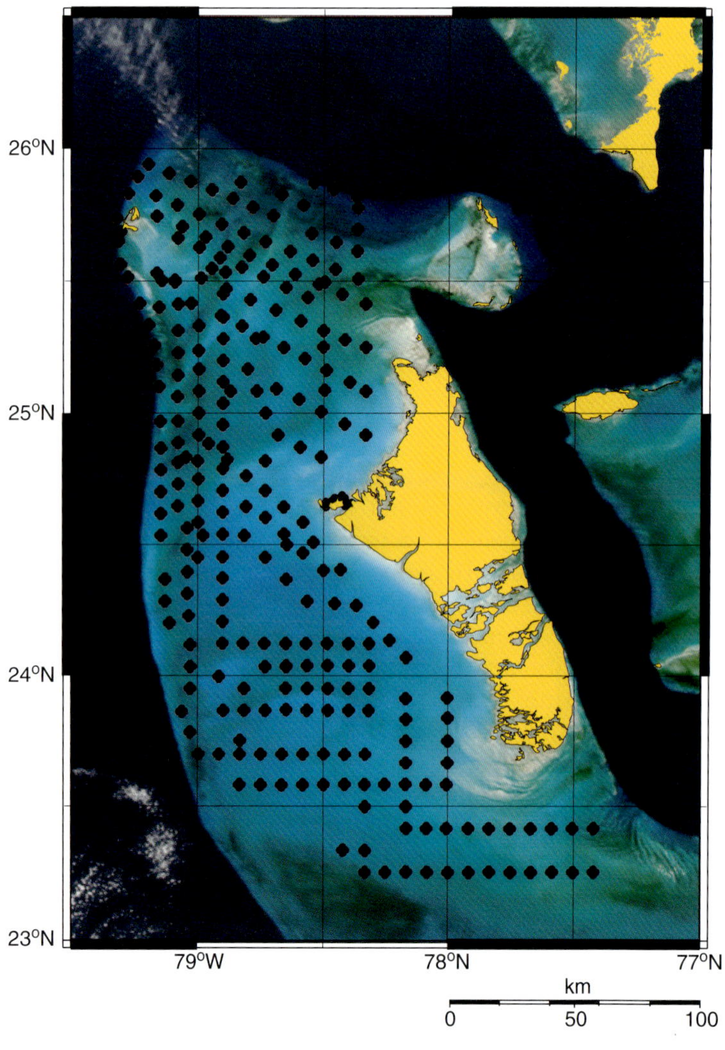

Fig. 1. Location map of the Bahamas and position of sample collection points (black diamonds).

<63 μm) and the relative percentages of aragonite, high-Mg calcite (HMC) and low-Mg calcite (LMC) were determined together with their carbon and oxygen-isotopic compositions. Each sample was also separated into a modified Dunham facies classification. At the same time water samples were taken for salinity and stable oxygen and hydrogen measurements. Samples for carbon isotopic analysis of the dissolved inorganic carbon (DIC) were filtered using a 1 μm filter and preserved in sealed crimped vials with excess mercuric chloride.

METHODS

X-ray mineralogy

The percentages of aragonite, HMC and LMC were determined using the method of Swart & Melim (2000). In this method the sample is assumed to be completely composed of aragonite, HMC and LMC. Areas of the appropriate peaks for each mineral are determined using a scan between 23 and 32° 2θ (CuKα radiation). These are then compared with standard relationships between peak area and percentage mineral and the percentage of the mineral in the sample determined.

Stable isotopes carbonates

The $\delta^{13}C$ and $\delta^{18}O$ of the carbonate materials was determined using dissolution in phosphoric acid using the common acid bath method (Swart *et al.*, 1991) at 90°C. The gas produced was analysed using a Finnigan-MAT 251. Later some samples were also analysed using a Kiel III device attached to a Finnigan Delta plus. Data produced in both methods were corrected for isobaric interferences using the procedures (Craig, 1957) modified for a triple collector mass spectrometer. Data are reported relative to Vienna Pee Dee Belemnite (V-PDB) using the conventional notation. Average standard deviation based on replicate analyses of internal standards is <0.1‰.

Water samples

The oxygen and hydrogen-isotopic composition ($\delta^{18}O$ and δD) were determined using the methods of Epstein & Mayeda (1953) and Coplen *et al.* (1991) respectively. All $\delta^{18}O$ and δD data for waters are reported relative to Vienna Standard Mean Ocean Water (VSMOW). Average standard deviation, based on replicate analyses of internal standards is <0.1‰ for $\delta^{18}O$ and 2.0‰ for δD. The $\delta^{13}C$ of the dissolved inorganic carbon (DIC) was determined by acidification of the sample followed by extraction using a flowing stream of He and analysis in a stable isotope ratio mass spectrometer (Europa 20–20). Data are reported relative to V-PDB using the conventional notation. Average standard deviation based on replicate analyses of internal standards is <0.1‰.

Facies classification

Immediately after collection, samples were assigned to a modified Dunham scheme (Dunham, 1962). In order to supplement the classification divisions were added between wackestone and mudstone (mud-rich wackestone), between a packstone and wackestone (mud-rich packstone) and finally between packstone and grainstone (mud-lean packstone). For display purposes the mudstone, wackestone, packstone, grainstone and rudstone classification were assigned numbers from 1 to 5, with the intermediary categories being assigned values of 1.5, 2.5 and 3.5. Samples from all cruises were preserved in cold storage and after the 2004 cruise all samples were re-examined to ensure consistency between cruises (Reijmer *et al.*, 2009).

RESULTS

Stable isotopes of carbonates

The mean $\delta^{13}C$ and $\delta^{18}O$ values for the carbonate sediments and the various size fractions are shown in Tables 1 and 2 and Fig. 2. The $\delta^{18}O$ of all fractions are statistically the same, but the $\delta^{13}C$ of the coarser material (>1000 μm) is statistically significantly more negative than the finer fraction and the bulk. Relative to the modified Dunham classification there is no statistically significant difference between the mean $\delta^{13}C$ and $\delta^{18}O$ of the coarser and finer sediment types (Table 3; Figs 3 and 4). However, the range of values of the modified Dunham categories increases with grain size (Table 3). The spatial distribution of $\delta^{13}C$ and $\delta^{18}O$ are shown in Figs 5 and 6.

Water samples

The $\delta^{18}O$ of the waters are shown in Fig. 7 and range between +0.75 and +2.93‰. There is an approximate positive correlation with salinity,

Table 1. Carbon isotopic composition of size fractions.

	Bulk	>1000 µm	1000–500 µm	500–250 µm	250–125 µm	125–63 µm	<63 µm
Mean	+4.8	+4.0	+4.6	+4.8	+4.9	+4.9	+4.4
SD	0.3	0.7	0.5	0.4	0.5	0.3	0.5
Minimum	+3.2	+2.0	+3.4	+3.3	+3.5	+3.9	+3.4
Maximum	+5.5	+5.1	+5.6	+5.9	+5.9	+5.4	+5.5

Table 2. Oxygen isotopic composition of size fractions.

	Bulk	>1000 µm	1000–500 µm	500–250 µm	250–125 µm	125–63 µm	<63 µm
Mean	+0.4	+0.3	+0.4	+0.5	+0.4	+0.5	+0.4
SD	0.4	0.4	0.3	0.4	0.44	0.4	0.4
Minimum	−0.8	−0.5	−0.5	−0.40	−0.3	−0.3	−0.5
Maximum	+1.8	+1.5	+2.0	+2.0	+1.7	+1.3	+2.0

Table 3. The $\delta^{13}C$ and $\delta^{18}O$ in different modified Dunham classification.

	1.5 Mud-rich wackestone	2 Wackestone	2.5 Mud-rich packstone	3 Packstone	3.5 Mud-lean packstone	4 Grainstone	5 Rudstone
$\delta^{13}C$	+4.8	+4.8	+4.8	+4.8	+5.0	+4.9	+3.7
SD	0.1	0.2	0.2	0.2	0.3	0.3	0.1
$\delta^{18}O$	+0.5	+0.6	+0.6	+0.6	+0.4	+0.2	−0.1
SD	0.3	0.4	0.4	0.3	0.5	0.4	0.2

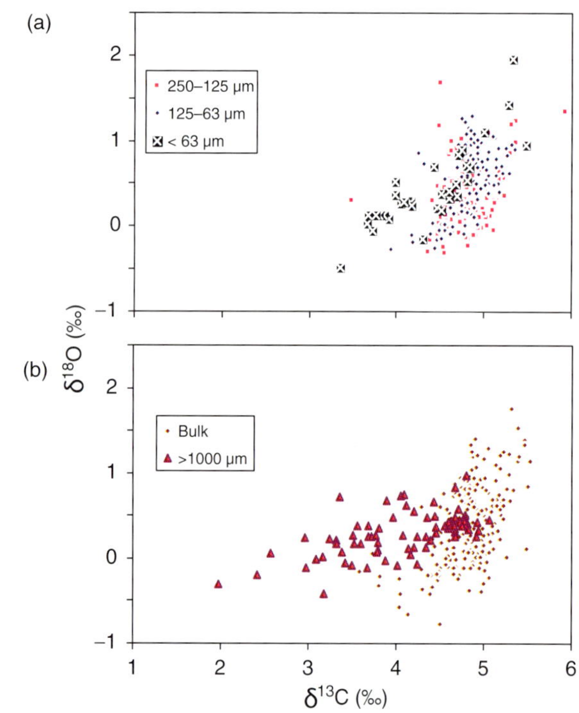

Fig. 2. Stable $\delta^{13}C$ and $\delta^{18}O$ of all samples and size fractions from 2001 to 2004 cruises: (a) data from <63, 63–125 and 125–250 µm size fractions, (b) data from bulk, 250–500, 500–1000 and >1000 µm size fractions.

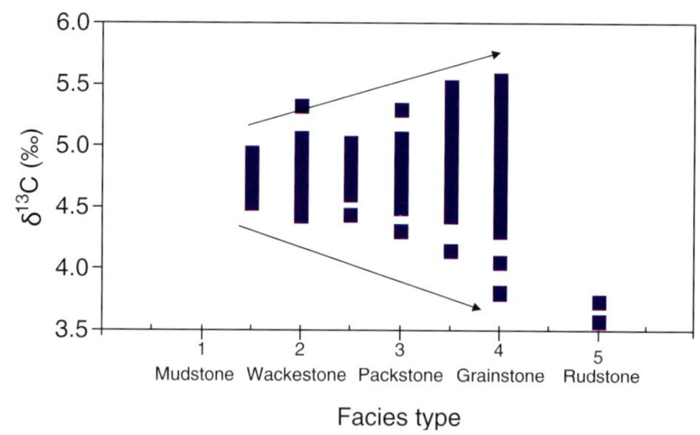

Fig. 3. Relationship between $\delta^{13}C$ and facies type (1 = mud, 1.5 = mud-rich wackestone, 2 = wackestone, 2.5 = mud-rich-packstone, 3 = packstone, 3.5 = mud-lean packstone, 4 = grainstone, 5 = rudstone).

similar to that shown previously (Lowenstam & Epstein, 1957; Shinn *et al.*, 1989). The $\delta^{13}C$ values of the DIC range from +0.4 to +2.4‰ and show no specific relationship with salinity (Fig. 8).

Sediment classification

The descriptions and distribution of these facies are fully described in a companion

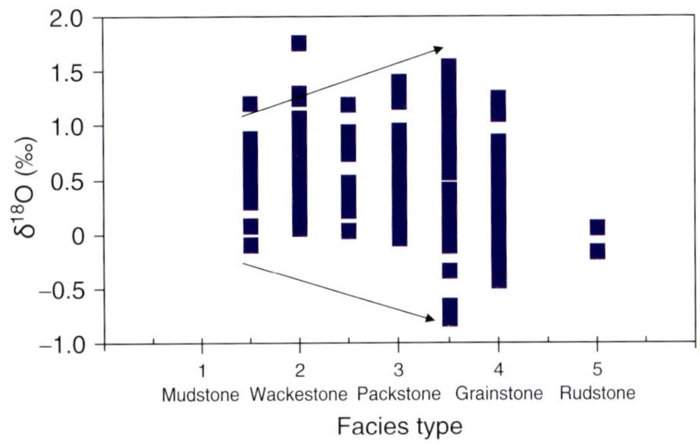

Fig. 4. Relationship between $\delta^{18}O$ and facies type (1 = mud, 1.5 = mud-rich wackestone, 2 = wackestone, 2.5 = mud-rich-packstone, 3 = packstone, 3.5 = mud-lean packstone, 4 = grainstone, 5 = rudstone).

paper (Reijmer *et al.*, 2009). The majority of the sediments are dominated by non-skeletal material (mud, pellets, peloids, ooids and grapestones), although all samples contained small amounts of skeletal debris. The percentage of mud in the samples increases rapidly moving away from the bank margin towards the interior of the platform with samples completely composed of skeletal material only present near the bank margin. The facies distribution pattern found is fairly similar to that shown in the facies distribution map of Purdy (1963a,b) and roughly agrees with the map of Traverse & Ginsburg (1966).

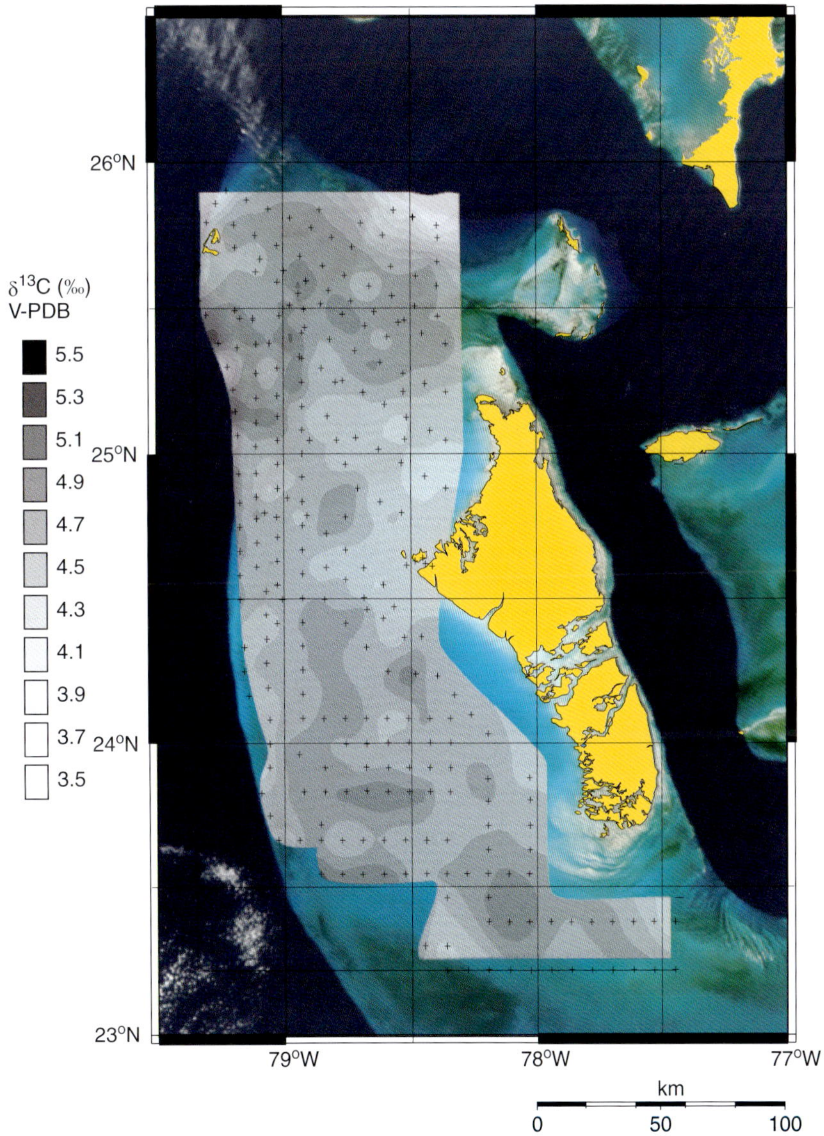

Fig. 5. Spatial composition map of the $\delta^{13}C$ of bulk carbonate sediments from GBB.

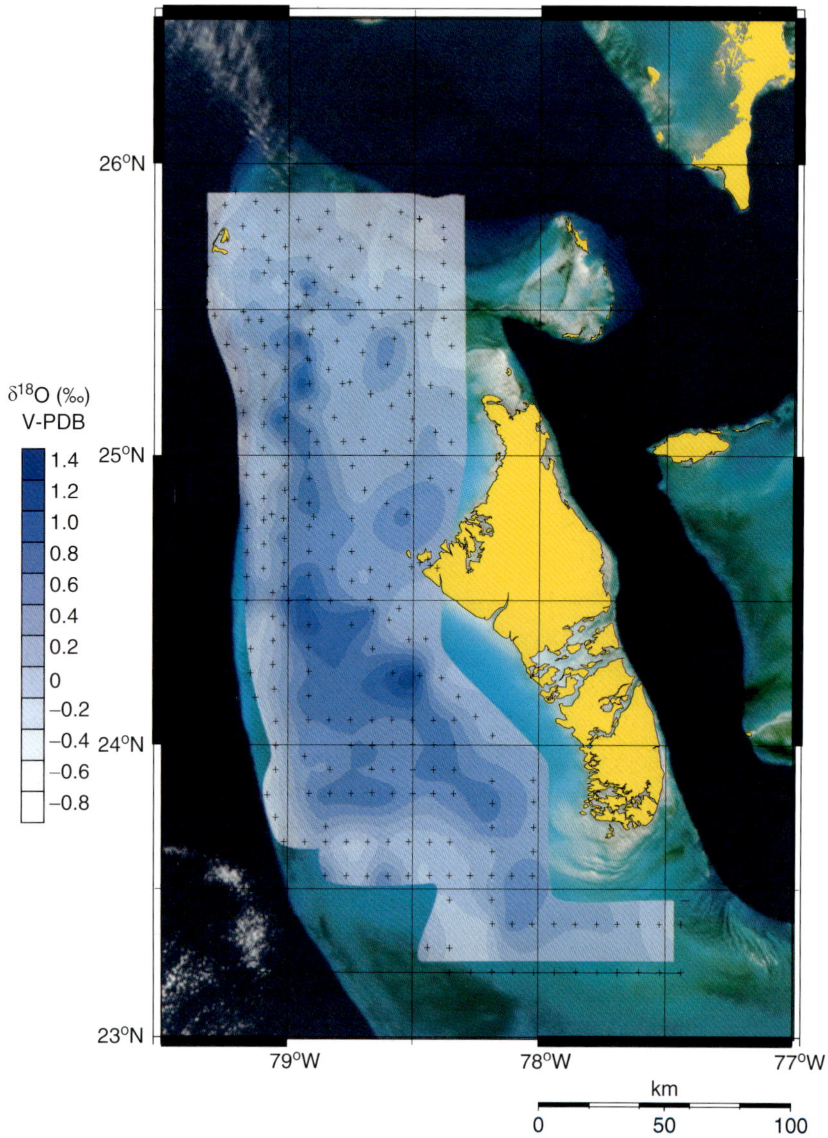

Fig. 6. Spatial composition map of $\delta^{18}O$ of bulk carbonate sediments from GBB.

DISCUSSION

Previous work on the variation of $\delta^{13}C$ on modern carbonate platforms

There have been relatively few previous studies on the distribution of $\delta^{13}C$ of sediments on modern carbonate platforms (Lowenstam & Epstein, 1957; Weber, 1967; Weber & Schmalz, 1968; Weber & Woodhead, 1969; Shinn *et al.*, 1989; Dix *et al.*, 2005). Studies such as the one carried out at Heron Island (Weber & Woodhead, 1969) examined variations over a relatively small area, a coral atoll approximately 5 × 20 km. The Heron Island study showed substantial variation in $\delta^{13}C$ progressing across the reef, with the margins being significantly more negative (up to 2‰) in $\delta^{13}C$ compared with portions of the interior (Fig. 9a). These changes are strongly correlated with the percentage of aragonite and arise because carbonate materials produced from the decomposition of algae, such as the various calcareous green algae commonly found on coral reefs (*Halimeda* sp., *Penicillus* sp., *Acetabularia* sp. and others), have very positive $\delta^{13}C$ values (typically > +4‰). In contrast, the $\delta^{13}C$ of most LMC and HMC components is close to or below 0‰ (Keith & Weber, 1965; Weber, 1965; Milliman, 1974; Land, 1989). Hence, a mixing line between aragonite and LMC or HMC will produce a covarying trend. However, a close inspection of the data (Fig. 9) shows two apparent trends, one at concentrations of

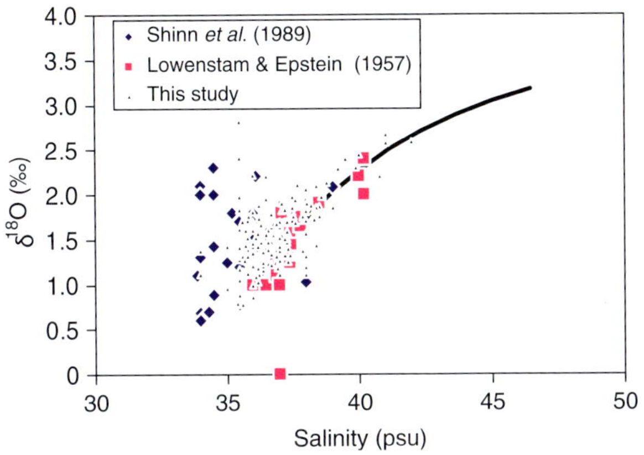

Fig. 7. Relationship between $\delta^{18}O$ and salinity for water samples from GBB measured by Shinn *et al.* (1989), Lowenstam & Epstein (1957) and this study. The line shows a model result of the evaporation of a water body with an initial salinity of 32 and $\delta^{18}O$ of 0‰ using the equations outlined by Gonfiantini (1986) and the model of Craig & Gordon (1965).

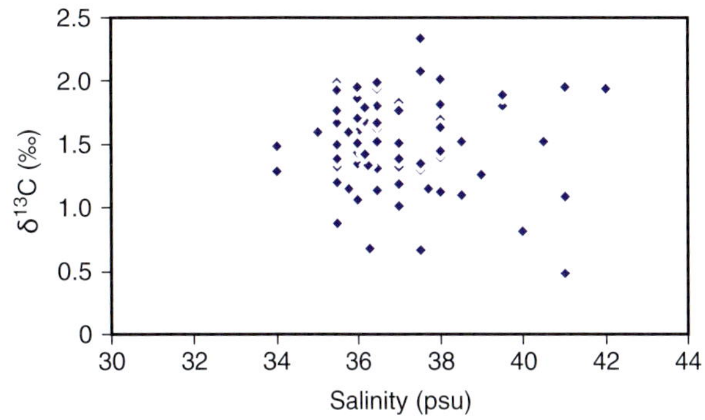

Fig. 8. Relationship between $\delta^{13}C$ of DIC and salinity for samples collected in this study. There is no statistically significant correlation in these data between $\delta^{13}C$ of DIC and salinity.

aragonite higher than 60%, where there is a strong positive association between percentage aragonite and $\delta^{13}C$, and one at concentrations below 60%, where there is apparently no correlation between the percentage of aragonite and $\delta^{13}C$ (Fig. 9b). This apparent difference probably arises because there are at least two sources of aragonite with very different $\delta^{13}C$ values on GBB. One source is the calcareous green algae exhibiting high $\delta^{13}C$ values, mentioned above. The other source is the aragonite produced by scleractinian corals and red calcareous algae which possess much lower $\delta^{13}C$ values (−1 to +1‰) (Keith & Weber, 1965; Milliman, 1974; Land, 1989). Hence, increasing the concentration of aragonite by adding material derived from corals and red calcareous algae will not result in a corresponding $\delta^{13}C$ increase and will not produce a correlation between the concentration of aragonite and $\delta^{13}C$. A similar result was found at Enewetok, with relationships between aragonite and $\delta^{13}C$ being either positive or negative depending upon the location sampled (Weber & Schmalz, 1968). Another significant study which examined a much larger spatial area was that by Dix *et al.* (2005). Here workers examined approximately 100 samples collected over an approximate 600 km^2 area on the North West Australian Shelf and also identified a positive relationship between the percentage of aragonite and $\delta^{13}C$, again produced by a mixing between isotopically positive aragonite and more depleted LMC. A similar explanation has been invoked for different correlations between C and O found in modern sediments collected from Belize and the Persian Gulf (Gischler *et al.*, 2007).

RELATIONSHIPS ON GREAT BAHAMA BANK

Salinity and stable isotopes

There have been no systematic temporal or spatial investigations of the water chemistry of Great Bahama Bank, but the analyses that have been made show waters increasing in salinity as one progresses across Great Bahama Bank from west to east (Black, 1933; Smith, 1940b; Lowenstam & Epstein, 1957; Cloud Jr., 1962; Shinn *et al.*, 1989). Very close to Andros Island, salinities have been shown to decrease as a result of runoff (Cloud Jr., 1962).

Oxygen-isotopic analyses of the waters generally show a weak positive relationship with salinity as do the data measured in this study (Lowenstam & Epstein, 1957; Shinn *et al.*, 1989) (Fig. 7). However, as a result of exchange between atmospheric water vapour and the water on GBB, the maximum $\delta^{18}O$ that can be attained is limited (Lloyd, 1964; Craig & Gordon, 1965; Gonfiantini, 1986) and therefore there is no clear relationship between salinity and $\delta^{18}O$. In fact the maximum $\delta^{18}O$ value measured in this study was ~+2.8‰ at a salinity of 35.75. Waters with higher salinity values had actually lower $\delta^{18}O$ values. The δD data also show clear evidence of a strong evaporative trend (Fig. 10) with original source water close to 0‰. As the samples were collected over a four-year time period during which the salinity and $\delta^{18}O$ of

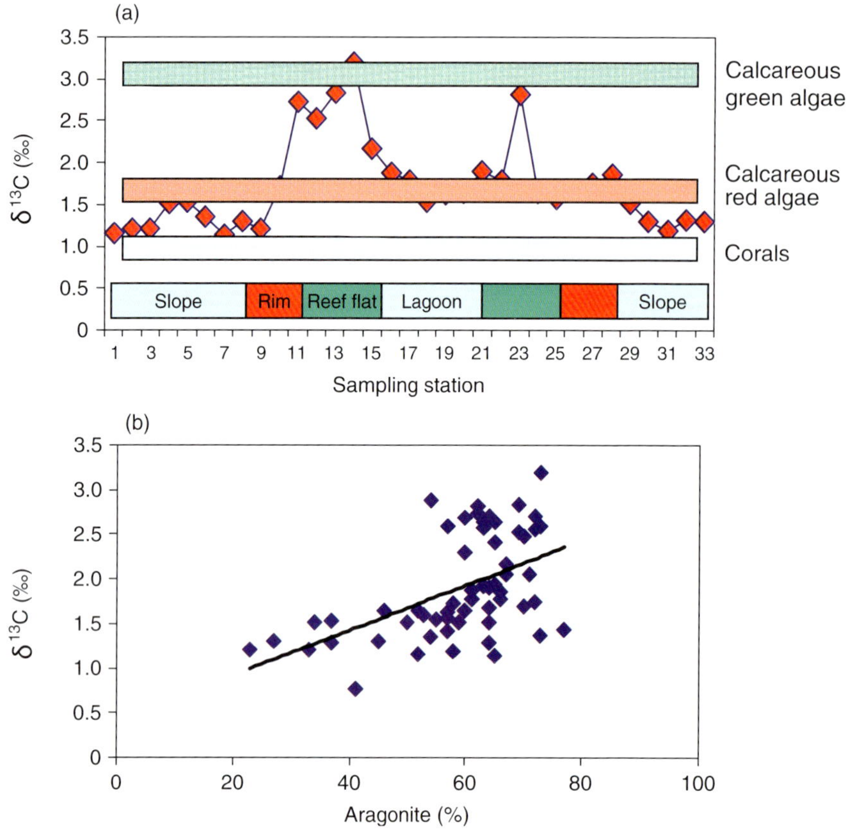

Fig. 9. (a) Sediment $\delta^{13}C$ data from Heron Island, Australia showing the $\delta^{13}C$ values in a south to north transect (low numbers to high numbers) across the reef. See Weber & Woodhead (1969) for sample location. Changes in $\delta^{13}C$ correspond to changes in facies with elevated values corresponding to areas dominated by calcareous green algae and lower values in areas where there are abundant red algae and scleractinian corals (Weber & Woodhead, 1969). (b) Sediment $\delta^{13}C$ data shown in (a) correlated with the percentage of aragonite.

the water varied as a result of weather conditions, it is not possible to combine the data and make a generalization about the distribution of $\delta^{18}O$ on GBB. However, it is probable that waters further from the margin might have elevated $\delta^{18}O$ values on a consistent basis. Hence carbonates formed further from the margin might also be expected to have higher $\delta^{18}O$ values than carbonates found on or near the margin.

Previous studies on the $\delta^{13}C$ of waters on GBB are limited to one, which showed large ranges in $\delta^{13}C$ ranging from ~−5 to +1‰ (Patterson & Walter, 1994). It was suggested that these variations were related to oxidation of organic material, photosynthesis and precipitation of aragonite and LMC. The $\delta^{13}C$ of the DIC measured in this study were significantly higher and varied from ~+0.5 to +2.5‰ (Fig. 8). In no instance over five years of analyses were waters with $\delta^{13}C$ values less than +0.5‰ recorded. The lower values measured in this study are typical of open marine values (Weber & Woodhead, 1971; Kroopnick, 1974; Swart *et al.*, 2005), while the higher ones suggest both fractionation of CO_2 during photosynthesis and/or fractionation of

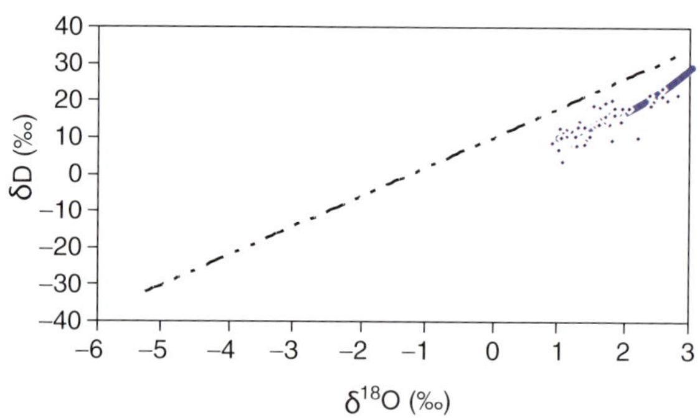

Fig. 10. Relationship between $\delta^{18}O$ and δD in the data measured in this study relative to the meteoric water line (dashed line). The solid line shows the evaporative trend of a water body using the same conditions as used in Fig. 7.

the HCO_3^- pool as CO_2 is produced during the precipitation of $CaCO_3$. Photosynthesis by the abundant algae and seagrasses on GBB preferentially removes the lighter isotope of carbon, causing the residual bicarbonate pool to become enriched in the heavier isotope of carbon, while precipitation of calcium carbonate lowers the pH, converting HCO_3^- into CO_2. As there is an isotopic fractionation of approximately 8‰

in this process (Emrich *et al.*, 1970; Romanek *et al.*, 1992), the residual bicarbonate becomes isotopically enriched.

Sediments

Perhaps the most startling results arising from this study are (i) the absence of significant patterns in the $\delta^{13}C$ across the GBB (Fig. 5) and (ii) the lack of a relationship between facies type and $\delta^{13}C$ (Fig. 3). The only exception to the latter is that very large grains (>1000 µm) are slightly less positive (+4.2 vs. +4.9‰). The relatively depleted values are a result of the fact that these grains are more likely to be composed of skeletal material, which generally have lower $\delta^{13}C$ values (see previous discussion). The origin of relatively high $\delta^{13}C$ values of the sediments compared with pelagic LMC arises from two phenomena. First, as mentioned above the $\delta^{13}C$ of the DIC averages +1.5‰ ranging from +0.4 to +2.4‰. Second, enrichment of $\delta^{13}C$ in aragonite is about 2.7 ± 0.6‰ relative to only about 1‰ for LMC (Romanek *et al.*, 1992). Hence equilibrium values for aragonite precipitated directly from surface waters of GBB should range from ~+2.6 to ~+5.8‰ (this range is calculated using the error of 0.6‰ and the $\delta^{13}C$ values measured on the surface water DIC (Fig. 8). This corresponds to a measured range of +3.6 to +5.5‰ (see Tables 1 and 2).

As regards the $\delta^{18}O$, the variation in water temperatures on GBB probably exceeds 10°C, but averages 25°C. Using the standard palaeotemperature equation (Epstein *et al.*, 1953) and the values reported in Table 2, the sediments if formed at equilibrium should lie between −1.2 and +0.8‰. This is a little more negative than the range of the sediments (−0.8 to +1.8‰) but still within the errors of the temperature estimate (Fig. 11). So if the predominantly aragonite sediments on GBB are in equilibrium with the ambient waters, then why do all the sediments, from the mud-dominated wackestones to the grainstones, have similar $\delta^{13}C$ and $\delta^{18}O$ compositions? The simple answer is that the majority of the sediments are genetically related; the exception being the small amount of skeletal material mentioned previously.

The origin of the finest carbonate sediment on GBB has been the subject of speculation for over 50 years (Black, 1933; Smith, 1940a,b; Cloud Jr., 1962). Authors have proposed that the material is either a result of direct or algal induced precipitation from seawater (Cloud Jr., 1962; Morse *et al.*, 1984; Shinn *et al.*, 1989; Robbins & Blackwelder, 1992), or that they are stirred up bottom sediments, ultimately originating from the decomposition of calcareous algae (Broecker & Takahashi, 1966; Stockman *et al.*, 1967; Morse *et al.*, 1984). This debate has been heated over the years, but regardless of the ultimate origin of the mud component, this fine-grained sediment seems to be the building material for most of the other sedimentary particles on GBB. Mud becomes ingested by benthic organisms forming faecal pellets which harden to form peloids, some of which become coated to form ooids or clumped together to form grapestones. This process is supported by the fact that the finest sediments have the narrowest range of C and O isotopic compositions (Fig. 2a), gradually increasing in range as diagenetic and cementation processes add carbonate material of differing generations and isotopic compositions to the grains.

RELATIONSHIPS BETWEEN $\delta^{13}C$ AND $\delta^{18}O$

The relationships between the $\delta^{13}C$ and $\delta^{18}O$ in the bulk samples and the separated size fractions are shown in Fig. 2a and b. In $\delta^{13}C$ and $\delta^{18}O$ space, the data define a triangle with the samples possessing the most positive $\delta^{18}O$ values also containing the narrowest range and the most elevated $\delta^{13}C$ values. These samples tend to be mud-rich wackestones. This pattern (Fig. 11) probably reflects the relationships between those areas on GBB where the water consistently possesses the highest $\delta^{13}C$ (DIC) and $\delta^{18}O$ values. These areas are probably fairly restricted in geographical extent. In contrast, the areas where there are wider ranges in $\delta^{13}C$ and $\delta^{18}O$ are more prevalent and account for the majority of the samples. The largest sized samples (500–1000 µm and >1000 µm) have a much shallower relationship between $\delta^{13}C$ and $\delta^{18}O$ reflecting a larger proportion of skeletal material in the samples (Fig. 11).

IMPLICATIONS FOR CARBON ISOTOPE STRATIGRAPHY

This study on the $\delta^{13}C$ and $\delta^{18}O$ of surface sediments of GBB shows that in spite of a large range in sediment sizes, there is a relatively small spread in the $\delta^{13}C$ and $\delta^{18}O$ values. The elevated $\delta^{13}C$ values are a result of a combination of isotopic

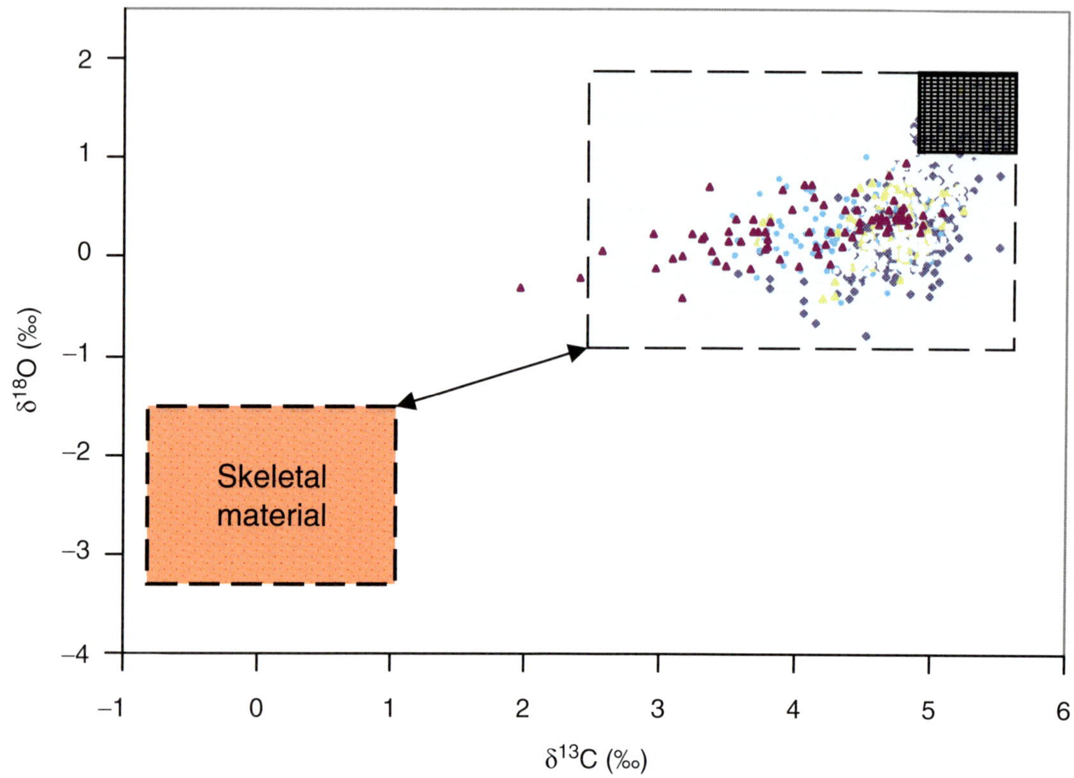

Fig. 11. Ranges of equilibrium estimates of $\delta^{13}C$ and $\delta^{18}O$ of sediments based on data shown in Figs 5 and 6 (region enclosed by cross-hatched symbol), compared with the mean $\delta^{13}C$ and $\delta^{18}O$ values measured on carbonates in this study (area enclosed by blue rectangular area). Carbonate data are the same as shown in Fig. 2. Estimates for skeletal data (red rectangular area) are taken from the literature (Weber, 1965; Milliman, 1974; Swart, 1983; Land, 1989).

fractionation of the DIC during photosynthesis and the natural enrichment of aragonite relative to LMC (see above). These sediments are exported from the platform and form periplatform sediments (Schlager & Ginsburg, 1981; Eberli *et al.*, 1997; Eberli, 2000) that drape the flanks of the platform and extend into the pelagic realm. It has been shown that the $\delta^{13}C$ of such sediments can be correlated between units of equivalent age, but that these $\delta^{13}C$ values are not related to pelagic $\delta^{13}C$ values (Swart & Eberli, 2005). The reason for the correlation between units of similar age arises as varying amounts of surface sediments enriched in $\delta^{13}C$, mix with more depleted carbonate derived from pelagic sources. As all sediment size fractions have similar $\delta^{13}C$ values, the mixing does not appear to change as a function of distance from the platform. This pattern may not be present surrounding carbonate buildups with high percentages of aragonite derived from calcareous organisms such as corals.

The question posed by this study is whether the results are at all applicable to carbonate platforms in previous geological time periods. For example, it has been proposed that the normal form of calcium carbonate precipitated from marine seawater has changed throughout geological time as a function of the Mg/Ca ratio of the oceans varying between 'calcite' and 'aragonite' seas (Sandberg, 1983; Hardie, 2003; Lowenstein *et al.*, 2003). During periods of 'calcite' seas, clearly the $\delta^{13}C$ of carbonates formed on the platform would probably not be as isotopically positive today because the only mechanisms that would enrich the platform carbonates would be the influences of photosynthesis and fractionation during the precipitation of calcium carbonate. The platform-derived carbonate would nevertheless be more isotopically positive than pelagic material formed at the same time, although not as enriched that as today. Hence it is proposed that there still might be a carbon isotopic effect as a result of precipitation on carbonate platforms during periods of 'calcite' seas, but that this might be reduced compared with periods of 'aragonite' seas.

CONCLUSIONS

1 The positive $\delta^{13}C$ values of surface sediments arises from a combination of the large fractionation between aragonite and bicarbonate (~2.7‰) and the elevated $\delta^{13}C$ of the DIC caused by the fractionation during photosynthesis of benthic algae, seagrasses and cyanobacteria and fractionation during the conversion of HCO_3^- into CO_2 during the precipitation of calcium carbonate.

2 All sediments formed upon GBB have similar $\delta^{13}C$ values suggesting a common origin. Mud, whether formed by direct precipitation or breakdown from algal carbonate, is ingested by benthic organisms forming pellets, which harden to produce peloids, which in turn are coated to form coated grains, grapestones and/or ooids. As a result of this paragenetic sequence, the range of $\delta^{13}C$ and $\delta^{18}O$ of the sediments increases with increasing grain size.

3 Skeletal material only forms a small proportion of sediment on GBB, but has more negative $\delta^{13}C$ and $\delta^{18}O$ values.

4 The $\delta^{18}O$ of the carbonate sediment is elevated in the interior bank relative to the margin as a result of the persistent elevated salinity and $\delta^{18}O$ of the water.

ACKNOWLEDGEMENTS

The authors would like to thank the staff and students of the stable isotope laboratory at the University of Miami (USA) and the carbonate sedimentology group of the Leibniz-Institut für Meereswisschaften–IFM-GEOMAR in Kiel (Germany). Writing of this paper was supported by a visiting professorship award to PKS from the Université de Provence (Aix-Marseille). The captain and crew of the RV *Bellows* are thanked for their support. This work was partially supported by the German Science Foundation (Re-1051/09) and Leibniz-price funds of Wolf-Christian Dullo, by grants from the Florida Institute of Oceanography, and the Comparative Sedimentology Laboratory. This paper benefited from reviews by Steve Burns and Adrian Immenhauser.

REFERENCES

Allan, J.R. and **Matthews, R.K.** (1982) Isotope signatures associated with early meteoric diagenesis. *Sedimentology*, **29**, 797–817.

Berner, R.A. (1987) Models for carbon and sulfur cycles and atmospheric oxygen: application to Paleozoic geologic history. *Am. J. Sci.*, **287**, 177–196.

Berner, R.A., Lasaga, A.C. and **Garrels, R.M.** (1983) The carbonate-silicate geochemical cycle and its effect on atmospheric carbon dioxide over the past 100 million years. *Am. J. Sci.*, **283**, 641–683.

Black, M. (1933) The precipitation of calcium carbonate on the Great Bahama Bank. *Geol. Mag.*, **70**, 455–466.

Broecker, W.S. and **Takahashi, T.** (1966) Calcium carbonate precipitation on the Great Bahama Banks. *J. Geophys. Res.*, **71**, 1575–1602.

Cloud Jr., P.E. (1962) Environment of calcium carbonate deposition west of Andros Island. *Bull. Geol. Soc. Am.*, **66**, 1542.

Coplen, T.B., Wildman, J.D. and **Chen, J.** (1991) Improvements in the gaseous hydrogen-water equilibration technique for hydrogen isotope ratio analysis. *Anal. Chem.*, **63**, 910–912.

Craig, H. (1957) Isotopic standards for carbon and oxygen and correction factors for mass-spectrometric analysis of carbon dioxide. *Geochim. Cosmochim. Acta*, **12**, 133–149.

Craig, H. and **Gordon, L.I.** (1965) Deuterium and oxygen-18 variations in the ocean and the marine atmosphere. *Stable Isotopes in Oceanographic Studies and Paleotemperatures Consiglio Nazionales delle Ricerche, Lab. Di Geo. Nucleare, Pisa* 9 p.

Dix, G.R., James, N.P., Kyser, T.K., Bonne, Y. and **Collins, L.B.** (2005) Genesis and dispersal of carbonate mud relative to late Quaternary sea-level change along a distally steepend carbonate ramp (Northwestern Shelf, Western Australia). *J. Sedimentary Res.*, **75**, 665–678.

Dunham, R.J. (1962) Classification of carbonate rocks according to depositional texture. In: *Classification of Carbonate Rocks* (Ed. W.E. Ham), pp. 108–121. American Association Petroleum Geologists, Tulsa, OK.

Eberli, G.P. (2000) The record of Neogene sea-level changes in the prograding carbonates along the Bahamas transect-Leg 166 synthesis. In: *Proceedings of the Ocean Drilling Program, Scientific Results* (Eds P.K. Swart, G.P. Eberli, M.J. Malone and J.F. Sarg), Vol. **166**, pp. 167–177.

Eberli, G.P., Swart, P.K. and **Malone, M.J.** (1997) *Proceedings of ODP* Initial Reports, **166**. Ocean Drilling Program, College Station, TX.

Emrich, K., Ehhalt, D.H. and **Vogel, J.C.** (1970) Carbon isotope fractionation during the precipitation of calcium carbonate. *Earth Planetary Sci. Lett.*, **8**, 363–371.

Enos, P. (1974) Surface sediment facies map of the Florida-Bahamas Plateau. Geological Society of America.

Epstein, S. and **Mayeda, T.** (1953) Variations in O^{18} content of waters from natural sources. *Geochim. Cosmochim. Acta*, **4**, 213–224.

Epstein, S., Buchsbaum, R., Lowenstam, H.A. and **Urey, H.C.** (1953) Revised carbonate-water isotopic temperature scale. *Bull. Geol. Soc. Am.*, **64**, 1315–1326.

Föllmi, K.B., Weissert, H., Bisping, M. and **Funk, H.** (1994) Phosphogenesis, carbon-isotope stratigraphy, and carbonate-platform evolution along the Lower Cretaceous northern Tethyan margin. *Geol. Soc. Am. Bull.*, **106**, 729–746.

Gale, A. (1993) Chemostratigraphy vs biostratigraphy: data from around the Cenomanian-Turonian boundary. *J. Geol. Soc.*, **150**, 29–33.

Gischler, E., **Swart, P.K.** and **Lomando, A.J.** (2009) Geochemical facies variation in Belize and the Persian Gulf. In: *Perspectives in Sedimentary Geology: A Tribute to the Career of Robert Nathan Ginsburg* (Eds P.K. Swart, G.P. Eberli and J. McKenzie). Blackwell, Oxford. *Spec. Publ. Int. Assoc. Sedimentol.*, **41**, 61–74.

Gonfiantini, R. (1986) Environmental isotopes in lake studies. In: *Handbook of Environmental Isotope Geochemistry* (Eds P. Fritz and J.C. Fontes), Vol. **1**, pp. 113–168. Elsevier, Amsterdam.

Hardie, L.A. (2003) Secular variations in Precambrian seawater chemistry and the timing of Precambrian aragonite seas and calcite seas. *Geology*, **31**, 785–788.

Illing, L. (1954) Bahaman calcareous sands. *Bull. Am. Assoc. Petrol. Geol.*, **38**, 1–95.

Immenhauser, A., **Della Porta, G.**, **Kenter, J.A.M.** and **Bahamonde, J.R.** (2003) An alternative model for positive shifts in shallow-marine carbonate $\delta^{13}C$ and $\delta^{18}O$. *Sedimentology*, **50**, 953–959.

Immenhauser, A., **Kenter, J.A.M.**, **Ganssen, G.**, **Bahamonde, J.R.**, **Van Vliet, A.** and **Saher, M.H.** (2002) Origin and significance of isotope shifts in Pennsylvanian carbonates (Asturias, NW Spain). *J. Sedimentary Res.*, **72**, 82–94.

Keith, M.L. and **Weber, J.N.** (1965) Systematic relationships between carbon and oxygen isotopes in carbonates deposited by modern corals and algae. *Science*, **150**, 498–501.

Kroopnick, P.M. (1974) Correlations between ^{13}C and TCO_2 in surface waters and atmospheric CO_2. *Earth Planetary Sci. Lett.*, **22**, 397–403.

Kump, L.R. and **Arthur, M.A.** (1999) Interpreting carbon-isotope excursions: carbonates and organic matter. *Chem. Geol.*, **161**, 181–198.

Land, L.S. (1989) The carbon and oxygen isotopic chemistry of surficial Holocene shallow carbonate sediment and Quaternary limestone and dolomite. In: *Handbook of Environmental Isotope Geochemistry* (Eds P. Fritz and J.C. Fontes), Vol. **3**, pp. 191–217. Elsevier, Amsterdam.

Lloyd, M.R. (1964) Variations in the oxygen and carbon isotope ratios of Florida Bay mollusks and their environmental significance. *J. Geol.*, **72**, 84.

Lowenstam, H.A. and **Epstein, S.** (1957) On the origin of sedimentary aragonite needles of the Great Bahama Bank. *J. Geol.*, **65**, 364–375.

Lowenstein, T.K., **Hardie, L.A.**, **Timofeeff, M.N.** and **Demicco, R.V.** (2003) Secular variation in seawater chemistry and the origin of calcium chloride basinal brines. *Geology*, **31**, 857–860.

Milliman, J.D. (1974) *Marine Carbonates*. Springer, Berlin, 375 pp.

Morse, J.W., **Millero, F.J.**, **Thurmond, V.**, **Brown, E.** and **Ostlund, H.G.** (1984) The carbonate chemistry of Grand Bahama Bank Waters – after 18 years another look. *J. Geophys. Res.-Oceans*, **89**, 3604–3614.

Newell, N.D., **Imbrie, J.**, **Purdy, E.G.** and **Thurber, D.L.** (1959) Organism communities and bottom facies, Great Bahama Bank. *Bull. Am. Museum Natural History*, **117**, 177–228.

Patterson, W.P. and **Walter, L.M.** (1994) Depletion in ^{13}C in seawater CO_2 on modern carbonate platforms: significance for the carbon isotopic record of carbonate. *Geology*, **22**, 885–888.

Purdy, E. (1963a) Recent calcium carbonate facies of the Great Bahama Bank.1. Petrography and reaction groups. *J. Geol.*, **71**, 334–335.

Purdy, E. (1963b) Recent calcium carbonate facies of the Great Bahama Bank. 2. Sedimentary facies. *J. Geol.*, **71**, 472–497.

Reijmer, J.J., **Swart, P.K.**, **Bauch, T.**, **Otto, R.**, **Reuning, L.**, **Roth, S.** and **Zechel, S.** (2009) A re-evaluation of facies variations on Great Bahama Bank 1: new facies maps of Western Great Bahama Bank. In: *Perspectives in Sedimentary Geology: A Tribute to the Career of Robert Nathan Ginsburg* (Eds P.K. Swart, G.P. Eberli and J.A. McKenzie) Blackwell, Oxford. *Spec. Publ. Int. Assoc. Sedimentol.*, **41**, 29–46.

Reuning, L., **Reijmer, J.J.G.**, **Betzler, C.**, **Timmerman, A.** and **Steph, S.** (2006) Sub-Milankovitch cycles in periplatform carbonates from the early Pliocene Great Bahama Bank. *Paleoceanography*, **21**, PA1017.

Robbins, L.L. and **Blackwelder, P.L.** (1992) Biochemical and ultrastructural evidence for the origin of whitings: a biologically induced calcium carbonate precipitation mechanism. *Geology*, **20**, 464–468.

Romanek, C.S., **Grossman, E.L.** and **Morse, J.W.** (1992) Carbon isotopic fractionation in synthetic aragonite and calcite: effects of temperature and precipitation rate. *Geochim. Cosmochim. Acta*, **56**, 419–430.

Saltzman, M.R. (2002a) Carbon and oxygen isotope stratigraphy of the Lower Mississippian (Kinderhookian-lower Osagean), western United States: Implications for seawater chemistry and glaciation. *Geol. Soc. Am. Bull.*, **114**, 96–108.

Saltzman, M.R. (2002b) Carbon isotope ($\delta^{13}C$) stratigraphy across the Silurian–Devonian transition in North America: evidence for a perturbation of the global carbon cycle. *Palaeogeogr. Palaeoclimatol. Palaeoecol.*, **187**, 83–100.

Saltzman, M.R. and **Young, S.A.** (2005) Long-lived glaciation in the Late Ordovician? Isotopic and sequence-stratigraphic evidence from western Laurentia. *Geology*, **33**, 109–112.

Saltzman, M.R., **Groessens, E.** and **Zhuravlev, A.V.** (2004) Carbon cycle models based on extreme changes in $\delta^{13}C$: an example from the lower Mississippian. *Palaeogeogr. Palaeoclimatol. Palaeoecol.*, **213**, 359–377.

Sandberg, P.A. (1983) An oscillating trend in Phanerozoic non-skeletal carbonate mineralogy. *Nature*, **305**, 19–22.

Schlager, W. and **Ginsburg, R.N.** (1981) Bahama carbonate platforms – the deep and the past. *Marine Geol.*, **44**, 1–24.

Shinn, E.A., **Steinen, R.P.**, **Lidz, B.H.** and **Swart, P.K.** (1989) Whitings, a sedimentologic dilemma. *J. Sedimentary Petrol.*, **59**, 147–161.

Smith, C.L. (1940a) The Great Bahama Bank. 2. Calcium carbonate precipitation. *J. Marine Res.*, **3**, 147–189.

Smith, C.L. (1940b) The Great Bahama Bank. 1. General hydrographic and chemical factors. *J. Marine Res.*, **3**, 1–31.

Stockman, K.W., **Ginsburg, R.N.** and **Shinn, E.A.** (1967) The production of lime mud by algae in south Florida. *J. Sedimentary Petrol.*, **37**, 633–648.

Swart, P.K. (1983) Carbon and oxygen isotope fractionation in scleractinian corals: a review. *Earth-Science Rev.*, **19**, 51–80.

Swart, P.K. and **Eberli, G.P.** (2005) The nature of the $\delta^{13}C$ of periplatform sediments: implications for stratigraphy and the global carbon cycle. *Sedimentary Geol.*, **175**, 115–130.

Swart, P.K. and **Melim, L.A.** (2000) The origin of dolomites in Tertiary sediments from the margin of Great Bahama Bank. *Sedimentary Res.*, **70**, 738–748.

Swart, P.K., Burns, S.J. and **Leder, J.J.** (1991) Fractionation of the stable isotopes of oxygen and carbon in carbon dioxide during the reaction of calcite with phosphoric acid as a function of temperature and technique. *Chem. Geol.*, **86**, 89–96.

Swart, P.K., Szmant, A., Porter, J.W., Dodge, R.E., Tougas, J.I. and **Southam, J.R.** (2005) The isotopic composition of respired carbon dioxide in scleractinian corals: implications for cycling of organic carbon in corals. *Geochim. Cosmochim. Acta*, **69**, 1495–1509.

Traverse, A. and **Ginsburg, R.N.** (1966) Palynology of the surface sediments of Great Bahama Bank, as related to water movement and sedimentation. *Marine Geol.*, **4**, 417–459.

Vahrenkamp, V.C. (1996) Carbon isotope stratigraphy of the Upper Kharaib and Shuaiba Formations: implications for the Early Cretaceous evolution of the Arabian Gulf region. *Am. Assoc. Petrol. Geol. Bull.*, **80**, 647–662.

Valladares, I., Recio, C. and **Lendinez, A.** (1996) Sequence stratigraphy and stable isotopes ($\delta^{13}C$, $\delta^{18}O$) of the late Cretaceous carbonate ramp of the western margin of the Iberian chain (Soria, Spain). *Sedimentary Geol.*, **105**, 11–28.

Veizer, J., Ala, D., Azmy, K., Bruckschen, P., Buhl, D., Bruhn, F., Carden, G.A.F., Diener, A., Ebneth, S., Godderis, Y., Jasper, T., Korte, G., Pawellek, F., Podlaha, O.G. and **Strauss, H.** (1999) $^{87}Sr/^{86}Sr$, $\delta^{13}C$ and delta $\delta^{18}O$ evolution of Phanerozoic seawater. *Chem. Geol.*, **161**, 59–88.

Weber, J. (1965) Fractionation of stable isotopes of O and C in marine calcareous organisms. *Geochim. Cosmochim. Acta*, **30**, 681–703.

Weber, J.N. (1967) Factors affecting the carbon and oxygen isotopic composition of marine carbonate sediments. Part 1, Bermuda. *Am. J. Sci.*, **265**, 586–608.

Weber, J.N. and **Schmalz, R.F.** (1968) Factors affecting the carbon and oxygen isotopic composition of marine carbonate sediments, Part III, Enewetok Atoll. *J. Sedimentary Petrol.*, **38**, 1270–1279.

Weber, J.N. and **Woodhead, P.M.J.** (1969) Factors affecting the carbon and oxygen isotopic composition of marine carbonate sediments-II. Heron Island, Great Barrier Reef, Australia. *Geochim. Cosmochim. Acta*, **33**, 19–38.

Weber, J.N. and Woodhead, P.M.J. (1971) Diurnal variations in the isotopic composition of dissolved inorganic carbon in seawater from coral reef environment. *Geochim. Cosmochim. Acta*, **35**, 891–902.

Weissert, H., Lini, A., Föllmi, K.B. and Kuhn, O. (1998) Correlation of Early Cretaceous carbon isotope stratigraphy and platform drowning events: a possible link? *Palaeogeogr. Palaeoclimatol. Palaeoecol.*, **137**, 189–203.

Int. Assoc. Sedimentol. Spec. Publ. (2009) **41**, 61–74

Stable isotopes of carbon and oxygen in modern sediments of carbonate platforms, barrier reefs, atolls and ramps: patterns and implications

EBERHARD GISCHLER*, PETER K. SWART† and ANTHONY J. LOMANDO‡

**Johann Wolfgang Goethe-Universität, Geowissenschaften, 60438 Frankfurt am Main, Germany (E-mail: gischler@em.uni-frankfurt.de)*
†University of Miami, RSMAS-MGG, Miami, FL 33149, USA
‡ChevronTexaco, Bahrain Pouch, San Ramon, CA 94583, USA

ABSTRACT

Stable isotopes of carbon ($\delta^{13}C$) and oxygen ($\delta^{18}O$) were determined in surface and subsurface sediments of modern carbonate platform and barrier reef (Belize), atoll (Maldives) and ramp (Kuwait) settings. The $\delta^{13}C$ values ranged from −1.6 to +5.3‰ and $\delta^{18}O$ values from −3.2 to +2.3‰. Among individual particle types, non-skeletal grains such as peloids and ooids exhibit highest values whereas coral fragments display relatively low $\delta^{13}C$ and $\delta^{18}O$ values. The most statistically significant correlations between abundance of carbonate grains, mineralogy and geochemistry occur in the ramp setting, where facies types may be defined using carbon and oxygen isotopic composition. In the platform, barrier and atoll examples, correlations are poor and there are no clear spatial trends in the carbon and oxygen isotopic values of sediments or facies. In locations with comparably high amounts of non-skeletal grains, $\delta^{13}C$ values correlate with the aragonite content of the samples. The difference between the ramp and other examples is probably a consequence of the continuous depositional energy gradient, which appears to be the major controlling factor of facies distribution in the ramp setting. In the atoll, barrier reef and platform examples, depositional energy gradients are more variable due to higher morphological and environmental variability. The $\delta^{13}C$ and $\delta^{18}O$ values of samples from Holocene cores also exhibit high total variability ranging from −1.9 to −0.5‰ in $\delta^{13}C$ and from −0.2 to +5.2‰ in $\delta^{18}O$. Strong excursions in $\delta^{13}C$ and $\delta^{18}O$ in individual cores are seen near the base where values become more negative as a consequence of diagenesis, i.e., the influence of a subaerial exposure horizon. Above the meteorically influenced section, however, variability along individual cores is comparably low (<1.2%). The enormous carbon and oxygen isotopic variability observed in modern shallow-water carbonates as compared with fossil sections, where $\delta^{13}C$ is used as stratigraphic marker, is probably a result of as yet poorly defined taphonomic and diagenetic filters, which significantly reduce isotopic variability.

Keywords Carbonates, sediment, isotopes, Belize, Kuwait, Maldives.

INTRODUCTION

The ratio of the stable isotopes of carbon (^{13}C and ^{12}C), in carbonate sediments and rocks, expressed as $\delta^{13}C$, is used in palaeoceanography as a proxy for changes in production, burial and preservation of organic matter (Broecker, 1982). Carbon isotope records are also used as a proxy for the concentration of atmospheric CO_2 through geological time. These records form a significant component of palaeoclimate models that are in turn used to predict future global warming. The $\delta^{13}C$ from bulk rock is also used as a stratigraphic tool (Scholle & Arthur, 1980; Weissert, 1989), especially in pelagic deposits. During the past decade, the stratigraphic method has also been applied to ancient shallow-water carbonate settings of ramps and carbonate platforms (Jenkyns, 1995; Vahrenkamp, 1996; Drzewiecki & Simo, 1997; Ferreri *et al.*, 1997; Mutti *et al.*, 1997; Grötsch *et al.*, 1998; Wissler *et al.*, 2003; Cozzi *et al.*, 2004). Immenhauser *et al.* (2002) discussed the large variability in carbon and oxygen isotope values in shallow-water carbonates and showed that there is a systematic offset between bank and off-bank isotope data in a Carboniferous example. A possible explanation is the depletion in ^{13}C of carbonate bank waters due to respiration of organic matter, which was demonstrated on the modern Great Bahama Bank by Patterson & Walter (1994a). Elevated ^{13}C may be observed during high-amplitude transgressions when isotopically depleted bank water mixes with open marine water (Immenhauser *et al.*, 2003). Swart & Eberli (2005) were able to correlate shallow-water bank, slope and basin sections using $\delta^{13}C$ in the Miocene to Holocene of the Bahamas. However, $\delta^{13}C$ values along their sediment cores did not correlate with the global $\delta^{13}C$ trends. Apparently, there are problems with the applicability of carbon isotopes as environmental and stratigraphic proxies in shallow-water depositional environments. In this study, considerable variability of carbon and oxygen isotopes in Holocene shallow-water carbonate sediments has been documented, the significance of which will be discussed in relation to the use of $\delta^{13}C$ as a palaeoceanographic and stratigraphic proxies in fossil shallow-water settings.

STUDY AREAS

Three areas formed the focus of the present study (Fig. 1). The Belize–Yucatan carbonate platforms include Glovers Reef, Lighthouse Reef, Turneffe Islands and Banco Chinchorro, which range in size from 200 to 550 km^2 (Fig. 1b). Surface-breaking reefs surround interior lagoons that are up to 18 m deep. Lagoonal patch reefs are common with the exception of Turneffe Islands, where interior lagoonal circulation is restricted as a result of dense mangrove growth. Reefal sediments are largely coralgal grainstones whereas lagoonal sediments are dominated by fragments of molluscs, foraminifera and non-skeletal grains (Glovers, Lighthouse, Chinchorro) as well as *Halimeda* (Turneffe) (Gischler & Lomando, 1999). Holocene platform lagoon cores usually show an up-core succession of Late Pleistocene/early Holocene soil, early Holocene mangrove peat and marine carbonates. Marine carbonate sediments often show indications of bioturbation and are predominantly composed of fragments of mollusc, foraminifera and *Halimeda*. The succession overlies Pleistocene reef limestone, which is usually altered diagenetically (Gischler, 2003).

The Belize Barrier Reef is 250 km long and forms the margin of the Belize shelf, which deepens from about 5 m in the north to some 50 m in the south (Fig. 1b). Surface sediments include reefal coralgal grainstones and *Halimeda*-rich wackestones and packstones in the back reef areas. Mollusc marls and siliciclastic sands are found on the southern shelf and nearshore areas, whereas molluscs, foraminifera and micritized grains predominate on the northern shelf (Purdy & Gischler, 2003).

Rasdhoo and Ari atolls are located in the western chain of the Maldives archipelago (Fig. 1c). Rasdhoo is a comparably small (60 km^2) atoll with an almost continuous reef margin and a 40 m deep lagoon. Ari covers 2300 km^2, the reef margin is rather discontinuous with numerous tidal channels and the lagoon is as deep as 80 m. Both lagoons contain patch reefs. Marginal and reefal sediments are coralgal grainstones. Lagoonal sediments are mollusc and foraminiferal wackestones and packstones, as well as mudstone (Gischler, 2006). Mudstone occurs only in Rasdhoo lagoon and most common grains >125 μm are cemented faecal pellets.

The ramp offshore Kuwait deepens more or less continuously to some 30 m over a distance of 20 km (Fig. 1d). Isolated patch reefs occur in the inner and outer ramps. Facies belts run more or less parallel to bathymetric contours. They include nearshore ooid–skeletal grainstone and quartz sand, mid-ramp mollusc grainstone and packstone, outer ramp mollusc marl wackestone and coralgal grainstone at reefs (Gischler & Lomando, 2005).

In all samples used in this study, carbonate mineralogy is predominated by aragonite, which usually ranges from 70 to 90%. High magnesium calcite normally reaches 10–20% and low-magnesium calcite <10% (Gischler & Lomando, 1999, 2005; Gischler, 2006).

METHODS

The $\delta^{13}C$ and $\delta^{18}O$ values were determined on a total of 559 surface sediment samples from the three modern carbonate areas. These include 326 samples from the Belize–Yucatan platforms, 102 samples from the Belize barrier reef and shelf, 38 samples from two Maldivian atolls and 93 samples from the Kuwait ramp. In addition, 69 carbonate particle end-members (coral, mollusc, algae and echinoderm fragments; ooids and peloids) were analysed with regard to C and O isotopes. From the Belize platforms, carbon and O isotopes along six Holocene lagoon cores were also

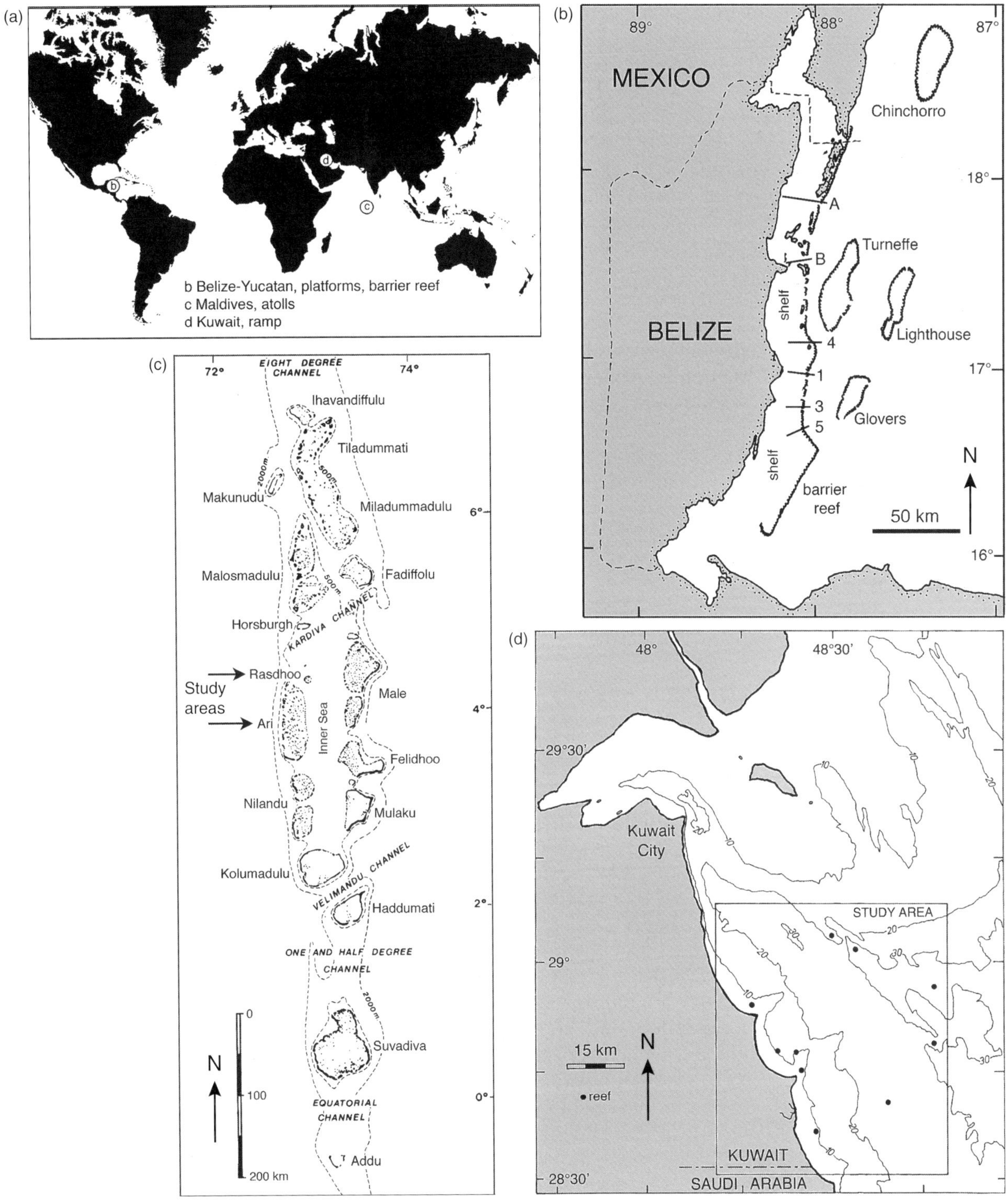

Fig. 1. Location maps of the study areas. (a) General geographical locations. (b) Map of the Belize–Yucatan area with carbonate platforms Glovers, Lighthouse, Turneffe (Belize) and Chinchorro (Mexico), as well as Belize Barrier Reef. For individual sample locations on platforms see Gischler & Lomando (1999). Barrier reef sample traverses indicated by labelled black lines. Platforms and barrier reef are bathymetrically bounded by 200 m contour. (c) Maldive archipelago. For location of individual sample points see Gischler (2006). (d) Kuwait ramp area. For individual sample locations see Gischler & Lomando (2005).

measured, totalling 139 samples. The extensive raw data are deposited in electronic form.

Samples were reduced with a sample splitter and ground with a mortar and pestle. Ground samples were analysed at the University of Frankfurt with a ThermoFinnigan MAT 253 mass spectrometer using the Gas Bench II (Finnigan Inc.) and at the University of Miami using an automated carousel attached to a Finnigan-MAT 251 mass spectrometer. Isotopic data are corrected using the procedures of Craig (1957) modified for a triple collector mass spectrometer. Data are quoted relative to Vienna Pee Dee Belemnite (V-PDB) standard according to the conventional notation. The analytical precision at both laboratories is approximately 0.1‰. Isotope data were compared with mineralogical composition, i.e., percentages of aragonite, high-magnesium calcite (HMC), low-magnesium calcite (LMC) determined using X-ray diffraction and to facies, i.e., amounts of constituent particles as quantified by point counting in thin-section.

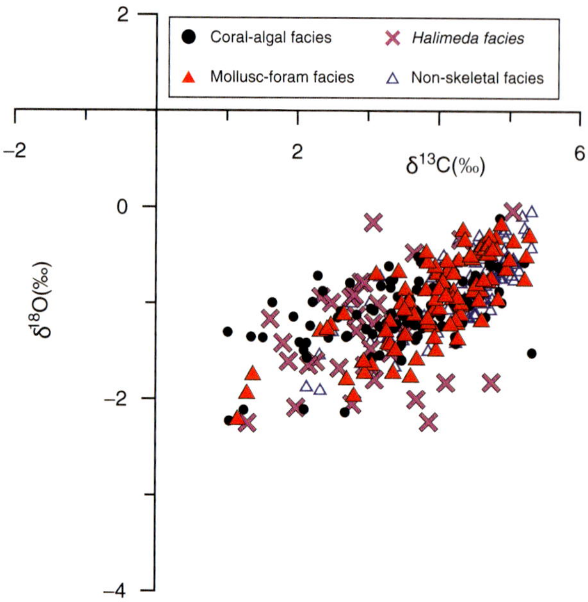

Fig. 2. C–O-isotope plot of sediments from Belize–Yucatan platforms. Note the overlap of isotopic composition of different facies.

RESULTS

Belize–Yucatan carbonate platforms, Caribbean Sea

Surface sediments

Carbon isotope values ($\delta^{13}C$) range from +1.0 to +5.5‰ and oxygen isotopes ($\delta^{18}O$) from −2.3 to 0‰ (Fig. 2). There are no isotopic differences between carbonate facies, i.e., it is not possible to delineate facies based on their $\delta^{13}C$ and $\delta^{18}O$ values. End-members of constituent particles exhibit even wider ranges (Fig. 3). As regards $\delta^{13}C$, corals usually have the lowest values, whereas non-skeletal grains such as peloids and green calcareous algae such as *Penicillus* show the highest values.

There are only a few cases of statistically significant correlations between isotopes and abundance of carbonate grains (Tables 1–4). In the shallow and open platforms of Lighthouse and Chinchorro there are negative correlations between $\delta^{13}C$, $\delta^{18}O$ and abundance of coral and coralline algal fragments. In the Glovers and Turneffe areas, no significant correlations between isotopes and constituent grains exist. Significant correlations between $\delta^{13}C$, $\delta^{18}O$ and the fine grain-size fraction (<125 µm) exist only in Lighthouse. Statistically significant correlations between $\delta^{13}C$, $\delta^{18}O$ and aragonite content occur only at Chinchorro.

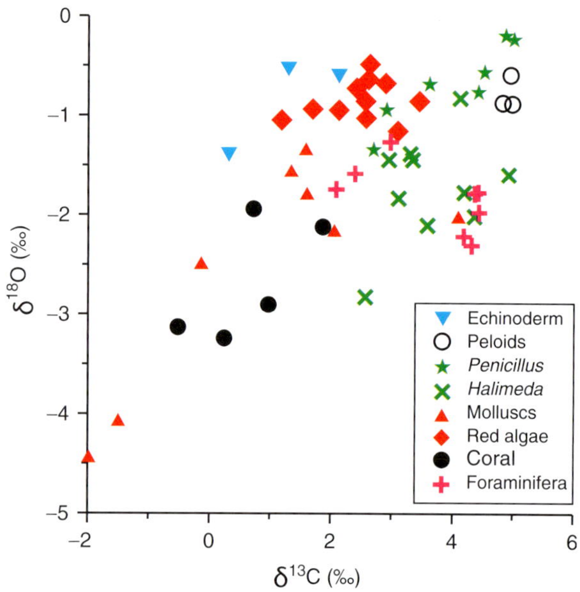

Fig. 3. C–O isotopes of end-members, Belize–Yucatan platforms. Isotopic variation is larger as compared with variation in bulk samples.

Correlations are positive. There are no significant correlations between isotopes and water depth.

Subsurface sediments

In six cores, $\delta^{13}C$ ranges from −0.2 to +5.2‰ (Fig. 4a). All cores show an up-core trend towards

Table 1. Correlation matrix of Chinchorro data. Bold numbers indicate statistically significant correlation ($p<0.05$).

	Depth	Mud	*Halimeda*	Coral	Red algae	Mollusc	Foram	Nonskel	$\delta^{13}C$	$\delta^{18}O$	ARA
Depth	1										
Mud	**.576**	1									
Halimeda	**–.327**	**–.287**	1								
Coral	–.208	**–.283**	–.032	1							
Red algae	**–.332**	**–.299**	–.003	**.805**	1						
Mollusc	.161	.042	.020	**–.261**	–.129	1					
Foram	.053	.015	.172	**–.341**	**–.281**	**.568**	1				
Nonskel	.038	–.039	**–.429**	**–.383**	**–.404**	**–.521**	**–.524**	1			
$\delta^{13}C$	.145	.020	.093	**–.228**	**–.267**	–.167	.023	.197	1		
$\delta^{18}O$	.059	.034	.013	**–.233**	**–.215**	–.188	.157	.166	**.740**	1	
ARA	–.015	–.165	–.014	-.005	–.052	**–.233**	–.067	**.212**	**.605**	**.424**	1

Table 2. Correlation matrix of Lighthouse data. Bold numbers indicate statistically significant correlation ($p<0.05$).

	Depth	Mud	*Halimeda*	Coral	Red algae	Mollusc	Foram	Nonskel	$\delta^{13}C$	$\delta^{18}O$	ARA
Depth	1										
Mud	**.274**	1									
Halimeda	–.067	**–.482**	1								
Coral	–.179	**–.693**	.068	1							
Red algae	–.156	**–.745**	**.291**	**.839**	1						
Mollusc	.143	–.141	.091	**–.281**	–.198	1					
Foram	.114	–.027	.257	**–.356**	–.160	**.474**	1				
Nonskel	–.117	**.350**	**–.531**	**–.485**	**–.585**	–.136	–.241	1			
$\delta^{13}C$	–.098	**.409**	–.262	**–.482**	**–.597**	.089	.198	**.425**	1		
$\delta^{18}O$	–.021	**.379**	**–.306**	**–.340**	**–.428**	.088	.219	.232	**.540**	1	
ARA	**–.291**	**–.542**	**.339**	**.414**	**.327**	–.127	–.177	–.098	–.107	–.136	1

Table 3. Correlation matrix of Turneffe data. Bold numbers indicate statistically significant correlation ($p<0.05$).

	Depth	Mud	*Halimeda*	Coral	Red algae	Mollusc	Foram	Nonskel	$\delta^{13}C$	$\delta^{18}O$	ARA
Depth	1										
Mud	–.030	1									
Halimeda	.126	**–.348**	1								
Coral	–.090	**–.428**	**–.508**	1							
Red algae	–.090	**–.379**	**–.482**	**.750**	1						
Mollusc	.078	–.177	.067	**–.342**	**–.291**	1					
Foram	–.007	**–.238**	–.062	–.116	–.088	**.241**	1				
Nonskel	–.015	.099	–.186	–.089	–.055	–.072	.049	1			
$\delta^{13}C$	.011	.091	–.220	.068	.073	–.034	.122	.130	1		
$\delta^{18}O$	.021	.054	–.190	.130	.194	–.106	–.021	.096	**.600**	1	
ARA	.081	.032	**.486**	**–.318**	**–.468**	.060	–.190	–.170	–.082	–.086	1

Table 4. Correlation matrix of Glovers Reef data. Bold numbers indicate statistically significant correlation ($p<0.05$).

	Depth	Mud	*Halimeda*	Coral	Red algae	Mollusc	Foram	Nonskel	$\delta^{13}C$	$\delta^{18}O$	ARA
Depth	1										
Mud	**.461**	1									
Halimeda	**–.310**	**–.537**	1								
Coral	**–.402**	**–.787**	**.350**	1							
Red algae	**–.355**	**–.770**	**.312**	**.799**	1						
Mollusc	.251	.159	–.108	**–.512**	**–.301**	1					
Foram	.230	.253	–.076	**–.475**	**–.509**	**.449**	1				
Nonskel	–.014	.056	**–.378**	**–.389**	**–.344**	.030	.080	1			
$\delta^{13}C$	.064	.249	–.082	–.165	–.247	–.037	.129	–.023	1		
$\delta^{18}O$	–.055	.019	.132	–.039	–.180	–.012	.156	–.003	**.723**	1	
ARA	.195	.073	–.042	**–.318**	–.222	**.339**	.199	**.262**	.172	.154	1

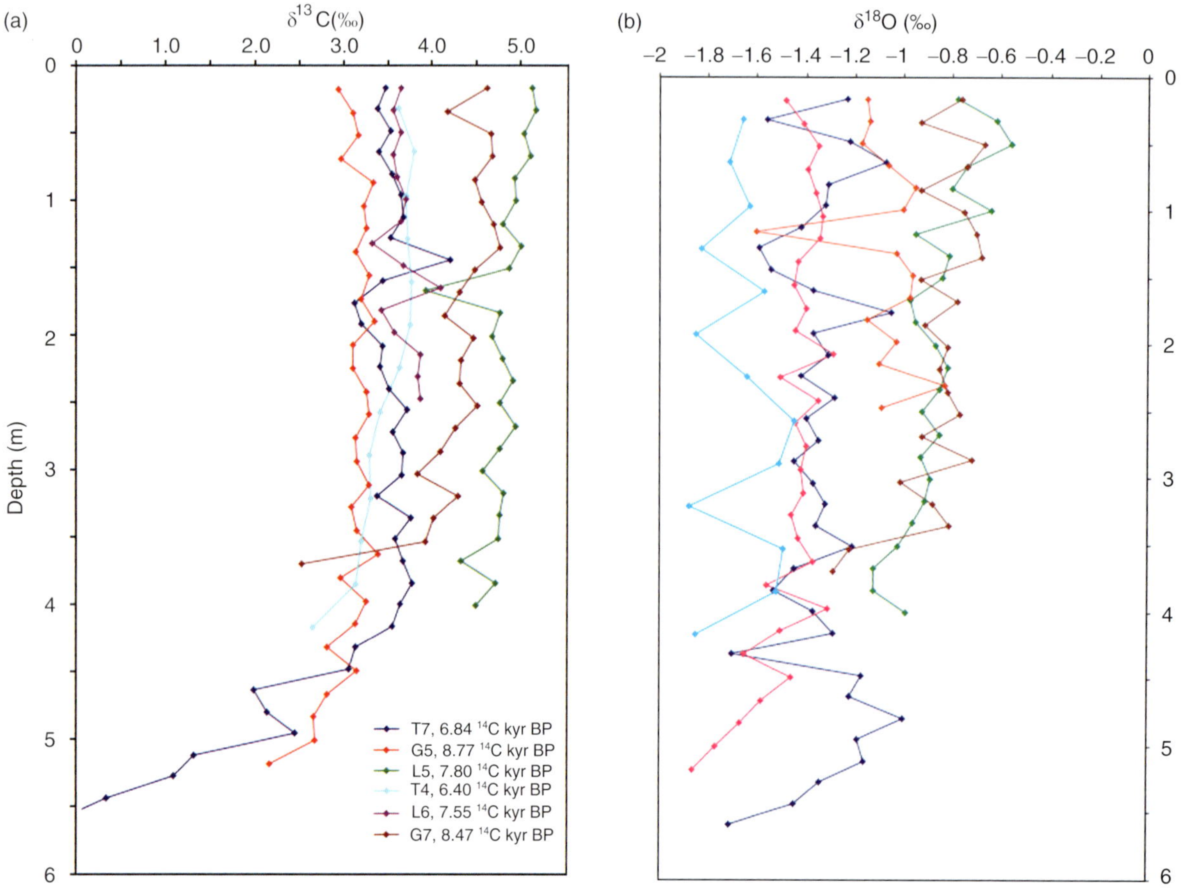

Fig. 4. C–O isotopes of sediments along Holocene lagoon cores, Belize platforms. Basal ^{14}C ages of carbonate sediments of individual cores are shown in the lower right corner. For location and description of cores see Gischler (2003).

higher $\delta^{13}C$ values. Very low values at core bases correlate with diagenetic alteration close to deposits of the meteoric realm (soil, altered reef limestone), which usually are characterized by light $\delta^{13}C$ values (Allan & Matthews, 1977). Above the meteorically influenced zone, i.e., above 4 m depth, $\delta^{13}C$ values of individual cores fluctuate only slightly, up to 1.2%. The $\delta^{13}C$ of cores L5 and G7 exhibit considerably higher values as compared with the other cores. These cores also have a high abundance of *Halimeda* (Gischler, 2003).

The values of $\delta^{18}O$ range from –1.9 to –0.6‰ (Fig. 4b). As with $\delta^{13}C$, there are mostly trends towards higher values up-core as a consequence of diagenesis, as discussed above. Again, cores L5 and G7 show higher values than observed in the other cores.

Belize Barrier Reef, Caribbean Sea

Within the barrier reef and shelf samples, $\delta^{13}C$ values range from –1.5 to +4.7%. The $\delta^{18}O$ values plot between –3.2 and –0.8% (Fig. 5). As in the platform samples, there is no clear separation of samples with regard to carbonate facies, however, non-skeletal grains are among the heaviest measured with regard to carbon isotopes.

A number of statistically significant correlations exist between carbon and oxygen isotopes and abundance of constituent grains such as coral, foraminifera, non-skeletal grains and *Halimeda* (Table 5). The $\delta^{18}O$ values also correlate with the amount of the fine grain-size fraction and with water depth. The $\delta^{18}O$ values also exhibit a statistically significant negative correlation with aragonite content.

Rasdhoo and Ari Atolls, Maldives, Indian Ocean

Stable isotope values range from +0.5 to +3.2% for $\delta^{13}C$ and from –2.9 to –1.4% for $\delta^{18}O$, respectively (Fig. 6). It is not possible to distinguish facies based on isotope values. Only the mudstone facies samples plot in a relatively narrow field.

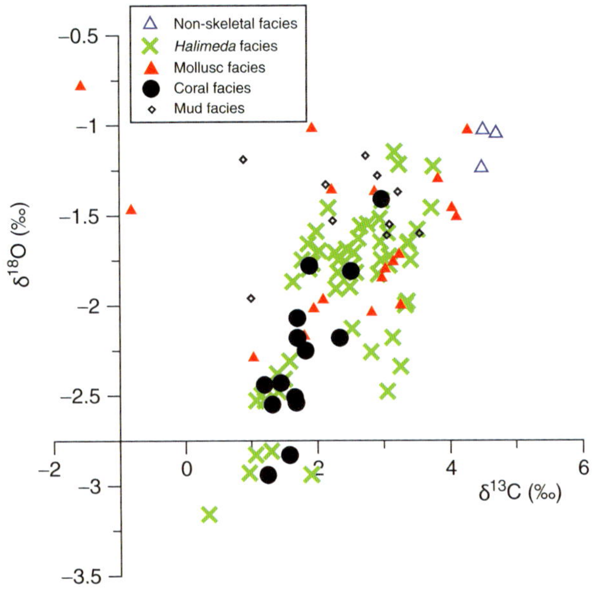

Fig. 5. C–O-isotope plot of sediments from Belize Barrier Reef. Samples from different facies largely overlap with the exception of the non-skeletal facies.

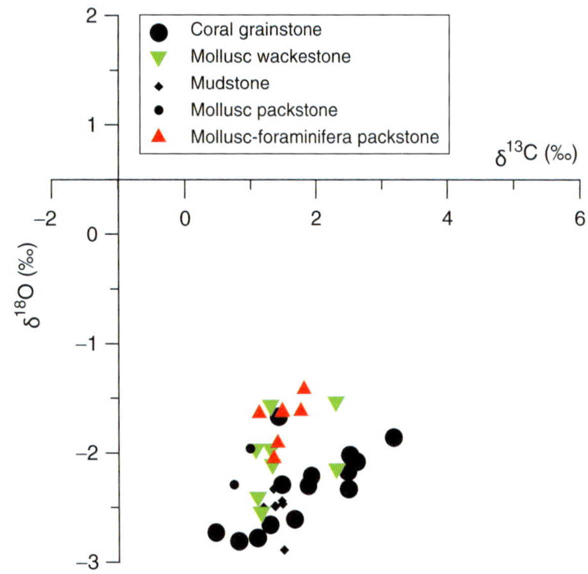

Fig. 6. C–O-isotope plot of sediments from Maldive atolls. Note the overlap of isotopic composition of facies.

Table 5. Correlation matrix of Belize Barrier Reef data. Bold numbers indicate statistically significant correlation ($p<0.05$).

	Depth	ARA	$\delta^{18}O$	$\delta^{13}C$	Mollusc	*Halimeda*	Coral	Red algae	Nonskel	Mud
Depth	1									
ARA	.003	1								
$\delta^{18}O$	**.214**	**–.463**	1							
$\delta^{13}C$	.112	.123	**.463**	1						
Mollusc	–.117	.123	.026	–.089	1					
Halimeda	–.089	**–.376**	**.549**	**.340**	.047	1				
Coral	**.261**	**.433**	**–.205**	.065	**–.442**	**–.397**	1			
Red algae	–.002	**.299**	**–.533**	**–.387**	–.112	**–.476**	–.015	1		
Nonskel	.069	–.086	.073	–.057	**–.300**	–.154	.181	**.322**	1	
Mud	.040	.232	**–.477**	**–.480**	.099	**–.450**	.079	**.402**	.157	1

In Rasdhoo Atoll, a number of significant correlations are seen between $\delta^{13}C$, $\delta^{18}O$ and the abundance of *Halimeda*, mollusc and foraminifera (Table 6). The $\delta^{13}C$ values also correlated with aragonite content and water depth. In Ari Atoll, no statistically significant correlations can be detected between $\delta^{13}C$, $\delta^{18}O$, abundance of constituent particles, fines, aragonite content and water depth (Table 7).

Kuwait Ramp, Arabian-Persian Gulf

Among surface sediment samples and constituent particle end-members, $\delta^{13}C$ values range from 0 to +4.5‰ and $\delta^{18}O$ from –2.1 to +1.5‰ (Fig. 7). Unlike in the barrier reef, atoll and platform examples, carbonate facies have characteristic isotope compositions.

There are statistically significant correlations between $\delta^{13}C$, $\delta^{18}O$ and the abundances of almost all constituent particles, with the fine grain-size fraction and with aragonite content (Table 8). The $\delta^{13}C$ pattern is also statistically correlated with water depth.

DISCUSSION

Patterns and comparisons

Whereas the range of bulk $\delta^{13}C$ (*c.* +1 to +5‰) is comparable among the sites studied, bulk $\delta^{18}O$

Table 6. Correlation matrix of Rasdhoo Atoll data. Statistically significant correlations are marked bold ($p < 0.05$).

	Depth	Coral	Red algae	*Halimeda*	Mollusc	Foram	Peloid	Aggreg	Echino	Mud	$\delta^{13}C$	$\delta^{18}O$	ARA
Depth	1												
Coral	**–.843**	1											
Red algae	–.440	**.512**	1										
Halimeda	**–.690**	**.656**	**.625**	1									
Mollusc	.350	–.253	.330	–.119	1								
Foram	.200	.170	**.588**	**.602**	**.464**	1							
Peloid	–.009	–.147	–.157	–.218	.131	–.113	1						
Aggreg	–.435	.233	.253	.297	.059	.305	.206	1					
Echino	.128	.026	.315	–.081	**.604**	.231	–.013	–.056	1				
Mud	**.614**	**–.730**	**–.827**	**–.750**	–.385	**–.660**	.066	–.353	–.352	1			
$\delta^{13}C$	**–.562**	.296	.333	**.782**	–.235	**.445**	.112	.385	–.381	–.388	1		
$\delta^{18}O$	.120	–.158	.374	.422	**.482**	**.641**	.102	.194	.001	–.374	**.591**	1	
ARA	**–.840**	**.813**	.424	**.725**	**–.460**	.232	–.084	.205	–.128	**–.543**	**.616**	.019	1

Table 7. Correlation matrix of Ari Atoll data. Statistically significant correlations are marked bold ($p < 0.05$).

	Depth	Coral	Red algae	*Halimeda*	Mollusc	Foram	Peloid	Aggreg	Echino	Mud	$\delta^{13}C$	$\delta^{18}O$	ARA
Depth	1												
Coral	**–.748**	1											
Red algae	**–.538**	.464	1										
Halimeda	**–.806**	**.573**	.325	1									
Mollusc	**.846**	**–.798**	–.394	**–.631**	1								
Foram	**.585**	**–.692**	–.068	–.458	**.702**	1							
Peloid	.406	–.232	–.154	–.204	.370	.134	1						
Aggreg	.388	–.460	.007	–.306	.443	**.817**	.012	1					
Echino	–.021	–.219	.003	.040	.285	.195	.009	–.189	1				
Mud	.434	**–.597**	**–.641**	**–.480**	.203	–.055	.036	–.158	–.056	1			
$\delta^{13}C$	–.091	–.004	–.158	.395	.141	.189	–.233	.155	.028	–.321	1		
$\delta^{18}O$	–.021	.082	.372	–.114	–.022	.060	.022	.024	–.070	–.163	.162	1	
ARA	**–.799**	.452	.125	**.729**	**–.712**	**–.640**	–.161	–.418	–.072	–.006	.073	–.118	1

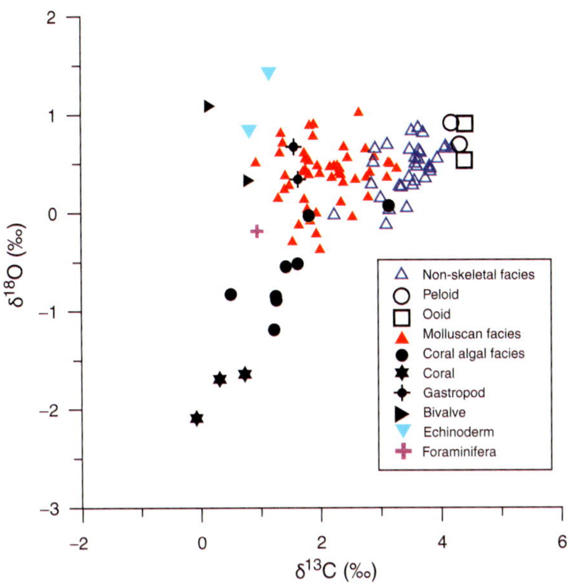

Fig. 7. C–O-isotope plot of sediments and end-members from Kuwait ramp. Note relatively clear separation of different facies and larger isotopic range of constituent particles as compared with bulk samples.

values (range *c.* −3 to +2.3‰) show considerable differences between individual sites. In the Belize Barrier Reef and Maldive atoll samples, values are in the same range. Belize–Yucatan platform samples show slightly heavier values and Kuwait ramp samples even more positive values. One possible explanation for these differences might be salinity, which is normal marine in the Maldives (around 35‰), normal marine to hyposaline (30–35‰) on the Belize shelf (Purdy *et al.*, 1975), elevated (35–42‰) in Belize–Yucatan platform lagoons (Gischler *et al.*, 2003) and hypersaline (39–43‰) on the Kuwait ramp (Gischler & Lomando, 2005).

For comparison, Weber (1967) and Weber & Woodhead (1969) measured ranges in $\delta^{13}C$ of bulk carbonate sediments of +1.6 to +4.5‰ and +0.75 to +3.2‰ in Bermuda and on Heron Reef off Australia, respectively. The $\delta^{18}O$ values from these settings ranged from −0.6 to +0.4‰ in Bermuda and from −2.1 to −0.4‰ on Heron Reef. Land (1989) compiled isotope data of shallow-marine carbonate sediments. These values range from about −3.5 to +5‰ for $\delta^{13}C$ and from −3.5 to +2‰ for $\delta^{18}O$.

The $\delta^{13}C$ and $\delta^{18}O$ values of constituent carbonate sediment particles such as corals, algae, molluscs, foraminifera and others from the Belize and Kuwait locations exhibit even larger variations as compared with the bulk samples. Land (1989) came to a similar conclusion, which can be explained by extreme isotope values cancelling each other out in a mixture of grain types (see e.g. Fig. 7). Corals usually show the most negative values, whereas non-skeletal grains such as peloids and ooids and calcareous algae are isotopically heaviest. Variation within skeletal grain groups may be explained by taxonomic differences in isotopic fractionation and also by seasonal isotopic variation within skeletons (Emiliani *et al.*, 1978; Wefer & Berger, 1991). As a consequence of the latter observation, small fragments of skeletons may not necessarily yield representative $\delta^{13}C$ and $\delta^{18}O$ signals.

Spatial trends in isotopic composition of bulk samples are visible in the Kuwait ramp example in that carbon isotopes decrease from nearshore to offshore. In the atoll, barrier reef and platform examples, however, no clear trends are discernible. In Glovers, Lighthouse and Chinchorro, $\delta^{13}C$ decreases from the platform interiors to margins. In the Belize Barrier Reef transects 1, 3, 5 and B there is an offshore increase in $\delta^{13}C$, however, in transects 4 and A, there is an offshore decrease.

Likewise, carbonate facies can be characterized by their $\delta^{13}C$ and $\delta^{18}O$ composition in the ramp example, but not in the other locations. Statistical analyses show that there are also more significant correlations between isotope composition and carbonate mineralogy, abundance of grain types or facies on the Kuwait ramp. Also, all measured parameters including $\delta^{13}C$ and $\delta^{18}O$ values correlate significantly with water depth, i.e., with depositional energy on the ramp example. This difference is probably a consequence of the fact that the continuous depositional energy gradient on the Kuwait ramp largely controls carbonate facies. In the atoll, barrier reef and platform examples, depositional energy gradients are more variable and morphological and environmental variability is much greater.

Several workers have found significant positive correlations between $\delta^{13}C$ and the concentration of aragonite on various carbonate platforms from numerous locations such as the Bahamas (Weber, 1967; Swart & Eberli, 2005; Swart *et al.*, 2009), Australia (Weber & Woodhead, 1969) and the NW Australian shelf (Dix *et al.*, 2005). In our study, such a correlation was only present in the Chinchorro platform, Rasdhoo Atoll and the Kuwait ramp samples, but not in the other examples investigated. A potential reason for this difference is the elevated content of isotopically heavy non-skeletal grains at these three locations. Among the Belize–Yucatan platforms, Chinchorro has 30% peloids on average as compared with

Table 8. Correlation matrix of Kuwait ramp data. Statistically significant correlations are marked bold ($p < 0.05$).

	Mollusc	Coral	Red algae	Foram	Ooid	Peloid	Aggreg	Echino	Mud	Depth	$\delta^{13}C$	$\delta^{18}O$	ARA
Mollusc	1												
Coral	**–.498**	1											
Red algae	**–.427**	**.873**	1										
Foram	**.290**	**–.280**	**–.272**	1									
Ooid	**–.601**	–.151	–.165	–.118	1								
Peloid	–.159	**–.318**	**–.324**	–.104	**.234**	1							
Aggreg	–.001	**–.210**	**–.216**	**–.341**	–.029	.012	1						
Echino	**.245**	–.041	.101	**.323**	**–.312**	–.165	**–.262**	1					
Mud	.088	–.178	–.177	**.303**	**–.300**	–.068	**–.332**	**.203**	1				
Depth	**.692**	**–.429**	**–.332**	**.358**	**–.468**	**–.256**	–.078	**.232**	**.478**	1			
$\delta^{13}C$	**–.438**	**–.292**	**–.338**	**–.334**	**.739**	**.536**	**.274**	**–.438**	**–.317**	**–.478**	1		
$\delta^{18}O$	**.264**	**–.568**	**–.510**	**–.205**	**.310**	**.303**	.197	**–.333**	**–.232**	.092	**.555**	1	
ARA	**–.288**	**.031**	–.001	**–.654**	**.413**	**.325**	**.317**	**–.466**	**–.527**	**–.519**	**.725**	**.557**	1

c. 10% in the other platform locations. In the Maldive examples, Rasdhoo Atoll sediments contain peloids in contrast to Ari Atoll. The Kuwait ramp is rich in non-skeletal grains such as ooids, peloids and aggregate grains, just like the above-mentioned example of Great Bahama Bank (Purdy, 1963; Swart & Eberli, 2005; Swart *et al.*, 2009).

Taphonomic and diagenetic filters

Subsurface results from the Belize platforms show that major fluctuations of several per mille in $\delta^{13}C$ along individual Holocene sections are associated with meteoric alteration near the Pleistocene/Holocene hiatus. However, isotopic variation along individual cores above the meteorically influenced horizon were rather low (<1.2%). Possible explanations for the latter observation include taphonomic processes such as bioturbation and early diagenetic effects.

Bioturbation by the burrowing shrimp *Callianassa* characterizes large parts of the lagoonal areas of the Belize platforms (Gischler & Lomando, 1999). The construction of large and complex burrows in unconsolidated sediment, *Callianassa* is able to rework and mix sediment to depths of almost 1 m (Scoffin, 1992). We speculate that during this process, isotopic variability is apparently reduced, as seen in the bulk of modern sediments. The fossil record of callianassids (Thalassinidean shrimps) extends to the Jurassic (Myrow, 1995) and *Callianassa*-type bioturbation is known from carbonate platforms as old as the Cretaceous (Sanders, 2001). The trace fossil *Thalassinoides* is reported from sedimentary rocks as old as the Cambrian (Myrow, 1995). Other organisms observed in bioturbation in the modern environments studied here include mollusc bivalves, echinoderms, polychaetes and fish. The fossil record of bioturbation for some of these organism groups goes as far back as the late Proterozoic, however, in general the Palaeozoic record is much more fragmentary as compared with the Meso-Cenozoic record (Bromley, 1990).

Early diagenetic effects also impact on the isotope composition of Holocene lagoonal carbonate sediments in this study, which are to a large part made up of the metastable minerals aragonite and HMC. A number of previous studies have shown scanning electron microscopy (SEM) evidence of partial dissolution of carbonate grains such as *Halimeda*, foraminifera and coralline algae in modern Caribbean reef settings (Perry, 2000). Partial dissolution occurs preferentially in back-reef or lagoonal settings, largely a result of the lowering of pH during organic matter oxidation. Sediment in *Callianassa*-rich settings is especially prone to dissolution (Sanders, 2001). There is also geochemical evidence of dissolution and reprecipitation of calcium carbonate in Holocene and modern platform sediments of Florida Bay (Burns & Swart, 1992; Patterson & Walter, 1994b). No major changes in mineralogy were observed, however, shifts in trace element (strontium, magnesium) and stable isotopes of carbon and oxygen were significant. Recrystallization of modern shallow-water skeletal grains from calcareous algae and foraminifera is widespread (Reid & Macintyre, 1998) and was demonstrated in the fine grain-size fraction of sediments of the Belize platforms (Gischler & Zingeler, 2002). Few geochemical studies indicate a lowering of minor element content, e.g., strontium (Reid & Macintyre, 1998; Gischler & Zingeler, 2002), however, it is largely unknown as yet to which degree early recrystallization processes also modify the isotopic composition of these carbonate grains.

For pre-Holocene settings, late diagenetic effects in the meteoric and burial environment have to be considered as well. Processes described include dissolution and recrystallization, neomorphism, LMC precipitation and dolomitization. In addition, lagoonal carbonate sediment can possibly become compacted and eventually develop features indicative of pressure dissolution. These processes are discussed at length in textbooks, and, because this study focuses on the Holocene, they will not be treated in detail. Melim *et al.* (2001) described a Neogene example based on the investigation of diagenesis in two cores in Great Bahama Bank.

Significance for carbon isotope stratigraphy

The application of $\delta^{13}C$ for stratigraphic correlation purposes has been applied for about a decade to ancient shallow-water carbonate settings of ramps and platforms spanning the time from the Precambrian to the Tertiary (Joachimski & Buggisch, 1993; Jenkyns, 1995; Vahrenkamp, 1996; Ferreri *et al.*, 1997; Mutti *et al.*, 1997; Grötsch *et al.*, 1998; Wissler *et al.*, 2003; Saltzman *et al.*, 2004). In these studies, $\delta^{13}C$ ranges from about −1 to +5‰ and individual stratigraphic sections show an absolute range from approximately 1‰ to 4‰. Only Valladares *et al.* (1996) reported a range of up to 8‰ for $\delta^{13}C$ along investigated Cretaceous sections. Likewise, Cozzi *et al.* (2004) presented large $\delta^{13}C$ ranges from −8 to +3‰

in a Proterozoic ramp. However, the reported ranges in $\delta^{13}C$ and $\delta^{18}O$ from the large majority of fossil ramp and platform sections fall well within the field of isotopic variability as seen in this study and in previous studies of $\delta^{13}C$ and $\delta^{18}O$ of modern shallow-water carbonate sediments (Weber, 1967; Weber & Woodhead, 1969; Land, 1989). As a consequence the isotopic variation along ancient rock sections that is used for stratigraphic correlation could easily be explained by changes in taxonomic composition within individual carbonate-producing organism groups, or simply by facies shifts.

Swart & Eberli (2005) showed that basin to bank correlations in the Miocene to Holocene of the Bahamas using $\delta^{13}C$ is possible. However, they also found that $\delta^{13}C$ trends through their sediment cores did not correlate with the global signal of $\delta^{13}C$. Compositional variation in periplatform settings was reported to be rather the expression of varying mixtures of carbonate grains produced on the platform top, which eventually are exported and mixed with pelagic carbonate. Similar processes were probably effective in the pre-Neogene, however, the abundance and isotopic compositions of ancient platform-dwelling, carbonate-producing organisms such as corals, calcareous sponges, molluscs, foraminifera and calcareous algae were not investigated in the above-mentioned bulk rock isotope studies.

In summary, there are a number of influences $\delta^{13}C$ and $\delta^{18}O$ values in shallow-water carbonate sediments. These include taxonomic variation, diagenesis and the fact that the majority of grains are composed of metastable aragonite and HMC. Taphonomic effects such as bioturbation are of importance as seen in the investigated Holocene cores. Therefore, results of bulk analyses of $\delta^{13}C$ and $\delta^{18}O$ of shallow water carbonate settings should be used with caution and in combination with additional sedimentological and paleontological data for stratigraphic and paleoenvironmental interpretations. Still, because carbon isotope stratigraphy appears reliable in the pre Holocene, it can be speculated that taphonomy and diagenesis act as isotopic 'filters', which level out isotopic variation seen in the modern time slice. These taphonomic and diagenetic filters, however, are only poorly defined and are as yet unquantified.

CONCLUSIONS

1. The variability of $\delta^{13}C$ and $\delta^{18}O$ values of sediments in modern and Holocene shallow-water carbonate settings such as platforms, barrier reefs, atolls and ramps is high, i.e., there are absolute ranges of 6.9‰ for $\delta^{13}C$ and 5.5‰ for $\delta^{18}O$, respectively.
2. Clear spatial trends in isotopic composition of sediments were found only in the ramp setting, where carbonate facies also have characteristic isotopic compositions.
3. The strongest isotopic excursions along Holocene sections are caused by diagenesis near a subaerial exposure horizon.
4. The much lower range of carbon and oxygen isotopes found in fossil shallow-water carbonates, e.g., in isotope stratigraphy, are presumably a consequence of taphonomic and diagenetic processes, which decrease isotopic variability.

ACKNOWLEDGEMENTS

We thank R. Gless, C. Schroeder, A. Saied and D. Harazim for help during sample preparation for stable isotope and X-ray-diffraction analyses. J. Fiebig ran the Frankfurt mass spectrometer and R. Petschick the Frankfurt diffractometer. The thoughtful reviews by W. Patterson (Saskatoon) and S. Hesselbo (Oxford) improved the paper and are gratefully acknowledged.

REFERENCES

Allan, J.R. and **Matthews, R.K.** (1977) Carbon and oxygen isotopes as diagenetic and stratigraphic tools: surface and subsurface data, Barbados. *Geology*, **5**, 16–20.

Broecker, W.S. (1982) Ocean chemistry during glacial time. *Geochim. Cosmochim. Acta*, **46**, 1689–1705.

Bromley, R.G. (1990) *Trace Fossils. Biology and Taphonomy*. Unwin Hyman, London, 280 pp.

Burns, S.J. and **Swart, P.K.** (1992) Diagenetic processes in Holocene carbonate sediments: Florida Bay mudbanks and islands. *Sedimentology*, **39**, 285–304.

Cozzi, A., **Allen, P.A.** and **Grotzinger, J.P.** (2004) Understanding carbonate ramp dynamics using $\delta^{13}C$ profiles: examples from the Neoproterozoic Buah Formation of Oman. *Terra Nova*, **16**, 62–67.

Craig, H. (1957) Isotopic standards for carbon and oxygen and correlation factors for mass-spectrometric analysis of carbon dioxide. *Geochim. Cosmochim. Acta*, **12**, 122.

Dix, G.R., **James, N.P.**, **Kyser, T.K.**, **Bone, Y.** and **Collins, L.B.** (2005) Genesis and dispersal of carbonate mud relative to late Quaternary sea-level change along a distally-steepened carbonate ramp (northwestern shelf, western Australia). *J. Sed. Res.*, **75**, 665–678.

Drzewiecki, P.A. and **Simo, J.A.** (1997) Carbonate platform drowning and oceanic anoxic events on a mid-Cretaceous carbonate platform, south-central Pyrenees, Spain. *J. Sed. Res.*, **67**, 698–714.

Emiliani, C., Hudson, J.H., Shinn, E.A., George, R.Y. and **Lidz, B.** (1978) Oxygen and carbon isotopic growth record in a reef coral from the Florida Keys and a deep-sea coral from the Blake Plateau. *Science*, **202**, 627–629.

Ferreri, V., Weissert, H., D'Argenio, B. and **Buoncunto, F.P.** (1997) Carbon isotope stratigraphy: a tool for basin to carbonate platform correlation. *Terra Nova*, **9**, 57–61.

Gischler, E. (2003) Holocene lagoonal development in the isolated carbonate platforms of Belize. In: *Late Quaternary Reef Development* (Eds. P. Blanchon and L.F. Montaggioni). *Sed. Geol.*, **159**, 113–132.

Gischler, E. (2006) Sedimentation on Rasdhoo and Ari atolls, Maldives, Indian Ocean. *Facies*, **52**, 341–360.

Gischler, E. and **Lomando, A.J.** (1999) Recent sedimentary facies of isolated carbonate platforms, Belize-Yucatan system, Central America. *J. Sed. Res.*, **69**, 747–763.

Gischler, E. and **Lomando, A.J.** (2005) Offshore sedimentary facies of a modern carbonate ramp, Kuwait, northwestern Arabian-Persian Gulf. *Facies*, **50**, 443–462.

Gischler, E. and **Zingeler, D.** (2002) The origin of carbonate mud in isolated carbonate platforms of Belize, Central America. *Int. J. Earth Sci.*, **91**, 1054–1070.

Gischler, E., Hauser, I., Heinrich, K. and **Scheitel, U.** (2003) Characterization of depositional environments in isolated carbonate platforms based on benthic foraminifera, Belize, Central America. *Palaios*, **18**, 136–255.

Grötsch, J., Billing, I. and **Vahrenkamp, V.** (1998) Carbon-isotope stratigraphy in shallow-water carbonates: implications for Cretaceous black-shale deposition. *Sedimentology*, **45**, 623–634.

Immenhauser, A., Della Porta, G., Kenter, J.A.M. and **Bahamonde, J.E.** (2003) An alternative model for positive shifts in shallow-marine carbonate $\delta^{13}C$ and $\delta^{18}O$. *Sedimentology*, **50**, 953–959.

Immenhauser, A., Kenter, J.A.M., Ganssen, G., Bahamonde, J.R., Van Vliet, A. and **Saher, M.H.** (2002) Origin and significance of isotopic shifts in Pennsylvanian carbonates (Asturias, Spain). *J. Sed. Res.*, **72**, 82–94.

Jenkyns, H.C. (1995) Carbon-isotope stratigraphy and paleoceanographic significance of the lower Cretaceous shallow-water carbonates of Resolution Guyot, mid-Pacific mountains. *Proc. Ocean Drilling Prog. Sci. Res.* (Eds E.L. Winterer, W.W. Sager, J.V. Firth and J.M. Sinton), Vol. 143, Ocean Drilling Program, College Station, Texas, pp. 99–104.

Joachimski, M.M. and **Buggisch, W.** (1993) Anoxic events in the late Frasnian – causes of the Frasnian-Famennian faunal crisis? *Geology*, **21**, 675–678.

Land, L.S. (1989) The carbon and oxygen isotopic chemistry of surficial Holocene shallow marine carbonate sediment and Quaternary limestone and dolomite. In: *Handbook of Environmental Isotope Geochemistry* (Eds P. Fritz and J.C. Fontes), Vol. 3, pp. 191–217. Elsevier, Amsterdam.

Melim, L.A., Swart, P.K. and **Maliva, R.G.** (2001) Meteoric and marine-burial diagenesis in the subsurface of Great Bahama Bank. In: *Subsurface Geology of a Prograding Carbonate Platform Margin* (Ed. R.N. Ginsburg). SEPM Spec. Publ., **70**, 137–161.

Mutti, M., Bernoulli, D. and **Stille, P.** (1997) Temperate carbonate platform drowning linked to Miocene oceanographic events: Maiella platform margin, Italy. *Terra Nova*, **9**, 122–125.

Myrow, P.M. (1995) *Thalassinoides* and the enigma of early Paleozoic open-framework burrow systems. *Palaios*, **10**, 58–74.

Patterson, W.P. and **Walter, L.M.** (1994a) Depletion of ^{13}C in seawater ΣCO_2 on modern carbonate platforms: significance for the carbon isotopic record of carbonates. *Geology*, **22**, 885–888.

Patterson, W.P. and **Walter, L.M.** (1994b) Syndepositional diagenesis of modern platform carbonates: evidence from isotopic and minor element data. *Geology*, **22**, 127–130.

Perry, C.T. (2000) Factors controlling sediment preservation on a north Jamaican fringing reef: a process-based approach to microfacies analysis. *J. Sed. Res.*, **70**, 633–648.

Purdy, E.G. (1963) Recent calcium carbonate facies of the Great Bahama Bank. 2. Sedimentary facies. *J. Geol.*, **71**, 472–497.

Purdy, E.G. and **Gischler, E.** (2003) The Belize margin revisited. 1. Holocene marine facies. *Int. J. Earth Sci.*, **92**, 532–551.

Purdy, E.G., Pusey, W.C. and **Wantland, K.F.** (1975) Continental shelf of Belize – regional shelf attributes. *Am. Assoc. Petrol. Geol. Stud. Geol.*, **2**, 1–52.

Reid, R.P. and **Macintyre, I.G.** (1998) Carbonate recrystallization in shallow marine environments: a widespread diagenetic process forming micritized grains. *J. Sed. Res.*, **68**, 928–946.

Saltzman, M.R., Groessens, E. and **Zhuralev, A.V.** (2004) Carbon cycle models based on extreme changes in $\delta^{13}C$: an example from the lower Mississippian. *Palaeogeogr. Palaeoclimatol. Palaeoecol.*, **213**, 359–377.

Sanders, D. (2001) Burrow-mediated carbonate dissolution in rudist biostromes (Aurisina, Italy): implications for taphonomy in tropical, shallow subtidal environments. *Paleogeogr. Palaeoclimatol. Palaeoecol.*, **168**, 39–74.

Scholle, P.A. and **Arthur, M.A.** (1980) Carbon isotopic fluctuations in pelagic limestones: potential stratigraphic and petroleum exploration tool. *Am. Assoc. Petrol. Geol. Bull.*, **64**, 67–87.

Scoffin, T.P. (1992) Taphonomy of coral reefs: a review. *Coral Reefs*, **11**, 57–77.

Swart, P.K. and **Eberli, G.P.** (2005) The nature of the $\delta^{13}C$ of periplatform sediments: implications for stratigraphy and the global carbon cycle. *Sed. Geol.*, **175**, 115–130.

Swart, P.K., Reijmer, J.J. and **Otto, R.** (2009) Variations in the $\delta^{13}C$, $\delta^{18}O$ and mineralogy of surface sediments on Great Bahama Bank: implications for reading the fossil record. In: *Perspectives in Carbonate Geology: A Tribute to the Career of Robert Nathan Ginsburg* (Eds P. Swart, G. Eberli and J. McKenzie), IAS/Wiley–Blackwell, London. *Spec. Publ. Int. Assoc. Sedimentol.*, 41, 47–60.

Vahrenkamp, V.C. (1996) Carbon isotope stratigraphy of the upper Kharaib and Shuaiba Formations: implications for the early Cretaceous evolution of the Arabian Gulf region. *Am. Assoc. Petrol. Geol. Bull.*, **80**, 647–662.

Valladares, I., Recio, C. and **Lendinez, A.** (1996) Sequence stratigraphy and stable isotopes ($\delta^{13}C$, $\delta^{18}O$) of the late

Cretaceous carbonate ramp of the western margin of the Iberian chain (Soria, Spain). *Sed. Geol.*, **105**, 11–28.

Weber, J.N. (1967) Factors affecting the carbon and oxygen isotopic composition of marine carbonate sediments – Part I, Bermuda. *Am. J. Sci.*, **265**, 586–608.

Weber, J.N. and **Woodhead, P.M.J.** (1969) Factors affecting the carbon and oxygen isotopic composition of marine carbonate sediments – II. Heron Island, Great Barrier Reef, Australia. *Geochim. Cosmochim. Acta*, **33**, 19–38.

Wefer, G. and **Berger, W.H.** (1991) Isotope paleontology: growth and composition of extant calcareous species. *Mar. Geol.*, **100**, 207–248.

Weissert, H. (1989) C-isotope stratigraphy, a monitor of paleoenvironmental change: a case study from the early Cretaceous. *Surv. Geophys.*, **10**, 1–61.

Wissler, L., **Funk, H.** and **Weissert, H.** (2003) Response of early Cretaceous carbonate platforms to changes in atmospheric carbon dioxide levels. *Palaeogeogr. Palaeoclimatol. Palaeoecol.*, **200**, 187–205.

Int. Assoc. Sedimentol. Spec. Publ. (2009) **41**, 75–90

A tale of two storms: an integrated field, remote sensing and modelling study examining the impact of hurricanes Frances and Jeanne on carbonate systems, Bahamas

STACY L. REEDER*† and EUGENE C. RANKEY*‡

*Comparative Sedimentology Laboratory and Center for Southeastern Tropical Advanced Remote Sensing Rosenstiel School of Marine and Atmospheric Science University of Miami, Miami, FL 33149, USA

† Schlumberger-Doll Research, One Hampshire Street, Cambridge, MA, 02139 (E-mail: sreeder@slb.com)

‡ Department of Geology, University of Kansas, Lawrence, KS, 66045

ABSTRACT

Tropical cyclones are commonly cited as being influential, or even dominant, controls on the geomorphological evolution of carbonate systems, and their influence is highlighted in many interpretations of ancient platforms. A unique opportunity to explore the effects of storms on platform systems occurred when two strong tropical cyclones (Frances and Jeanne) passed directly over the shoals and reefs of the Abacos, northern Bahamas, in 2004. Observations of the influence of these storms coupled with hydrodynamic observations and wave models provide insights into the nature and extent of geomorphological change that might be related to the passage of such storms.

Comparisons of high resolution (2.4 m pixels) QuickBird remote sensing images and modelling lead to a conclusion that storms have spatially limited effects on the subtidal platform and shoals, and that storms are neither sufficient to alter the system nor necessary for lobate shoals to form. Instead, daily processes (winds, waves and tides) produce conditions adequate to explain geomorphological features.

The effects of these two storms are not enough to explain the geomorphological evolution of this tidally influenced system. Other studies show that storms had comparable impacts, suggesting that this is not an exceptional case. These observations are consistent with a quantitative wave model showing the distribution of energy across carbonate systems with different geometries. These observations and model results highlight the difficulty of generating grainy 'storm deposits' of comparable extent to those interpreted in some ancient successions. Collectively, these observations and models question the paradigm of storm influence on shallow-water carbonate platform geomorphology.

Keywords Storms, hurricanes, waves, carbonate sedimentology, geomorphology, Bahamas, tidal deltas, ooid shoals.

INTRODUCTION

Tropical cyclones are commonly invoked as significant geomorphological and geological forces that shape carbonate sedimentary systems (Ball, 1967; Perkins & Enos, 1968; Wanless *et al.*, 1988; Scoffin, 1993; Wanless & Tedesco, 1993) due to their powerful forces including excessive wind speeds, increased current velocities, storm surges and a broader higher-energy wave spectrum. In many ancient carbonate successions, thin, but apparently laterally extensive, grainy beds have been interpreted as storm-generated deposits (Kreisa, 1981; Seilacher & Aigner, 1991; Aigner, 1982, 1985; Kreisa & Bambach, 1982; Jennette & Pryor, 1993). Additionally, many stratigraphic studies highlight interpretations of 'storm-dominated platforms' and 'storm wave base' (reviews by Ager, 1981; Aigner, 1985; Pomar, 2001). At a smaller scale, many lobate features on Holocene platforms were originally described as 'spillover lobes' (Ball, 1967) with a morphology interpreted to be driven mainly by storms (Hine, 1977). Nonetheless, although storms frequently traverse carbonate platforms, direct observations of major storm impacts on carbonate platforms are relatively rare (Major *et al.*, 1996). In fact, some studies have illustrated the lack of impact of storms on carbonate systems (Boss & Neumann, 1993; Rankey *et al.*, 2004).

The passage of two tropical cyclones (Hurricanes Frances and Jeanne) over the same area on Little

Bahama Bank, Bahamas, within one month in September 2004, provided a unique opportunity to examine the impact of strong tropical storms on active Holocene carbonate systems. If storm-generated morphological changes were ever expected to occur, they should be readily apparent after the passage of *two* storms over the same area within a 1 month period.

The purpose of this paper is to describe the impacts of these storms, explicitly exploring the hypothesis that storms are not a major influence on the geomorphology of the shoals in this region. To further explore the effects of storms, a wave model studies the depths of influence of storm waves on more simplified carbonate bathymetries such as a gentle ramp or a flat, unrimmed carbonate platform. The results of the modelling suggest that larger storms are not necessarily more influential on the geomorphology of platform-top carbonate shoals since the larger waves are more likely to break over the platform edge. The results are important in understanding the fundamental processes involved in shaping the geomorphology of carbonate systems. Additionally, they raise questions regarding the realism of interpretations of widespread storm deposits in analogous systems in the ancient geological record.

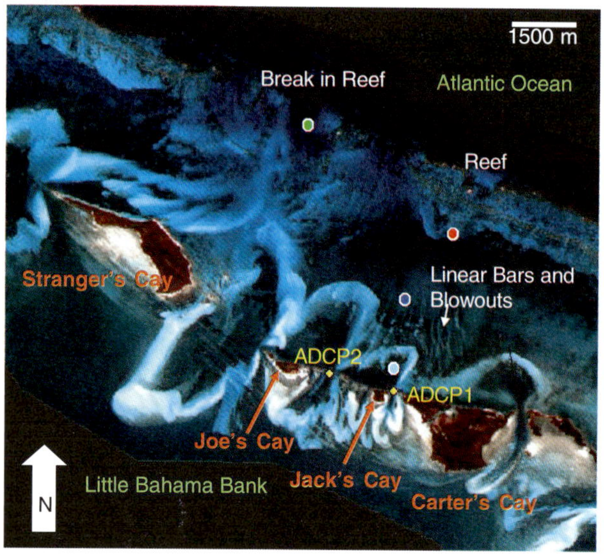

Fig. 1. QuickBird image of the study region in the Northern Abacos, Bahamas. QuickBird satellite image courtesy: DigitalGlobe. The yellow dots indicate the locations of the bottom mounted ADCPs ('ADCP1' in August and 'ADCP2' in September). The green, red, dark blue and light blue dots indicate the locations of the Dobie wave gauges. There was one at the dark blue dot during the entire sampling period for calibration purposes, and the other rotated through the other three locations to indicate differences in the wave field during this time period.

BACKGROUND

Study area

The study focuses on observations from the northwestern portion of the Abaco Island chain, on the northern edge of Little Bahama Bank. Little Bahama Bank is an isolated carbonate platform bound by the Atlantic Ocean to the north and east, the Straits of Florida to the west, and Northwest Providence Channel to the south. A discontinuous barrier reef is located at the northern shelf break (Fig. 1). The western portion of the northern boundary of the study area contains a 5-km-wide breach in the reef complex, through which many open-ocean waves can pass without being affected by the bottom. This area is deeper (~10+ m deep) than the adjacent parts of the shelf located behind the reef (~6–7 m deep), and the sandy bottom includes many areas with symmetrical skeletal sand ripples (~0.15 m high, spaced about 0.30 m apart), although large (up to 5 m relief) patch reefs are common. East of this break, the barrier reef is continuous for 10.5 km, and is flanked by a reef apron that extends ~1 km onto the platform. The apron is generally ~4–5 m deep, and it consists mainly of a skeletal sandy bottom. Between the reef apron and the island system (a distance of ~4.5–5 km), the sea-bottom consists mostly of peloid-skeletal packstone to grainstone largely stabilized by seagrass, and abundant patch reefs. (In this paper, although these sediments are not rocks, for ease, the Dunham classification based on grain support and mud content is used to describe the modern depositional texture.)

Rocky islands formed by Pleistocene aeolianite are separated by inlets that focus the tidal flow. Flood and ebb tidal deltas in the area consist of generally lobate forms, convex away from the islands with endpoints near the inlets (Fig. 1). The ebb deltas extend up to 2 km from the islands towards the reef, but the flood deltas mostly remain within 1.3 km of the islands. Although the inner portions of these lobes are mainly seagrass-stabilized fine-sand-sized peloid–skeletal packstone with a few small local patch reefs, the main channels are generally hard bottomed (assumed to be Pleistocene bedrock) with growth of hard and soft corals and sponges. Outwards from the inlet centre, this hard bottom passes into coarse lag and eventually into a seagrass-stabilized area before reaching the active oolitic shoals, which form the crest of

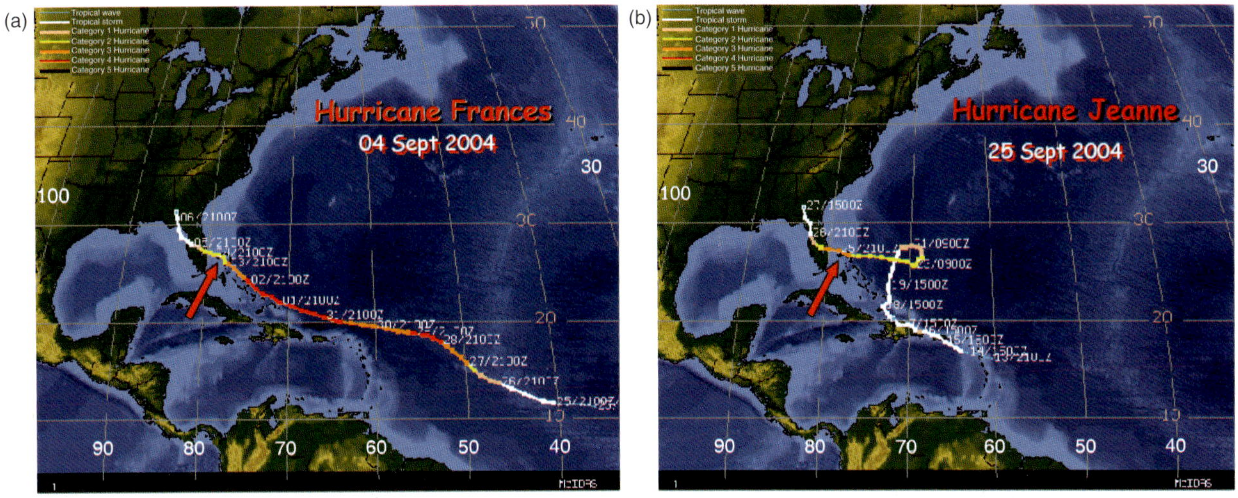

Fig. 2. Hurricane tracks and intensity plots for Hurricanes Frances (a) and Jeanne (b) from September 2004. The red arrow indicates the location of the study area in the northern Abacos, Bahamas. Figures are modified from the National Hurricane Center/Tropical Prediction Center's website (www.nhc.noaa.gov). The colours of the storm tracks indicate the intensity of the storm: tropical storm (white), category 1 hurricane (pink), category 2 hurricane (yellow), category 3 hurricane (orange), category 4 hurricane (red) and category 5 hurricane (black).

the deltas. Many flood deltas include rock ridges parallel to the island trend, and not all facies are represented in every flood lobe due to the different sizes and shapes. South of the Abaco island chain, the surface of Little Bahama Bank consists largely of peloid–skeletal wackestone to packstone and is stabilized by seagrass.

Oceanographic and meteorological setting

The physical setting of the study area dictates the daily oceanographic forces to which it is exposed. The tidal deltas are exposed to the open Atlantic Ocean to the north and are therefore prone to larger swells from this direction than from the shallower bank to the south, although the continuous sections of the barrier reef reduce the impact of these larger swells. Similarly, the islands shield the bank from incoming open-ocean swells, creating a mixed tidally dominated system that is predominately semidiurnal. The tidal range in the region averages around 1 m (cf. Reeder, 2007).

The area is influenced by several winter cold fronts annually. Such events have been documented as a significant cause of erosion in areas such as the northwest coast of Andros Island, Bahamas (Shinn *et al.*, 1969; Rankey & Morgan, 2002). These fronts are accompanied by stronger than average winds from the north and northeast that differ both in direction and in magnitude from the otherwise dominating easterly trade winds.

The Abacos are prone to tropical cyclones from June through to November. In 2004 alone, two major tropical cyclones passed directly over the study area. On 4 September 2004, Hurricane Frances (Fig. 2a) passed through this region as a category 3 storm on the Saffir–Simpson scale (sustained winds of 50–58 $m\,s^{-1}$), decreasing in intensity to a category 2 (sustained winds of 43–49 $m\,s^{-1}$) upon departure (Beven, 2004). Less than a month later, on 25 September, Hurricane Jeanne (Fig. 2b) passed through the same area as a strong category 3 storm. These storms did considerable damage to the Bahamian communities in the area.

Observations

An ultra-high-resolution (2.4 m^2 pixel size) QuickBird remote sensing image acquired on 27 November 2003, prior to the passage of these two storms, provides information on pre-storm morphology. For comparison, a second QuickBird image was obtained on 16 March 2005, after the passage of the storms. These data, calibrated by field observations, provide an important foundation for this study; however, differences in boundary conditions for each image acquisition (different atmospheric conditions, time frames, wave climate, tide level, etc.) limited quantitative comparisons between images. Bathymetry derived from satellite data (Stumpf *et al.*, 2003) is not accurate enough to measure centimetre-scale changes. Therefore, it cannot be determined if sediment

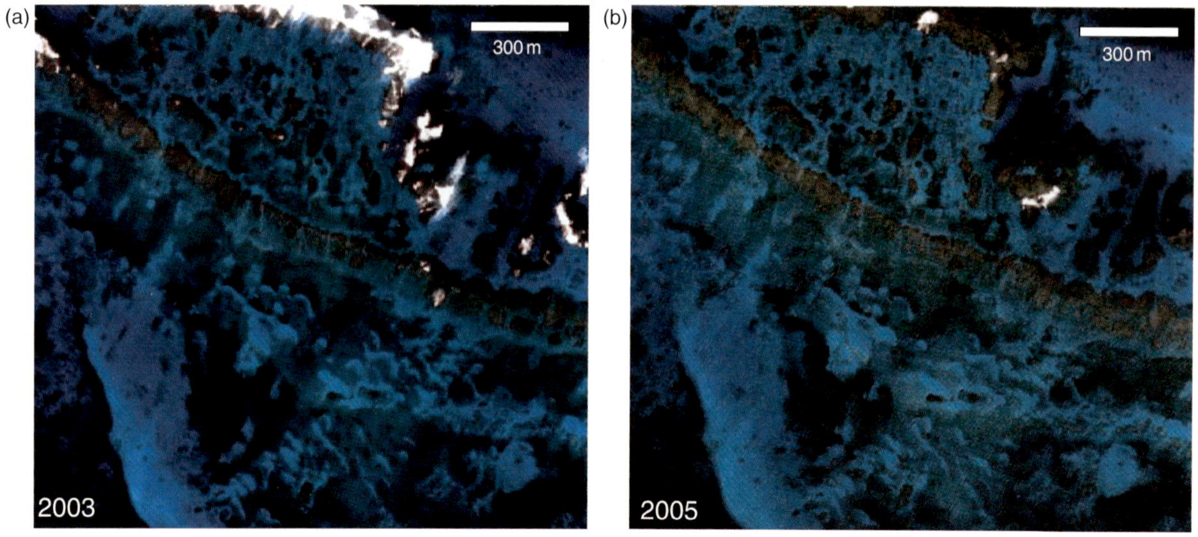

Fig. 3. High-resolution remote sensing image of pre- (a) and post- (b) storm morphology of a portion of the barrier reef and its reef apron. This figure demonstrates the absence of significant geomorphological change (e.g. no expanded reef apron) during a period that included the passage of hurricanes Frances and Jeanne in 2004. Although there are some expanded sand patches around some patch reefs, no major geomorphological changes occur. (The whitish colour areas on the pre-storm image are waves breaking on the reef crest, and the darker patches in the apron on the pre-storm image are most probably due to seasonal differences in seagrass colonization rather than depth differences.) Image courtesy of DigitalGlobe.

was either eroded from or deposited to the existing system through the available datasets.

What changes resulted from the passage of the two hurricanes?

Comparisons between these two datasets demonstrate very little geomorphological change in the study area. In many areas, the two images appear more-or-less identical (Fig. 3), illustrating little to no change. For example, the reef aprons show minimal morphological change after the passage of the two storms, and the majority of the shelf behind the reef does not illustrate any noticeable change in the remote sensing images. There are, however, some minor modifications to the platform after the storm passage. In the shelf region behind the reef, just seaward of the deltas, two of the linear bars and blowouts (Fig. 1) experienced some seagrass removal and slight (<10 m) eastward expansion of the blowouts.

The most pronounced changes in the study area occur in the crests of the tidal deltas (Fig. 4) where some sand wave crests migrated ~30–35 m, but none of the crests moved more than 50 m. On the ebb lobes, some shoal crests migrated to the southeast. The largest change in the ebb lobes occurred in the southwestern portion of the area illustrated in Fig. 4. In the 2003 image, there is a small (~35 m) seagrass-stabilized area on the shoal (black dashed box in Fig. 4c). In the 2005 image, this small region has been 'filled in' with sand. The crest of the northern lobe in the area of the black dashed box of Fig. 4c migrated ~30 m westward, whereas the adjacent crest southwest of this one did not move to the same extent. The linear bar (Fig. 1) perpendicular to this region also broadened.

The flood deltas showed the most obvious changes over this 2-year period. The sandy extent of some deltas expanded ~30–60 m out onto the platform and ~10–20 m laterally (relative to the central channel). Many of the sand wave crests of the flood deltas followed similar trends to those on the ebb deltas and migrated to the west-southwest. Some of the crests, however, move islandward (solid black box in Fig. 4c), and others move to the east.

These observations indicate that the storms did modify the carbonate platform, at least to some extent. The crests of some sand waves migrated 30+ m. Such modifications to the shoals may be due to the passage of the two storms. With the direction of approach of both Hurricane Frances and Hurricane Jeanne, the strongest winds would have come from the northeast, which may account for the southwestern migration patterns. However, storms should not be considered the only possible explanation for these morphological adjustments.

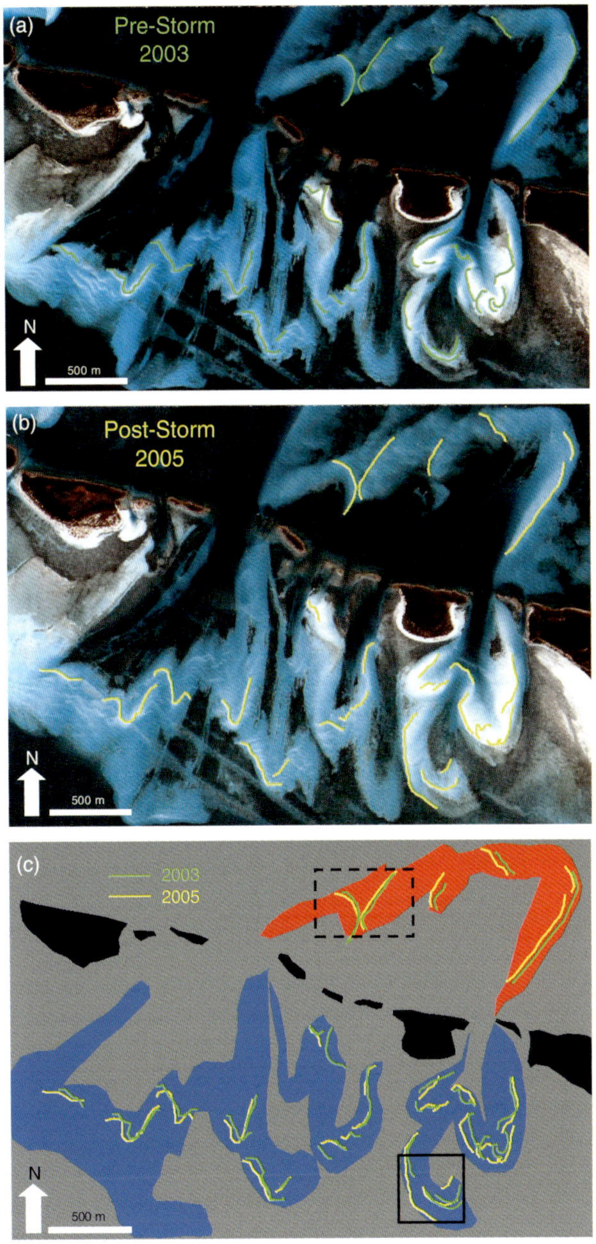

Fig. 4. Extent of the geomorphological changes that occurred over the 1.5-year period between remote sensing image acquisitions, during which the two storms (Hurricane Frances and Hurricane Jeanne) passed. The majority of the study area experienced few to no modifications, but the crests of some tidal deltas did move. This figure highlights key pre-storm crest locations (a, and green lines in c) and post-storm crest locations (b and yellow lines in c), selected to illustrate morphological adjustments between the remote sensing images. The modifications are compared in c. Most of the crests on both flood and ebb deltas shifted to the southwest, but this is not the case for all the crests highlighted. See text for more details. Images copyright DigitalGlobe.

Sand waves can migrate ~4 m in a month under fair-weather conditions in a tidal channel (Gonzalez & Eberli, 1997). The observed sand wave migration rate of $20\,\mathrm{m\,yr^{-1}}$ at the tidal deltas is lower than this documented rate, but these rates are highly dependent on the tidal velocity. On the tidal deltas, the current velocities range between 0.24 and $0.30\,\mathrm{m\,s^{-1}}$, so the rates of sand wave migration may be much smaller here.

The lobes themselves remained generally stationary, and modifications occurred mainly in sand wave crests. Overall, on the tidal deltas, there was no major reorganization of the shoal geometries (e.g. no new 'spillover lobes'). Only minor changes are apparent, such as removal of small patches of seagrass (Fig. 5, red boxes), smearing of sand waves on shoal crests, and slight expansion of the shoal by deposition of a thin layer of sand ~60 m outward in some areas (Fig. 5, orange and green boxes). In the context of the entire carbonate platform, these are small perturbations, and, although it is possible that such small perturbations may be further amplified by daily processes, this does not appear to occur here. Observations (in summer 2005) of sand waves reforming and seagrass re-establishing suggest that the system was returning to its pre-storm state.

Field observations led to further questions about the influence of storms on the platform. No major discernable sediment erosion or deposition was observed in the field.

Many of the areas of 'shoal widening' on the remote sensing images proved to be a thin layer of sediment that was transported off the shoal crest onto adjacent grassy areas rather than a large addition to the lobe, and these areas were beginning to stabilize by seagrass in summer 2005. These observations indicate that the daily tides, winds and waves of the region are restoring the system from the minor perturbations brought about by storms. Although storms may modify the system in the short term, it appears unlikely that such perturbations will be preserved.

MODELLING STORM WAVES: WHAT MIGHT THEY DO?

The field and remote sensing observations suggest relatively little impact of the storms on this carbonate system. In the context of the often assumed geological importance of storms, this

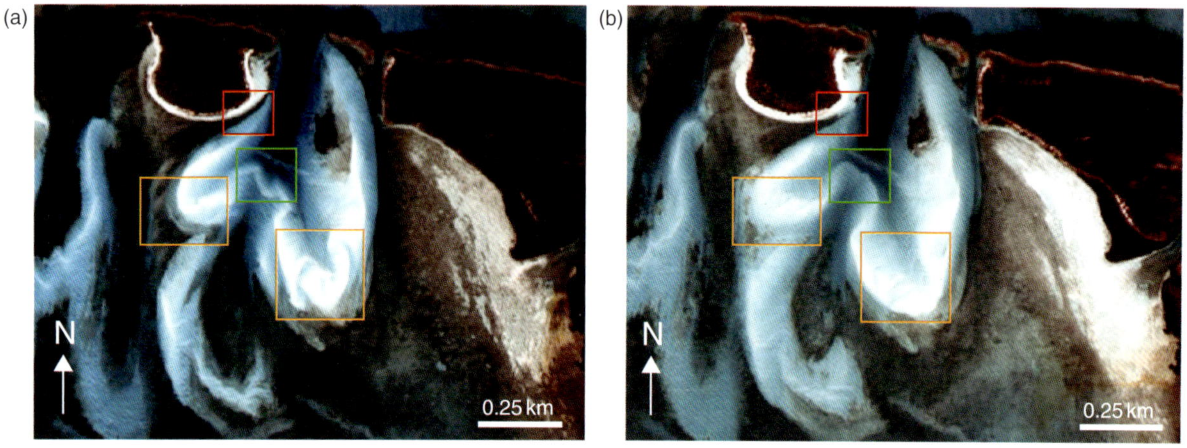

Fig. 5. QuickBird high-resolution remote sensing images of a flood lobe of a tidal delta near Carter's Cay, Bahamas, acquired in November 2003 (a) and in March 2005 (b). This time interval surrounds the active 2004 hurricane season that affected the regions, and shows one of the regions more affected by the storms, with erosion of seagrass (red box), migration of bar crests (green box) and sedimentation (orange boxes). Satellite image courtesy: DigitalGlobe.

result is not trivial, given the apparent strength of the storms, and the fact that two passed over the area in a short time period. To further investigate the potential influence of storms, a simple wave model was used to explore the forces that might be exerted by storm waves on a carbonate platform.

The model consists of two main parts: a wave prediction module and a wave transformation module. The wave prediction module calculates the deep water significant wave height (the average wave height of the highest 1/3 of the measured waves in a location) and peak wave period for a given wind speed. As the waves approach shallow water, the wave transformation module of the model calculates the adjusted significant wave heights and the bottom orbital velocities produced by these waves. Breaker criteria are used to calculate the maximum height of a wave before it will break and are used to determine the depth at which the wave will break. Once the wave breaks, the heights and orbital velocities are set to zero, terminating the propagation of the wave as well as any sediment transport that may have been initiated.

The wave model was run for two different bathymetric configurations (Fig. 6). The simplest bathymetry was a ramp sloping at an angle of 0.3° from horizontal, spanning 10 km (Fig. 6a). The second bathymetric configuration that was studied simulated a simplified flat-topped platform. In this case, a 5-m-deep shelf dropped steeply (14°) to deep water at the shelf margin, mimicking some Bahamian platforms (Fig. 6b).

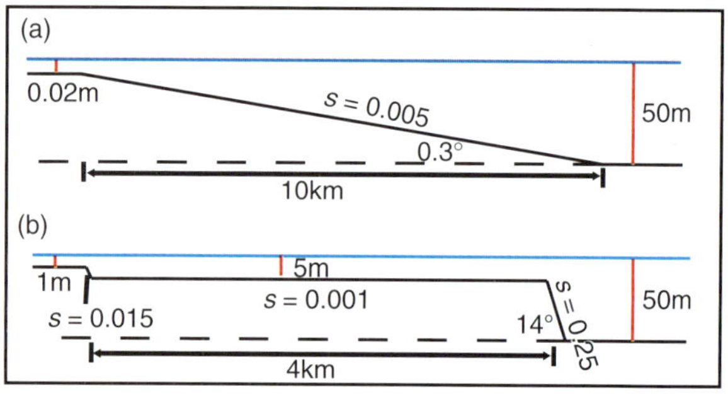

Fig. 6. Bathymetric configurations for the ramp (a) and platform (b) situations modelled with the wave model. The platform bathymetry (b) consists of three different slopes used to determine the breaking criteria for the waves.

Deep water wave model

Initially, the deep-water wave heights (H_0) are calculated for a given wind speed measured 10 m above sea level (u_{10}) using the deep water fetch-limited wave equations (Komar, 1976)

$$\frac{gH_0}{u_*^2} = 1.6 \times 10^{-3} \left(\frac{gF}{u_*^2}\right)^{1/2} \quad (1)$$

where g is the acceleration due to gravity (~9.81 m s^{-2}), F is the fetch, and u_* is the wind stress factor. Given a known u_{10}, the wind stress factor is

$$u_* = 0.71 u_{10}^{1.23}$$

Wind waves are initially dependent on the time the wind has been blowing (duration) as well as the distance over which it has blown (fetch). After a certain period of time, the waves are no longer time dependent and become fetch-limited. To ensure that the waves are past the limit for a duration-limited sea state and that the fetch-limited equation given above is adequate for calculating the wave heights, the model calculates this time limit (t_d) using

$$\frac{t_d g}{u_*} = 68.8\left(\frac{gF}{u_*}\right)^{2/3} \tag{2}$$

The model then assumes the wind has been blowing steady longer than this limit to the duration-limited sea state. The model then generates waves from a fetch of 16 km (based on an assumption of a 10-mile scale for daily land or sea breezes). Model runs with other fetch values (25 and 50 km) yielded similar results to the 16 km fetch, except the larger waves produced by a longer fetch broke in deeper water such as at the platform margin, limiting the waves that pass over a platform. The calculations with the 16 km fetch produced waves that propagated over the platform rather than breaking at the platform edge, for most wind speeds.

The model calculates the peak wave period for the given wind conditions using

$$\frac{gT_p}{u_*} = 0.286\left(\frac{gF}{u_*^2}\right)^{1/3} \tag{3}$$

The effects of four different wind conditions representing the daily trade winds, winter North Easters, tropical storm winds, and hurricane (Category 1) strength winds are explored with the model. Table 1 gives the results of the deep water wave prediction module.

Wave transformation and breaking

Waves generated in deeper water can propagate towards the coast (carbonate ramp) or across a platform. As waves enter relatively shallow water, the waves shoal, indicating that the wave height increases, the wave length (L) decreases, and steepness (ratio of wave height to wave length) increases. The wave height (H) at a given depth (h) is calculated using the equations for wave shoaling:

$$H = H_0 k_s \tag{4}$$

$$k_s = \frac{H}{H_0} = \left(\frac{1}{2n}\frac{c_o}{c}\right)^{1/2} \tag{5}$$

where k_s is the shoaling coefficient (relates the deep water wave height to the wave height at a given water depth, h), c is the wave phase velocity (the speed of individual waves), c_o is the deep-water wave phase velocity

$$c_o = \frac{g}{2\pi f} \tag{6}$$

$$f = \frac{1}{T_p} \tag{7}$$

and n is defined as

$$n = \frac{1}{2}\left[1 + \frac{2kh}{\sinh(2kh)}\right] \tag{8}$$

where h is the water depth and k is the wave number. For simplicity, the model assumes shore-perpendicular waves and shore-parallel depth contours; therefore, wave refraction can be neglected. Storm surges were incorporated by adding the average surge estimated by the Tropical Prediction Center/National Hurricane Center for each type of storm (Table 2) to the bathymetrical profile (in the above equations, substitute (h + surge) for h), but the current velocities produced by a storm surge were not modelled. These currents, however, may be important to the overall impact of the storm and cause downslope

Table 1. Results of the deep water wave module computations. This table gives the values of the wind speed at 10 m above sea level (u_{10}), the wind stress factor (u_*), the deep water significant wave height (H_0), the peak period of the wave spectrum (T_p), and the time limit to the duration-limited sea state (t_d).

u_{10} (m s^{-1})	u_* (m s^{-1})	H_0 (m)	T_p (s)	t_d (h)
6	6.43	0.42	2.93	10.5
12	15.09	0.98	3.89	14.0
18	24.89	1.61	4.59	16.6
34	54.32	3.51	5.96	21.5

Table 2. Wind speed (u_{10}) and surge values used in the numerical wave model for the different types of winds: daily winds, storm winds (winter storm), and tropical cyclones of both Tropical Storm strength and Category 1 Hurricane strength (on the Saffir–Simpson Scale).

Category	Daily winds	Storm winds	Tropical storm	Category 1
Wind speed (m s^{-1})	6	12	18	34
Surge (m)	0.0	0.5	1.0	1.6

Table 3. Model results for the case of a gently sloping carbonate ramp: H_b is the wave height when the wave breaks, h_b is the water depth at which the wave breaks, h_t is the depth where sediment begins to move ($u > u_t$), and u_{max} is the maximum orbital velocity obtained on the ramp prior to the wave breaking. See Table 1 for the other parameter definitions.

u_{10} (m s^{-1})	H_0 (m)	T_p (s)	H_b (m)	h_b (m)	h_t (m)	u_{max} (m s^{-1})
6	0.42	2.93	0.38	1.00	1.75	0.54
12	0.98	3.89	0.90	1.99	4.12	0.78
18	1.61	4.59	1.45	3.28	6.55	0.86
34	3.51	5.96	3.11	7.30	12.37	0.80

transport during surge retraction. Since the point of the model is to explore the effects of storm waves on a sedimentary system, the model calculates the bottom orbital velocities for the shoaled waves through linear wave theory. For each point along the bathymetric profile, the bottom orbital velocity (u) is calculated by

$$u = \frac{\pi H}{T_p \sinh(2\pi h/L)} \quad (9)$$

To determine the effects of the waves on the sediments, the bottom orbital velocity was compared to the threshold velocity (u_t) for entrainment of sediment of which 50% of the grains have a diameter smaller than d_{50}.

$$u_t = \sqrt{8g\left(\frac{\rho_s}{\rho_w} - 1\right)d_{50}}, \quad (10)$$

where ρ_s and ρ_w are the densities of the sediment and water, respectively. In the model, the threshold velocity for sediment entrainment of sediments of 0.5 mm in diameter (approximately the average size of the Abaco tidal delta sands) with an average density of 2.75 g cm^{-3} (Incze, 1998) is 0.26 m s^{-1}. This velocity, however, is calculated for bare sand. If the bottom contains seagrass or is partially cohesive due to algae or cementation, the threshold velocity could be much higher. The depths at which sediment transport begins are therefore most probably smaller than those calculated with u_t = 0.26 m s^{-1}.

As the wave steepness increases through shoaling, it can reach a point where the wave breaks. To determine the wave height at the point where it breaks as well as the depth of the water when the wave breaks, the model includes a breaker criterion (Kaminsky & Kraus, 1993). Once a wave exceeds the breaking condition

$$H_b = \gamma_b h_b \quad (11)$$

$$\gamma_b = 1.20\xi^{0.27} \quad (12)$$

$$\xi = \frac{S}{\sqrt{H_0/L_0}} \quad (13)$$

(H_b is the critical wave height at which point the wave breaks,

$$L_0 = \frac{gT_p^2}{2\pi} \quad (14)$$

is the deep water wave length and S is the bottom slope), the wave breaks, the wave height is set to 0.00 m, the bottom orbital velocity is set to 0.0 m s^{-1}, and the waves do not regenerate in this model.

Application to a carbonate ramp

To investigate the physical potential of laterally extensive storm deposits that have been interpreted in the stratigraphic record, the wave model was first run for a gently sloping ramp (Fig. 6a). Table 3 and Fig. 7 provide the results of these calculations.

For the tropical storms, the waves can break in 3.90 m of water, while the category 1 hurricane can produce breaking waves in 8.21 m of water depth which move the sediments beginning in depths of 12.37 m prior to breaking. For these stronger storms in the real world, the waves lose energy to white capping and non-linear interactions, so the cases illustrated in Fig. 7 represent waves larger than would actually occur. The depths of influence, therefore, are also larger than should be expected. Overall, with the modelled waves from the peak of the wave frequency spectrum, the maximum depth of impact on a ramp occurs during a Category 1 Hurricane. In these conditions, the waves can begin to move bare sediments in 12.37 m of water (these depths include the surge values, so this depth corresponds to a depth of 10.87 m below mean sea level). These results are consistent with the results of detailed

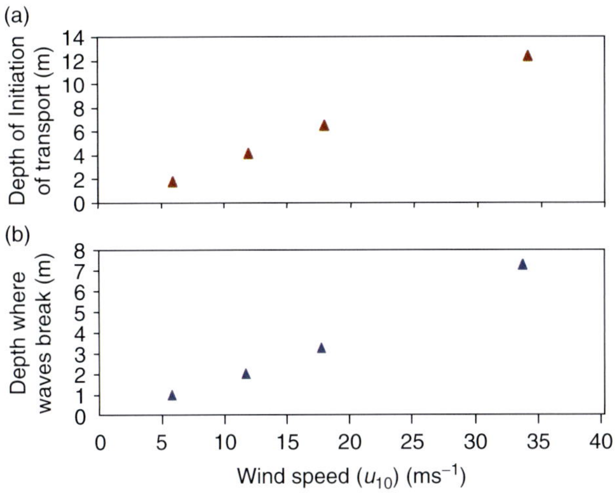

Fig. 7. Wave model results for the simple ramp bathymetry at the peak frequency of the wave spectrum for four different wind speeds: daily $6.0\,m\,s^{-1}$ winds, the average winter cold front winds ($12\,m\,s^{-1}$), tropical storm winds ($18\,m\,s^{-1}$), and Category 1 Hurricane strength winds (based on the Saffir–Simpson scale). (a) This plot illustrates the depth at which the wave's bottom orbital velocity exceeds the threshold velocity of sediment entrainment. At depths shallower than these plotted values, bare sediment is being transported. Once a wave breaks, however, the energy disperses and the sediment transport is terminated (for that wave). The depths at which the waves break are shown in (b). If the wave begins to transport sediment at 6.5 m water depth, but breaks in 4 m, it is only responsible for sediment transport along a short distance of the ramp.

studies of wave impact on exposed coasts, which indicate that significant sediment transport due to waves will not occur below 20 m of water depth (cf. Dietz & Fairbridge, 1968; Duane, 1976; Gordon & Roy, 1977). A similar study on a coast with low wave energy (Sapelo Island, Georgia; Coastal Engineering Research Center, 1984) further limited this depth to less than 10 m on coasts exposed to lower wave energy (~0.25 m annually averaged significant wave height), and the model results are also fairly consistent with these results.

These data illustrate the difficulty of storm-generated waves impacting broad expanses of ramps, given the model assumptions. It appears that in shallower areas (e.g. <5 m), such wind-generated waves could, however, have an impact. At these depths, bottom effects cause the waves to disperse their energy through sediment transport and eventual breaking.

Application to a carbonate platform

To investigate the effects of storm waves on a platform, the model calculated the wave heights, bottom orbital velocities and breaking limits for a simple, unrimmed carbonate platform (Fig. 6b). The platform configuration does not include shoals such as the tidal deltas of the Northern Abacos, but understanding the potential influence of storm waves on a flat platform is useful in understanding the forces that could possibly have an impact on such a shoal system. The results of the wave model as applied to a simplified platform are summarized in Table 4.

Table 4. Results of the wave model for the simplified carbonate platform bathymetry. Refer to Table 3 for parameter definitions. If h_b > platform depth (5 m) + surge (refer to Table 2 for the surge value), the wave breaks before reaching the platform. If h_t < platform depth (5 m) + surge, the sediment on the platform top is not transported. The values for the platform differ from the results for the ramp (Table 3) due to the different bathymetries. Similarly, the different values in u_{max} within this table are due to the different bathymetrical gradients influencing the waves. Recall that u_{max} is the maximum orbital velocity obtained prior to breaking, not the orbital velocity produced on the flat platform.

u_{10} ($m\,s^{-1}$)	H_b (m)	h_b (m)	h_t (m)	u_{max} ($m\,s^{-1}$)
6	0.38	1.01	1.75	0.53
12	0.93	1.57	4.11	1.03
18	1.49	2.58	6.50	1.17
34	3.12	6.60	12.35	0.99

The case of the platform margin (Fig. 8) provides insights on how a more complex bathymetry might be impacted by storm waves. Although sediment transport on the platform top can be initiated from waves produced by $18\,m\,s^{-1}$ storm winds, the bottom orbital velocities remain below $1.2\,m\,s^{-1}$ for even the strongest of storms on the platform. From the results presented here (Fig. 8; Table 4), the most influential storm would be similar to the winter storm ($12\,m\,s^{-1}$ winds). The waves produced from winds around $12\,m\,s^{-1}$ (under the assumed duration and fetch) do not produce bottom orbital velocities large enough to initiate sediment transport on the platform. Although the maximum orbital velocities are larger than the threshold velocity, the depth of initiation of motion is shallower than the platform depth, so the sediment does not move along the platform (the $12\,m\,s^{-1}$ winds initiate sediment motion at 4.1 m depth, but the platform depth is 5.5 m with the surge set-up for this storm). Instead, motion is initialized at the shallower-end of the bathymetrical profile, which could indicate the location of a carbonate shoal body (Fig. 6b). For winds stronger than $18\,m\,s^{-1}$, the waves break

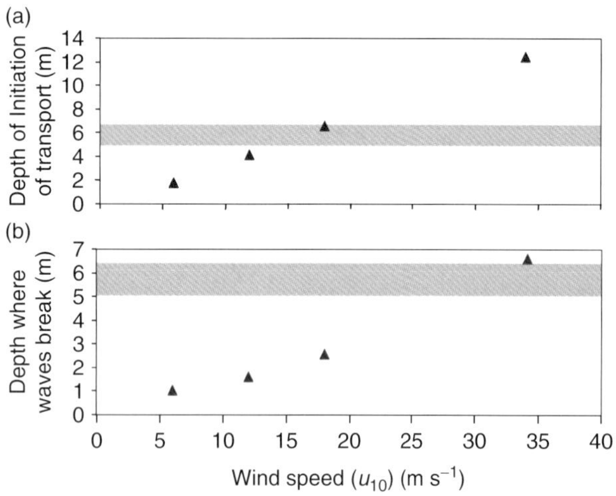

Fig. 8. The results of the wave model for five frequencies under four different wind strengths ($6\,m\,s^{-1}$, $12\,m\,s^{-1}$, $18\,m\,s^{-1}$ (Tropical Storm), $34\,m\,s^{-1}$ (Category 1 Hurricane)) over a shallow, unrimmed platform. Since the platform is flat and bottom friction is not accounted for, if the wave does not break on the reef, it will persist throughout the platform. This figure shows the depths at which each wave produced by the different wind conditions begins to move sediment at the bottom (a) and the depths at which the waves break and would therefore stop influencing the bottom. The grey box in both figures indicates the depth of the platform (which varies according to the storm surge). If the wave breaks above this grey box (b), the wave does not influence the platform. Similarly, if the depth of transport initiation is below this box (a), no sediment is transported on the platform. For model runs of storms of the order of Category 1 and greater, the waves break at the platform edge, so there is effectively less influence on the platform.

at the edge of the platform, so the platform sediments will not be affected by these waves. There remains a small window of storm intensities that can influence the platform interior and, hence, the carbonate shoal bodies.

Model assumptions and limitations

This simple model fails to show the post-breakage effects. As waves enter the platform over the shelf margin, they are modified by bottom effects. The larger swells break at the platform edge upon hitting shallow water, since the maximum wave that can occur is a fraction of the water depth. The waves that break can reform, but due to the short fetch and shallow depths of the region between the reefs and the islands in the study area, these regenerated waves cannot grow to the size and force of the original swells that broke at the platform edge. These dynamics introduce an interesting paradox. Although larger storms produce bigger waves offshore, a category 4 or 5 storm may lead to less wave-influenced transport on the platform than a less intense storm such as a winter cold front. The reefs, however, will probably feel more of an impact from these large storms. The relationship between storm strength and storm impact to the carbonate system, can therefore vary spatially across a platform.

Another limitation to this wave model is that it only considers the peak wave period of the wind-wave spectrum. Although this wave may break, there are some wave height and period combinations that would penetrate into shallower waters, either on the ramp or along the platform. Although the model may indicate that the waves have broken on the reef, it is possible that some waves produced by a particular wind speed may yield sediment transport beyond the point of breaking for the peak period waves.

Furthermore, wave generation by wind on the platform was not explored in this simple wave model. As a strong wind traverses a carbonate platform, it will produce fetch and duration-limited wind-generated waves. These waves will not have to pass over a steep slope, so they may be able to produce more sediment transport on the platform than the waves coming from the deep ocean.

Finally, the estimated values of wave heights, bottom orbital velocities and depths of initiation of motion or breaking from the model represent the maximum values for several reasons. In the case of a reef-rimmed platform such as the Abaco tidal delta region, the majority of the waves would break over the reef, so the conditions depicted in the flat-topped platform model correspond to the highest waves and orbital velocities on the reef-rimmed platform.

Storm forces versus day-to-day processes

Storms have an effect on sediment distribution. The modelling illustrates how tropical cyclones can yield large enough bottom orbital velocities to move sediment, and observations after hurricanes Frances and Jeanne illustrated limited redistribution of sediment, with the most pronounced changes occurring in the flood tidal delta lobes. However, the question remains as to how much effect the storms have on shaping geomorphological bodies and, eventually, the sedimentary record. Although they may have a slight impact on geomorphology in the short term, are they enough to counteract the daily processes?

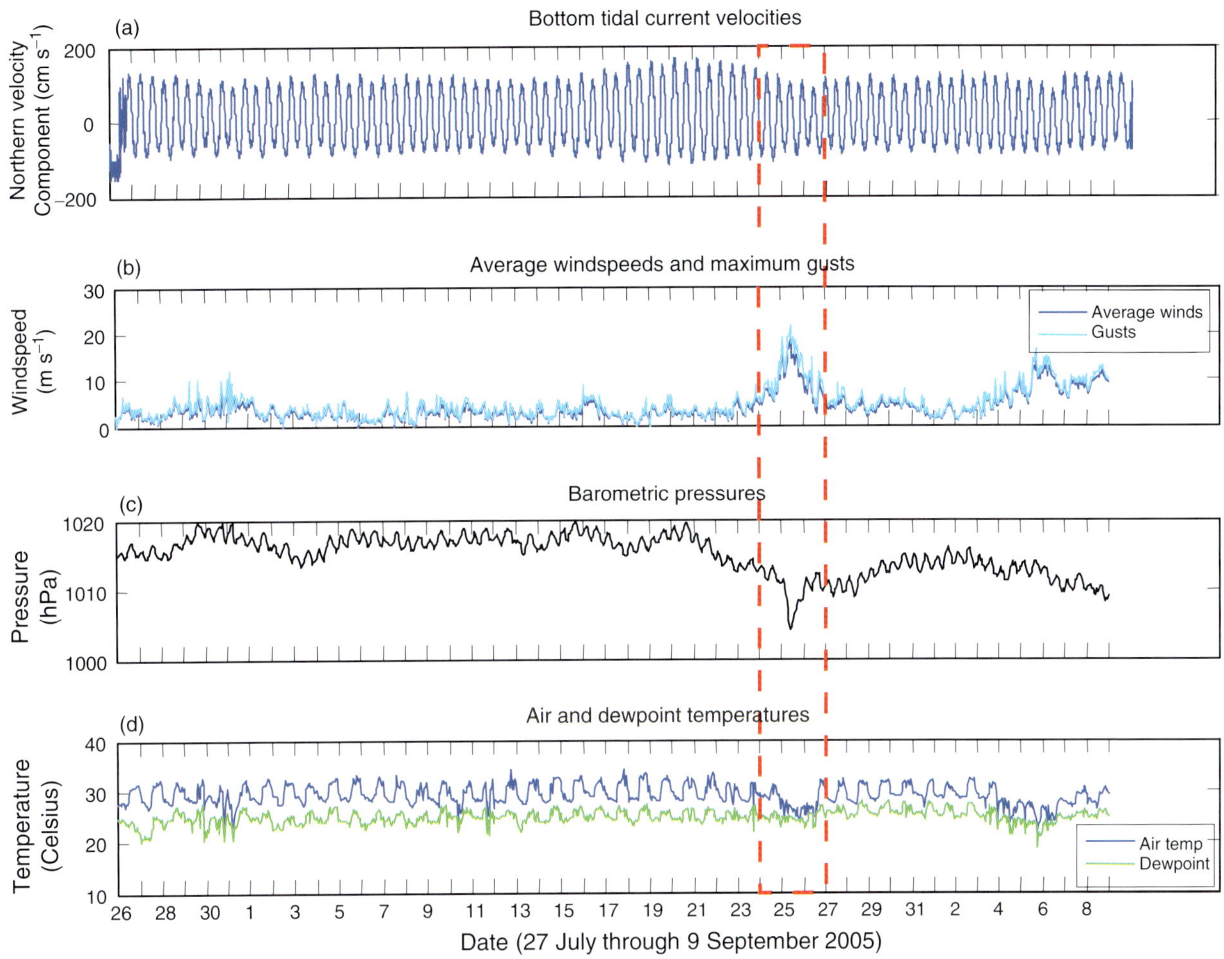

Fig. 9. Comparison of the northern component of the tidal forcing at the bottom of a tidal delta (top) with atmospheric effects (NOAA buoy, bottom three plots) in late summer, 2005. The measurements were obtained by an ADCP located at ADCP1 in Fig. 1. Although the presence of the storm (Hurricane Katrina, red dashed box) is obvious in the atmospheric data, it does not appear as obvious in the tidal measurements. Note that the average tidal velocities exceed $1\,m\,s^{-1}$, daily.

To quantify the daily forces that act on the platform in the study area, waves and currents were measured during the summer and autumn of 2005. Two bottom-mounted, pressure-sensitive wave gauges (see Fig. 1 for location) measured the sea state for August and September. One wave gauge was kept stationary throughout the sampling time at a location between an ebb lobe and the barrier reef (dark blue dot in Fig. 1) while the second gauge spent 2 weeks inside the ebb lobe (light blue dot in Fig. 1) and was then transferred to the deeper water near the break in the barrier reef (green dot in Fig. 1). Simultaneously, a bottom-mounted Acoustic Doppler Current Profiler (ADCP) was placed for one month at a time in two different tidal inlets (Fig. 1, located at ADCP1 in August and ADCP2 in September), and measured the three-dimensional velocity profile in 4 minute intervals, averaging over 3 minutes. These data were compared with atmospheric conditions monitored at the nearest National Oceanographic and Atmospheric Administration (NOAA) buoy, located at Settlement Point on Grand Bahama Island, ~105 km to the southeast (Fig. 9). Although these data may not represent the exact atmospheric conditions of the study area, the winds measured at the buoy and the waves measured at the study site appear to correspond well.

The storm of 24–27 August (Fig. 9, red dashed box) is Hurricane Katrina, which passed ~160 km south of the study area. In the Abacos, winds were from the north as the storm first approached, then from the east as it passed south of the area. The waves produced by this storm persisted for two days, but significant wave heights remained smaller than 1 m (Fig. 10). The bottom currents

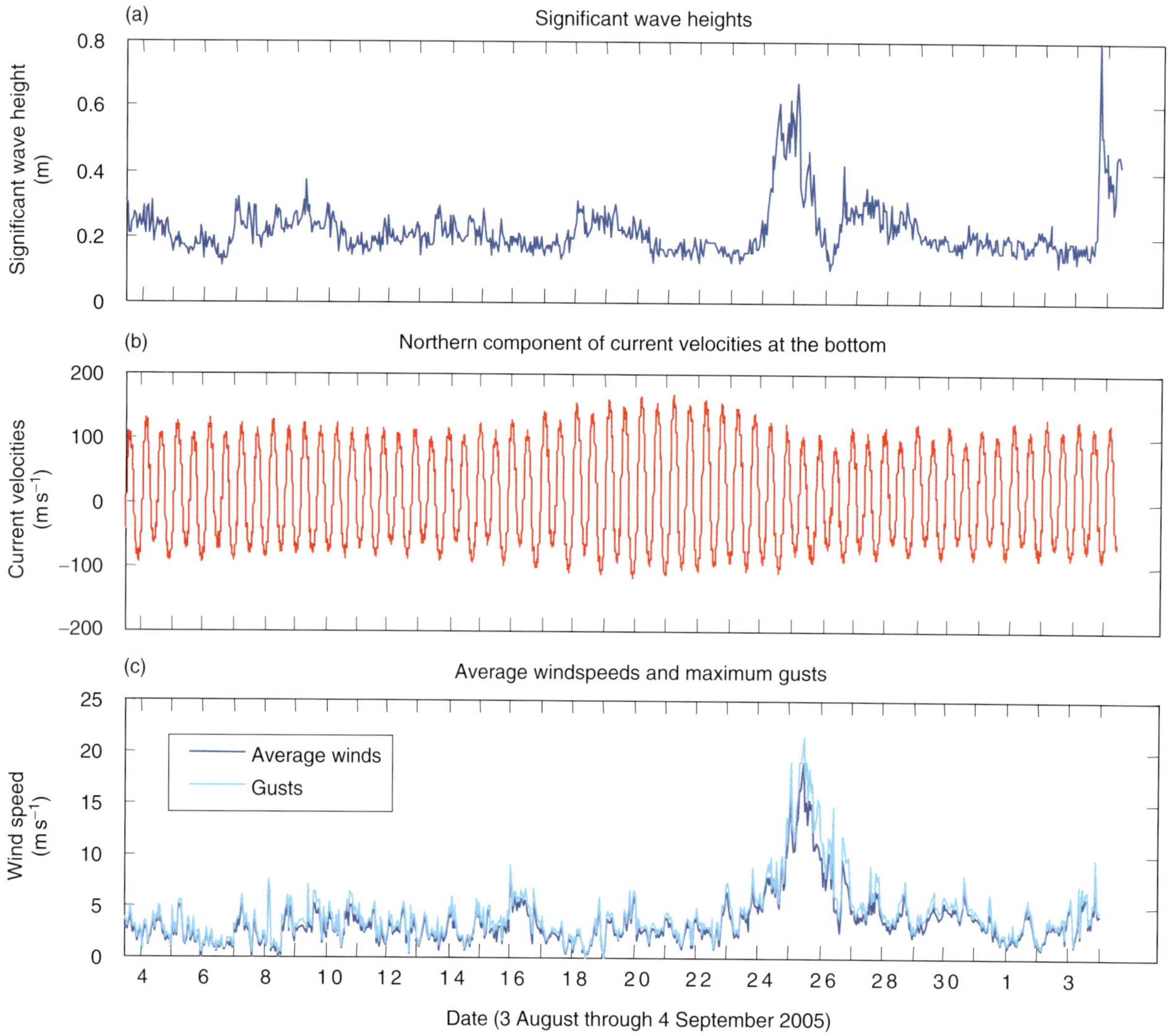

Fig. 10. Wave gauge measurements (top) compared with the tidal velocities (middle) and atmospheric data from the NOAA buoy (bottom). The wave gauge was located between the reef and the apex of an ebb tidal delta (dark blue in Fig. 1). The ADCP was located in the inlet between Jack's Cay and Carter's Cay (ADCP1 in Fig. 1). The wave height increases in response to Hurricane Katrina (red dashed box, Fig. 9) prior to the decrease in wind speeds. The waves in this region, however, generally remain between 0.2 and 0.4 m high.

were only slightly modified from the rest of the tidal cycle's magnitudes. From the simplified storm wave calculations described previously, this corresponds to conditions analogous to the 12 $m s^{-1}$ winds, which may produce waves that could move sediment in waters shallower than 4.8 m. Therefore, the combination of the currents and waves produced by this storm could have impacted the tidal deltas, albeit for a short time period, and changed them somewhat.

In comparison, the daily bottom tidal currents of the tidal deltas approach and can exceed 1 $m s^{-1}$ in some areas (Fig. 10), much greater than the threshold velocity of sediment entrainment of the average diameter (0.5 mm) for ooids found in the tidal delta region (0.26 $m s^{-1}$). Although the velocities produced during a storm may be greater than these tidal velocities, they are short-lived. In contrast, tides move in and out, day after day, reapplying the same forces to the system. Therefore, although sediment may be transported and redeposited during storms and periods of higher-than-average velocities, the duration of storms may not be enough for significant morphological adjustments. Likewise, based on our continuing observations (described above) it is possible that the daily forces may be enough to readjust the system back to its previous morphological state.

DISCUSSION

Why only subtle changes?

Strong storms are destructive. Many lives have been lost to tropical cyclones (e.g. the Galveston storm of 1900), and they can cause extensive property damage (e.g. Hurricane Andrew, 1992; Hurricane Katrina, 2005). In regions containing sandy barrier islands and beaches, a single storm can cause considerable damage along the shore due to storm surges and waves (Hill *et al.*, 2004; Keen *et al.*, 2004; Stone & Orford, 2004; Stone *et al.*, 2004). However, impacts on tidally influenced subtidal seascapes such as the oolitic tidal deltas and shelf margin of the Northern Abacos appear much less pronounced.

The remote sensing images clearly demonstrate that although there were no major lasting geomorphological changes (e.g. new 'spillover lobes', or extensive movement of sands) as a result of passage of these tropical cyclones, there are indeed some geographically limited changes to this system. In the study area, between the 2003 and 2005 image acquisitions, crests of sand waves appear to have moved, and sand was transported off the shoal crest over adjacent grassy areas. If these changes were not due to day-to-day processes (the observations do not permit this possibility to be eliminated), these small apparent changes imply either of the following.

(a) Changes were simply too small to observe on the ultra-high-resolution remote sensing data, or there simply were no changes. This could be due to the presence of a storm surge, which diminishes the impact of the storm waves on the platform by increasing the water depth. This result indicates that, although storms may be powerful, they are not the most important forces shaping geomorphology in this and analogous depositional systems.

(b) There were alterations due to storms, but they returned to their original state after the storms and before acquisition of the second image and field observations. If this situation is valid, it would emphasize the role of daily processes in shaping the overall morphology of the shelf.

Tropical cyclones are powerful enough to produce large waves and storm surges, both of which can generate strong currents. The models (Figs 7 and 8) illustrate that although not all storms (from winter storms to strong cyclones) produce waves strong enough to have an impact on the bottom in all areas, they appear to have important roles in two regions. The first area is at the shoreline. In all cases, energy is expended on the shoreline due to breaking; this energy would be greater than that due to the daily (non-storm) waves. In these areas, more pronounced changes are expected. In the Abacos study area, many of the islands are rocky, and do not have sandy beaches, which may explain the lack of change at the shoreline. The second important area is the 'outer shelf' between the reef and the shore. Some, but not all, storms produce waves large enough to be impacted by the bottom in this area. These deeper areas (between 5 and 12 m deep) are beyond the penetration depths of daily waves (Fig. 8a). Tide- and wind-generated currents, which are weaker in these less-restricted regions, become much stronger in the presence of high winds and storm surges. Thus, in these areas, some sediment may have been suspended and transported.

Another possible interpretation for the lack of significant change is that the geomorphological elements formed in response to storms earlier in the geological history of the region, and are now in 'equilibrium' with such high-energy conditions. Many tropical cyclones and storm fronts pass through this area (Landsea *et al.*, 2003) and the tidal delta region of the Abacos must have been subjected to a very large quantity of storms ranging from tropical cyclones to winter cold fronts throughout its history. However, this hypothesis is unlikely for several reasons. Visual observations at the study site revealed active sediment transport with the daily tidal cycles. These observations suggest that the forms are not relict, being modified only by high energy storm events, but they are changing daily. Also, if they were to be formed by storms, as 'spillover lobes', it would be difficult to obtain such well-defined lobes on both sides of the islands. The storm patterns as discussed previously would most likely lead to larger 'flood' lobes than 'ebb' lobes, whereas the observations illustrate the opposite relationship. Therefore, it is unlikely that these geomorphological bodies were formed by storms, and it is more likely that they are the result of daily forces.

Perspectives and limitations

Based on his modelling study of wave-dominated shorelines, Storms (2003) wrote that '... the shallow-marine stratigraphic record is built predominantly by high magnitude–low frequency processes rather than low magnitude–high frequency processes'. Similarly, Quiquerez *et al.* (2004) wrote that '... observations therefore suggest that ... storm events and reworking of sediments contribute significantly to the sedimentary record on ancient continental shelves at long time scales ...'. Wanless *et al.* (1988) summarized 25 years of research saying that 'a large body of literature has suggested that catastrophic events are a dominant influence on sedimentation in shallow-marine carbonate environments'. These types of statements form the basis of interpretations of storm deposits of great lateral continuity in ancient successions, such as the beds interpreted by Jennette & Pryor (1993) to be continuous across an area greater than their 32 × 35 km study area.

In stark contrast, the observational and modelling results of this study illustrate that storms are neither sufficient nor necessary to explain geomorphological features and subtle changes in the Abacos carbonate system. Similarly, evidence for aerially extensive tempestite deposits, such as those commonly attributed to storms (Aigner, 1985), remains to be observed in many Holocene carbonate systems (Gagan *et al.* (1988) is a notable exception; cf. Wanless *et al.*, 1988), although it is not for lack of trying. For example, Robert Ginsburg (personal communication, 2006) related how he had an associate examine the shelf off Key Biscayne, FL, immediately following the passage of a Category 5 storm, Hurricane Andrew, directly over the area. The person found merely a one-cm-thick layer of mud in some places. How might this apparent paradox be reconciled?

To some extent it must be conceded that the answer, of course, is that storms can, in some instances, have an impact. Carbonate systems are heterogeneous, and interpretations of the dominance or irrelevance of storms cannot be applied universally. Subsequently, perhaps it is better to ask, in what settings might we expect to find the effects of storms to occur?

Similar to siliciclastic shorelines, some carbonate systems are wave-dominated whereas others are controlled primarily by tides. In wave-dominated systems such as the east coast of the USA, it is possible that the added velocities from the storm surge and the increased wave energy of the storm waves may produce greater forces on the carbonate system than those experienced daily. These may drastically affect the system.

Likewise, it is possible that storm surges have an impact on parts of the geomorphological system. Although the models included the increased water depths that would be produced by surge, they did not explicitly include the currents that would be generated by similar storm surges. These could have an effect on the transport of sediment on- or offshore, but this was not apparent in the shallow areas considered by this study.

In contrast, the modelling of this study highlights elevated velocities on the platform edge and shoals, yet there are relatively few changes in these regions. The measured tidal velocities at the sediment–water interface (Figs 9 and 10) in these deltas are not obviously modified by the passage of the storms, even in the presence of significantly larger storm waves (Figs 9 and 10). Thus, although the passage of storms may slightly modify velocity patterns, the daily tidal velocities are of a strong enough magnitude not to be affected by these minor perturbations. It may be that features shaped by tidal flows (tidal deltas and sandy shoals) are less susceptible to changes due to elevated velocities brought about by storm surges or storm waves (cf. Rankey *et al.*, 2006). Similarly, the reefs include rigid frameworks that face tides and waves on a daily basis and appear to have been sturdy enough to not be damaged notably by the storms.

CONCLUSIONS

Interpretations of storms and hurricanes are implicitly tied to the concepts of wave base and storm wave base (Gulliver, 1899; Fenneman, 1902; see review in Dietz, 1963). The latter is generally taken to be 'the greatest depth to which wave action ceases to stir sediments' (Dietz, 1963, p. 973) and is often cited as a sedimentologically important water depth, even if its value is never quite known. Its importance has extended to concepts of shelf profiles [the classic 'undaform' of Rich (1951) and the 'wave-base razor' of Cross and Lessenger (1997)] and even so far as to be used to estimate amplitudes of sea-level change (Immenhauser & Scott, 2002).

The observations and modelling of this study illustrate that storms are neither sufficient nor necessary to shape the geomorphological patterns observed in this shelf margin system. Although it is likely that tide-influenced and wave-influenced systems respond to storms differently, it still remains that there is precious little direct observational data providing examples of the impact of storms of an extent comparable to those interpreted from the geological record. This work highlights the need to understand basic physical processes behind sedimentation in carbonate systems and integrate this knowledge to develop new interpretations of ancient successions.

ACKNOWLEDGEMENTS

We are grateful to Robert Ginsburg, for his support and always keen insights. Numerous individuals have offered support and insights during field work and field trips to the area, including: Barry Albury, Scott Ritter, Kristine Sanchez, Lowell Waite and participants in Comparative Sedimentology Laboratory and Pioneer Natural Resources field trips to the area. Jacobus van de Kreeke was an invaluable resource for the modelling section of this work, and Gregor Eberli provided thorough reviews which greatly enhanced the logic and clarity of this paper. Peter Burgess, Peter Winefield, and Vic Hitchings provided support in the final throws of the project. Brian Haus graciously provided use of the ADCP, and provided training in its use. DigitalGlobe provided some of the imagery for this report. Thanks to Roland Albury and the Department of Fisheries of The Bahamas for permission to study in the area, and to Miguel McKinney for facilitating permits. The research was supported in part by the contributors to the Comparative Sedimentology Laboratory, the Center for Southeastern Tropical Advanced Remote Sensing (CSTARS) at University of Miami, the National Science Foundation (Grant # EAR-0418815) and the Army Research Office (Grant # W911NF0510005).

REFERENCES

Ager, D.V. (1981) *The Nature of the Stratigraphic Record*, 2nd ed. Macmillan Press, London, 122 pp.

Aigner, T. (1982) Calcareous tempestites: storm-dominated stratification. In: *Upper Muschelkalk Limestones* (Middle Trias. SW-Germany) (Eds G. Einsele and A. Seilacher), pp. 180–198. *Cyclic and Event Stratification*. Springer-Verlag, Berlin.

Aigner, T. (1985) *Storm Depositional Systems: Dynamic Stratigraphy in Modern and Ancient Shallow-Marine Sequences*. Springer, Berlin, 174 pp.

Ball, M.M. (1967) Carbonate sand bodies of Florida and the Bahamas. *J. Sed. Petrol.*, **37**, 556–591.

Beven, J.L. (2004) Tropical Cyclone Report: Hurricane Frances. 25 August–8 September. http://www.nhc.noaa.gov/2004frances.shtml.

Boss, S.K. and **Neumann, A.C.** (1993) Impacts of Hurricane Andrew on carbonate platform environments, northern Great Bahama Bank. *Geology*, **21**, 897–900.

Coastal Engineering Research Center (1984) Chapter 4: Littoral processes. *Shore Protection Manual*. Vicksburg, MS, **1**, 214 pp.

Cross, T.A. and **Lessenger, M.A.** (1997) Correlation strategies for clastic wedges. In: *Innovative Applications of Petroleum Technology in the Rocky Mountain Area* (Eds E.B. Coalson, J.C. Osmond and E.T. Williams). *Rocky Mountain Association of Geologists*, Denver, CO, pp. 183–203.

Dietz, R.S. (1963) Wave-base, marine profile of equilibrium, and wave-built terraces: a critical appraisal. *Geol. Soc. Am. Bull.*, **74**, 971–990.

Dietz, R.S. and **Fairbridge, R.W.** (1968) Wave base. In: *The Encyclopedia of Geomorphology* (Ed. R.W. Fairbridge). Reinhold, New York, pp. 1224–1228.

Duane, D.B. (1976) Sedimentation and coastal engineering: beaches and harbors. In: *Marine Sediment Transport and Environmental Management* (Eds D.J. Stanley and D.J.P. Swift). Wiley, New York, pp. 493–517.

Fenneman, N.M. (1902) Development of a profile of equilibrium of the subaqueous shore terrace. *J. Geol.*, **10**, 31.

Gagan, M.K., Johnson, D.P. and **Carter, R.M.** (1988) The cyclone Winifred storm bed, Central Great Barrier Reef, Australia. *J. Sedimentary Petrol.*, **58**, 845–856.

Gonzalez, R. and **Eberli, G.P.** (1997) Sediment transport and bedforms in a carbonate tidal inlet; Lee Stocking Island, Exumas, Bahamas. *Sedimentology*, **44**, 1015–1030.

Gordon, A.D. and **Roy, P.S.** (1977) Sand movements in Newcastle Bight. *Proceedings of the 3rd Australian Conference on Coastal and Ocean Engineering*, Melbourne, Australia, pp. 64–69.

Gulliver, F. (1899) Shoreline topography. *Am. Acad. Art and Sci. Proc.*, **34**, 151–258.

Hill, H.W., Kelley, J.T., Belknap, D.F. and **Dickson, S.M.** (2004) The effects of storms and storm-generated currents on sand beaches in Southern Maine, USA. *Marine Geol.*, **210**, 149–168.

Hine, A.C. (1977) Lily Bank, Bahamas: History of an active oolite sand shoal. *J. Sed. Res.*, **47**, 1554–1581.

Immenhauser, A. and **Scott, R.W.** (2002). An estimate of Albian sea-level amplitudes and its implications for the duration of stratigraphic hiatuses. *Sedimentary Geology/ExpresSed*, **152**, 19–28.

Incze, M.L. (1998) *Petrophysical properties of shallow-water carbonates in modern depositional and shallow sub-surface environments*. PhD dissertation, University of Miami, 405 pp.

Jennette, D.C. and **Pryor, W.A.** (1993) Cyclic alteration of proximal and distal storm facies; Kope and Fairview formations (Upper Ordovician), Ohio and Kentucky. *J. Sed. Petrol.*, **63**, 183–203.

Keen, T.R., Bentley, S.J., Vaughan, W.C. and **Blain, C.A.** (2004) The generation and preservation of multiple hurricane beds in the northern Gulf of Mexico. *Marine Geol.*, **210**, 79–105.

Komar, P.D. (1976) *Beach Processes and Sedimentation.* Prentice-Hall Inc., Englewood Cliffs, NJ.

Kreisa, R.D. (1981) Storm-generated sedimentary structures in subtidal marine facies with examples from the Middle and Upper Ordovician of southeastern Virginia. *J. Sed. Petrol.*, **51**, 823–848.

Kreisa, R.D. and **Bambach, R.K.** (1982). The role of storm processes in generating shell beds in Paleozoic shelf environments. In: *Cyclic and Event Stratification* (Eds G. Einsele and A. Seilacher). Springer-Verlag, Berlin, pp. 200–207.

Landsea, C.W., Anderson, C.,Charles, N., Clark, G., Dunion, J., Fernandez-Partagas, J., Hungerford, P., Neumann, C. and **Zimmer, M.** (2003) The Atlantic Hurricane Database Reanalysis Project – Documentation for 1851–1910 alterations and additions to the HURDAT Database. *Hurricanes and Typhoons: Past, Present, and Future* (Eds R.J. Murnane and K.B. Liu). Columbia University Press, Columbia, 2004, pp. 178–221.

Major, R.P., Bebout, D.G. and **Harris, P.M.** (1996) Recent evolution of a Bahamian ooid shoal: effects of Hurricane Andrew. *Geol. Soc. Am. Bull.*, **108**, 168–180.

Perkins, R.D. and **Enos, P.** (1968) Hurricane Betsy in the Florida-Bahama area; geologic effects and comparison with Hurricane Donna. *J. Geol.*, **76**, 710–717.

Pomar, L. (2001) Types of carbonate platforms: a genetic approach. *Basin Res.*, **13,** 313–334.

Quiquerez, A. Allemand, P., Dromart, G. and **J-P. Garcia.** (2004) Impact of storms on mixed carbonate and siliciclastic shelves: insights from combined diffusive and fluid-flow transport stratigraphic forward model. *Basin Res.*, **16**, 431–449.

Rankey, E.C. and **Morgan, J.J.** (2002) Quantified rates of geomorphic change on a modern carbonate tidal flat, Bahamas. *Geology*, **30,** 583–586.

Rankey, E.C., Enos, P. Steffen, K. and **Druke, D.** (2004) Lack of impact of Hurricane Michelle on tidal flats, Andros Island, Bahamas: integrated remote sensing and field observations. *J. Sed. Res.*, **74**, 654–661.

Rankey, E.C., Riegl, B. and **Steffen, K.** (2006), Form and function in a tidally dominated ooid shoal, Bahamas. *Sedimentology*, **53**, 1191–1210.

Reeder, S.L. (2007) *Comparative morphology and dynamics of Holocene carbonate systems, northwestern Abaco Islands, Bahamas.* Unpublished doctoral dissertation, University of Miami, Coral Gables, FL, USA.

Rich, J.L. (1951) Three critical environments of deposition and criteria for recognition of rocks deposited in each of them. *Geol. Soc. Am. Bull.*, **62**, 1–20.

Shinn, E.A., Lloyd, R.M. and **Ginsburg, R.N.** (1969) Anatomy of a modern carbonate tidal-flat, Andros Island, Bahamas. *J. Sed. Petrol.*, **39**, 1202–1228.

Scoffin, T.P. (1993) The geological effects of hurricanes on coral reefs and the interpretation of storm deposits. *Coral Reefs*, **12**, 203–221.

Seilacher, A. and **Aigner, T.** (1991) Storm deposition at the bed, facies, and basin scale: The geological perspective. In: *Cycles and Events in Stratigraphy* (Eds G. Einsele, W. Ricken and A. Seilacher). Springer-Verlag, Berlin, pp. 249–268.

Stone, G.W. and **Orford, J.D.** (2004) Storms and their significance in coastal morpho-sedimentary dynamics. *Marine Geol.*, **210**, 1–5.

Stone, G.W., Baozhu, L., Pepper, D.A. and **Wang, P.** (2004) The importance of extratropical and tropical cyclones on the short-term evolution of barrier islands along the northern Gulf of Mexico, USA. *Marine Geol.*, **210**, 63–78.

Storms, J.E.A. (2003) Event-based stratigraphic simulation of wave-dominated shallow-marine environments. *Marine Geol.*, **199**, 83–100.

Stumpf, R.P., Holderied, K. and **Sinclair, M.** (2003) Determination of water depth with high-resolution satellite imagery over variable bottom types. *Limnology and Oceanography* **48**, 547–556.

Wanless, H.R. and **Tedesco, L.P.** (1993) Comparison of oolitic sand bodies generated by tidal vs. wind-wave agitation. In: *Mississippian Oolites and Modern Analogs* (Eds B.D. Keith and C.W. Zuppmann). *AAPG Stud. Geol.*, **35**, 199–225.

Wanless, H.R., Tyrrell, K.M., Tedesco, L.P. and **Dravis, J.J.** (1988) Tidal-flat sedimentation from Hurricane Kate, Caicos Platform, British West Indies. *J. Sed. Petrol.*, **58,** 724–738.

Int. Assoc. Sedimentol. Spec. Publ. (2009) **41**, 91–112

Rapid recycling of organic-rich carbonates during transgression: a complex coastal system in southwest Florida

BRIGITTE M. VLASWINKEL*[1] and HAROLD R. WANLESS†

**Marine Geology and Geophysics, University of Miami, 4600 Rickenbacker Causeway, Miami, FL 33149, USA*
†Department of Geological Sciences, University of Miami, 1301 Memorial Drive, Coral Gables, FL 33124, USA

ABSTRACT

Coastal and shallow marine environments are actively responding to an accelerated rise in sea level, eroding in some areas and rapidly accreting in others. Cape Sable in southwest Florida, with several natural and anthropogenic triggering events in the past century, illustrates the nearly instantaneous response that can occur in a sediment-rich coastal system. Results are presented of a field study that documents a shallowing-upwards sediment package as a response to a transgressive phase. Patterns and rate of sedimentation are reported, the different sediment sources identified and the governing processes that control sedimentation style are determined. The study integrates sedimentological and geochemical data with hydrodynamic time-series measurements of water level, currents, suspended sediment concentration and salinity. Results show a rapid, sequential infilling of the intertidal zone from the most seaward marine, to transitional marine–freshwater subenvironments, as accommodation space becomes available due to sea level rise, increased flood tidal volume and collapse of interior freshwater marshes. Average *in situ* sedimentation rates of 6.2 cm yr^{-1} are reported on the intertidal mudflats and daily tides are the most important agent responsible for the erosion, transportation and deposition of the fine-grained sediment. A significant amount of sediment is sourced from one coastal compartment and transported to another within the intracoastal system. The shallowing-upwards sediment package contains organic-rich carbonates with 15–35% total organic matter. In contrast to common stratigraphic wisdom, the shallowing-upwards peritidal sediments as recorded in southwest Florida are the depositional response to several small, rapid pulses of sea-level rise instead of being diagnostic of a single high-stand lithofacies such as commonly described in ancient epeiric sequences.

Keywords Recycling, organic-rich carbonates, shallowing-upward, process dynamics, sea-level rise, southwest Florida.

INTRODUCTION

Low-energy coasts represent dynamic systems that respond to a number of forcing mechanisms, including relative changes in sea level and changes in sediment supply. Subtle relative changes in sea level can result in pronounced sedimentological, geomorphological and ecological responses (Stapor *et al.*, 1991; Allen, 2000; Rankey & Morgan, 2002). Specifically, rising sea level commonly is accompanied by substantial sediment release, transport and redeposition (Scholl, 1964; Fairbridge, 1974; Carter, 1988; Wanless *et al.*, 1994). Stratigraphically, this recycling and deposition during transgression can lead to facies complexity within littoral and shallow-marine sediments (Lobo *et al.*, 2005). In addition to sea level, sediment availability exercises a strong control over changes in coastal morphology and the preservation potential of a sedimentary package (Hine *et al.*, 1988; Wanless & Tagett, 1989; Otvos, 2004). Contrasting modes of coastal evolution are commonly linked to variations in sediment supply (Hine *et al.*, 1988).

Southwest Florida includes a low-wave-energy coastal system that is responding to changes in sea level and sediment supply, as well as minor anthropogenic modifications. The late Holocene

[1]Current address: Shell International Exploration and Production B.V., Kessler Park 1, 2288 GS, Rijswijk, The Netherlands (E-mail: B. Vlaswinkel@shell.com).

stratigraphic record of Florida Bay and the southwest coast of Florida includes a transgressive sediment package overlain by a regressive package (Scholl, 1964; Enos & Perkins, 1979; Parkinson, 1987). This transgressive package has been interpreted to have been deposited during a period of rapid relative sea level (RSL) rise (~23 cm 100 yr^{-1}) in the mid-Holocene (5500–3200 yr BP) (Parkinson, 1987). The regressive package formed during a period with a slower average rate of RSL rise, ~4 cm 100 yr^{-1} over the past 3200 years (Scholl *et al.*, 1969; Robbin, 1984; Parkinson, 1987; Wanless *et al.*, 1994). This decelerated rate of RSL rise, in combination with abundant sediment supply, resulted in lateral accretion of islands in Florida Bay (Gorsline, 1963; Enos & Perkins, 1979; Cottrell, 1989; Wanless & Tagett, 1989) and progradation of coastal facies along the southwest coast of Florida (Evans *et al.*, 1985; Hine *et al.*, 1988; Parkinson, 1989; Stapor *et al.*, 1991). Over the past 75 years, south Florida has faced an accelerated rate of RSL rise of >23 cm 100 yr^{-1} (Wanless, 1982; Wanless *et al.*, 1988; Douglas, 1991). Hence, interpretation of earlier Holocene deposits might suggest that the existing coastal system should switch from aggradational/progradational to retrogradational.

In order to understand how rapid RSL rise and sediment availability might influence the evolution of a low-energy coastal system, an integrated study of a carbonate/organic system within Cape Sable, southwest Florida, is presented. The purposes of this paper are to (1) document the patterns and rates of sedimentation in the shallowing-upwards facies succession; (2) identify the different sediment sources and sinks, and (3) determine the process dynamics which govern the sediment redistribution and accumulation style. An integrated approach is adopted that uses sedimentological, hydrodynamic and geochemical data, to link process dynamics to the associated sedimentary products. The aim of the investigation is to fill a gap in the knowledge of quantitative relations between processes and products, a field not well explored in mixed organic–carbonate sedimentary systems.

The results of this research illustrate the complicated nature of processes and related products in a low-energy setting. Recycling of organic–carbonate muds during rapid RSL rise can result in a shallowing-upwards succession of tidal deposits and a complete spatial reorganization of facies. Thus, ancient metre-scale, shallowing-upwards lithologic units need not necessarily be the result of a stable or slow sea-level rise, such as suggested by several models of tidal flat progradation (Ginsburg, 1971; Goodwin & Anderson, 1985; Pratt & James, 1986; Burgess, 2001), but instead can be the depositional expression of a single rapid rise of sea level during a high-stand.

AREA OF STUDY

Cape Sable (900 km^2) forms the southwestern tip of the Florida Peninsula, an emergent low-relief carbonate platform (Fig. 1). The area is situated at the southwest end of the gentle Everglades depression between a Pleistocene limestone ridge to the northeast and emergent Pliocene limestone to the northwest. The karstified surface of the Pleistocene Miami Limestone lies at a depth of 3.5–4 m in the Cape Sable area (Fig. 2; Roberts *et al.*, 1977).

Cape Sable is located at the intersection of Florida Bay and the Gulf of Mexico. Characteristics of both the southwest Florida coast (tide-dominated, sand-starved, continuous mangrove belt) and Florida Bay (shallow coastal lagoon, carbonate mudbanks, mangrove-capped islands) are present in Cape Sable. Cape Sable has shell beach or mangrove shorelines along its western (Gulf of Mexico) coast, a storm-built marl ridge on its southern (Florida Bay) coast and mangroves bordering the interior coast on Whitewater Bay (Fig. 2). The southern interior of Cape Sable includes a series of at least three linear, emergent ridges (Fig. 1) composed of calcium carbonate mud (marl), interpreted as ancient shorelines (Roberts *et al.*, 1977). Landward of the youngest (most seaward) marl ridge is a vast wetland historically dominated by freshwater marsh species such as sawgrass and cordgrass. Today, salt-tolerant species including mangroves occur throughout this region. Seaward of the marl ridge lies Lake Ingraham (Fig. 1), formerly a shallow freshwater lake; more seaward of the lake lies a shelly beach ridge complex. This shell beach is just a thin veneer over a 4-m-thick marl sequence (Fig. 2), except at the three capes, where more than 2 m of sand has accumulated. Southeast of the coastal lagoon Lake Ingraham is an intricate complex of ponds, mudflats and tidal creeks, hereafter called the Southern Lakes (Fig. 1).

In the past century, this region has been modified by humans. Through the early to mid-1920s, several narrow canals (less than 4 m in width) were constructed connecting the different

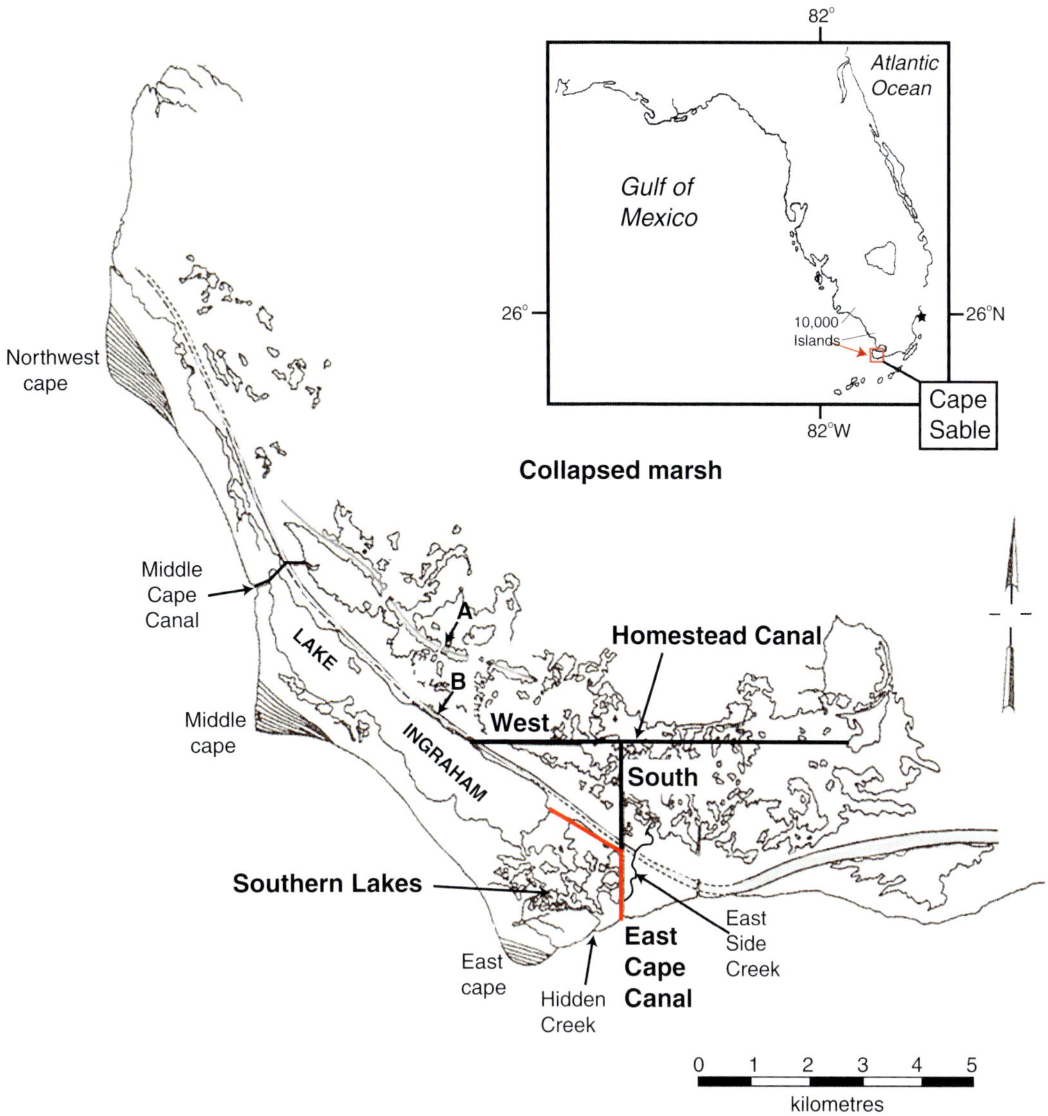

Fig. 1. Location map of the Cape Sable study area along the southwest coast of the Florida Peninsula, showing (1) prominent geomorphical features, (2) relict coastal levees in grey tints (A and B refer to dates in text), (3) Homestead Canal (HSC) West and South and East Cape Canal (ECC) respectively accentuated in solid black and red. Modified from Roberts *et al.* (1977).

subenvironments of the study area: Homestead Canal (HSC) connects Lake Ingraham to the inland marshes as it cuts through the slightly higher (<0.5 m above mean sea-level) marl ridge; Middle Cape Canal (MCC) provides tidal exchange between the Gulf of Mexico and the northern portion of the coastal lagoon; East Cape Canal (ECC) joins Lake Ingraham to Florida Bay as well as to HSC (Fig. 1). One natural creek, East Side Creek (ESC) (Fig. 1), cuts through the marl ridge, connecting the inland marsh with ECC. A natural creek, Hidden Creek (HC), connects Florida Bay with the Southern Lakes (Fig. 1). These canals have all played an important role in the initial and ongoing saline intrusion into the southern interior of Cape Sable.

CLIMATE, WIND AND TIDES

South Florida's climate is subtropical and the seasons are defined by rainfall amounts. Precipitation averages 120–160 cm annually, of which 60% falls in the summer months June to September (Duevier *et al.*, 1994). Cape Sable is affected only minimally by freshwater discharge

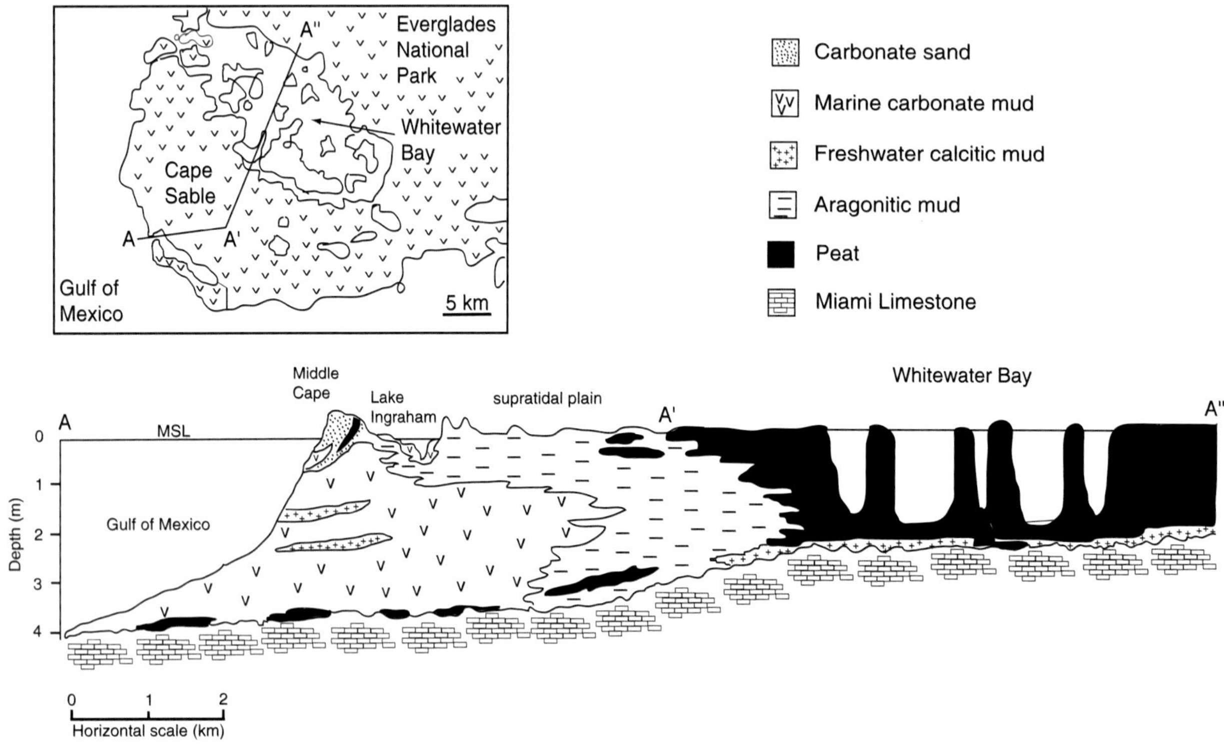

Fig. 2. Schematic cross-section of Holocene stratigraphy through Cape Sable. Modified from Roberts *et al.* (1977). The Pleistocene Miami Limestone lies at a depth of 3.5–4 m. The carbonate facies (aragonitic mud is also called 'marl') switch landward to peat, which in some places is underlain by freshwater calcitic mud on bedrock.

from the Everglades, but seasonal variation in rainfall results in substantial salinity oscillations in Cape Sable's interior.

The southwest coast of Florida has an intricate landscape with abundant tidal inlets, bays and islands, which results in an uncommonly large tidal prism relative to the wave regime (Goodbred *et al.*, 1998). This large tidal prism, coupled with low prevailing wave energy and low platform relief, results in a tide-dominated coastal system. Astronomical tides at Cape Sable are mixed, predominantly semi-diurnal with a strong diurnal inequality. Mean predicted tide range at East Cape is 0.9 m, minimum neap tide range is approximately 0.6 m, whereas maximum spring tide range is 1.7 m.

Prevailing winds are from the southeast and have average wind velocities of less than 4.5 m s^{-1} (Peng *et al.*, 1999). From November to May, approximately 25 cold fronts pass across South Florida (Hardy & Henderson, 2003). During passage of a cold front, the wind direction shifts clockwise from south to west to northeast and can carry sustained wind velocities of up to 11 m s^{-1}. The westerly winds passing during cold fronts have the greatest potential to develop waves that can affect the study area, because the trend of the coastline places the region in the lee of all other wind directions. Hurricanes can set up strong wind wave systems as well. On average, hurricanes occur once in every five to seven years (Neumann *et al.*, 1993), but there is great variability from year to year, as attested by comparing 2005 (seven major hurricanes) with 2006 (one weak hurricane). Two major hurricanes have played an important role in geomorphological changes in the study area (Perlmutter, 1982; Bischof, 1995), a Labor Day Hurricane in 1935 and Hurricane Donna in 1960.

PREVIOUS WORK

Pioneering work by Davis (1940) in the Ten Thousand Island region (Fig. 1) established a basic understanding of the evolution of southwest Florida's coastal zone and highlighted the importance of the biological production of coastal peat. Spackman *et al.* (1964) cored basal freshwater peats 2.5 km offshore and concluded that the southwest coast of Florida had undergone significant flooding over the past 5000 years. In this area, a transgressive facies succession is present, and consists of basal freshwater peat, overlain by transitional mangrove peat and marine carbonate mud (Spackman *et al.*, 1964; Scholl & Stuiver, 1967).

Scholl *et al.* (1969) generated the first Holocene sea-level curve for south Florida with an inflection point between 3500 and 3200 yr BP. An initial rapid rise, which averaged about 26 cm 100 yr^{-1}, was followed by a much slower rise of about 4 cm 100 yr^{-1}. Parkinson's (1987) subsurface work in the coastal zone of southwest Florida focused on evaluating the sedimentological succession deposited as a result of this changing rate of sea-level rise. He recognized a 6-m-thick Holocene package in the Ten Thousand Islands region, in which he defined a lower transgressive and upper regressive sediment package, and interpreted this shift to RSL rise. The lower sediment succession reflects shoreline retreat and subsequent accumulation of subtidal sediments. The upper sequence consists of biogenic shallowing-upwards buildups and coastal mangrove peats thicker than the present tidal range. As the rate of rise slowed, the rate of biological sediment production and accumulation began to outpace the rate of sea-level rise, initiating shoreline stabilization and island emergence. More recent studies along the southwest coast of Florida (Frederick, 1994; Gelsanliter, 1996), northern shore of Florida Bay (Huang, 1990) and within Florida Bay (Cottrell, 1989) have all documented a regressive package as a direct result of a declining rate of RSL rise.

Roberts *et al.* (1977) improved the understanding of the regional transgressive–regressive succession, identifying the major sedimentary units in a transect across Cape Sable and Lake Ingraham (Fig. 2). Sediment thickness of carbonate mud on the seaward portion of the tidal plain is 3.5–4.0 m. The carbonate facies switch landward to peat, which is underlain in some places by freshwater calcitic mud above bedrock. Roberts *et al.* (1977) also dated the late Holocene progradation of the shoreline and the three sandy capes after which Cape Sable is named. The oldest (most inland) ridge is dated at 2280 ± 100 yr BP (^{14}C years) (A in Fig. 1) and the youngest marl ridge at 2000 ± 80 yr BP (B in Fig. 1). Each one of these ridges was interpreted by Roberts *et al.* (1977) to be an ancient shoreline.

The major depositional environments in the Cape Sable region were briefly described by Roberts *et al.* (1977), and extensively by Gebelein (1977). Gebelein's work focused entirely on the surface geology, as he analysed the sedimentological and biological processes and patterns in the subtidal, intertidal and supratidal carbonate sediments in and around Lake Ingraham.

METHODS AND MATERIALS

Sediment sampling

Thirty sediment cores, ranging in length from 80 to 150 cm, were collected within Lake Ingraham, the Southern Lakes and the interior marsh to document the stratigraphic succession, including its texture, composition and sedimentary structures. Sediments are classified using Dunham's (1962) carbonate textural terminology for rocks.

Between May 2004 and January 2005, 34 sediment reference markers were deployed on the intertidal mudflats of Lake Ingraham and the Southern Lakes to measure *in situ* sedimentation rates. A series of four, 40 × 40 cm wide carpet tiles were pinned down on the sediment bed during low tide, acting as artificial marker horizons. One carpet tile from each location was removed after one, two, four and six months (Fig. 3). The sediment weight on top of the carpet was converted to vertical sedimentation rate using bulk densities of the wet mud. Precision of the weight measurements is ± 50 g.

Several 12-hour experiments were carried out throughout ECC to determine the spatial variability of suspended sediment concentration in the water column: water samples (250 mL) were collected 1 m below the water level each hour at Stations 1, 3 and 4. Earlier experiments, in which sediment concentrations were measured at 50, 100 and 150 cm above the sediment bed, displayed a homogeneous vertical turbidity profile. The carbon isotopic value of suspended sediment was measured simultaneously at the same three stations throughout one full tidal cycle in January 2005. Stable carbon isotopes ($\delta^{13}C$) are widely used to differentiate organic matter sources in estuarine sediments, because there is a significant difference in the ^{13}C content of the primary producers in question (Zieman *et al.*, 1984; Kennicutt *et al.*, 1987; Fleming *et al.*, 1990; Chmura & Aharon, 1995). Suspended sediment samples and several sediment reference marker samples were analysed with an ANCA GSL mass spectrometer to determine the isotopic composition of the

Fig. 3. (a) 40 × 40 cm carpet tiles as artificial marker horizons. Thirty-five sediment reference markers have been deployed on the intertidal mudflats along East Cape Canal, in Lake Ingraham and the Southern Lakes (for location see Fig. 8); (b) Example of rapid sedimentation along the levees of East Cape Canal: 8.5 cm sedimentation from 4 July to 5 January (6 months) at sediment marker 107.

particulate organic carbon fraction (δ^{13}C POC) and the organic carbon in the sediment. All data are reported relative to Vienna-Peedee Belemnite (V-PDB) standard.

Sediment characteristics

Grain size was determined with an electro-resistance particle size analyser. Sediment samples were wetted to disaggregate grains and then added to the fluid module on a Coulter Counter analyser, which measures particles from 0.04 μm to 2 mm.

Percentages of total organic matter (TOM), calcium carbonate and insoluble residue were measured on core, sediment reference marker and suspended sediment samples. Sediments were dried and ground with a mortar and pestel, and approximately 5 g of sediment was combusted at 550°C for 4 hours to determine the organic matter fraction. Results were accurate to 1–2% for organic matter. Subsequently 50 mg of the combusted sample was leached with 10% HCl to determine an estimate of percentage carbonate loss (Van Iperen & Helder, 1985). The residue after combustion and dissolution contained mainly quartz, opaline silica and clay minerals.

Hydrodynamic data acquisition

Continuous measurements of current velocity, water level, suspended sediment concentration and salinity were taken during several measurement campaigns between July 2003 and March

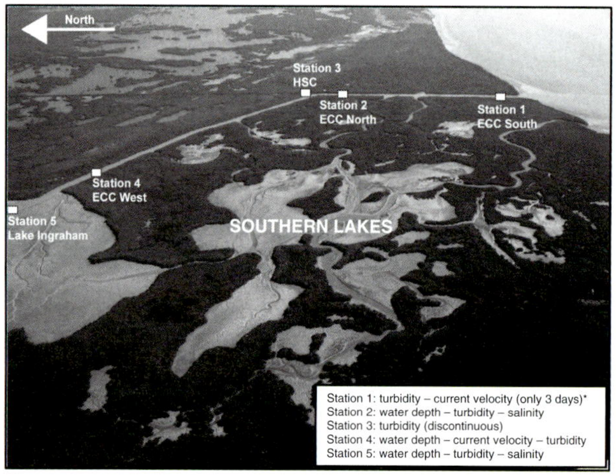

Fig. 4. Oblique aerial photograph during low tide of East Cape Canal (ECC), Homestead Canal South (HSC), East Side Creek (ESC), the Southern Lakes and part of the southeastern delta in Lake Ingraham with the positions of the measurement stations and parameters measured at each station. Width of photograph is 2 km. *Turbidity and current velocity are not measured simultaneously (Station 1).

2005 (Fig. 4). All instruments were mounted on weighted frames placed on the channel bottoms. Data were collected at 1 m above the bottom every 10 minutes, and over time periods ranging from days to 4 weeks. Current velocity and direction were measured with a Sontek Acoustic Doppler Current Profiler (ADCP), which generates a vertical velocity profile with depth cell intervals of 0.25 m (accuracy is 0.5 cm s^{-1}). Water-level fluctuations were indirectly determined from the ADCP,

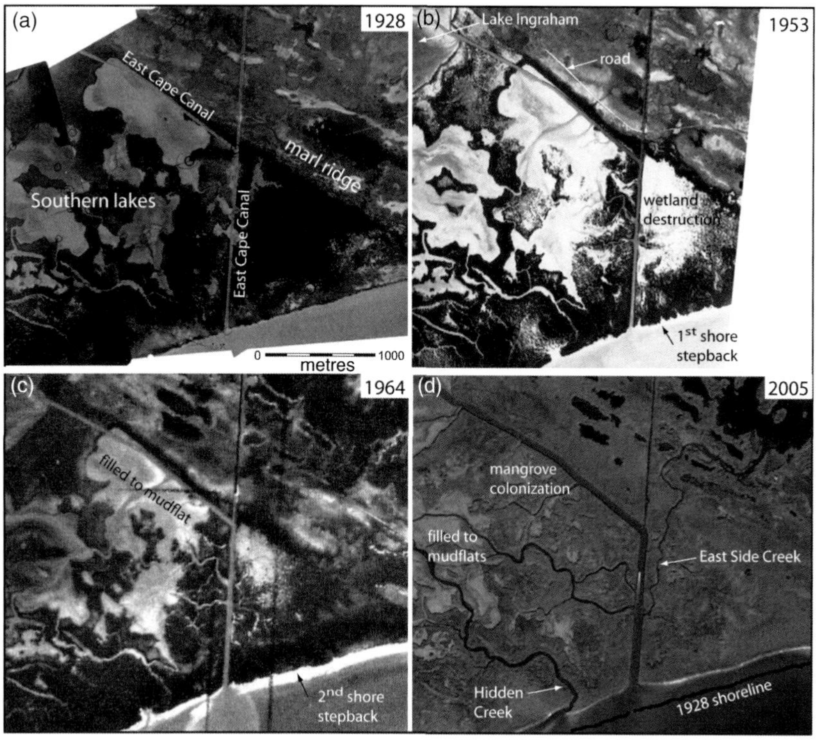

Fig. 5. (a–d) Sequential aerial photographs (1928–1953–1964) and satellite image (2005-Ikonos © GeoEye) of the East Cape Canal area. Shoreline erosion occurred between 1928–1953 and 1953 and 1964, interpreted to be the direct result of hurricanes. 1928 shoreline has moved inland approximately 180 m. Ponds on the 1928 aerial photograph have filled up and mangroves cap the intertidal mudflats at present. Hidden Creek is connected to Florida Bay by 1953, East Side Creek first appears on 1964 aerial photograph.

as well as directly measured in combination with salinity with a conductivity–temperature–depth (CTD) recorder (resolution is 0.02 psu for salinity and 0.03 m for water level).

The amount of suspended sediment in the water column was measured with a turbidity sensor. A total of 65 *in situ* water samples were collected for validation analysis between turbidity sensor-derived values and *in situ* measured values of suspended sediment. The correlation coefficient is 0.83 and there is a 95% confidence level that the compared values are correlated.

RESULTS

Historical stresses to study area

Cape Sable has undergone large geomorphological, ecological and sedimentological changes over the past 80 years, caused by a combination of factors: (1) small but significant human modifications, mainly the dredging of narrow canals in the 1920s, (2) major hurricanes and (3) historical relative rise in sea level.

Canals

Through the early to mid-1920s, a series of dredging projects created a network of canals across the interior freshwater marshes of southern Cape Sable, and between Cape Sable and the Gulf of Mexico and Florida Bay. Three canals are important for this study: ECC, HSC and MCC (Fig. 1). These canals cut through the marl ridge and connect the interior freshwater marsh with the coastal lagoon, Lake Ingraham and Florida Bay. Before canal construction, the water depth in the enclosed fresh- to brackish-water lagoon was 1–1.5 m (Gebelein, 1977). Initially, the canals were less than 4 m wide and did not reach limestone bedrock (Davis, 1972). The volume of water passing through the canals was small at first, and in fact, a hurricane in 1935 choked MCC with trees and overwash sediments (Simons & Ogden, 1998). However, tidal flow through ECC and MCC caused a continual and persistent deepening and widening in those channels at rates of 60 and 120 cm per year respectively (Davis, 1972), widening that continues to the present (Fig. 5d). Presently ECC measures 70 m in width, MCC 130 m in width, and both canals cut to the Pleistocene limestone surface at 3.5 m depth. Two natural creeks, Hidden Creek and East Side Creek (Fig. 5d) have also been widening steadily. HC naturally connected Florida Bay to the Southern Lakes and Lake Ingraham in 1950 and has since been widening at a rate of 60 cm per year (Davis, 1972) and is now 35 m wide.

Prior to canal construction, Lake Ingraham and the Southern Lakes (Figs 1 and 5a) were isolated fresh- to brackish-water lakes, receiving saline water only during storm surges (Simons & Ogden, 1998). Connecting Lake Ingraham and the Southern Lakes to the marine realm permitted saltwater intrusion during the dry months, and a combination of saltwater intrusion and freshwater discharge during the rainy months. The opening of the canals is interpreted to have triggered collapse of the interior freshwater marsh and initiated a phase of increasingly rapid marine sedimentation in areas reached by flood tidal waters. The change from fresh–brackish to marine sediment provides a visible marker horizon in the sediment sequence, permitting calculations of thickness and rate of historical sedimentation.

Hurricanes

Two intense hurricanes profoundly modified Cape Sable: the Labor Day Hurricane of 1935, a category 5 storm on the Saffir–Simpson scale, and Hurricane Donna in September 1960, a strong category 4 storm. Since the 1928 aerial photograph, the southern coast facing Florida Bay has eroded about 180 m (Fig. 5). As documented by aerial photographs, erosion occurred in two nearly equal steps of approximately 75 m, one between 1928 and 1953 (Fig. 5a and b) and a second between 1953 and 1964 (Fig. 5b and c). These two steps in erosion are interpreted to be the result of the 1935 and 1960 hurricanes. Only a very small amount of erosion has occurred on the south shore since 1964.

Sea-level rise

Since 1932, South Florida has experienced an accelerated rate of sea-level rise of ~23 cm 100 yr^{-1} (Wanless, 1982; Wanless *et al.*, 1988; Douglas, 1991), almost six times greater than that of the previous 2400 yr (~4 cm 100 yr^{-1}) (Wanless, 1982). This rate of sea-level rise is determined from a continuous series of tide gauge records starting in 1913 from Key West (Maul & Martin, 1993) and it does not take into account the estimated Florida subsidence rate of ~0.01 cm yr^{-1} (Rona & Clay, 1966).

At present, the slightly elevated marl ridge, ~0.5 m above MSL, is significantly overtopped by saline flood tide waters that have a (predicted) tidal level of 1.28 m, which occurs during 80 high tides a year. During these events the extent of the tide, and thus the tidal volume, expands. It is expected that with steadily rising sea level, marine water will spill more frequently across the ridge, feeding a progressively larger tidal volume.

Geomorphological patterns and rate of sediment accumulation

Where connected to the ocean, the coastal bays and lakes within Cape Sable are rapidly shallowing. First to infill have been the Southern Lakes adjacent to ECC (Fig. 5b). By 1953 these lakes were filled to form an intertidal mud flat dissected by tidal creeks, features that carried sediment-laden waters to the lakes' inner recesses. Red mangrove forests subsequently spread across these flats, capping a shallowing-upwards succession (mangroves overlying organic-rich mud, overlying medium grey, shelly subtidal mud, with indicative brackish water mollusc species).

As the Southern Lakes filled and ECC and HC widened, sediment delivery became focused inward into Lake Ingraham (Fig. 6), illustrated by the rapid expansion of the flood tidal delta and infilling of Lake Ingraham since 1953 (Fig. 6b and c). Aerial photographs since 1990 indicate that the rate of delta growth has progressively increased over the past two decades. At present, the southern Lake Ingraham flood delta forms a sediment body over 4 km long by 1–1.5 km wide and 50–90 cm in thickness. The delta elevation is highest near the southeastern entrance to the lagoon and along the margin of the axial channel. The primary channel-margin levee remains emergent even at the lower high tides and is becoming colonized with mangrove seedlings towards its southeastern (ECC) end.

The channel-margin levee constriction has promoted numerous side creeks extending perpendicular to the main channel towards the lake margins. These secondary channels have discrete sediment lobes that broaden the delta as a whole. One pronounced side creek connects HSC West to the axial delta channel and is maintained by ebb tidal discharge from HSC West (Fig. 6c). As waters overflow the relict marl ridge from Lake Ingraham into the interior marsh regularly (at least 80 times a year), muddy overwash sediments are also extending from the ridge into the shallow waters of the collapsed interior marsh (Fig. 7).

The nature of these spatial changes is refined by observations of *in situ* sedimentation patterns in Lake Ingraham. Short-term sedimentation

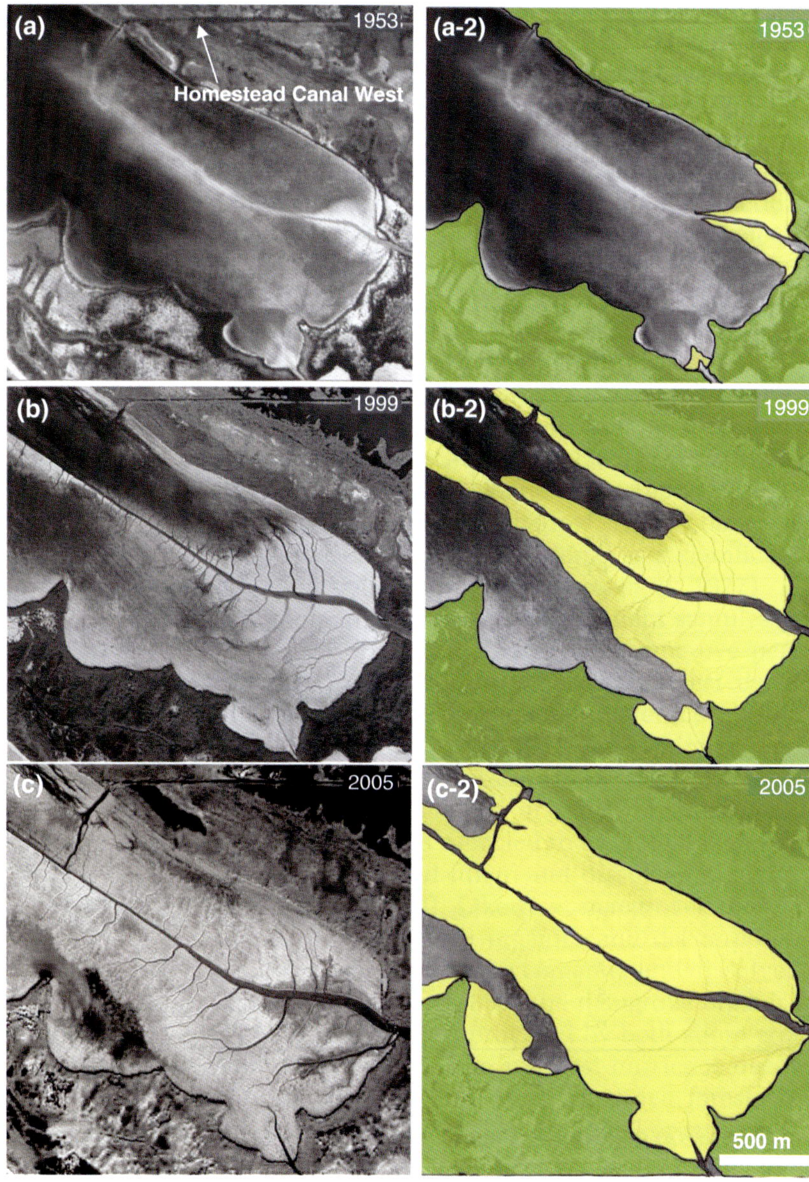

Fig. 6. Aerial photographs (1953, 1999) and satellite image (2005-Ikonos © GeoEye) with interpretative maps of southern Lake Ingraham showing infilling history. Yellow areas are out of water at most low tides; green is wetland and non-shaded is subtidal. The darker shade along the main channel in (c) is much higher and thus drier than the surrounding white colour. The elongate form and single axial channel is the result of motorboat traffic.

Fig. 7. Aerial photograph (2002) displaying overwash sediments into interior collapsed marsh. Water to the left is the northern extent of Lake Ingraham.

rates, measured with sediment reference markers along ECC, in Lake Ingraham and the Southern Lakes, range from 3 to 14 cm yr^{-1} (Table 1; Fig. 8a). The highest sedimentation rates (11–14 cm yr^{-1}) are measured along the banks of tidal creeks that connect ECC to the Southern Lakes (Fig. 3b). These observations suggest that large amounts of sediment are still being fed into the Southern Lakes. Within Lake Ingraham, the lower parts of the delta are accumulating rapidly at 5–7 cm yr^{-1}, whereas the oldest (and presently highest) parts of the delta are accreting at 3–4 cm yr^{-1}. The lower sedimentation rates and

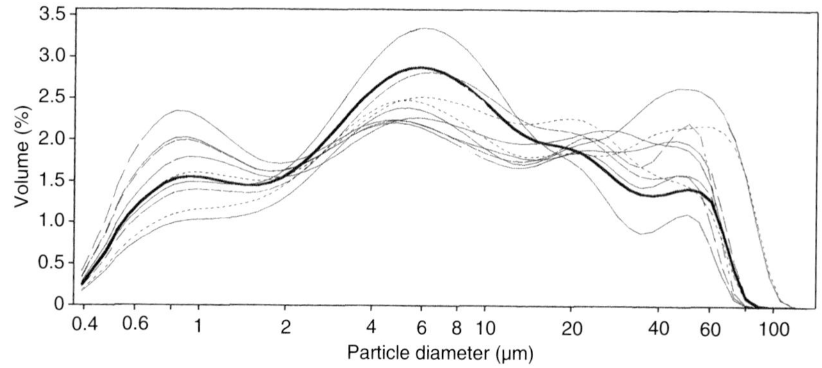

Fig. 9 Trimodal grain size distribution for selected sediment reference markers ($N = 26$) with peaks at 0.8 μm, 6 μm and 50 μm. Solid line indicates the mean grain size distribution for all sediment samples.

are doubled (150 cm s^{-1}), whereas maximum ebb velocities are only slightly higher (93 cm s^{-1}) (Fig. 10a). Suspended sediment concentration in the study area is very high compared with other estuarine settings in south Florida: ~20 mg L^{-1} at Marco Island (Barron, 1976) and <20 mg L^{-1} in western Florida Bay (Lutz, 1997; Jones & Boyer, 1999). During calm weather, average sediment concentration off East Cape is around 50 mg L^{-1}. Much higher concentrations, with peaks ranging between 200 and 1100 mg L^{-1}, are measured at the various stations within ECC (Fig. 11).

Time-series observations of suspended particulate matter at ECC entrance (Station 1) and ECC North (Station 2) show a distinct signal that mimics the daily inequality of the locally predicted tides (Fig. 11a–c). Large sediment peaks are recorded at the end of every 'high-amplitude' ebbing tide. During spring tide, a secondary smaller sediment peak coincides with the 'low-amplitude' ebbing tide (Fig. 11a and b). Importantly, notice the low levels of suspended sediment in incoming tidal waters, whether spring or neap tide, on a day-to-day basis.

On the western end of ECC, close to Lake Ingraham (Station 4), time-series measurements of suspended material do not show the same pattern as at Stations 1 and 2 (Figs 10a and 11d). On the contrary, sediment concentration is irregular through time and does not display a spring/neap tide effect. Many sediment peaks coincide with local high water level (Fig. 10a). Due to the fact that current lags water level (Fig. 10b), a phenomenon commonly recognized in estuaries (Van Veen, 1950), the current is still directed inward at high water and carries sediment towards Lake Ingraham. However, sediment peaks also appear at low water, and every so often at times of highest (ebb or flood) velocities. A possible explanation for this complicated sediment pattern is the presence of several little creeks that exit close to Station 4.

To ascertain the origin of the water mass that carries the observed sediment peaks, time-series measurements of salinity were made in combination with water level and turbidity, first at ECC North (Station 2) and two weeks later at Lake Ingraham (Station 5) (Fig. 4). Data from both stations show a clear, but contrasting, salinity pattern related to the tide. Incoming flood waters have normal seawater values of 35, whereas outgoing waters are either hypersaline (up to 38) (Fig. 12a) or brackish (25) (Fig. 12b). The suspended sediment peak concentrations, appearing more regularly at station 2 than at 5, are carried seaward with these waters. These brackish or very saline waters drain from the interior collapsed marsh, where local rainfall and evaporation rates can rapidly change the hydrological signature of the water mass: 20–25 February 2005 was a dry week, resulting in high salinities measured at Station 2; 8–13 March 2005 was a rainy week, thus resulting in low salinities measured at Station 5.

To determine whether the suspended particulate matter is related to the (brackish/very saline) interior waters that it is carried by, organic carbon isotopic values of the sediment in suspension were measured simultaneously at Stations 1, 3 and 4 throughout a tidal cycle. End-member isotopic values (Fig. 13a) include $\delta^{13}C$ values of around −26 to −27‰ for a peat dominated by sawgrass (*Cladium* sp.) and mangrove (*Rhizophora* sp.) (core 68 and 37), and ca. −21‰ for Florida Bay water reflecting mainly a marine phytoplankton signal (Zieman *et al.*, 1984; Lutz, 1997; Mead, 2003; Xu *et al.*, 2006). Throughout the entire ECC, the ebb tide carries isotopically lighter suspended sediments than a flood tide (Fig. 13b). The large sediment peaks that are observed at the end of the ebb feature the lightest $\delta^{13}C$ signal

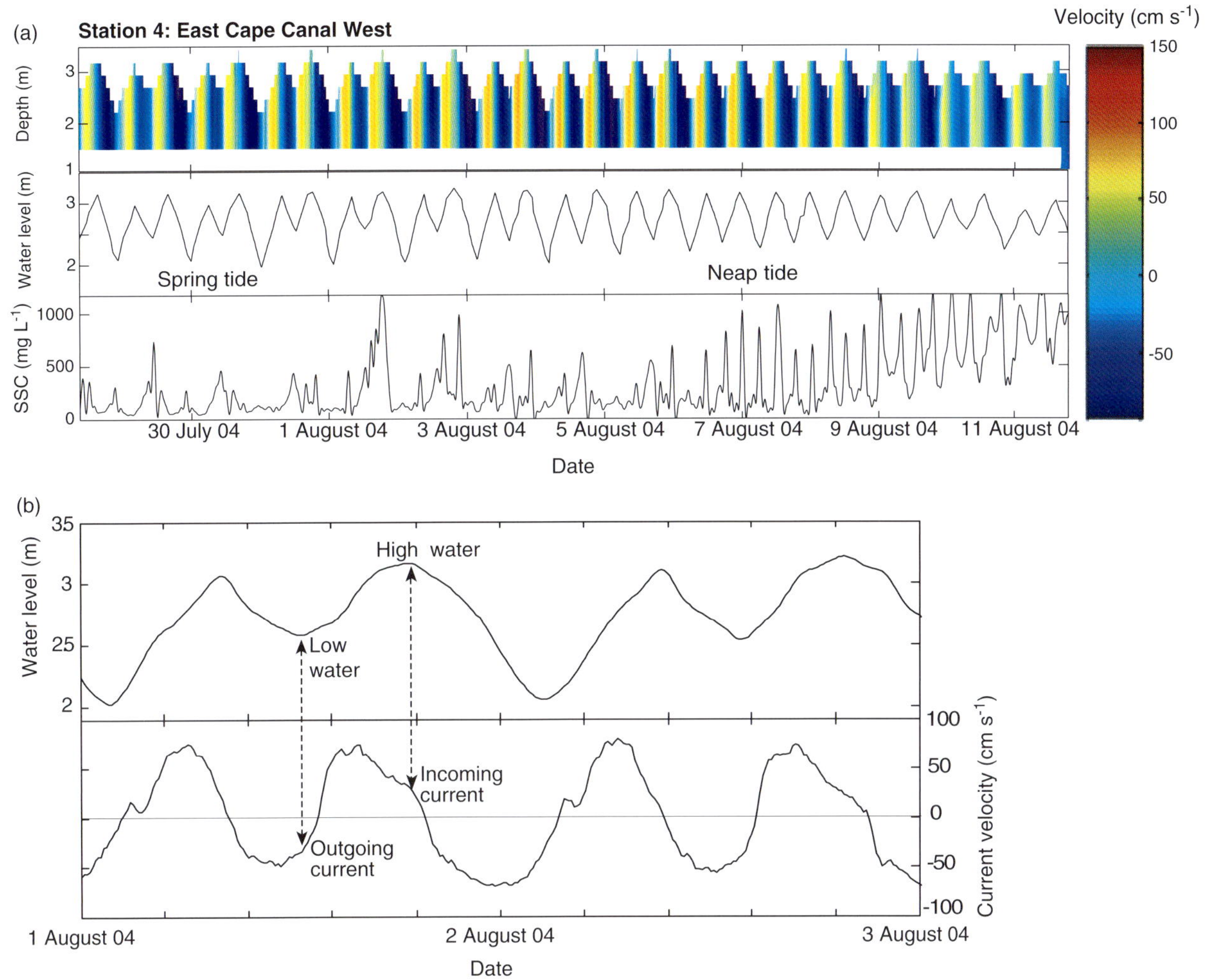

Fig. 10. (a) Time series of velocity, water level and suspended sediment concentration, measured at Station 4 in East Cape Canal in August 2004 with an ADCP and turbidity meter. The velocity time series shows an undulating vertical velocity profile, reflecting velocities throughout the entire water column as the water level goes up and down. The colours refer to the strength of the current: red and yellow tints being inflow, blue and green tints being outflow. The darker the colour, the stronger the velocities; (b) Detailed time series of water level and current velocities, demonstrating the time difference between maximum (minimum) water level and the reversal of the current flow (up to 1 hour), a phenomenon commonly observed in estuaries.

(sawgrass/mangrove); the incoming Florida Bay water has a heavier $\delta^{13}C$ signal (marine plankton and seagrass). Station 1 displays the largest isotopic range (from −23.7 to −21.7‰). The least varying carbon isotopic signal throughout the tide is seen at Station 4 (−22.9 to −21.9‰). These observations confirm that the brackish to hypersaline waters and the suspended sediment that it is transporting have the same origin: the collapsed interior marsh.

The mechanisms for sediment input to the major sediment sink of Lake Ingraham are provided by several field observations. Low and high water within HSC South (Station 3, Fig. 13b) lags approximately 2 hours behind Stations 1 and 4. As seawater comes in on a rising tide, the waters from the interior marsh are still draining from nearby HSC South and ESC (Fig. 4). Plumes of dark brown water can be observed to drain from HSC South into less turbid waters of ECC, creating swirl pools and vortices as they mix with incoming Florida Bay water on its way to Lake Ingraham. Time-series measurements combined with these field observations suggest a sediment pathway model, explained in the following synthesis.

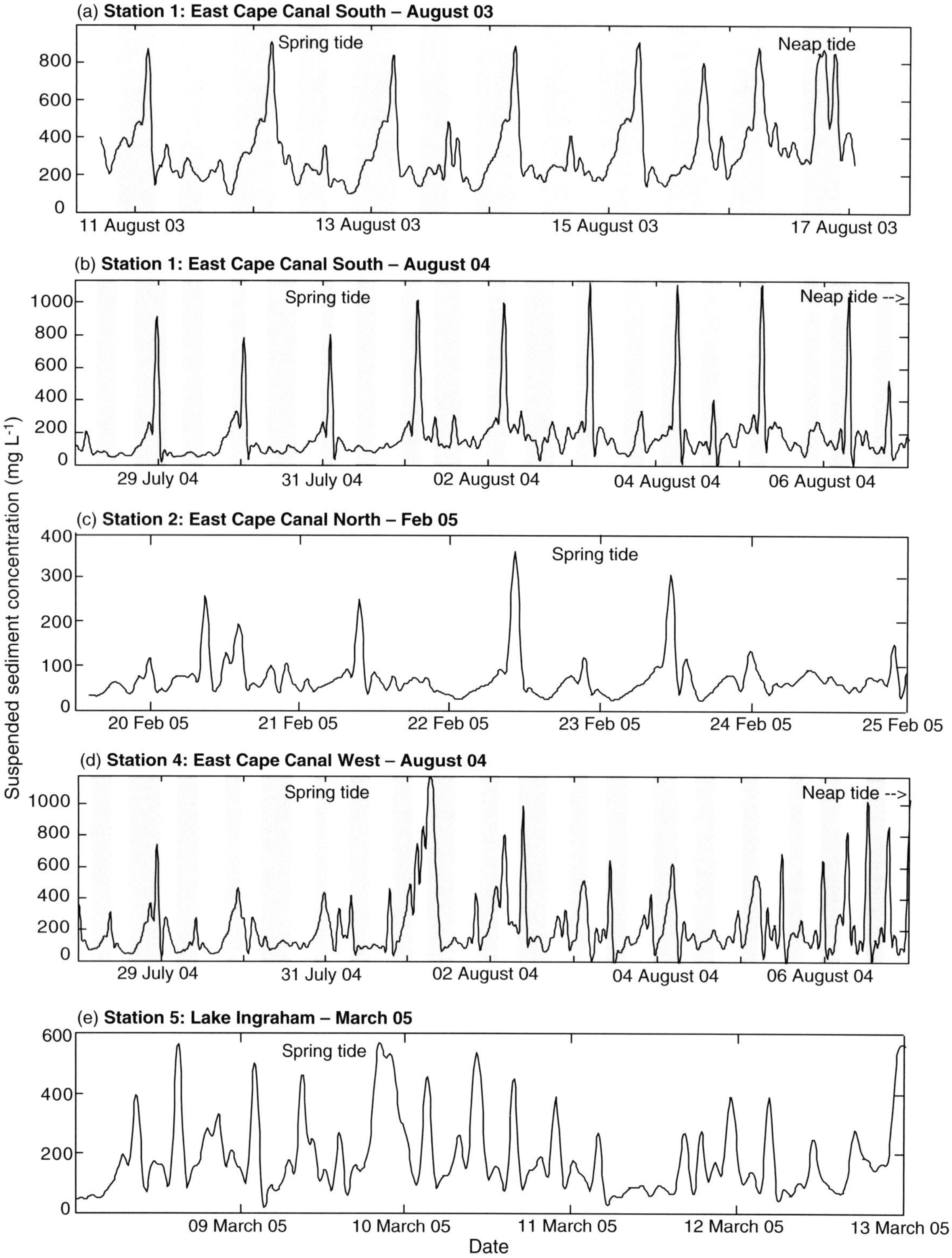

Fig. 11. (a–e) Suspended sediment concentrations for all measurement stations in East Cape Canal, measured at different time periods between August 2003 and March 2005. Grey shaded areas indicate the ebbing phase of the tide, measured with current meters.

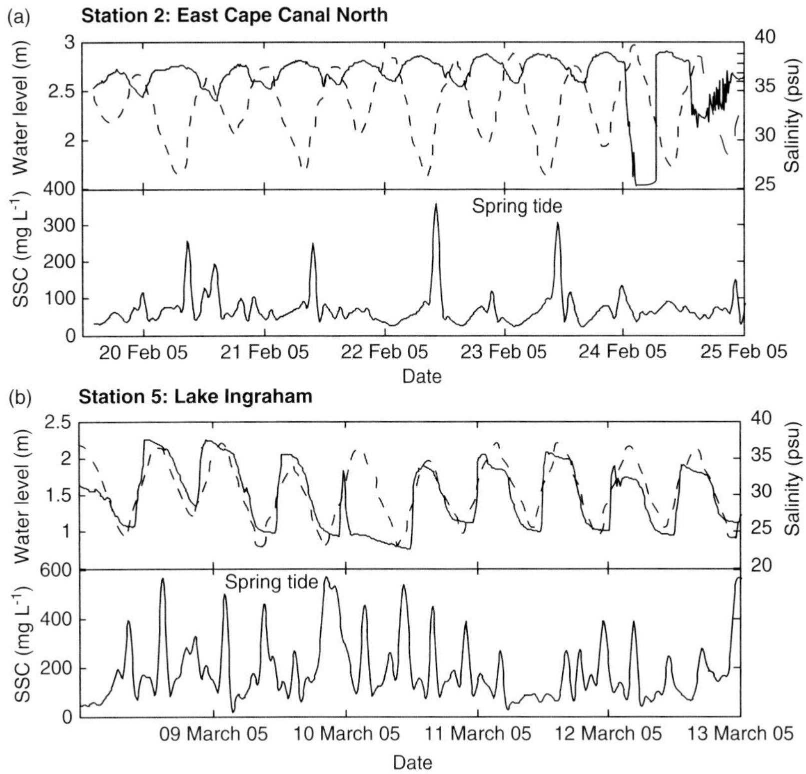

Fig. 12. Time series of water level (dashed) and salinity on top and suspended sediment concentration on bottom, measured with a CTD (conductivity–temperature–depth) and turbidity meter: (a) Station 2 in East Cape Canal (February 2005) and (b) Station 5 in Lake Ingraham (March 2005). The drop in salinity on February 24 measured at Station 2 was caused by extensive rainfall that day.

Synthesis and interpretation

The large amounts of suspended sediment during the last phase of the ebb tide contain an important component of carbonate and particulate organic matter, interpreted to originate from the collapsed and decaying marsh behind the marl ridge. A typical sediment sample in the collapsed freshwater marsh contains >80% TOM (decaying peat) at the sediment–water interface. This widespread subtidal surface of loosely compacted, low-density material can easily erode with the slightest flow. Saline water that flows over the marl ridge is trapped behind the marl ridge and can escape only through the canals and natural creeks that cut across this slightly higher, impermeable ridge. Large volumes of trapped tidal and rain water flush out from this subtidal back-barrier basin through HSC West and South and ESC, and into Lake Ingraham and ECC (Fig. 14a). As these waters 'flush' the collapsed freshwater marsh, they carry large quantities of decayed, organic-rich sediment. Every time the water level in the collapsed marsh is higher than the water level in ECC, and this situation happens basically with every ebbing tide, large volumes of water drain through the narrow canals. As the tide reverses and Florida Bay waters come in, heavy sediment plumes are still flowing out of HSC South and ESC (Fig. 14b), where they meet the incoming flood waters (Fig. 14c) and mix in the bend of ECC (Fig. 14d). Vortices of organic-rich sediment plumes are carried by the incoming tide towards Lake Ingraham (Fig. 14e), building a rapidly accreting delta (Fig. 14f).

DISCUSSION

Sea-level and coastal dynamics

Recent global stratigraphic and radiometric studies indicate the existence of metre-scale (0.5–2 m) high-frequency sea-level oscillations after the mid-Holocene climatic optimum around 6000 years ago (DePratter & Howard, 1981; Goodbred *et al.*, 1998; Angulo *et al.*, 1999; Islam & Tooley, 1999; Martin *et al.*, 2003). As a result, the late Holocene sea-level trend is suggested (Dominguez & Wanless, 1991; Stapor *et al.*, 1991; Baker & Haworth, 2000; Banerjee, 2000; Morton *et al.*, 2000) to have been stepwise or with embedded oscillations, instead of asymptotically reaching its present position.

A sea-level oscillation recognized throughout the Atlantic between 3200 and 2400 yr BP

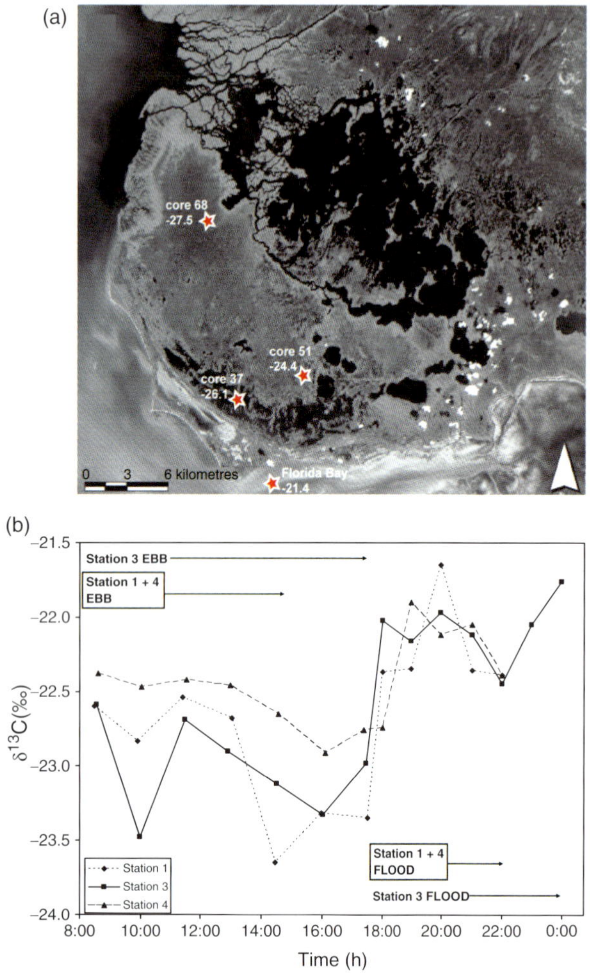

Fig. 13. (a) Stable carbon isotopic ($\delta^{13}C$) values around Cape Sable from the top 5 cm of sediment cores 37 (–26.1‰), 51 (–24.4‰) and 68 (–27.5‰) and $\delta^{13}C$ POC value of Florida Bay suspended sediment sample (–21.4‰); (b) $\delta^{13}C$ POC values for water samples at Stations 1 (ECC South), 3 (HSC) and 4 (ECC West), measured during one tidal cycle on 16 January 2005 (see Fig. 4 for station locations). Notice 2 hour delay at Station 3 (compared with Station 1 + 4) for low and high water (arrows indicate end of ebb and flood). In general, the ebb tide carries isotopically lighter suspended sediment than a flood tide. Station 1 displays largest isotopic variability (Δ 2‰), Station 4 displays least isotopic variability (Δ 1‰) throughout tidal cycle.

(DePratter & Howard, 1981; Dominguez & Wanless, 1991) has also been recognized in Florida (Gelsanliter, 1996) and has dramatically influenced coastal stratigraphy and sedimentation. A global cooling starting around 3200 yr BP and peaking at 2800 yr BP caused relative sea-level in south Florida to stabilize at –1.8 m and possibly fall (Wanless *et al.*, 1994). A small, quick rise of RSL (<1 m) followed between 2500 and 2400 yr BP and raised sea level up to about –1.2 m (Wanless *et al.*, 1994). This rise destabilized coastal environments in southwest Florida (Gelsanliter, 1996) and initiated a 400-year period of recycling of unstable sediment bodies. This transgressive, recycling phase triggered rapid resedimentation, as sediment bodies were reworked and sediment was transferred to more stable sites. A vast, subtidal to supratidal carbonate flat, extending for more than 100 km along the southwest coast, was deposited. This mud flat succession is up to 2 m thick and 8 km wide. Coastal mud deposition choked and blocked the palaeodrainage channels of the Everglades outflow, which had been through the Lake Ingraham area of Cape Sable (Jackson & Wanless, 2004). The emergent carbonate marl ridges that exist today behind the present shoreline of Lake Ingraham and along the north shore of Florida Bay were built in response to this 2500 yr BP rapid sea-level rise (Gelsanliter, 1996).

More recently, rapid RSL rise (IPCC, 2001) is again changing the rates and patterns of coastal response. The present rise is seven times faster than the average rate of rise during the previous 2400 years, during which littoral and shallow marine environments shallowed and prograded. The coastal and shallow marine environments in southwest Florida are actively responding to the accelerated rate of rise in sea level, eroding in some areas and rapidly accreting in others. Cape Sable, with several natural and anthropogenic triggering events in the past century, illustrates the nearly instantaneous response that can occur in a sediment-rich coastal system.

Rapid recycling via complex sediment transport pathways

The data presented in this paper suggest that sediment is recycled and stored within the Cape Sable coastal system. The resulting shallowing-upwards sediment package in Lake Ingraham and the Southern Lakes contains high percentages of TOM (15–35%) and accumulates with an average rate of 6.2 cm yr^{-1} ± 2 cm. In comparison, typical carbonate muds in the coastal bays of south Florida do not exceed accumulation rates of 2 cm yr^{-1} (Bosence *et al.*, 1985; Wanless & Tagett, 1989; Holmes *et al.*, 2001; Strasser & Samankassou, 2003) and contain 2–10% TOM (Lutz, 1997; Wanless *et al.*, 2005). The high organic content and accumulation rate, acting in conjunction with high erosion rates observed both in the marine and transitional freshwater environments (Wanless & Vlaswinkel, 2005), demonstrate the major recycling phase this coast is undergoing.

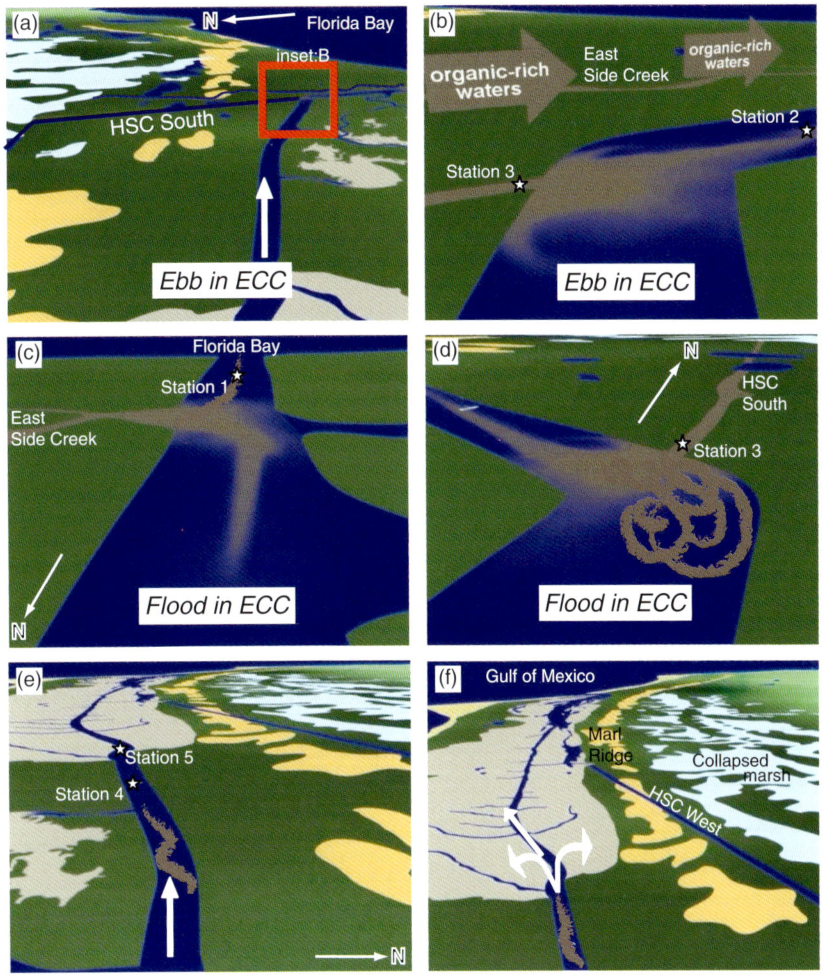

Fig. 14. 'Fly-over' cartoons visualizing the movement of sediment though a tidal cycle (12 hours). Cartoons show a birds-eye view above East Cape Canal (ECC), first moving away from Lake Ingraham eastward (a and b), then southward (c), turning around to north (d) and then moving westward back to Lake Ingraham (e and f). Waters draining from Lake Ingraham through ECC on an ebbing tide (a) join with organic-rich density plumes from Homestead Canal (HSC) South (b). Large sediment peaks are measured at Station 1 at the end of the ebb (c), but as the tide reverses, Florida Bay water carries organic-rich waters that are still draining from East Side Creek (c) and Homestead Canal South (d) towards Lake Ingraham (e), where flow diverges, and mixed carbonate–organic sediment settles to form a rapidly accreting delta (f).

Daily tidal currents are the most important agent responsible for the high sedimentation rates, as revealed by *in situ* sedimentation and time-series measurements made during 2004 and 2005. Past hurricanes have surely been the most important contributor to large, sudden morphological changes such as stepwise shoreline erosion and breaching of canals. However, since 1960 there has not been a category 4 or 5 hurricane affecting the study area directly; smaller hurricanes passing in 2005 had minimal effect on the sediment patterns and rates. Thus, the rapid sedimentation is interpreted to be a response to accelerated rate of RSL rise, responsible for the large sedimentological and ecological changes set into motion behind the ridge, probably given a hand by small human modifications (canals) acting as temporal catalysts.

Field observations, geochemical and salinity data suggest that the large volumes of sediment filling Lake Ingraham and the Southern Lakes are derived from three sediment sources that feed suspended sediment into ECC and the lakes. Source 1 is a carbonate-rich component that comes from flood tidal waters entering from Florida Bay through ECC and Hidden Creek; Source 2 is a mixed carbonate/organic component, derived from erosional widening of creeks and canals; and Source 3 is a dominantly organic component, of material discharging from the collapsed interior marsh out through HSC South and West and ESC. Because flood tide waters are washing across the marl ridge into the interior wetlands, prolonged discharge occurs from HSC South and West and ESC, persisting well after the flood tide has come in through ECC. This organic-rich, sediment-laden water from the interior (Source 3) thus largely moves with the flood tide into Lake Ingraham and the Southern Lakes rather than out to sea. It is well established that offshore sediment (Source 1) commonly moves landward into coastal bays (Postma, 1961), and it is expected that erosive widening of tidal channels will provide sediment (Source 2) to interior bays. One important result of this study is the recognition that organic-rich sediments are transported from one compartment to another within the intracoastal system.

The shallowing-upwards sediment package follows the increase in accommodation space that becomes available initially close to shore, in the coastal lakes, and later, further inland in the marshes. The first geomorphical features to fill in were small ponds directly behind the beach ridges (the Southern Lakes). Second was the large coastal lagoon that lies slightly more inland (Lake Ingraham). Finally, with sea-levels overtopping the marl ridge, the interior former freshwater marsh first collapsed to become a back-barrier lagoon, and recently started to infill with muds carried across the marl ridge by flood tides and storms. With steadily rising sea level, marine floodwaters probably will spill more frequently across the ridge, feeding a progressively larger tidal volume (prism). Generally, a larger tidal prism results in a larger cross-sectional area of the channels through which the water flows (O'Brien, 1931), and thus increased channel margin erosion can be expected, as the main channels already reach to bedrock.

Geological significance

It has proved difficult to determine from the stratigraphic record whether shallowing-upwards carbonate rock associations are driven by autocyclicity or allocyclicity (Lehrmann & Goldhammer, 1999). Mechanisms responsible for formation of peritidal lithological successions have been a focus of debate between supporters of the two end-member models (Burgess *et al.*, 2001). The main issue is whether the stacked shallowing-upwards successions were deposited due to external forcing such as relative sea-level oscillations (cf. Goodwin & Anderson, 1985; Goldhammer *et al.*, 1990; Osleger & Read, 1991; Yang *et al.*, 1995), or whether intrinsic dynamics, such as changing rates of carbonate productivity or sediment supply, could be a plausible alternative to create such stratal patterns (cf. Ginsburg, 1971; Pratt & James, 1986; Drummond & Wilkinson, 1993; Burgess, 2001). However, in both end-member models, the shallowing-upwards successions represent a distinct high-stand cycle of sedimentation with the peritidal deposits representing coastal progradation (Laporte, 1967; James, 1979; Wilkinson, 1982; Pratt & James, 1986).

However, the general assumption that each shallowing-upwards cycle represents one single sea-level excursion may well be incorrect. During a single episode of platform flooding, multiple upward-coarsening carbonate cycles, deposited during the infilling of Florida Bay by repeated aggradation of mudstone to packstone successions, have been described by Tedesco & Wanless (1991). Drummond & Wilkinson (1993) also create, in a one-dimensional computational forward model of carbonate accumulation, multiple shallowing-upwards cycles during a single sea-level fluctuation. These results from both field and modelling studies serve as a caution that a simple one-to-one relationship between sea-level change and cycle response may not always exist, particularly with scenarios of low-amplitude sea-level change.

Recent research on late Holocene sea level suggests small sea-level oscillations or steps embedded within the overall sea-level rise. This study of sedimentation in response to the latest small rise along the southwest coast of Florida shows that these embedded small, accelerated sea-level jumps can trigger rapid sedimentation and generation of a shallowing-upwards cycle within the coastal system. This recent rise, combined with the sea-level oscillation between 3200 and 2400 yr BP, occurs within a single overall sea-level high-stand. These embedded oscillations are generating multiple shallowing-upwards sediment successions. These multiple successions can either be vertically stacked or spatially offset, and completely, partially or not eroded by the subsequent succession. Importantly, the nature of each embedded succession may be quite different, controlled by the nature and pattern of pre-existing sedimentary environments and topography, and the details of the sea-level oscillation.

The data and interpretations described in this paper provide new insights into the generation of small-scale shallowing-upwards successions. Observations of modern environments such as Cape Sable reveal that each pulse of sea-level rise can result in a deposition of a shallowing-upwards facies succession and, since transgression is stepwise, multiple shallowing-upward successions may develop. Applying these findings to the rock record, a composite set of metre-scale peritidal carbonates, in which each unit commonly is interpreted to be deposited during a prolonged period of stable or slowly rising sea level, might indeed have been the depositional result of a series of rapid, small pulses of sea-level rise, within an overall high-stand.

CONCLUSIONS

The present depositional buildups of Cape Sable and the late Holocene geometry and stratigraphy of southwest Florida are the result of complex interplay between relative sea-level fluctuations, sediment production, distribution and accumulation, antecedent topography and climate. They preserve an excellent record of the principles governing accumulation and progradation of a low-energy mixed carbonate system. The following conclusions are based mainly on the results of this field study but also well-founded on many years of work in south Florida by the second author and his former students.

1 Rapid relative sea-level rise (>20 cm 100 yr^{-1}) results in a nearly instantaneous destabilization of the coastal system, erosion and transport of sediment in the nearshore zone and intracoastal redeposition of a shallowing-upwards facies succession.

2 Transgressive recycling in tropical, low-energy coastal systems can release substantial amounts of organic material leading to deposition and potential preservation of organic-enriched carbonates.

3 Intertidal mudflats can accumulate fine-grained material very rapidly, provided that sediment supply is abundant and transport processes are favourable (daily tides in this study).

4 A shallowing-upwards facies succession can be the product of a small sea-level oscillation, which is embedded within an overall rising sea-level trend. A set of multiple shallowing-upwards peritidal cycles, commonly interpreted in the stratigraphic record as an integral part of coastal progradation during high-stand, might be the depositional expression of a transgression.

ACKNOWLEDGEMENTS

This paper forms part of a PhD dissertation written at the University of Miami under the supervision of Harold Wanless and Eugene Rankey. The work was supported by a research grant from Everglades National Park and park personnel are acknowledged for granting permits, boats and a helicopter to collect within the park boundaries. Additional financial support was provided by the Comparative Sedimentology Laboratory at the University of Miami. All field assistants are thanked for their time in the field and Tali Babila is thanked for her help in the laboratory. Peter Swart is thanked for providing stable isotope data shown in Figure 13. The first author is especially grateful to Laurent Chérubin for continued inspiration and thought provoking discussions throughout the various field and data analysis stages of the project. Robert Ginsburg is thanked for his encouragement to always go after scientific challenges and to try answering each time the question: So What?

REFERENCES

Allen, J.R.L. (2000) Morphodynamics of Holocene salt marshes: a review sketch from the Atlantic and Southern North Sea coasts of Europe. *Quaternary Sci. Rev.*, **19**, 1155–1231.

Angulo, R.J., Giannini, P.C.F., Suguio, K. and **Pessenda, L.C.R.** (1999) Relative sea-level changes in the last 5500 years in southern Brazil (Laguna-Imbituba region, Santa Catarina State) based on vermetid 14C ages. *Marine Geol.*, **159**, 323–339.

Baker, R.G.V. and **Haworth, R.J.** (2000) Smooth or oscillating late Holocene sea-level curve? Evidence from the palaeo-zoology of fixed biological indicators in east Australia and beyond. *Marine Geol.*, **163**, 367–386.

Banerjee, P.K. (2000) Holocene and late Pleistocene relative sea-level fluctuations along the east coast of India. *Marine Geol.*, **167**, 243–260.

Barron, E.J. (1976) *Suspended Sedimentation Processes, Marco Island, Florida*. Technical Report TR76–9, University of Miami, RSMAS, FL, 183 pp.

Bathurst, R.G.C. (1975) *Carbonate Sediments and their Diagenesis: Developments in Sedimentology* (No .12), 2nd edn. Elsevier, Amsterdam, 658 pp.

Bischof, B.G. (1995) *Aerial Photographic Analysis of Coastal and Estuarine Mangrove System Dynamics of the Everglades National Park, Florida, in Response to Hurricanes: Implications for the Continuing Sea-level Rise*. MS thesis, University of Miami, Coral Gables, FL, 151 pp.

Bosence, D.W.J. (1995) Anatomy of a recent biodetrital mud-mound, Florida Bay, USA. In: *Carbonate Mud-mounds, their Origin and Evolution* (Eds C.L.V. Monty *et al.*). *Int. Assoc. Sedimentol. Special Publ.*, **23**, 475–493.

Bosence, D.W.J., Rowlands, R. and **Quine, M.** (1985) Sedimentology and budget of a recent carbonate mound, Florida Keys. *Sedimentology*, **32**, 317–343.

Burgess, P.M. (2001) Modeling carbonate sequence development without relative sea-level oscillations. *Geology*, **29**(12), 1127–1130.

Burgess, P.M., Wright, V.P. and **Emery, D.** (2001) Numerical forward modelling of peritidal carbonate parasequence development: implications for outcrop interpretation. *Basin Res.*, **13**, 1–16.

Carter, R.W.G. (1988) *Coastal Environments: An Introduction to the Physical, Ecological and Cultural Systems of Coastlines*. Academic Press, London, 617 pp.

Chmura, G.L. and **Aharon, P.** (1995) Stable carbon isotope signatures of sedimentary carbon in coastal wetlands as indicators of salinity regime. *J. Coastal Res.*, **11**, 124–135.

Cottrell, D.J. (1989) *Holocene evolution of the coast and nearshore islands, northeast Florida Bay, Florida.* PhD dissertation, University of Miami, Coral Gables, Florida, 194 pp.

Davis, G.E. (1972) *A Review of the Man-made Canals in the Cape Sable Region of Everglades National Park, Florida.* Report to superintendent Everglades National Park, 15 pp.

Davis, J.H. (1940) *The Ecology and Geologic Role of Mangroves in Florida.* Papers from the Tortugas Laboratory, **32**, 302–412.

DePratter, C.B. and **Howard, J.D.** (1981) Evidence for a sea-level lowstand between 4500 and 2400 years B.P. on the Southeast coast of the United States. *J. Sedimentary Petrol.*, **51**(4), 1287–1295.

Dominguez, J.M.L. and **Wanless, H.R.** (1991) Facies architecture of a falling sea-level strandplain, Doce River coast, Brazil. *Special Publication Int. Assoc. Sedimentol.*, **14**, 259–281.

Douglas, B.C. (1991) Global sea-level rise. *J. Geophys. Res.*, **96**(C4), 6981–6992.

Drummond, C.N. and **Wilkinson, B.H.** (1993) Carbonate cycle stacking patterns and hierarchies or orbitally forced eustatic sealevel change. *J. Sedimentary Petrol.*, **63**(3), 369–377.

Duevier, M.J., Meeder, J.F., Meeder, L.C. and **McCollom, J.M.** (1994) The climate of South Florida and its role in shaping the Everglades Ecosystem. In: *Everglades, the Ecosystem and its Restoration* (Eds S.M. Davis and J.C. Ogden). St Lucie Press, Delray Beach Florida, pp. 225–248.

Dunham, R.J. (1962) Classification of carbonate rocks according to depositional texture. *Am. Assoc. Petrol. Geol. Mem.*, **1**, 108–121.

Enos, P. and **Perkins, R.D.** (1979) Evolution of Florida bay from island stratigraphy. *Geol. Soc. Am. Bull.*, part 1, **90**, 59–83.

Evans, M.W., Hine, A.C., Belknap, D.F. and **Davis, R.A.** (1985) Bedrock controls on barrier island development: west-central Florida coast. *Marine Geol.*, **63**, 263–283.

Fairbridge, R.W. (1974) The Holocene sea-level record in south Florida. In: *Environments of South Florida: Past and Present* (Ed. P.J. Gleason), 2nd edn. Miami Geological Society, Miami, pp. 223–232.

Fleming, M., Lin, G., da-Silveira, L. and **Steinberg, L.** (1990) Influence of mangrove detritus in an estuarine ecosystem. *Bull. Marine Sci.*, **47**, 663–669.

Frederick, B.C. (1994) *The Holocene stratigraphy of the Broad-Lostman's River region in response to rising sea-level, SW Florida coast.* MS thesis, University of Miami, Coral Gables, Florida, 173 pp.

Gebelein, C.D. (1977) *Dynamics of Recent Carbonate Sedimentation and Ecology, Cape Sable, Florida.* E.J. Brill, Leiden, 120 pp.

Gelsanliter, S. (1996) *The Holocene stratigraphy of the Chatham river region, Southwest Florida; with a reevaluation of the late Holocene sea-level curve.* MS thesis, University of Miami, Coral Gables, Florida, 182 pp.

Ginsburg, R.N. (1956) Environmental relationships of grain size and constituent particles in some south Florida carbonate sediments. *Bull. Am. Assoc. Petrol. Geol.*, **40**, 2384–2427.

Ginsburg, R.N. (1971) Landward movement of carbonate mud: new model for regressive cycles in carbonates. *Am. Assoc. Petrol. Geol.* Convention Abstract, **55**, 340.

Goodbred, S.L., Hine, A.C. and **Wright, E.E.** (1998) Sea-level change and storm surge deposition in a late Holocene Florida salt marsh. *J. Sedimentary Res.*, **68**, 240–252.

Goodwin, P.W. and **Anderson, E.J.** (1985) Punctuated aggradational cycles: a general hypothesis of episodic stratigraphic accumulation. *J. Geol.*, **93**, 515–533.

Goldhammer, R., Dunn, P. and **Hardie, L.** (1990) Depositional cycles, composite sea-level changes, cycle stacking patterns, and the hierarchy of stratigraphic forcing: examples from Alpine Triassic platform carbonates. *Geol. Soc. Am. Bull.*, **102**, 535–562.

Gorsline, D.S. (1963) Environments of carbonate deposition, Florida Bay and the Florida Straits. In: *Shelf Carbonates of the Paradox Basin* (Ed. R.O. Bass). Four Corners. Geol. Soc. Symp., 4th Field Conf., pp. 130–143.

Gunderson, L.H. (1994) Vegetation of the Everglades: Determinants of community composition. In: *Everglades, the Ecosystem and its Restoration* (Eds S.M. Davis and J.C. Ogden). St Lucie Press, Delray Beach, FL, pp. 323–340.

Hardy, J.W. and **Henderson, K.G.** (2003) Cold front variability in the southern United States and the influence of atmospheric teleconnection patterns. *Phys. Geogr.*, **24**, 120–137.

Hine, A.C., Belknap, D.F., Hutton, J.G., Osking, E.B. and **Evans, M.W.** (1988) Recent geological history and modern sedimentary processes along an incipient, low-energy, epicontinental-sea coastline: northwest Florida. *J. Sedimentary Petrol.*, **58**, 567–579.

Holmes, C.W., Robbins, J., Halley, R. and **Bothner, M.** (2001) Sediment dynamics of Florida bay mud banks on a decadal time scale. *Bull. Am. Paleontol.*, **361**, 31–40.

Huang, H. (1990) *Holocene environmental history in a marginal marine area of the Everglades of South Florida.* MS thesis, University of Miami, Coral Gables, FL, 132 pp.

IPCC (2001) *Climate Change 2001: the Scientific Basis – International Panel on Climate Change.* Cambridge University Press, Cambridge, 881 pp.

Islam, M.S. and **Tooley, M.J.** (1999) Coastal and sea-level changes during the Holocene in Bangladesh. *Quat. Int.*, **55**, 61–75.

Jackson, K.L. and **Wanless, H.R.** (2004) Shift of Everglades discharge in response to late Holocene coastal buildup, southwest Florida. *Geol. Soc. Am. Abstracts with Programs*, **36**(5), 192.

James, N.P. (1979) Shallowing-upward sequences in carbonates. In: *Facies Models* (Ed. R. Walker). Geoscience Canada Reprint Series 1, Ottowa, pp. 109–119.

Jones, R.D. and **Boyer, J.N.** (1999) *An Integrated Surface Water Quality Monitoring Program for South Florida Coastal Waters.* Cumulative report to Everglades National Park and South Florida Water Management District. http://serc.fiu.edu/wqmnetwork/

Kennicutt II, M.C., Barker, C., Brooks, J.M., DeFreitas, D.A. and **Zhu, G.H.** (1987) Selected organic matter source indicators in the Orinoco, Nile and Changjiang deltas. *Org. Geochem.*, **11**, 41–51.

Laporte, L. (1967) Carbonate deposition near mean sea-level and resultant facies mosaic: Manlius Formation

(Lower Devonian) of New York State. *AAPG Bull.*, **51**(1), 73–101.

Lehrmann, D.J. and **Goldhammer, R.K.** (1999) Secular variation in parasequence and facies stacking patterns of platform carbonates: a guide to application of stacking pattern analysis in strata of diverse ages and settings. In: *Advances in Carbonate Sequencestratigraphy: Applications to Reservoirs, Outcrops and Models* (Eds P.M. Harris, A.H. Saller and J.A. Simo). Society for Sedimentary Geology Special Publication **63**, pp. 187–225.

Lobo, F.J., Fernández-Salas, L.M., Hernández-Molina, F.J., González, R., Dias, J.M.A., Días del Río, V. and **Somoza, L.** (2005) Holocene highstand deposits in the Gulf of Cadiz, SW Iberian Peninsula: a high resolution record of hierarchical environmental changes. *Marine Geol.*, **219**, 109–131.

Lutz, M.J. (1997) *A carbon isotope study of the flux of organic material in a subtropical carbonate estuary, Florida Bay.* MS thesis, University of Miami, Coral Gables, FL, 144 pp.

Macintyre, I.G. and **Reid, R.P.** (1992) Comment on the origin of aragonite needle mud: a picture is worth a thousand words. *J. Sedimentary Petrol.*, **62** (6), 1095–1097.

Martin, L., Dominguez, J.M.L. and **Bittencourt, A.C.S.P.** (2003) Fluctuating Holocene sea-levels in Eastern and Southeastern Brazil: evidence from multiple fossil and geometric indicators. *J. Coastal Res.*, **19**, 101–124.

Maul, G.A. and **Martin, D.M.** (1993) Sea-level rise at Key West, Florida, 1846–1992: America's longest instrument record? *Geophys. Res. Lett.*, **20** (18), 1955–1958.

Mead, R.N. (2003) *Organic matter dynamics in the Florida coastal Everglades: a molecular marker and isotopic approach.* Ph.D. dissertation, Florida International University, Miami.

Morton, R.A., Paine, J.G. and **Blum, M.D.** (2000) Responses of stable bay-margin and barrier-island systems to Holocene sea-level highstands, western Gulf of Mexico. *J. Sedimentary Res.*, **70**, 478–490.

Nelsen, Jr., J.E. and **Ginsburg, R.N.** (1986) Calcium carbonate production by epibionts on Thalassia in Florida Bay. *J. Sedimentary Petrol.*, **56**(5), 622–628.

Neumann, C.J., Jarvinen, B.R. McAdie, C.J. and **Hammer, G.R.** (1993) *Tropical Cyclones of the North Atlantic Ocean, 1871–1998.* NOAA Historical Climatology Series 6–2. The National Climatic Data Center, Asheville, NC, (http://www.ncdc.noaa.gov), 180 pp.

O'Brien, M.P. (1931) Estuary tidal prisms related to entrance areas. *Civil Eng.*, 1(8), 738–739.

Osleger, D. and **Read, J.F.** (1991) Relation of eustacy to stacking patterns of meter-scale carbonate cycles, Late Cambrian, USA. *J. Sedimentary Petrol.*, **61**, 1225–1252.

Otvos, E.G. (2004) Holocene Gulf levels: recognition issues and an updated sea-level curve. *J. Coastal Res.*, **20**(3), 680–699.

Parkinson, R.W. (1987) *Holocene sedimentation and coastal response to rising sea-level along a subtropical low energy coast. Ten thousand Islands, Southwest Florida.* PhD Dissertation, University of Miami, Coral Gables, FL, 225 pp.

Parkinson, R.W. (1989) Decelerating Holocene sea-level rise and its influence on the southwest Florida coastal evolution; a transgressive/regressive stratigraphy. *J. Sedimentary Petrol.*, **59**(6), 960–972.

Peng, G., Mooers, C.N.K. and **Graber, H.C.** (1999) Coastal winds in South Florida. *J. Appl. Meteorol.*, **38**, 1740–1757.

Perlmutter, M.A. (1982) *The recognition and reconstruction of storm sedimentation in the near-shore, Southwest Florida.* PhD dissertation, University of Miami, Coral Gables, FL, 230 pp.

Postma, H. (1961) Transport and accumulation of suspended matter in the Dutch Wadden Sea. Netherlands. *J. Sea Res.*, **1**, 148–190.

Pratt, B.R. and **James, N.P.** (1986) The St George Group (Lower Ordovician) of western Newfoundland: tidal flat island model for carbonate sedimentation in shallow epeiric seas. *Sedimentology*, **33**, 313–343.

Rankey, E.C. and **Morgan, J.J.** (2002) Quantified rates of geomorphic change on a modern carbonate tidal flat: Andros Island , Bahamas. *Geology*, **30**, 583–587.

Robbin, D.M. (1984) A new Holocene sea-level curve for the upper Florida Keys and Florida reef tract. In: *Environments of South Florida: Past and Present* (Ed. P.J. Gleason), 2nd edn. Miami Geological Society, Miami, pp. 437–458.

Roberts, H.H., Whelan, T. and **Smith, W.G.** (1977) Holocene sedimentation at Cape Sable, South Florida. *Sedimentary Geol.*, **18**, 25–60.

Rona, P.A. and **Clay, C.S.** (1966) Continuous seismic profiles of the continental terrace of southeast Florida. *Bull. Geol. Soc. Am.*, **77**, 31–44.

Scholl, D.W. (1964) Recent sedimentary record in mangrove swamps and rise in sea-level over the southwestern coast of Florida. *Marine Geol.*, **1**, 344–367.

Scholl, D.W. and **Stuiver, M.** (1967) Recent submergence of southern Florida: a comparison with adjacent coasts and other eustatic data. *GSA Bull.*, **78**, 437–454.

Scholl, D.W., Craighead, F.C. and **Stuiver, M.** (1969) Florida submergence curve revisited: its relation to sedimentation rates. *Science*, **163**, 562–564.

Simons, G. and **Ogden, L.** (1998) *Gladesmen: Gator Hunters, Moonshiners, and Skiffers.* University Press of Florida, Gainesville, FL, 191 pp.

Spackman, W., Scholl, D.W. and **Taft, W.H.** (1964) Environments of coal formation in South Florida. *Geological Society of America Pre-Convention Guidebook*, Miami Geological Society, Miami, 67 pp.

Stapor Jr., F.W., Mathews, T.D. and **Lindfors-Kearns, F.E.** (1991) Barrier-island progradation and Holocene sea-level history in Southwest Florida. *J. Coastal Res.*, **7**, 815–838.

Stockman, K.W., Ginsburg, R.N. and **Shinn, E.A.** (1967) The production of lime mud by algae in South Florida. *J. Sedimentary Petrol.*, **37**, 633–648.

Strasser, A. and **Samankassou, E.** (2003) Carbonate sedimentation rates today and in the past: Holocene of Florida Bay, Bahamas, and Bermuda vs. Upper Jurassic and Lower Cretaceous of the Jura Mountains (Switzerland and France). *Geol. Croatica*, **56/1**, 1–18.

Tedesco, L.P. and **Wanless, H.R.** (1991) Generation of sedimentary fabrics and facies by repetitive excavation and storm infilling of burrow networks, Holocene of South Florida and Caicos Platform, B.W.I. *Palaios*, **6**(3), 326–343.

Twilley, R.R. (1985) The exchange of organic carbon in basin mangrove forests in a southwest Florida estuary. *Estuarine, Coastal and Shelf Sci.*, **20**, 543–557.

Van Iperen, J. and **Helder, W.** (1985) A method for the determination of organic carbon in calcareous marine sediments. *Marine Geol.*, **64**, 179–187.

Van Veen, J. (1950) Ebb and flood channel systems in the Netherlands Tidal Waters. "Waddensymposium". *Tijdschr. Kon. Ned. Aardrijksk. Genoot.*, 43–65. Delft University Press, ISBN 90-407-2338-9.

Wanless, H.R. (1982) Sea-level is rising – so what? *J. Sediment. Petrol.*, **52**, 1051–1054.

Wanless, H.R. and **Tagett, M.G.** (1989) Origin, growth and evolution of carbonate mudbanks in Florida Bay. *Bull. Marine Sci.*, **44**, 454–489.

Wanless, H.R. and **Vlaswinkel, B.M.** (2005) Coastal landscape and channel evolution affecting critical habitats at Cape Sable, Everglades National Park, Florida. Final report to Everglades National Park, National Park Service, U.S. Department of Interior, 196 pp.

Wanless, H.R., Tedesco, L.P. and **Parkinson, R.W.** (1994) Sea-level control on the stability of the Everglades wetlands. In: *Everglades, the Ecosystem and its Restoration* (Eds S.M. Davis and J.C. Ogden). St Lucie Press, Delray Beach Florida, pp. 199–224.

Wanless, H.R., Tedesco, L.P., Rossinsky Jr., V. and **Dravis, J.J.** (1988) Carbonate environments and sequences of Caicos platform with an introductory evaluation of south Florida. *AGU 28 Int. Geol. Congr. Trip Guidebook*, Vol. T374, 75 pp.

Wanless, H.R., Vlaswinkel, B.M. and **Jackson, K.L.** (2005) Transgressive recycling produces organic-rich carbonate muds. *Abstract with Programs, 2005 AAPG Annual Convention*, Calgary, A-149.

Wilkinson, B.H. (1982) Cyclic cratonic carbonates and Phanerozoic calcite seas. *J. Geol. Educ.*, **30**, 189–203.

Xu, Y., Mead, R.N. and **Jaffe, R.** (2006) A molecular marker based assessment of sedimentary organic matter sources and distributions in Florida Bay. *Hydrobiologia*, **569**, 179–192.

Yang, W., Harmsen, F. and **Kominz, M.A.** (1995) Quantitative analysis of a cyclic peritidal sequence, the Middle and Upper Devonian Lost Burro Formation, Death Valley, California – a possible record of Milankovitch climatic cycles. *J. Sedimentary Res.*, **B65**, 3306–3322.

Zieman, J.C., Macko, S.A. and **Mills, A.L.** (1984) Role of seagrasses and mangroves in estuarine food webs: temporal and spatial changes in stable isotope composition and amino acid content during decomposition. *Bull. Marine Sci.*, **35**, 380–392.

Int. Assoc. Sedimentol. Spec. Publ. (2009) **41**, 113–122

The paradoxical occurrence of oolitic limestone on the eastern islands of Great Bahama Bank: where do the ooids come from?

PASCAL KINDLER* and ALBERT C. HINE†

**Section of Earth Sciences, University of Geneva, Maraîchers 13, 1205 Geneva, Switzerland (E-mail: Pascal.Kindler@terre.unige.ch)*

†Coastal Planning and Engineering Inc, Tampa Bay Regional Office, 101 16th Ave So. Suite 4, St Petersburg, FL 33701, USA

ABSTRACT

The origin of the ooids forming the bulk of many large Pleistocene and Holocene ridges on the windward islands of Great Bahama Bank is not well understood. A formation locus situated to the east of these islands would be consistent with present-day sediment fluxes, but outer platforms in these areas are extremely narrow today and appear unsuitable for large-scale ooid production. The four models presented suggest that ooids could originate (1) from these narrow outer platforms, (2) from wider outer platforms that no longer exist, (3) from the bank interior via tidal channels, and (4) directly from the bank interior by means of a westerly sediment flux. With the exception of option 3, all models are validated by field observations in Eleuthera and in the Exumas. Corroboration of models 2 (outer platform erosion) and 4 (westerly flux) further indicates that past depositional environments and sedimentary processes on Great Bahama Bank could have been significantly different from those operating today.

Keywords Great Bahama Bank, Eleuthera, Exumas, Quaternary, ooids, oolites, aeolianites.

INTRODUCTION

The main lesson that the first author learned from R.N. Ginsburg during his tenure at the University of Miami in the early 1990s was that the fundamental duty of a scientist is to constantly pose pertinent questions to himself and to others. One of the questions raised by Ginsburg, at that time, was about the paradoxical occurrence of extensive Pleistocene oolites on the windward islands of Great Bahama Bank (GBB; Fig. 1). The petrographic composition of these ridges is, indeed, peculiar because there is no obvious source of ooids, unless sediment fluxes were opposite to those operating today. This paradox is addressed in the following sections after a brief review of the formation of ooids on GBB, of the sediment transport vectors in the Bahamas, and of the surficial geology of the windward islands.

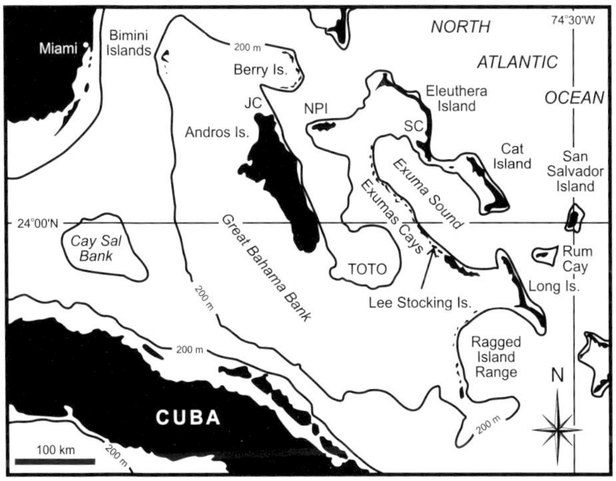

Fig. 1. Geographical situation of the Great Bahama Bank (GBB). The occurrence of islands (Eleuthera, Cat, Exumas, Long) on the eastern margin of GBB is probably related to the prevailing easterly winds. JC = Joulters Cays, NPI = New Providence Island, TOTO = Tongue of the Ocean, SC = Schooners Cays.

SETTING

Ooid production on Great Bahama Bank

The vast majority of authors (Newell *et al.*, 1960; Richter, 1983; Scholle & Ulmer-Scholle, 2003) agree that conditions required for the optimal production of Bahamian ooids (i.e. aragonitic ooids with a tangential crystal structure) are: (1) the presence of sedimentary particles that can serve as nuclei; (2) a flat topography where these particles can

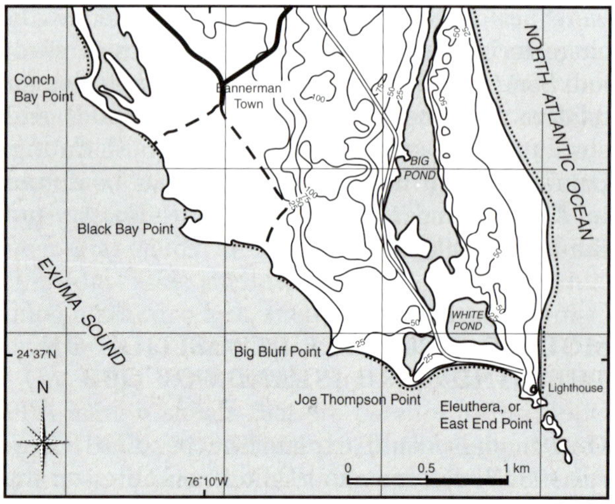

Fig. 3. Topographic map of East End Point, southern Eleuthera showing the NNW–SSE-trending, middle Holocene ridge between Big Pond and the North Atlantic shoreline. The ridge crest is located at about 4 km to the west of the bank margin.

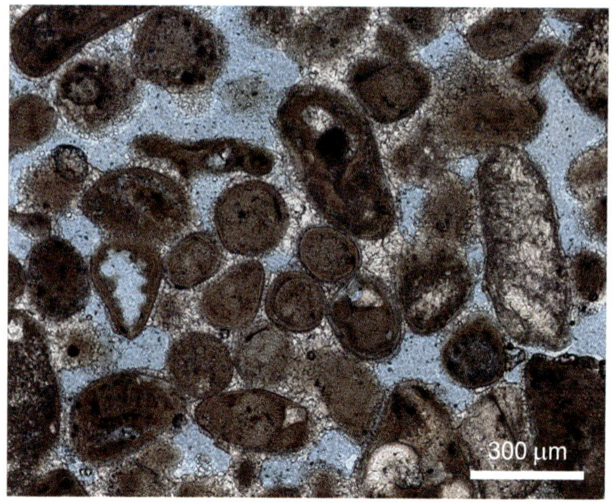

Fig. 5. Oolitic grainstone collected from the East End Point ridge (sample EL 201). Abundant superficial ooids, good particle preservation, and minor low-Mg calcite meteoric cement characterize this middle Holocene unit.

Fig. 4. Partial view of the East End Point oolitic ridge looking towards the south. The large, westward-dipping, aeolian foresets disappear beneath modern, bioclastic, beach sands, indicating that (1) the dune was deposited when sea level was slightly lower than present, and that (2) there was an ooid factory to the east of the island.

just offshore, as is the case near the Joulters Cays today. Ooid production stalled during the late Holocene, probably because of increasing water depth on the outer platform due to the ongoing transgression. Nonetheless, this depositional system was able to build up a 15-m-high oolitic ridge in about 1700 years, i.e., between the onset of open marine conditions on the bank about 4.7 ka ago (Hine, 1983) and the end of ooid production about 3 ka ago (Kindler, 1992, 1995). This oolitic ridge was then preserved due to the rapidity and efficiency of freshwater cementation processes (Halley & Harris, 1979; Kindler & Mazzolini, 2001).

To determine if it was possible to generate sufficient oolitic sand on a narrow shelf during a high-frequency sea-level event, the following simple assumptions and calculations are presented. Based upon field observations, we assume that each Pleistocene cycle would produce a 6 m thick aeolian deposit that would extend 1 km across an island such as Eleuthera. This produces a volume of 6000 m^3 if we make our cross-section to be 1 m wide. Now let us assume that the adjacent shelf was 1 km wide and oolitic sediments were generated over a 6000 yr time-frame before production shuts down. Sea-level curves indicate that the narrow (1 km wide) shelves fronting much of Eleuthera Island, for example, were flooded since 6 ka or earlier. Thus, 6000 m^3 of sediment would have to be produced across a strip on the shelf that is 1 km by 1 m or 1000 m^2 over 6000 years, or 1000 m^3 1000 yr^{-1} over this same strip of shelf. This converts to 1 $m^3\,yr^{-1}$ over 1000 m^2 and further converts to 0.001 $m^3\,yr^{-1}\,m^{-2}$. So, how thick is 0.001 m^3 when spread over 1 m^2? By converting to centimetres, it is 0.1 cm. So, we need a production rate of 0.1 cm yr^{-1} over a 1 km wide shelf for 6000 years. Can oolitic sediments be generated this fast? According to Schlager (1981), such sediments can be produced at rates ranging from ~40 to ~110 cm 1000 yr^{-1}. To make the calculations simpler, let us choose 100 cm 1000 yr^{-1} or 0.1 cm yr^{-1} – the same number we just calculated.

So, based upon this exercise, it is possible for a 1 km wide shelf to produce sufficient oolitic sediment to generate a realistically sized dune deposit on the adjacent island, assuming efficient transport mechanisms are in place. However, if the oolitic generation phase was significantly condensed temporally and spatially to only 1000 years and 100 m wide, then a thickness of 6 cm yr^{-1} needs to be generated. This is about 60 times faster than the fastest rate reported by Schlager (1981). Perhaps, this scenario is unrealistic? Perhaps, Schlager's (1981) numbers are incorrect? Perhaps, there is another model that needs to be considered? Nevertheless, it does appear possible to create the sediment from the immediately adjacent shelf.

Ooid production on a wider platform that has since collapsed

In the second model, based on the present-day example along the Berry Islands where the outer platform is over 15 km wide (Fig. 1), ooids were formed on a wide outer shelf, which no longer exists, and carried onto the islands by the easterly trade winds. Such a carbonate factory could have been more efficient than that of the first model because of a more extensive production area. It would thus better explain the occurrence of the large oolitic ridges forming the backbone of the windward islands.

One good example is the Gregory Town ridge in the northern and central part of Eleuthera (Fig. 6). This massive, NW–SE-trending crest is 8 km long, 1.5 km wide and culminates at about 55 m. It is parallel to the Eleuthera shoreline and its top stands less than 2 km away from the present-day platform margin, which shows a scalloped (i.e. convex bankward) morphology in this area (Fig. 7 Mullins & Hine, 1989; Mullins *et al.*, 1991). On the ocean side of the island, the ridge comprises well-exposed subtidal and beach deposits up to 4 m above mean sea level. In contrast, only aeolian sediments, which locally display rainfall-induced fenestral porosity (Bain & Kindler, 1994; Kindler & Strasser, 2000), can be seen at higher elevations and on the bank side.

The Gregory Town ridge consists of fine-grained, well-lithified oolitic grainstone characterized by the predominance of 'normal' (i.e. thickly coated) ooids (Fig. 8). Amino-acid ratios obtained from ridges of identical lithology and stratigraphic position on Eleuthera (Hearty, 1998) and San Salvador (Hearty & Kindler, 1993), the occurrence of an overlying *terra rossa* palaeosol, and the presence of associated marine deposits at about 4 m all indicate that this ridge was deposited during the last interglacial period, i.e. Marine Oxygen Isotope Stage (MIS) 5e. The primary morphology of the dune is well preserved, with a lee side facing towards the bank interior and large aeolian foresets dipping towards the southwest (Fig. 6). The orientation of the lee face of the dune and the dip of the foresets clearly show that the constituent particles originated from the outer platform to the northeast. Today, this outer platform

Fig. 6. View from the top (45 m at this particular location) of the Gregory Town ridge towards the west. This ridge consists of oolitic limestone of Late Pleistocene age. Note preserved dune morphology, with the lee side facing the bank interior. Large westward-dipping foresets (arrow) further indicate that the constituent sedimentary particles originated from the east. Road cut is 4 m high.

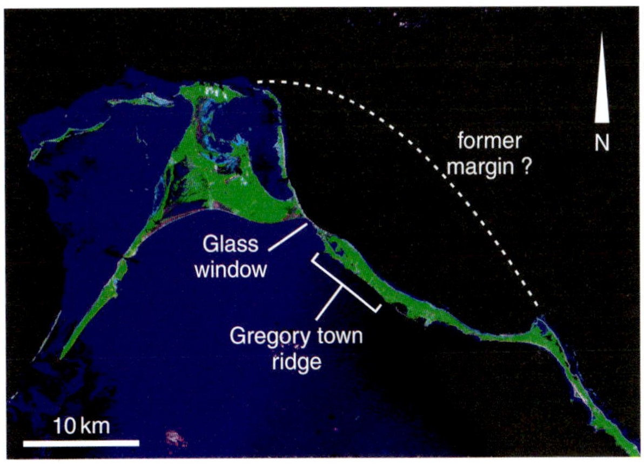

Fig. 7. Satellite view of northern Eleuthera showing scalloped (i.e. convex bankward) margin. The extreme narrowness of the outer platform in this area is probably related to bank-edge erosion or collapse. The missing part of the platform was probably the site of ooid production during MIS 5e. Image from: https://zulu.ssc.nasa.gov.

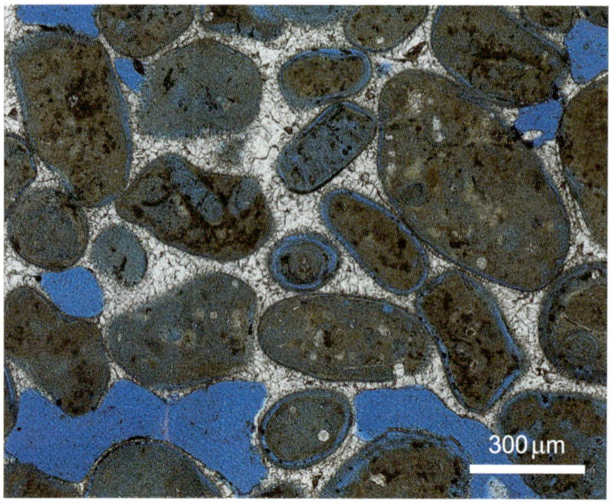

Fig. 8. Oolitic–peloidal grainstone collected from the Gregory Town ridge (sample EL 159). Note the greater abundance of normal ooids and meteoric sparry calcite cement compared with the younger East End Point oolite.

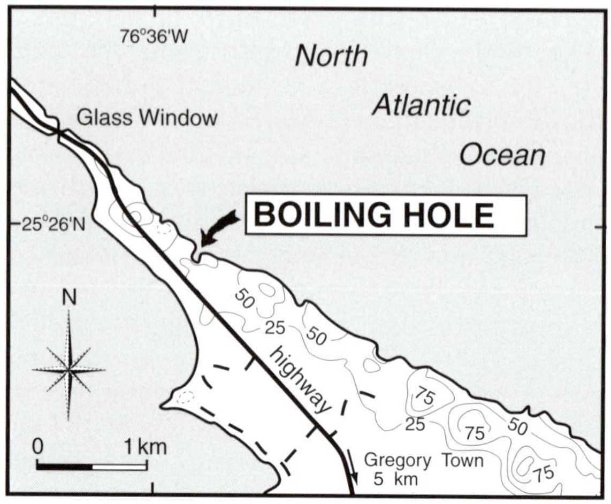

Fig. 9. Topographic map of the Boiling Hole–Glass Window area. The modern windward shoreline truncates a NW–SE trending aeolian ridge of Middle Pleistocene age, indicating that a substantial amount of regressive erosion occurred since that time (modified from the topographic map Eleuthera No. 4 published by the Department of Lands and Surveys, Nassau; contour lines are in feet).

is extremely narrow (Fig. 7), and does not seem extensive enough to support an ooid factory, even of the Joulters Cays type. However, the scalloped shape of the bank margin, which could be related to platform collapse (Mullins & Hine, 1989; Mullins *et al.*, 1991), and the truncation of middle Pleistocene deposits in the Boiling Hole area, just north of Gregory Town (Fig. 9), both suggest that the North Eleuthera bank has experienced a significant amount of recessive erosion since the late Pleistocene. Similar bank-margin retreats have been documented from the Cat Island platform (Freeman-Lynde & Ryan, 1985) and from Silver, Navidad and Mouchoir Banks in the SE Bahamas (Mullins *et al.*, 1992) for older geological periods. Furthermore, seismic data from the Bahamas/Blake Escarpment north of Little Bahama Bank (LBB) indicate about a 5 km retreat since the Early Cretaceous. It is not clear whether the recess of these bank margins results from a few major mass-wasting events or from a constant process of chipping away. In the former case, large-scale collapse of platform margins, well away from a plate boundary, would appear as a recent (i.e. late Pleistocene), if not an on-going phenomenon that could represent serious tsunami hazards for the low-lying coastal zones in the Caribbean and southeastern USA (H.T. Mullins, personal communication). Regardless, part of the large ooid accumulations exposed on the windward islands could be derived from shoals that covered an outer portion of the platform that no longer exists due to recessive erosion processes (Fig. 7).

Ooid production on the bank side and exportation to the outer platform

According to a third model, suggested to the first author by R.N. Ginsburg some years ago, ooids were formed on the bank side of the islands, transported eastward by tidal currents to the ocean side, and then back onto the islands by the trade-wind action. An illustration of this model can perhaps be observed near Lee Stocking Island (LSI, Exumas, Fig. 1). In this area, kilometre-sized ooid shoals formed by flood-tidal delta bars and levees occur at the bankward end of inter-island channels (Fig. 10; Kendall *et al.*, 1990; Gonzalez & Eberli, 1997). Indeed, one might expect that ooids could be carried seaward by ebb currents, deposited on ocean-facing beaches by waves, and finally blown onto dunes by the prevailing easterly winds. However, the modern, ocean-facing beaches on LSI are composed of bioclastic sand probably derived from offshore reefs. In addition, ebb-tidal deltas do not occur at the seaward end of the channels (Fig. 10), indicating that sediments are either swept off to deeper depositional environments (Dill *et al.*, 1989) or confined to the tidal channel by a circular hydrodynamic pattern (Reeder & Rankey, 2005).

Trade winds further provide a persistent set-up by which water is piled up against the bank margin and flows through the inter-island gaps, preferentially augmenting the daily flood-tidal

Fig. 10. Oblique aerial photograph of the southern Exumas showing Norman Ponds Cay (NPC) and Lee Stocking Island (LSI). Note extensive ooid shoal on the bankward side of the tidal channel between the two islands. Despite the presence of this ooid factory, beaches on the windward side of LSI are characterized by bioclastic sand. Photograph courtesy Paul J. Hearty.

currents over the ebb-tidal currents. For example, Lily Bank on the northern edge of LBB (Hine, 1977) clearly shows a flood or bankward dominance of bedforms, and sand-body geometric features such as bankward-oriented spillover lobes. Given an appropriate track, a hurricane could produce much stronger easterly flows that would transport lagoonal sediments eastward through the inter-island gaps and place these sediments in the windward environment for dune development. However, such an event is probably rare. Hurricanes pass by quickly, and there is simply not enough time to transport significant quantities of sand over long distances (at least several kilometres, but probably tens of kilometres) in a 3–4 hour time frame. This model does not appear to be the solution to explain the presence of extensive Pleistocene oolites on the windward islands.

Bank-side production and eastward transport of ooids

In the fourth and last model, ooids were formed on the bank interior and transported directly onto the islands by a westerly energy flux. The Boiling Hole section in northern Eleuthera (Fig. 11) represents a good illustration of such a case. At this location, oolitic limestones dating from the last interglacial (Kindler & Hearty, 1995; Hearty, 1998) partly fill a large depression between two ~20-m-high aeolianites of Middle Pleistocene age, and are capped by MIS 5a dunal deposits (Fig. 3b in Kindler & Hearty, 1997). New observations at the southeast end of this swale reveal that (1) the MIS 5e oolites consist of two shallowing-upward sequences including subtidal, beach and aeolian deposits, separated by an erosional surface (Figs 12 and 13), and (2) both beach and aeolian bed sets dip towards the southwest, i.e., the bank interior.

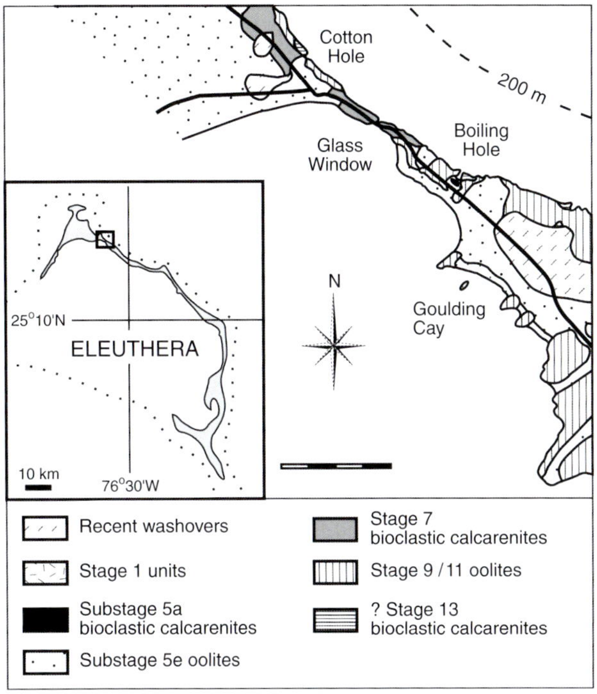

Fig. 11. Geological map of the Boiling Hole area (modified from Kindler & Hearty, 1997). Dotted line in inset is the 200 m isobath.

Fig. 12. View of the southeastern end of the Boiling Hole outcrop showing two stacked Late Pleistocene oolites (units 2 and 3) onlapping a skeletal aeolianite of Middle Pleistocene age (unit 1). Unit 3 beach deposits (upper right-hand side) show low-angle cross beds that dip towards the southwest indicating that constituent particles were formed on and transported from the bank interior.

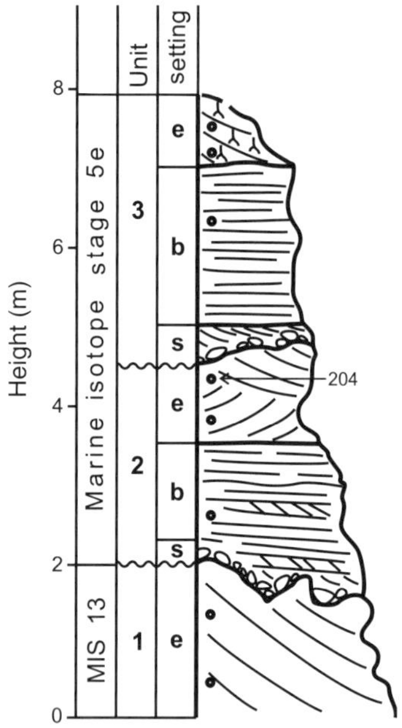

Fig. 13. Stratigraphic/sedimentological section logged on the southeastern end of the Boiling Hole site. In the setting column, s = subtidal, b = beach, e = aeolian. Dots indicate sample location. 204 marks the location of sample EL 204 illustrated in Fig. 14.

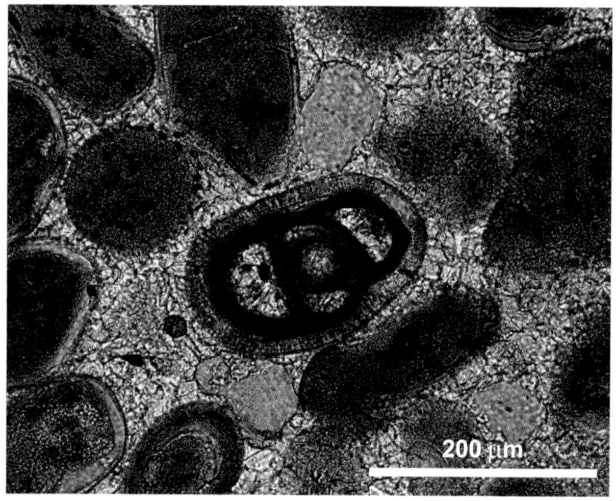

Fig. 14. Radially coated ooid from Late Pleistocene oolite at Boiling Hole (sample EL 204). Radial cortex suggests that ooid formation occurred in a relatively low-energy setting. Sample location is indicated in Fig. 13.

Further, as already noted by Kindler & Hearty (1995), the MIS 5e deposits include some peculiar ooids with a radial cortex (Fig. 14).

The dip of the beach beds clearly shows that the constituent grains of these sediments originated from the interior of the bank, to the southwest. The eolian strata dip in the same direction. However they are not foresets, as suggested by earlier authors (Kindler & Hearty, 1995; Hearty, 1998), but backsets draped over the pre-existing middle Pleistocene topography. Thus, their component particles were also brought up from the bank interior by a westerly flux, and not from the open ocean by the prevailing easterly winds, as previously interpreted. Further, the unusual occurrence of radial ooids (Fig. 14) in these deposits suggests a relatively low-energy production locus (Land *et al.*, 1979; Flügel, 2004), thus strengthening a bank-side source for these deposits. The case is even clearer at Cotton Hole, 1 km to the northwest (Fig. 11), where bankward-dipping beach sediments exposed on the ocean-facing cliffs can be walked on all the way to the bank-facing shoreline of the island. The Boiling Hole outcrop thus validates the fourth model. Moreover, ooids were not carried eastward during a single and catastrophic depositional event, such as a storm or a northwesterly gale, but by a sustained westerly flux that lasted long enough for the shoreline to prograde significantly towards the southwest.

CONCLUSIONS

The response to Ginsburg's question about the curious occurrence of large oolite accumulations on the windward islands of GBB is the following. Most likely, ooids have been produced on a very narrow outer platform between the bank edge and the islands, as is the case today along the Yucatan Peninsula (Ward & Brady, 1979), and were brought onto the islands by transport vectors (i.e. wind and waves) similar in orientation to those operating today. This model is supported by theoretical calculations. In addition, ooids may have been generated on a wide outer platform that has since disappeared due to recessive erosion. In some cases, for example the Boiling Hole area in Eleuthera, ooids have been formed on the bank side and transported by energy fluxes that were opposite to modern fluxes. The important point to consider from this 'ooid story', in particular from the fourth model, is that past sedimentary features may not necessarily be explained by the observation of modern processes, or in other words, that the present is not always a key to the past.

ACKNOWLEDGEMENTS

PK would like to thank again R.N. Ginsburg for accepting him as a post-doctoral student in his laboratory some 15 years ago, and for asking good questions. Many thanks also to F. Gischig and J. Metzger (University of Geneva) for their technical help, and to the Swiss National Science Foundation for supporting PK's travel expenses to Salt Lake City (Grant No. 200020–107436). We further thank A. Strasser, H.T. Mullins and the editors for reviewing and revising our manuscript.

REFERENCES

Bain, R.J. and **Kindler, P.** (1994) Irregular fenestrae in Bahamian eolianites: a rainstorm induced origin. *J. Sed. Res.*, **A64**, 140–146.

Ball, M.M. (1967) Carbonate sand bodies of Florida and the Bahamas. *J. Sed. Petrol.*, **37**, 556–591.

Bathurst, R.B.G. (1975) *Carbonate Sedimentology and Diagenesis*. Elsevier, Amsterdam.

Boss, S.K. and **Neumann, A.C.** (1993) Impacts of Hurricane Andrew on carbonate platform environments, northern Great Bahama Bank. *Geology*, 21, 897–900.

Carew, J.L. and **Mylroie, J.E.** (1987) A refined geochronology for San Salvador Island, Bahamas. In: *Proceedings of the 3rd Symposium on the Geology of the Bahamas* (Ed. H.A. Curran), pp. 35–44. Bahamian Field Station, San Salvador.

Carew, J.L. and **Mylroie, J.E.** (2001) Quaternary carbonate eolianites of the Bahamas: useful analogs for the interpretation of ancient rocks? In: *Modern and Ancient Carbonate Eolianites: Sedimentology, Sequence Stratigraphy, and Diagenesis* (Eds. F.E. Abegg, D.B. Loope and P.M. Harris), *SEPM Spec. Publ.*, 71, 33–45.

Department of Lands and Surveys (1975) Eleuthera, sheet 4. *Topographic Maps of the Bahamas 1: 25,000*, Nassau, Bahamas.

Dill, R.F., **Kendall, C.G.St.C.** and **Shinn, E.** (1989) Giant subtidal stromatolites and related sedimentary features. *Field Trip Guidebook T373, 28th International Geological Congress*, 33 pp., American Geophysical Union, Washington.

Flügel, E. (2004) *Microfacies of Carbonate Rocks*. Springer Verlag, Berlin, 976 pp.

Freeman-Lynde, R.P. and **Ryan, W.B.F.** (1985) Erosional modification of Bahama escarpment. *Geol. Soc. Am. Bull.*, **96**, 481–494.

Ginsburg, R.N. (2005) Disobedient sediments can feedback on their transportation, deposition and geomorphology. *Sed. Geol.*, **175**, 9–18.

Gonzalez, R. and **Eberli, G.P.** (1997) Sediment transport and bedforms in a carbonate tidal inlet; Lee Stocking Island, Exumas, Bahamas. *Sedimentology*, **44**, 1015–1030.

Haak, A.B. and **Schlager, W.** (1989) Compositional variations in calciturbidites due to sea-level fluctuations, late Quaternary, Bahamas. *Geol. Rundsch.*, **78**, 477–486.

Halley, R.B. and **Harris, P.M.** (1979) Fresh-water cementation of a 1,000-year-old oolite. *J. Sed. Petrol.*, **49**, 969–988.

Harris, P.M. (1979) *Facies Anatomy and Diagenesis of a Bahamian Ooid Shoal. Sedimenta* **VII**, University of Miami, Florida, 163 pp.

Hearty, P.J. (1998) The geology of Eleuthera Island, Bahamas: a rosetta stone of Quaternary stratigraphy and sea-level history. *Quatern. Sci. Rev.*, **17**, 333–355.

Hearty, P.J. and **Kaufman, D.S.** (2000) Whole-rock aminostratigraphy and Quaternary sea-level history of the Bahamas. *Quatern. Res.*, **54**, 163–173.

Hearty, P.J. and **Kindler, P.** (1993) New perspectives on Bahamian geology: San Salvador Island, Bahamas. *J. Coastal Res.*, **9**, 577–594.

Hearty, P.J., **Kindler, P.**, **Cheng, H.** and **Edwards, L.** (1999) A +20 m middle Pleistocene sea-level highstand (Bermuda and The Bahamas) due to partial collapse of Antarctic ice. *Geology*, **27**, 375–378.

Hine, A.C. (1977) Lily Bank, Bahamas: history of an active oolite sand shoal. *J. Sed. Petrol.*, **47**, 1554–1581.

Hine, A.C. (1983) Relict sand bodies and bedforms of the Northern Bahamas: evidence of extensive early Holocene sand transport. In: *Coated Grains* (Ed. T.M. Peryt). Springer Verlag, Berlin, pp. 116–131.

Kendall, C.G.St.C., **Dill, R.F.** and **Shinn, E.** (1990) *Guidebook to the Marine Geology and Tropical Environments of Lee Stocking Island, The Southern Exumas, Bahamas*. KenDill Publishers, San Diego, 33 pp.

Kindler, P. (1992) Coastal response to the Holocene transgression in the Bahamas: episodic sedimentation versus continuous sea-level rise. *Sed. Geol.*, **80**, 319–329.

Kindler, P. (1995) New data on the Holocene stratigraphy of Lee Stocking Island (Bahamas) and its relation to sea-level history. In: *Terrestrial and Shallow Marine Geology of the Bahamas and Bermuda* (Eds H.A. Curran and B. White), *Geol. Soc. Am. Spec. Pap.*, 300, pp. 105–116.

Kindler, P. and **Hearty, P.J.** (1995) Pre-Sangamonian eolianites in the Bahamas? New evidence from Eleuthera Island. *Mar. Geol.*, **127**, 73–86.

Kindler, P. and **Hearty, P.J.** (1996) Carbonate petrography as an indicator of climate and sea-level changes: new data from Bahamian Quaternary units. *Sedimentology*, **43**, 381–399.

Kindler, P. and **Hearty, P.J.** (1997) Geology of The Bahamas: architecture of Bahamian Islands. In: *Geology and Hydrogeology of Carbonate Islands* (Eds H.L. Vacher and T.M. Quinn), *Dev. Sedimentol.*, **54**, 141–160.

Kindler, P. and **Hearty, P.J.** (2000) Elevated marine terraces from Eleuthera (Bahamas) and Bermuda: sedimentological, petrographic and geochronological evidence for important deglaciation events during the middle Pleistocene. *Global Planet. Change*, **24**, 41–58.

Kindler, P. and **Mazzolini, D.** (2001) Sedimentology and petrography of dredged carbonate sands from Stocking Island (Bahamas). Implications for meteoric diagenesis and aeolianite formation. *Palaeogeogr. Palaeoclimatol. Palaeoecol.*, **175**, 369–379.

Kindler, P. and **Strasser, A.** (2000) Palaeoclimatic significance of co-occurring wind- and water-induced sedimentary structures in the last-interglacial coastal deposits from Bermuda and the Bahamas. *Sed. Geol.*, **131**, 1–7.

Land, L.S., **Behrens, E.W.** and **Frishman, S.A.** (1979) The ooids of Baffin Bay, Texas. *J. Sed. Petrol.*, **49**, 1269–1278.

Mackenzie, F.T. (1964) Bermuda Pleistocene eolianites and paleowinds. *Sedimentology*, **3**, 52–64.

McKee, E.D. and **Ward, W.C.** (1983) Eolian environment. In: *Carbonate Depositional Environments* (Eds P.A. Scholle, D.G. Bebout and C.H. Moore), *AAPG Mem.*, **33**, 132–170. American Association of Petroleum Geologists, Tulsa.

Mullins, H.T. and **Hine, A.C.** (1989) Scalloped bank margins: beginning of the end for carbonate platforms? *Geology*, **17**, 30–33.

Mullins, H.T., Breen, N., Dolan, J., Wellner, R.W., Petruccione, J.L., Gaylord, M., Andersen, B., Melillo, A.J., Jurgens, A.D. and **Orange, D.** (1992) Carbonate platforms along the southeast Bahamas-Hispaniola collision zones. *Mar. Geol.*, **105**, 169–209.

Mullins, H.T., Dolan, J., Breen, N., Andersen, B., Gaylord, M., Petruccione, J.L., Wellner, R.W., Melillo, A.J. and **Jurgens, A.D.** (1991) Retreat of carbonate platforms: response to tectonic processes. *Geology*, **19**, 1089–1092.

Newell, N.D., Purdy, E.G. and **Imbrie, J.** (1960) Bahamian oolitic sand. *J. Geol.*, **68**, 481–496.

Purdy, E.G. (1963) Recent calcium carbonate facies of the Great Bahama Bank. 2. Sedimentary Facies. *J. Geol.*, **71**, 472–497.

Reeder, S.L. and **Rankey, E.** (2005) Stuck on a spin cycle: interactions between tidal flows and ooid shoals, Bahamas. *Geol. Soc. Am. Ann. Meeting. Abstracts with Programs*, **37**, 401.

Reijmer, J.J., Swart, P.K., Bauch, T., Otto, R., Reuning, L. Roth, S. and **Zechel, S.** (2009) A re-evaluation of facies variations on Great Bahama Bank 1: new facies maps of Western Great Bahama Bank. In: *Perspectives in Sedimentary Geology: A Tribute to the Career of Robert Nathan Ginsburg, IAS Special Publication* (Eds P.K. Swart, G.P. Eberli and J.A. McKenzie). Blackwell, Oxford. *Int. Assoc. Sedimentol. Spec. Publ.*, **41**, 29–46.

Richter, D.K. (1983) Calcareous ooids: a synopsis. In: *Coated Grains* (Ed. T.M. Peryt). Springer Verlag, Berlin, pp. 71–99.

Schlager, W. (1981) The paradox of drowned reefs and carbonate platforms. *Geol. Soc. Am. Bull.*, **92**, 197–211.

Scholle, P.A. and **Ulmer-Scholle, D.S.** (2003) A color guide to the petrography of carbonate rocks: grains, textures, porosity, diagenesis. *AAPG Mem.*, **77**, 474 pp.

Sealey, N.E. (1994) *Bahamian Landscapes. An Introduction to the Geography of the Bahamas.* Media Publishing, Nassau, Bahamas, 128 pp.

Siewers, F.D. (2003) Oolite and coated grains. In: *Encyclopedia of Sediments and Sedimentary Rocks* (Ed. G.V. Middleton). Kluwer Academic Press, Dordrecht, pp. 502–506.

Swart, P.K., Reijmer, J.J. and **Otto, R.** (2009) A re-evaluation of facies on Great Bahama Bank II: variations in the $\delta^{13}C$, $\delta^{18}O$ and mineralogy of surface sediments. In: *Perspectives in Sedimentary Geology: A Tribute to the Career of Robert Nathan Ginsburg, IAS Special Publication* (Eds P.K. Swart, G.P. Eberli and J.A. McKenzie). Blackwell, Oxford. *Int. Assoc. Sedimentol. Spec. Publ.*, **41**, 47–60.

Ward, W.C. and **Brady, M.J.** (1979) Strandline sedimentation of carbonate grainstones, Upper Pleistocene, Yucatan Peninsula, Mexico. *Am. Assoc. Petrol. Geol. Bull.*, **63**, 362–369.

Int. Assoc. Sedimentol. Spec. Publ. (2009) **41**, 123–148

Calcareous epiphyte production in cool-water carbonate seagrass depositional environments – southern Australia

NOEL P. JAMES*, YVONNE BONE‡, KIRSTY M. BROWN†‡ and ANTHONY CHESHIRE‡

**Department of Geological Sciences and Geological Engineering, Queen's University, Kingston, Ontario, Canada K7L 3N6 (E-mail: james@geol.queensu.ca)*

‡School of Earth and Environmental Sciences, University of Adelaide, South Australia, Australia 2005 (E-mail: yvonne.bone@adelaide.edu.au)

ABSTRACT

The southern continental margin of Australia, the largest area of cool-water carbonate sedimentation on the globe, is characterized by extensive marine grassbeds in many inshore environments. The most important seagrasses in terms of calcareous epiphyte production are *Posidonia sinuosa, P. angustifolia, P. australis, Amphibolis antarctica* and *A. griffithii.* The predominant control on relative abundance of calcareous epiphytes is seagrass biomass. These grasses have a biomass of 50–500 $g\,m^{-2}$, which peaks at 2–4 m water depth. The most abundant calcareous epiphytes are geniculate (articulated) and non-geniculate (encrusting) coralline algae that together comprise ~38–80% of the epiphyte carbonate. The only other significant epiphytes are bryozoans and benthic foraminifera, which contribute roughly equal amounts (~8–33% each) of carbonate. Unlike the seagrass biomass, calcareous epiphyte abundance peaks at water depths of ~10 m. The rates of epiphyte production are roughly similar to those from epiphytes in tropical environments, averaging $210 \pm 26\,g\,m^{-2}\,yr^{-1}$. *Posidonia* is morphologically similar to tropical seagrasses (e.g. *Thalassia*) and produces largely carbonate mud from the disintegration of blade-encrusting corallines. *Amphibolis*, on the other hand, has an extensive upright, exposed shoot system that is much longer lived (biennial) and so is encrusted with prolific articulated corallines thus producing ~3× more carbonate in terms of $g\,kg^{-1}$ from the stems than the blades. Average accumulation rates of epiphytic carbonate are calculated to be ~7.4 $cm\,kyr^{-1}$. This accounts for a major proportion of the carbonate sequestered in grass beds in this cool-water realm, and probably accounts for much of the nearshore and supratidal carbonate mud. Thus, the nearshore, grass-covered habitat is a cool-water carbonate factory surprisingly similar to the shallow-water tropical system, except that the sediment produced is poorly sorted Mg-calcite carbonate with little or no aragonite.

Keywords Cool-water carbonate, epiphyte, southern Australia, seagrass, sedimentation, coralline algae.

INTRODUCTION

The nature of a carbonate sedimentary rock reflects the environment in which it formed and the time in which it was generated. This axiom, as articulated by Robert Ginsburg, is one of the founding principles of comparative sedimentology. The modern seafloor is a prism through which the ancient rock record is viewed, and so the better modern sedimentary processes are understood, the more confident is our interpretation of the ancient sedimentary rock record. Anyone who has been guided through the spectacular shallow-water environments of the Caribbean by Robert Ginsburg cannot help, egged on by his enthusiasm, but be surprised by the abundance and diversity of the calcareous benthos and the ubiquity of the marine grasses. The lush meadows seem to be a hallmark of the brightly lit, warm, shallow, marine carbonate seafloor. In his seminal studies of carbonate sedimentology from the Florida–Caribbean region, Ginsburg has repeatedly stressed the importance of these plants to the depositional system (Ginsburg, 1956; Nelsen & Ginsburg, 1986). Their function as sediment binders, as refuges for infaunal and epifaunal, mobile and sessile

†Deceased.

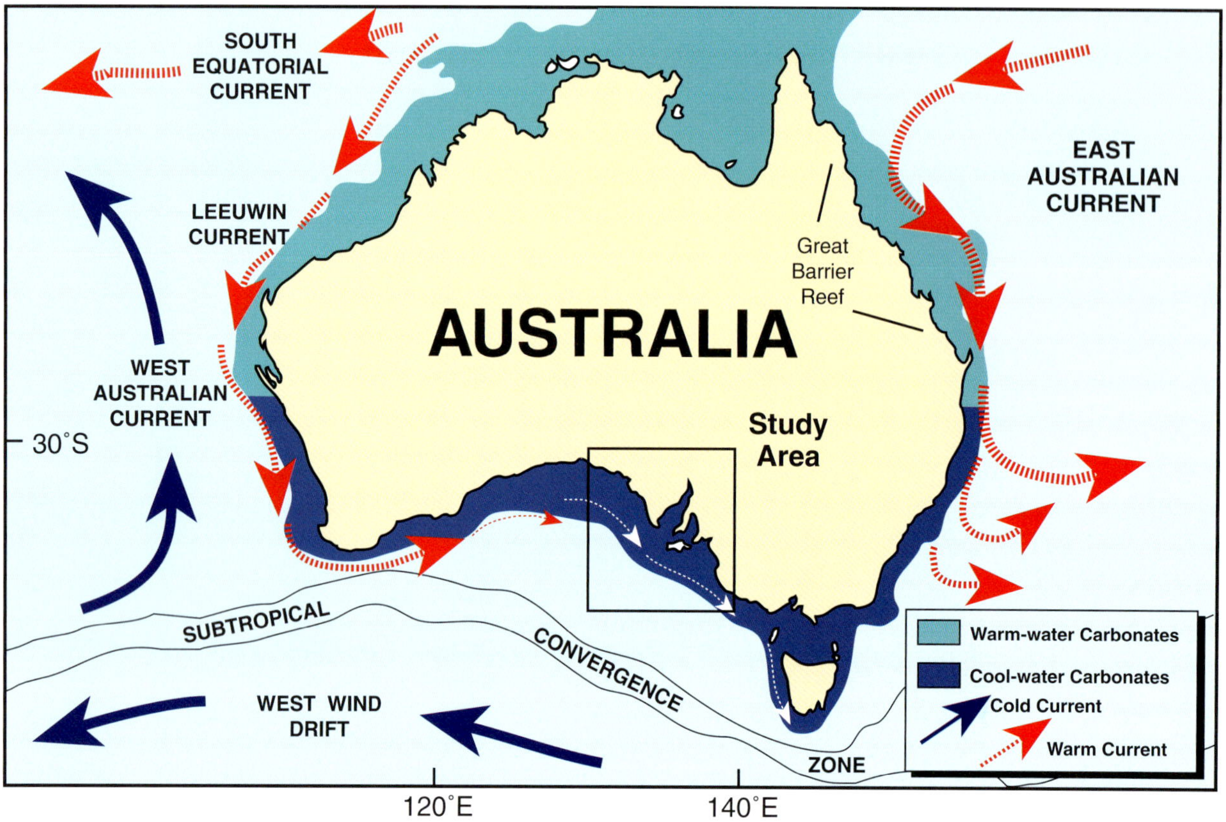

Fig. 1. Map of Australia showing major ocean currents, areas of cool- and warm-water carbonate deposition, and location of study area.

calcareous organisms, and as substrates for calcareous epiphytic growth is profound (Patriquin, 1972; Brasier, 1975; Wanless, 1981; Perry & Beavington-Penny, 2005; Corlett & Jones, 2007).

The continental shelf of southern Australia, south of 30°S and facing the Indian, Southern and Pacific oceans (Fig. 1), is the largest area of cool-water carbonate sedimentation on the globe (James, 1997). Vast areas of the inner shelf seafloor here are covered with seagrasses. It has long been known that the rates of sedimentation on these cool-water, marine grass banks are much higher than in contemporaneous offshore marine environments (Gostin *et al.*, 1988; James, 1997), and yet little is known about how these sediments are generated and the precise role of seagrass in sediment production.

The purpose of this study is to document the nature of such sediment production in a suite of nearshore environments in southern Australia, from the Great Australian Bight east to the Otway shelf a distance of ~1300 km (Fig. 2). This paper represents part of a PhD research project undertaken by KMB, who was tragically killed in Antarctica before the study could be finished. The problem was conceived by NPJ and YB who, together with AC, supervised the study, assembled and completed the thesis, and wrote this paper.

SEAGRASSES IN SOUTHERN AUSTRALIA

Introduction

Seagrasses are aquatic monocotyledons that continuously produce new blades, while shedding old ones. Calcareous epiphytes that encrust this renewable substrate are also continuously growing and being removed as the seagrass blades are discarded. Southern Australia and Western Australia are one of the world's most extensive known areas of temperate seagrass (Kirkman & Kuo, 1990). South Australia alone has 9612 km^2 (Edyvane, 1999a). The most extensive of these seagrass meadows occur in the clear, shallow, sheltered gulf waters of the Spencer Gulf (5520 km^2) and Gulf St. Vincent (1530 km^2) (Fig. 2) (Lewis *et al.*, 1997; Edyvane, 1999b). Twelve species of

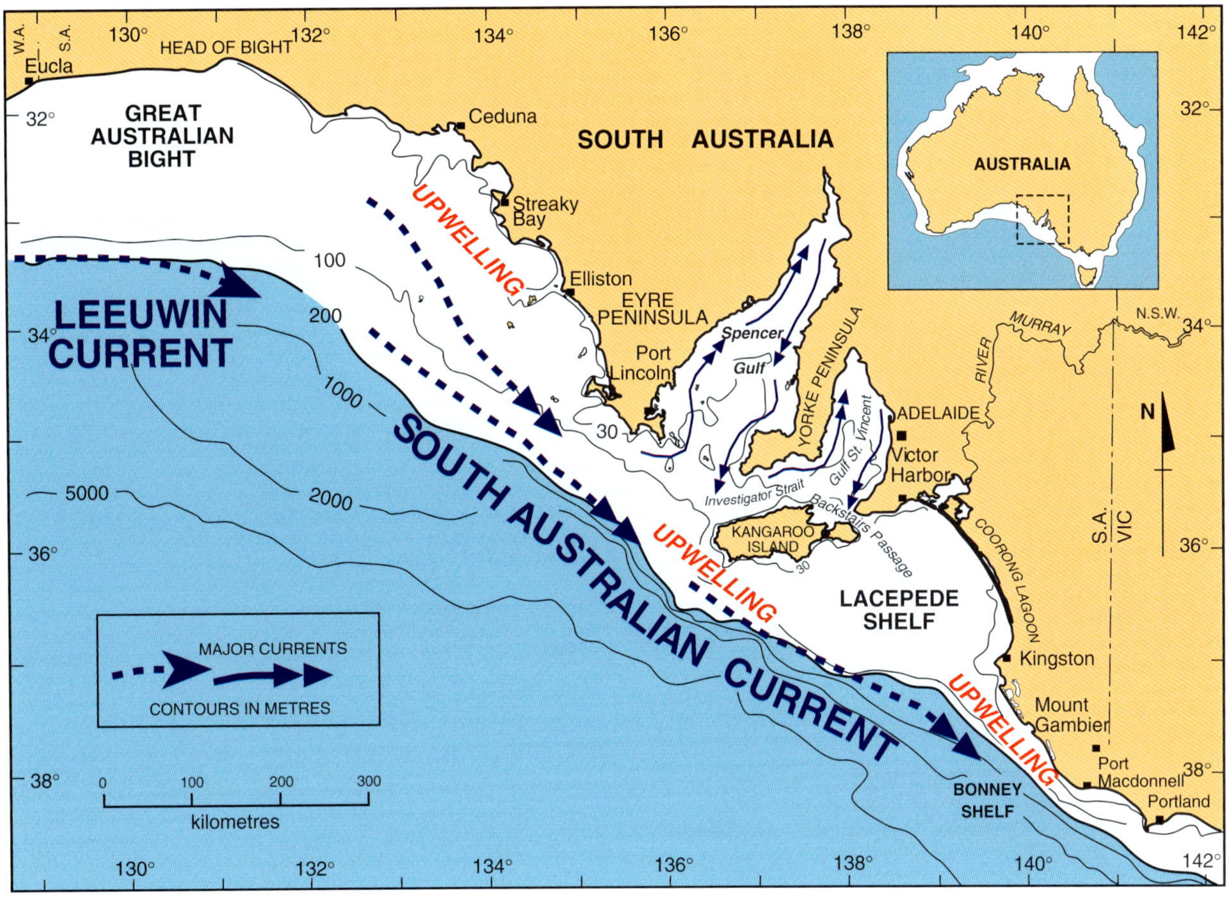

Fig. 2. Chart of the central part of southern Australia and location of the study area, main ocean circulation patterns, areas of upwelling and location of major population centres.

seagrass are known in South Australian coastal waters (Edyvane, 1999b).

Seagrasses present off South Australia are mainly from the genera *Posidonia, Amphibolis, Heterozostera, Zostera* and *Halophila*, of which only the first two are significant as hosts for calcareous epiphytes. Three species of *Posidonia* (*P. sinuosa, P. angustifolia*, and *P. australis*) and two species of *Amphibolis* (*A. antarctica* and *A. griffithii*) were studied. The shelf off South Australia, the region of this study, is, however, near the southern limit of their range (Robertson, 1984, 1986; Shepherd & Robertson, 1989; Kirkman, 1997; Edgar, 2000, 2001). *Posidonia* sp. and *Amphibolis* sp. species generally decline in abundance to the east. This has been attributed to their affinity with warm temperate waters that decrease to the east with the reduced effects of the warm water Leeuwin Current (Kirkman, 1997).

These marine angiosperms consist of a well-developed rhizome that runs in sediment beneath the seabed and has regularly spaced nodes, each bearing roots below and an erect stem or shoot with strap-like leaves above. The structure of these two grasses is somewhat different. Whereas *Posidonia* has a shoot with blades growing upward from close to the sediment–water interface, *Amphibolis* has a relatively long stem extending well above the sediment from which clusters of blades emerge (Fig. 3). Both have strong root systems that form a dense mat within the sediment and both shed their leaves over a period of about 2 months.

Posidonia

Posidonia australis is found in sheltered sand and mud environments, in 0–15 m water depth (m w.d.), from Shark Bay in Western Australia across southern Australia to Lake Macquarie on the east coast and along the northern coast of Tasmania. The maximum leaf length is 450 mm and maximum width is 20 mm. Rhizomes lie horizontally at a depth of 10–20 cm below the sediment surface. This, the most widely dispersed form

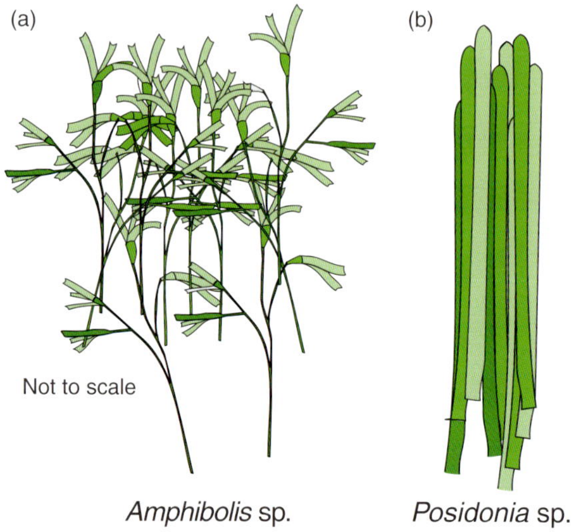

Fig. 3. The main attributes of *Posidonia* (right) with numerous strap-like leaves above the sediment substrate, much like the tropical grass *Thalassia*, compared to *Amphibolis* (left) that consists of a stem above the sediment surface from which emerge clusters of leaves.

seems to tolerate greater extremes in temperature and salinity compared with related species.

Posidonia angustifolia grows in moderately exposed sand between 2 and 35 m w.d. from the Houtman Abrolhos in Western Australia to Port Moorowie, South Australia (Fig. 2), in the middle of this study area. The leaves, although as long as those of *P. australis*, are thinner, and up to only 6 mm wide. These leaves are easily broken, and typically coated with dense epiphytes. Rhizomes lie just below the surface at 5–15 cm. It is often the dominant species of *Posidonia* in deeper water.

Posidonia sinuosa lives in moderately exposed sand and sheltered sand from 0 to 15 m w.d. It is found from Shark Bay Western Australia to Kingston, South Australia (Fig. 2) near the eastern margin of the study area. The leaves are similar to those of *P. australis*, a maximum of 1.2 m long and 11 mm wide. Rhizomes are firmly rooted in sand and can be removed only with difficulty.

Amphibolis

The two *Amphibolis* species have woody stems arising from the rhizome, the stems have a regular arrangement of leaf scars that are tough and wiry. The leaves are small relative to the stems. Both species are restricted to southern Australia.

Amphibolis antarctica grows in moderately exposed sand from 0 to 23 m w.d. and extends from Carnarvon, Western Australia to Wilson's Promontory, Victoria and halfway down the coast of Tasmania. Leaves are up to 50 mm long and the stem height to 1.5 m, while leaves occur in clusters of 8–10. It mostly grows by itself in very dense beds and is particularly extensive in Shark Bay where it forms the dominant cover over 3700 km^2.

Amphibolis griffithii is found on moderately exposed sand and rock in 0–40 m w.d., from Champion Bay Western Australia to Victor Harbor (Fig. 2) again near the centre of the study area. Leaves are a maximum of 100 mm, twice the length of *A. antarctica,* and lower in number (4–5) whereas the maximum stem height is similar, 1.1 m.

SETTING

Oceanography

The shelf of southern Australia is storm-dominated with high (>2.5 m) modal deep-water wave heights (Davies, 1980; Short & Hesp, 1982; Wright *et al.*, 1982). Long period (>12 s) swell waves are common, and wavelengths of 200 m have been reported. The oceanography is otherwise complex; shelf water masses, which vary in character throughout the year are mixtures of cold Southern Ocean Water, warm Indian Ocean water, saline waters from the adjacent gulfs, and water formed in the Great Australian Bight, all of which flow to the southeast. Nearshore water temperatures are generally less than 20°C, except in the large gulfs, and become progressively cooler southeastward.

The seasonal Leeuwin Current (Fig. 2) is a shallow-water shelf-edge stream of warm (17–19°C), low-salinity (35.7–35.8‰), low nutrient, tropical surface water that during the winter flows southward along the west coast of the continent and eastward into the Great Australian Bight (Cresswell, 1991). This shelf-edge flow continues eastward as the South Australian Current (Fig. 2) (Ridgeway & Condie, 2004), a flow derived from a warm (17–19°C in winter; 19–22°C in summer) and saline (35.9–36.4‰) water mass in the central and western Great Australian Bight (Rochford, 1986). The generally southeast flow of these surface waters is matched by westward geostrophic flow of a cooler intermediate depth, upwelling-favourable boundary current (Bye, 1972, 1983; Middleton & Cirano, 2002) called the Flinders Current. Coast-directed winds result in significant upwelling (Fig. 2) off Kangaroo Island, the west coast of Eyre Peninsula, and the Bonney Shelf (Schahinger, 1987; Griffin *et al.*, 1997; Middleton & Platov, 2003).

Spencer Gulf and Gulf St. Vincent (including Investigator Strait) are inverse estuaries (Nunes Vaz *et al.*, 1990), where seawater is concentrated by evaporation exceeding precipitation. Salinities at the head of Spencer Gulf remain above 40‰ year-round while water temperatures fall to ~12°C in the winter and rise to 24°C in summer (Bye, 1983). Temperature and salinity gradients in Gulf St. Vincent are not as strong as those in Spencer Gulf. Although the River Murray (Fig. 2) is the largest river in Australia its mouth is generally sealed by longshore drift and so it has little influence on Lacepede Shelf waters, which are normal marine throughout.

Sedimentology

The seafloor throughout the area is covered by heterozoan carbonate sediment (cf. James, 1997). The carbonate fraction of the deposits is almost wholly biogenic and produced mainly by coralline algae, foraminifera, molluscs, bryozoans and echinoderms (Gostin *et al.*, 1988; James *et al.*, 1992, 1997, 2001; Fuller *et al.*, 1994). In some inboard areas they are augmented by tests of large symbiont-bearing foraminifera (particularly *Peneroplis* sp.) and locally in the gulfs there is sporadic growth of the zooxanthellate coral *Pleiseastrea* sp. Thus, overall the sedimentary environment is at the warmer end of the cool-water spectrum (cf. Betzler *et al.*, 1997).

TERMINOLOGY, PREVIOUS STUDIES

Terminology

The following groups of calcareous organisms are known to grow on seagrasses; coralline algae, bryozoans, foraminifera, serpulids, spirorbids, ostracods and bivalves (Humm, 1964; Ducker *et al.*, 1977; Ducker & Knox, 1978; Harlin, 1980). The name for organisms growing on a variety of marine hosts is the subject of an extensive literature (Harlin, 1980; Borowitzka & Lethbridge, 1989; Taylor, 1990; Frankovich & Zeiman, 1994; Hageman *et al.*, 1996; Womersley, 1996; Jernakoff *et al.*, 1996; English *et al.*, 1997). The term calcareous epiphyte is herein defined as 'an organism, plant or animal, that secretes a calcareous skeleton and that, for most of its life is attached to the living outer tissues of the plant: it is assumed that it does not derive food or nutrients from the host'.

Previous studies

The first studies evaluating the quantity of carbonate from epiphytes were carried out by Land (1970), Patriquin (1972) and Smith (1972). Nelsen & Ginsburg (1986) concluded that the epibionts (epiphytes) could account for the entire high-magnesium calcite (HMC) and aragonite mud-sized fraction within eastern Florida Bay. There are only three other sedimentological studies of epiphyte carbonate productivity in tropical waters (Boscence, 1989; Frankovich & Zieman, 1994; Corlett & Jones, 2007) and one in the subtropical realm (Perry & Beavington-Penney, 2005).

In South Australia there has been only one examination of seagrass calcareous epiphytes and this is only at one site (Thomas & Clarke, 2001). In contrast, there are numerous studies in Western Australia (Smith & Atkinson, 1983; Searle, 1984; Horner, 1987; Walker & Woerkerling, 1988; Walker *et al.*, 1991; Sim, 1991; Lord, 1998; Lavery & Vanderklift, 2000).

METHODS

Site selection

Twenty study sites were chosen to represent the wide variety of settings in which seagrass is found (Fig. 4). They ranged from the head of reverse estuaries (Gulf St. Vincent and Spencer Gulf) to environments exposed to the Southern Ocean. Water depths ranged from 0.4 to 15.5 m w.d., across a salinity range of 32.2–43.7‰ with surface-water temperatures varying from 14.1 to 25.8°C. West Island was chosen for detailed study because it faced the Southern Ocean and because there was a marine laboratory on the island that facilitated diving and analysis. Chinaman Creek was selected for additional study because it represented the other extreme, a restricted, seasonally saline and high-temperature environment.

Sample collection

Marine study was undertaken by snorkeling or SCUBA diving where appropriate. Each site was visited at least once. West Island (Fig. 4) was studied in detail at one shallow-water site

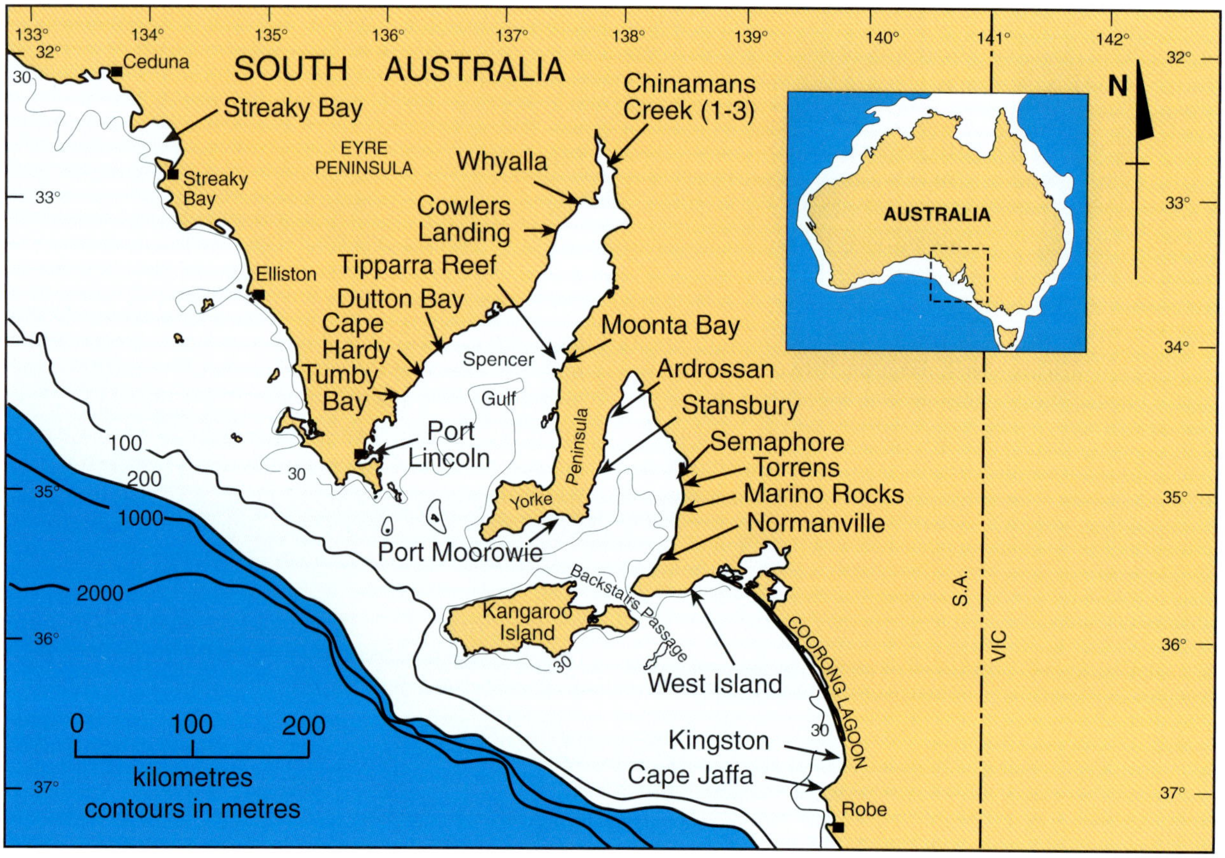

Fig. 4. A map of the study area illustrating location of the 20 study sites.

(4–6 m w.d.) and a deep water site (10 m w.d.) every three months over a 2.5 year period. Each site was sampled along two transects perpendicular to the coastline, 30 m in length and 10 m apart using 50 × 50 cm quadrats, 5 m apart as described in English *et al.* (1997). All plant matter was removed with a blade or scissors and placed in a shade cloth bag. Water samples were collected just above the quadrat. Collected samples were air dried.

Temperature, salinity and conductivity were measured, samples were frozen as soon as possible for nutrient analysis and depth measurements were made using a dive computer. Climatological data for two weeks surrounding the sampling dates were obtained from the Australian Meteorological Office.

Laboratory methods

Samples were analysed using the following procedure: (1) samples were rinsed in fresh water; (2) they were then air dried, extraneous material removed, and subsequently weighed; (3) different seagrass species were then separated and the blades and stems for each species further separated; (4) organic matter was removed via several treatments with 10% H_2O_2; (5) samples were then placed in an ultrasonic bath for 3 hours, partly to remove remaining organic material; (6) the sample was poured through 3.5 mm and 65 μm sieves, and further epiphytes gently rubbed off and carbonate retained on a 65 μm sieve; (7) this material was further treated with 35% H_2O_2; and (8) the carbonate was finally air dried and weighed.

Data analysis

Carbonate values have been calculated as the amount of carbonate per area (m^2) and per kilogram weight of seagrass. The implications of using these two methods of calculating carbonate values are described below.

The amount of carbonate per m^2 is calculated for each quadrat or plant species using the following equation:

$$C\ \mathrm{gm}^2 = \left[\left(\frac{C}{\sum P_i}\right) \times P\right] \times 4 \tag{1}$$

where C is $CaCO_3$ (g), P is total dry seagrass (includes the carbonate) (g) and P_i is the weight of individual seagrass groups used (weight includes the carbonate) (g). As the quadrat size is 50 × 50 cm, then a multiplication of 4 is used to bring the value up to 1 m^{-2}.

The value C in g m^{-2} is dependant on the weight of the seagrass. As the weight of the seagrass is measured by how much is present within a certain area, this weight is also expressed as density. Density depends on the number of plants, the type of plants, and the amount of carbonate on those plants. An increase in plant density can be attributed to either an increase in the number of plants or an increase in the amount of carbonate. Both an increase in plant weight and carbonate value will increase the C g m^{-2} value. Hence with increasing plant density it would also be expected that there would be an increase in C g m^{-2}.

A disadvantage of the carbonate g m^{-2} values is that this is very subjective to the density of the seagrass and not the amount of carbonate that is actually found on each plant. To compare carbonate values between localities it is necessary to calculate the carbonate quantity independent of the various seagrass densities from site to site. To do this the amount of carbonate found upon a known weight of seagrass is calculated, i.e. $CaCO_3$ per kg of seagrass. This is simply calculated by

$$\left(\frac{C}{P_i}\right) \times 1000 \tag{2}$$

The relationship between C kg^{-1} seagrass and seagrass sample size is a matter of ratio. The greater the ratio, the lower the C kg^{-1} seagrass value and vice versa.

Calcareous epiphyte productivity is achieved by multiplying standing stock value by the number of crops per year. Biomass (standing stock) is the dry weight of all biomass in each quadrant ×4; seagrass biomass which Σ = seagrass biomass in g m^{-2}.

Factors affecting biomass and carbonate quantity were analysed using Analysis of Variance (ANOVA) with SPSS 10.0 software and JMP3.0.2. Variance heterogeneity was checked using Leven's test and normal distribution was tested using the Kolmogorov–Smirnov test where the number of samples was >20, otherwise the Shapiro–Wilk test was used. Data were transformed where necessary, the appropriate transformation estimated using Taylor's power law and the ladder of powers. Most often log 10 transformation was found to be appropriate. The probability level of $\sigma = 0.05$ was assumed, unless transformation data failed Levene's test for variance homogeneity in which case a more conservative alpha value was adopted (i.e $\alpha = 0.01$) (Underwood, 1981). *Post hoc* comparison of means determining where significant differences lie was carried out using Tukey's HSD test. Non-parametric Kruskall–Wallace analysis of variance was used where gross violations of normal distribution and failure of homogeneity occurred (Underwood, 1981; Zar, 1996). Significant differences were identified using the *post hoc* Tukey style, Nemenyi comparison of means. Seagrass species distribution was analysed by cluster analysis (JMP3.0.2) and multidimensional scaling (MDS) using the Bray–Curtis association measure for the similarity matrix (Pcord4). Student's t-test was used to test differences in carbonate abundance between genera. All averages given are means and standard errors are provided where possible.

The method used to derive qualitative estimates of the relative abundances from ranked data is adapted from Saito and Atobe (1970) (cited in English *et al.*, 1997). The formula is

$$A = \frac{\sum(M_i \times f_i)}{\sum f} \tag{3}$$

where M_i is the mid point value of each class and f_i is the frequency of each class (number of blades with same class).

Estimation of the amount of carbonate derived from each species was calculated using the abundance ratio of each species, the standing stock values of calcareous epiphyte abundance from each site on each seagrass species present.

$$\frac{\%sp_i}{100} C \tag{4}$$

where $\%spi$ is the percentage of species 'i' present on seagrass species a and site b and C is the carbonate abundance (g kg^{-1} of seagrass) on the same seagrass species at the same site. These calculations assume that the density (or specific gravity) of each form of the coralline algae is the same.

RESULTS AND INTERPRETATION

Seagrass biomass

The seagrasses are variably distributed across the region (Fig. 5). *Posidonia* and *Amphibolis*

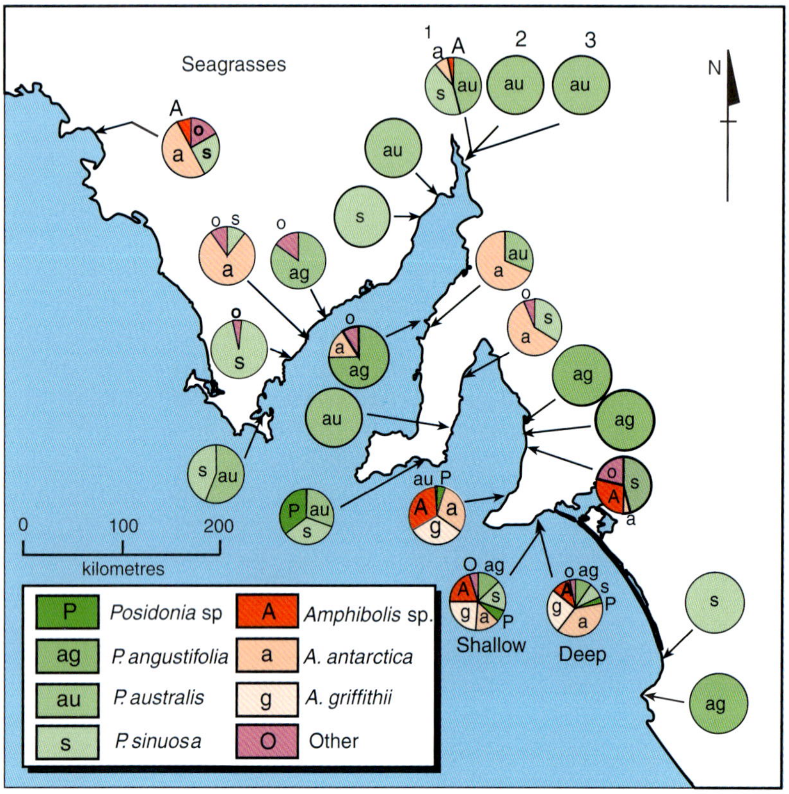

Fig. 5. A map showing the relative proportions of seagrasses at each study site.

occur throughout. *Posidonia angustifolia* is more abundant in Gulf St. Vincent and towards the eastern side of the area and grows largely in monotypic beds. *Posidonia australis* reaches its furthest eastern extent along the southern edge of Yorke Peninsula and also grows mostly in monotypic beds but locally with *P. sinuosa*, *A. antarctica*, and other species. *Amphibolis antarctica* grows along the entire coastline and always with *P. australis* and *P. sinuosa* and locally with *A. griffithii*. *Amphibolis griffithii* occurs only in the east and always in association with *A. antarctica* and *P. angustifolia*.

On the assumption that the biomass of the marine grasses is in some way related to the amount of calcareous epiphyte production, the attributes of seagrasses are assessed across the areas. The average dry weight of seagrass per site is 322.5 $g\,m^{-2}$, with >74% of the areas ranging from 50 to 500 $g\,m^{-2}$ (Table 1). Kruskall–Wallis analysis of variance with water depth indicates that the biomass decreases from 504 ± 34.5 $g\,m^{-2}$ at 2–4 m w.d. to 155.7 ± 89.7 $g\,m^{-2}$ at >10 m w.d.. Furthermore, biomass is greater at 2–4 m w.d. than at 0–2 m w.d. (Fig. 6).

Different genera and species have high biomass (Fig. 7). *Amphibolis*-dominated quadrats consistently show higher biomass values (522.9 $g\,m^{-2}$) than *Posidonia*-dominated quadrats (280.9 $g\,m^{-2}$). This is due to the higher density of shoots/m^2. On the other hand, the surface area for a typical blade cluster of 3–4 blades for *A. griffithii* is 273–515 cm^2, versus 3–4 blades of *P. sinuosa* which is 599–1019.2 cm^2.

Although no clear trends are apparent, *Amphibolis* shows no significant variation with temperature and salinity whereas *Posidonia* does show variation with changing temperature and salinity. There is also no correlation between biomass and species type and grain-size of the rooting sediment.

Calcareous epiphytes

General attributes

A total of 1544 blades and stems were examined to assess the distribution and importance of calcareous epiphytes (Table 2); *P. sinuosa* (333 blades), *P. australis* (179), *P. augustofolia* (130), *A. antarctica* stems (58), *A. antarctica* blades (435), *A. griffithii* stems (53), *A. griffithii* blades (356). The relative importance of epiphyte taxa are (Figs 7 and 8); corallines, 53.6%; benthic foraminifera, 17.4%; bryozoans, 16.4%; spirorbids, 8.0%;

Table 1. Summary of total (•bio) and seagrass biomass (sg bio) averages (g dw m^{-2}) from all sites.

	Total		*A. antarctica*		*A. griffithii*		*Amphibolis* sp.		*P. augustifolia*		*P. australis*		*P. sinuosa*		*Posidonia* sp.		Other	
Site	•bio	sg bio	•bio	sg bio	•bio	sg bio	•bio	sg bio	•bio	sg bio	•bio	sg bio	•bio	sg bio	•bio	sg bio	•bio	sg bio
Ardrossan	660.3	492.7	572.9	447.8	0.0	0.0	0.0	0.0	0.0	0.0	0.0	0.0	321.7	287.0	0.0	0.0	62.5	0.1
Cape Hardy	451.6	338.5	379.0	313.5	0.0	0.0	0.0	0.0	0.0	0.0	0.0	0.0	52.6	49.6	0.0	0.0	48.0	2.8
Cape Jaffa	540.5	484.1	0.0	0.0	0.0	0.0	0.0	0.0	518.8	462.6	0.0	0.0	0.0	0.0	0.0	0.0	0.0	0.0
Chinaman Ck-1	244.2	206.9	38.6	33.2	0.0	0.0	18.0	14.8	0.0	0.0	244.0	212.9	233.3	194.3	0.0	0.0	0.0	0.0
Chinaman Ck-2	455.3	442.9	0.0	0.0	0.0	0.0	0.0	0.0	0.0	0.0	455.3	442.9	0.0	0.0	0.0	0.0	0.0	0.0
Chinaman Ck-3	227.4	203.9	0.0	0.0	0.0	0.0	0.0	0.0	0.0	0.0	227.4	203.9	0.0	0.0	0.0	0.0	0.0	0.0
Cowlers Landing	206.6	168.9	0.0	0.0	0.0	0.0	0.0	0.0	0.0	0.0	0.0	0.0	206.6	168.9	0.0	0.0	0.0	0.0
Dutton Bay	99.9	85.4	0.0	0.0	0.0	0.0	0.0	0.0	97.0	85.4	0.0	0.0	0.0	0.0	0.0	0.0	17.4	0.0
Kingston	334.0	298.3	0.0	0.0	0.0	0.0	0.0	0.0	0.0	0.0	0.0	0.0	334.0	298.3	0.0	0.0	0.0	0.0
Marino Rocks	423.8	397.8	39.3	36.7	0.0	0.0	242.5	228.0	0.0	0.0	0.0	0.0	394.3	377.5	0.0	0.0	184.7	164.3
Moonta Bay	746.0	583.1	724.1	563.6	0.0	0.0	0.0	0.0	0.0	0.0	329.0	265.9	0.0	0.0	0.0	0.0	0.0	0.0
Normanville	976.0	599.0	852.8	528.1	932.7	571.9	928.0	551.6	140.2	126.3	0.0	0.0	0.0	0.0	0.0	0.0	29.4	25.2
Port Lincoln	354.3	310.3	0.0	0.0	0.0	0.0	0.0	0.0	0.0	0.0	365.4	325.5	287.3	219.1	0.0	0.0	0.0	0.0
Port Moorowie	461.3	392.0	0.0	0.0	0.0	0.0	0.0	0.0	0.0	0.0	421.8	385.6	454.4	387.2	485.1	403.3	0.0	0.0
Semaphore	57.7	42.9	0.0	0.0	0.0	0.0	0.0	0.0	66.0	49.2	0.0	0.0	0.0	0.0	0.0	0.0	0.0	0.0
Stansbury	259.8	230.6	0.0	0.0	0.0	0.0	0.0	0.0	0.0	0.0	259.8	230.6	0.0	0.0	0.0	0.0	0.0	0.0
Streaky Bay	781.5	557.5	697.7	576.7	0.0	0.0	0.0	0.0	0.0	0.0	237.9	210.3	349.0	294.1	0.0	0.0	113.1	0.0
Tipparra Reef	110.0	93.3	19.4	16.3	0.0	0.0	0.0	0.0	91.2	81.0	0.0	0.0	0.0	0.0	0.0	0.0	11.3	3.8
Torrens outlet	177.1	162.9	0.0	0.0	0.0	0.0	0.0	0.0	177.1	162.9	0.0	0.0	0.0	0.0	0.0	0.0	0.0	0.0
Tumbly Bay	487.1	352.5	0.0	0.0	0.0	0.0	0.0	0.0	0.0	0.0	0.0	0.0	457.1	328.4	0.0	0.0	21.8	0.0
Whyalla	214.1	203.8	0.0	0.0	0.0	0.0	0.0	0.0	0.0	0.0	214.1	203.8	0.0	0.0	0.0	0.0	0.0	0.0
West Island 6m	544.7	413.4	205.0	150.0	338.7	240.2	286.0	176.1	184.8	160.1	0.0	0.0	246.9	232.9	95.6	78.8	72.5	29.4
West Island 10m	374.5	257.5	578.8	377.0	376.4	281.4	168.8	111.5	147.5	138.5	0.0	0.0	126.5	114.5	57.1	53.6	48.7	9.2

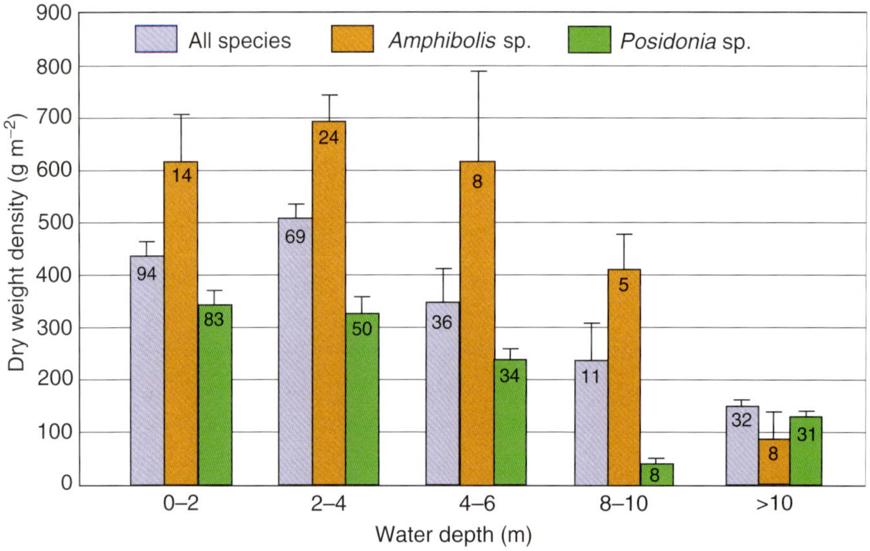

Fig. 6. Graphical distribution of seagrass biomass with depth. Values grouped at 2 m water depth intervals, combined by species and by genera. Number at top of column refers to number of samples collected. Mean ± SE.

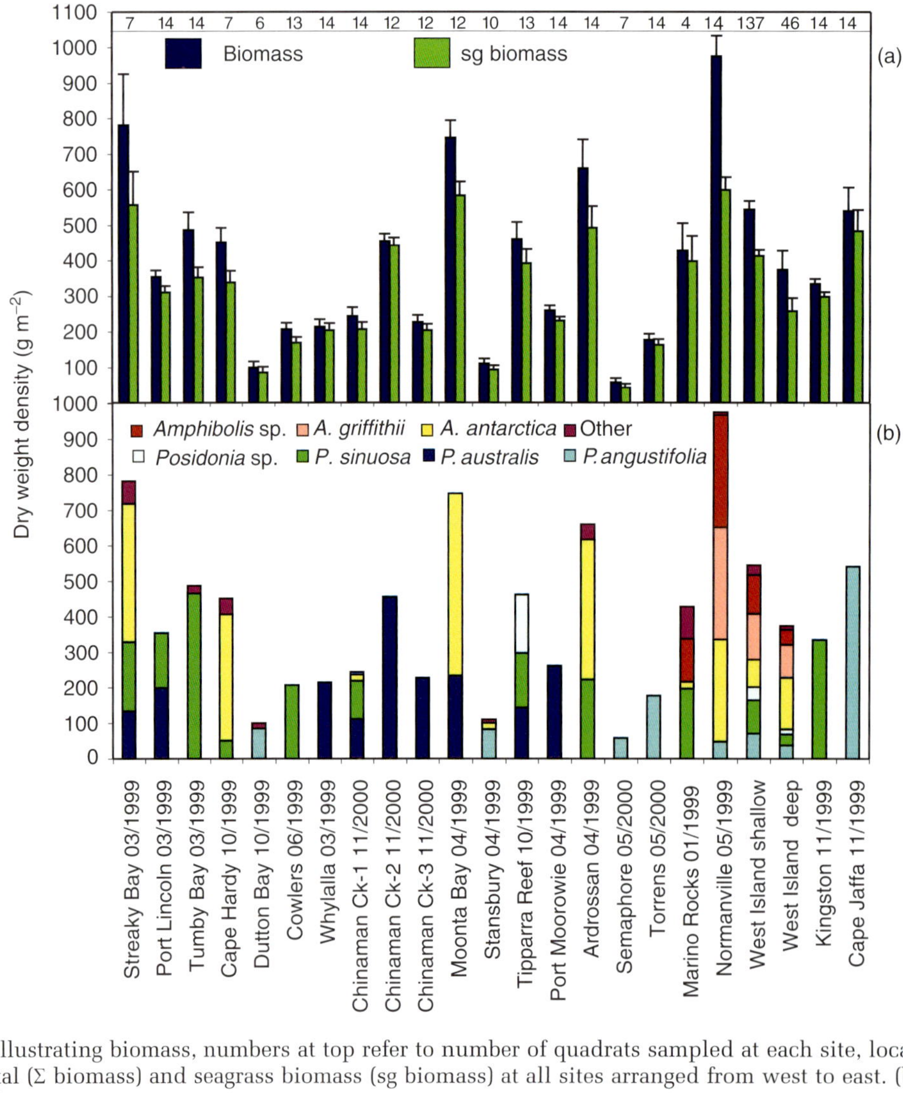

Fig. 7. Graphs illustrating biomass, numbers at top refer to number of quadrats sampled at each site, locations at bottom on Fig. 4. (a) total (Σ biomass) and seagrass biomass (sg biomass) at all sites arranged from west to east. (b) Grass species contribution to biomass.

Table 2. Calcareous epiphyte components from each site, sampled from west to east. Percentage values (%) are ratios of the relative abundance of each taxa from each site. Carbonate values are the actual amounts of carbonate derived from each taxa ($CaCO_3$ columns are: A = $g\,m^{-2}$ and B = $g\,kg^{-1}sg^{-1}$). West Island (shallow site) derived from samples collected in February, May and September, 1999. • = total % estimation, N = total number of blades and stems analysed for each site.

	Coralline algae			Bryozoans			Spirorbids			Serpulids			Foraminifera			Ostracods			Bivalves			Other				
Site	%	A	B	%	A	B	%	A	B	%	A	B	%	A	B	%	A	B	%	A	B	%	A	B	•	*N*
Streaky Bay	55.1	123.3	144.8	0.4	0.8	1.0	6.2	13.8	16.2	0.1	0.2	0.2	33.3	74.5	87.5	0	0	0	0.42	0.9	1.1	0	0	0	95.3	96
Port Lincoln	69.4	30.5	88.8	3.2	1.4	4.1	0.2	0.1	0.2	0	0	0	18.6	8.2	23.8	0.01	0	0	0.17	0.1	0.2	0	0	0	91.6	29
Tumby Bay	56.6	76.2	143.8	19.6	26.3	49.7	11.6	15.7	29.6	0.1	0.1	0.2	9.3	12.5	23.7	0	0	0	0	0	0	0.61	0.8	1.5	97.9	29
Cape Hardy	54.9	62.1	138.4	31.8	36.0	80.1	0.3	0.3	0.6	0.1	0.2	0.3	6.3	7.1	15.8	0.04	0	0.1	0.14	0.2	0.3	0	0	0	93.6	72
Dutton Bay	60.8	8.8	94.9	8.3	1.2	12.9	8.9	1.3	13.9	0	0	0	14.9	2.2	23.2	0	0	0	3.13	0.5	4.9	0	0	0	95.9	20
Cowlers Landing	30.0	11.3	56.1	15.4	5.8	28.9	0	0	0	0.2	0.1	0.5	48.4	18.3	90.5	0.01	0	0	0.09	0	0.2	0	0	0	94.2	30
Whyalla	85.8	8.8	40.7	6.0	0.6	2.8	0	0	0	0	0	0	2.4	0.3	1.2	0	0	0	0	0	0	0	0	0	94.3	30
Chinaman Ck-1	62.7	23.4	92.0	21.3	7.9	31.3	2.4	0.9	3.6	0.8	0.3	1.1	12.8	4.8	18.9	0.01	0	0	0.09	0	0.1	0	0	0	100.1	29
Chinaman Ck-2	65.5	8.1	18.2	28.3	3.5	7.9	0	0	0	0	0	0	0	0	0	0.01	0	0	0	0	0	0	0	0	93.9	30
Chinaman Ck-3	26.2	6.2	27.3	43.9	10.3	45.7	7.1	1.7	7.4	0	0	0	19.9	4.7	20.7	0.01	0	0	0.09	0	0.1	0	0	0	97.2	30
Moonta Bay	27.6	44.9	59.9	15.9	25.9	34.5	18.6	30.4	40.5	0.7	1.1	1.4	29.6	48.2	64.3	0.16	0.3	0.4	0.20	0.3	0.4	0	0	0	92.7	126
Tipparra reef	41.9	7.0	63.0	31.6	5.3	47.5	7.1	1.2	10.7	1.1	0.2	1.6	24.6	4.1	36.9	0	0	0	0	0	0	0	0	0	106.2	33
Port Moorowie	56.7	39.3	84.6	0	0	0	29.3	20.3	43.8	0	0	0	7.3	5.0	10.8	0	0	0	0	0	0	0	0	0	93.2	41
Stansbury	72.7	21.2	79.7	3.8	1.1	4.2	7.9	2.3	8.6	0	0	0	12.3	3.6	13.5	0.01	0	0	0	0	0	0	0	0	96.7	27
Ardrossan	66.8	111.9	171.2	5.7	9.5	14.6	0.5	0.9	1.4	0.1	0.1	0.2	22.1	37.1	56.7	0.01	0	0	0.44	0.7	1.1	0	0	0	95.	86
Semaphore	2.2	0.3	7.2	1.7	0.3	5.6	0.2	0	0.6	0.3	0.1	1.1	87.5	13.0	284.1	0	0	0	1.22	0.2	3.9	0	0	0	93.22	25
Torrens	40.7	5.8	33.1	29.3	4.1	23.8	7.8	1.1	6.3	1.0	0.1	0.8	28.4	4.0	23.2	0.01	0	0	0.02	0	0	0	0	0	107.1	30
Marino Rocks	39.1	12.1	29.1	17.0	5.3	12.6	35.1	10.9	26.1	0	0	0	3.2	1.0	2.4	0	0	0	0	0	0	0	0	0	94.4	117
Normanville	70.2	256.0	262.2	10.7	39.0	40.0	3.3	12.2	12.5	0	0.1	0.1	5.5	20.1	20.5	0	0	0	0	0	0	0	0	0	89.8	171
West Island shallow	59.5	78.1	136.4	18.6	24.5	42.8	14.2	18.6	32.5	0.2	0.2	0.4	0.4	0.5	0.8	0	0	0	0	0	0	0	0	0	92.8	356
West Island deep	41.9	49.0	75.4	39.9	46.6	71.8	6.5	7.6	11.7	0.1	0.1	0.2	0.4	0.4	0.7	0	0	0	0	0	0	0	0	0	88.7	65
Kingston	58.7	20.9	61.8	24.5	8.7	25.8	16.5	5.9	17.4	0	0	0	12.4	4.4	13.1	0	0	0	0.08	0	0.1	0	0	0	112.2	30
Cape Jaffa	87.5	49.3	94.5	0.4	0.2	0.5	0	0	0	0	0	0	0.3	0.2	0.4	0	0	0	0	0	0	0	0	0	88.3	25

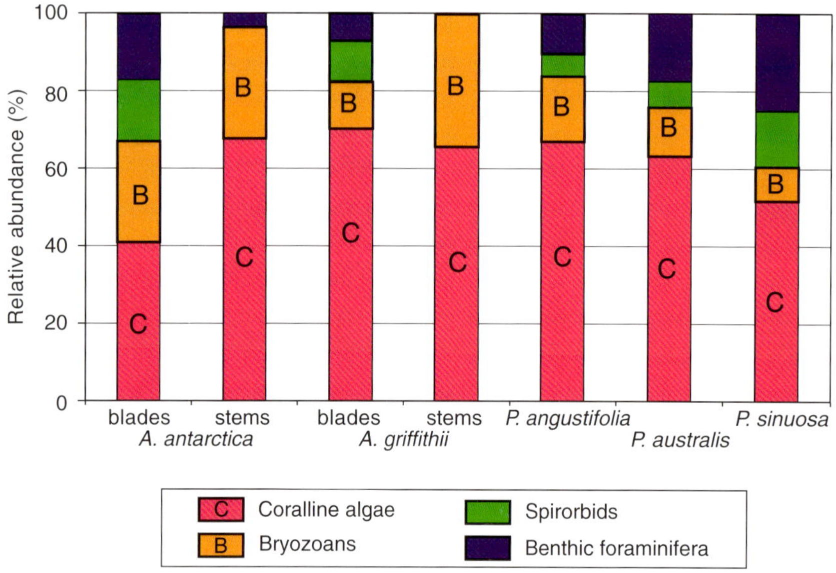

Fig. 8. Relative percentage of different epiphyte taxa from each seagrass species summarized from all sites.

bivalves, 0.3%; serpulids, 0.2%; ostracods and others <0.1%. Corallines have the highest abundance at nearly every site (Fig. 8). Only corallines show any geographic trend, decreasing northward up Spencer Gulf.

Epiphytes on all seagrasses are dominated by corallines 38.2–67.8% (Table 3 and Fig. 9), although dominance on *A. antarctica* is least pronounced. *Amphibolis* stems show the highest proportion of corallines (average 65.3%) and bryozoans (average 30.8%), but they have few other calcareous epiphytes. *Posidonia sinuosa* has a relatively high abundance of spirorbids and foraminifera in comparison to other *Posidonia* species (Fig. 8).

No differences are found for taxon abundance with depth, water temperature or salinity using in ANOVA and regression analysis. Nevertheless, there are trends; corallines decrease with increasing depth, salinity and temperature whereas bryozoans increase with increasing depth, temperature and salinity, but foraminifera show no change.

There are specific differences in the ratio of calcareous epiphytes between seagrass species. Corallines are relatively sparse on *A. antarctica* blades, bryozoans are numerous on *A. antarctica* stems, spirorbids are abundant on *A. antarctica* and *P. sinuosa*, benthic foraminifera are high on *P. sinuosa* blades but low on *Amphibolis* stems and *A. griffithii* blades (Fig. 8).

Corallines

These algae (Fig. 10) are composed of magnesium calcite. They produce 38.2–67.8% of all epiphytic carbonate and are present on 1523 of the blades and stems analysed; thick encrusting sheet (type 1 – *Hydrolithon* sp., *Pneophyllum* sp., *Synarthrophyton* sp.) = 2.3%; encrusting sheet (type 2 – *Hydrolithon* sp., *Pneophyllum* sp.) = 68.7%; erect, thinly branching, filamentous (type 3 – *Jania* sp.) = 20.9%; erect, multiple branching, narrowly segmented (type 4 – *Corallina* sp., *Arthocardia* sp., *Jania* sp., *Haliptilon* sp.) = 7.8%; erect, simple branching, broadly segmented (type 5 – *Metagoniolithon radiatum*) = 0.3% (Table 4). Encrusting forms produce 71.0% whereas rigid branching and articulated branching forms comprise 29.0%. This is the same for all except *A. antarctica* where erect narrow segmented and thick encrusting algae are greater. The total amount of carbonate produced ranges from <0.01 to 232 g kg^{-1} of grass.

Detailed study at West Island shows that 80% of epiphytic carbonate on *Amphibolis* is associated with the stems. *Amphibolis* stems and blades, however, share similar epiphytic species, in particular geniculate corallines (e.g. *Pneophyllum-Fosliella* group) (den Hartog, 1970; Ducker *et al.*, 1977; Sim, 1991). Temperature and light have the greatest effect on growth rates of the corallines (Bressan & Tomini, 1982; Jones & Woerkling, 1983), in particular *Fosliella cruciata* (a common species on *A. antarctica*) does not germinate at temperatures below 10°C.

There is a seasonal variation in epiphytes. A decline in autumn/winter biomass is attributed to a simultaneous decrease in seagrass biomass, hence recruitment space, largely brought about by

Table 3. Percentage of epiphyte taxa on different seagrass species, and their various components. • = total %, *N* = number of samples.

Seagrass species	Coralline algae	Bryozoans	Spirorbids	Serpulids	Foraminifera	Ostracods	Bivalves	Other	•	*N*
A. antarctica blade	38.2	24.3	14.7	0.1	15.9	0.03	0.22	0.00	93.5	425
A. antarctica stem	66.0	27.9	0.5	0.4	2.7	0.03	0.01	0.00	97.6	54
A. griffithii blade	67.8	9.8	7.6	0.1	4.8	0.00	0.00	0.00	90.5	357
A. griffithii stem	64.5	33.6	0.1	0.1	0.1	0.00	0.00	0.00	98.4	53
P. angustifolia	63.0	15.9	5.3	0.2	9.3	0.00	0.52	0.00	94.1	120
P. australis	60.4	12.0	6.4	0.1	16.7	0.00	0.01	0.00	95.6	179
P. sinuosa	50.4	8.5	14.1	0.2	24.1	0.00	0.14	0.04	97.4	334

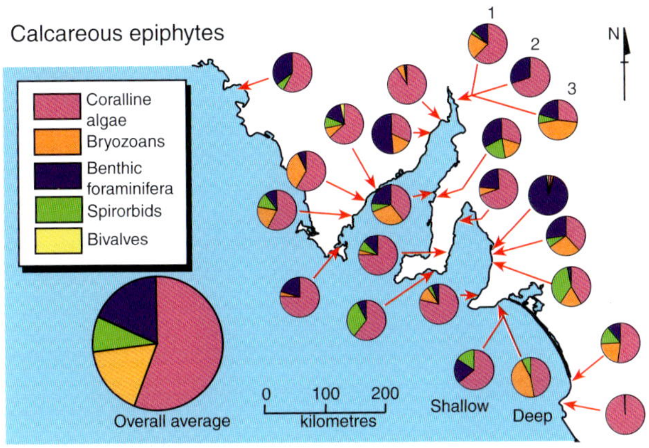

Fig. 9. A map showing the relative proportions of calcareous epiphytes on all grasses at each site.

decreased light, lower temperatures, and shorter day lengths during that time of year.

Bryozoans

Bryozoans (Fig. 10) occur at all sites and on all seagrasses (Tables 2–4). They produce 8.5–33.6% of all epiphytic carbonate and are present on 916 (~2/3) of the blades and stems analysed. The biota comprises 61 species from 47 genera. The most frequently occurring form is the cheilostome *Heterooecium* sp. (19 of 23 sites). The next is another cheilostome *Thairopora* sp., closely followed by the cyclostomes *Diaperocecia australis*, *Crisia acropora*, *Disporella pristis* and *Favospira* sp. All of the foregoing bryozoans are composed of Mg calcite (Bone & James, 1993).

The most common morphotype (cf. Bone & James, 1993) is ‘encrusting delicate sheet’ whereas the most common erect bryozoan is the ‘delicate branching type’ and then ‘articulated zooidal’ type. When bryozoan carbonate is averaged from all sites, the most important species is the erect fenestrate *Iodictyum* sp. (21% of all carbonate generated from bryozoan epiphytes), followed by the erect arborescent *Celleporaria* sp. (8%), encrusting robust sheet *Thairopora cincta*, (6.9%) and *Mychopletra pocula* (6.3%). The most common species *Heterooecium*, however, contributes only 1.2% of the carbonate. The most important carbonate-producing bryozoans per unit area are *Iodictyum* (14.8%), *Thairopora* (13.6%) and *Mychopletra pocula* (10.1%). The most important producing morphotype is ‘erect fenestrate’, which produces an average of 26.7 g m^{-2} or 44.1 g kg^{-1} of seagrass, but this form is relatively uncommon amongst the epiphytic bryozoans.

Benthic foraminifera

These protists (Fig. 11) occur at all sites and on all seagrass species and generated 0.1 to 33.6% of all epiphytic carbonate (Table 3). They are present on 939 (~2/3) of the blades and stems analysed. They comprise 30 species from 18 genera, mostly ‘clinging’ types followed by ‘adhesive’ and ‘encrusting’ types. The only encrusting form is *Nubecularia* sp. All of these tests are composed of calcite with varying Mg contents.

The maximum amount of carbonate is 27.6 g m^{-2} or 282 g kg^{-2} of seagrass (average 1 g m^{-2} or 4 g kg^{-1}). About 45% is from *Discorbis dimidiatus*, and 33% from *Nubecularia* sp. *Discorbis dimidiatus* grows on all seagrass species and occurs at every site. The second most widely distributed genera is *Elphidium*. The total amount of carbonate production does not have any particular relationship with specific geographical region nor any correlation with temperature, salinity and depth.

The most frequently occurring, in terms of their distribution on different seagrass species, are *D. dimidiatus*, *N. lucifuga*, *Nubecularia* sp. and *Elphidium* sp., which are present on all seagrasses, whereas *Spioloculina antillaum* grows

Fig. 10. Images of sea grasses and their calcareous epiphytes. (a) A lush meadow of *Posidonia* sp. in 3m w.d. at Chinaman Creek in which many blades are coated with encrusting coralline algae (white) (image = 60 cm wide); (b) *Posidonia* sp. blades washed up on the beach illustrating white coralline algal encrustations and white branching coralline algae (image = 20 cm wide), (c) *Posidonia* sp. blade heavily encrusted with coralline algae (scale in mm), (d) *Posidonia* sp. blade entwined with branching coralline algae (scale in mm), (e) *Posidonia* sp. blade heavily encrusted with coralline algae and a few spirorbids (top) (scale in mm), (f) stem of *Amphibolis* sp. encrusted with coralline algae (scale in mm), (g) blades of *Amphibolis* encrusted with coralline algae (scale in mm), (h) *Amphibolis* stems encrusted with bryozoans; 1 = *Iodictum phoenicium* (fenestrate), 2 = *Orthoscuticella ventricosa* (articulated zooidal), 3 = *Cellaria* sp. (articulated branching), 4 = *Celleporaria cristata* (foliose).

Table 4. Average abundance of $CaCO_3$ g kg^{-1} seagrass from each coralline algae type on each seagrass species/components. Average ± s.e. (number of occurrences).

Seagrass	Type 1 – thick encrusting sheet	Type 2 – encrusting sheet	Type 3 – articulated, erect thin branching	Type 4 – articulate, erect, narrow branching	Type 5 – non-articulated erect, simple branching
A. antarctica blade	0.6	67.0	31.7	0.4	0
A. antarctica stem	0.0	18.8	36.0	42.6	5.6
A. griffithii blade	3.4	88.8	26.5	1.5	0.0
A. griffithii stem	9.6	25.9	0.1	12.0	0.0
P. angustifolia	0.6	42.1	21.6	0.2	0.0
P. australis	0.0	59.5	15.2	24.0	0.0
P. sinuosa	0.8	19.5	15.5	3.8	0.0
Average	3.4 ± 1.3 (12)	42.7 ± 5.5 (62)	21.9 ± 4.1 (37)	14.5 ± 5.9 (37)	6.0 ± 0.9 (3)

only on *Posidonia* blades. *Quinqueloculina bradyana* and *Q. subpolygona* are found only on *Amphibolis* blades.

Ostracods

These arthropods are found on 48 samples (1/30) of the blades and stems analysed. They do not contribute significantly to epiphyte production, but are useful as environmental indicators. *Xestolibris* was the most common genera.

Annelids

Spirorbids occur on 741 (~1/2) of the blades and stems whereas serpulids are found on only

Fig. 11. Scanning electron microscope images of benthic foraminifera and spirorbid epiphytes. (a) *Discorbis dimidiatus*, (b) *Nubecularia lucifuga*, (c) *Elphidium fichtellianum*, (d) *Spiroloculina antillarium*, (e) *Peneropolis planatus*, (f) *Spirorbis* sp.

73 (~1/20). Spirorbids produce ~5% of the epiphytic carbonate and are most abundant on *A. antarctica* and *P. sinuosa* blades (never on *Amphibolis* stems). Serpulids do not produce significant carbonate. There is no relationship to temperature, salinity or depth. The skeleton of these annelids is aragonite and/or Mg-calcite.

Bivalves

Attached bivalves are present on 47 (~1/30) of the blades and stems and comprise five species, all of which are attached by byssal threads.

Epiphyte standing stock

Standing stock of epiphyte carbonate averages 79 g m^{-2} or 162 g kg^{-1} of seagrass (Fig. 12; Table 4). Each of the 21 sites has different abundances of calcareous epiphytes. Quadrat values range from 3.06 to 655.6 g m^{-2} (average 78.7 g m^{-2}; SD = 105.1 g m^{-2}) or 11.21 to 767.32 g kg^{-1} of seagrass (average 164; SD = 107.8 g kg^{-1}). Chinaman Creek (Fig. 5) at the head of Spencer Gulf, although having the highest biomass of seagrass, has the lowest abundance of epiphytes per plant (27.8 g kg^{-1} seagrass), probably the result of the extreme salinity values during summer months. Comparison with other areas shows a wider range in values from the minimum–maximum abundance for individual seagrass species data, but average values lie within the range of previous studies. Standing stock values from the tropical grass *Thalassia*, for example, tend to lie between average values of *Posidonia* and *Amphibolis*, e.g. 0.2–182.9 g m^{-2} (Frankovitch & Ziemnan, 1994), 1.19–140.95 g m^{-2} (Nelsen & Ginsburg, 1986) and 7.2–48 g m^{-2} (Land, 1970).

Predominant control on the abundance of epiphytes is seagrass biomass (Fig. 13), i.e. the more seagrass the more surface area for recruitment, the greater abundance of epiphytes. The relationship is allometric, i.e there is a constant relationship between the variables (Fig. 13). This is in contrast to the findings of Sim (1991). Regression analysis

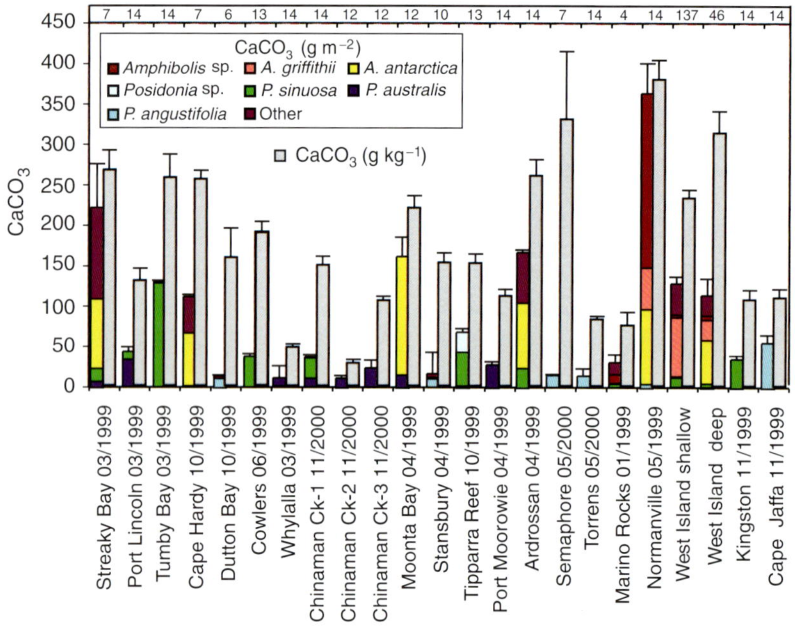

Fig. 12. Average epiphyte abundance from all sites studied. Values of $CaCo_3$ are in a $g\,kg^{-1}$ seagrass (grey bars) and $g\,m^{-2}$ (coloured). Numbers at top of bars refer to number of quadrats sampled at each site, with locations at the bottom given on Fig. 4. Data are grouped by seagrass species.

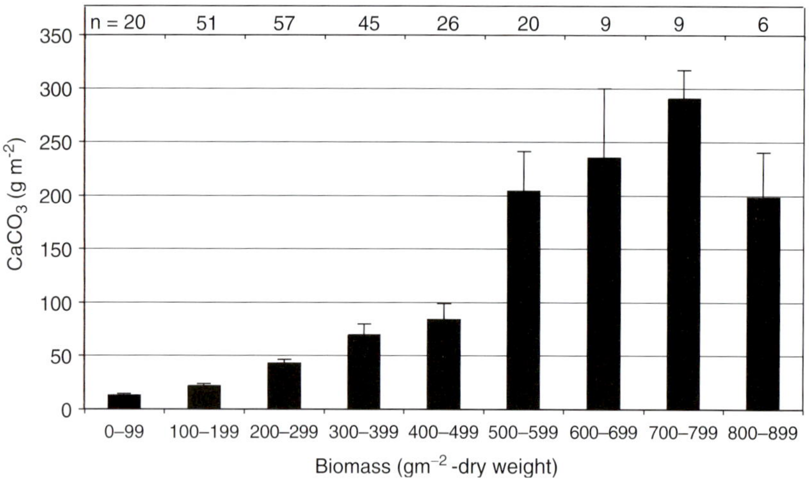

Fig. 13. Relationship between epiphyte $CaCO_3$ and seagrass biomass. Epiphyte $CaCO_3\,g\,m^{-2}$ by 100 g seagrass biomass bins.

(Fig. 13) shows that calcareous epiphyte abundance per unit area is closely associated with the abundance of seagrass substrate (equivalent biomass) available within that area. The fact that the relationship is not linear indicates that some other process is involved.

This exponential increase in epiphyte abundance is herein attributed to the proximity of the conspecific adults, reduced current velocity allowing higher settlement, success rates, and light availability below the canopy, which reduces competition for space with non-calcareous epiphytes.

Epiphyte biomass is also patchy; where patchiness of biomass is high, patchiness of epiphytes is also high. The $g\,kg^{-1}$ of seagrass data shows a decrease in epiphyte abundance per shoot above a biomass of $700\,g\,m^{-2}$ (Fig. 13).

Greater abundance of calcareous epiphytes, independent of seagrass biomass, is associated more with *Amphibolis* ($238.5\,g\,kg^{-1}$) than *Posidonia* ($127.2\,g\,kg^{-1}$) due to the accumulation of calcareous epiphytes over time on the relatively long-lived *Amphibolis* stems. This is confirmed by the Student's *t*-test ($P \leq 0.001$). These results are similar to studies in Western Australia (Lavery *et al.*, 1998). It must also be remembered that *Amphibolis* has the largest biomass of epiphytic (calcareous and non-calcareous) organisms of any species of seagrass (Borowitzka & Lethbridge, 1989). There is also a correlation between low carbonate quantities/high dominance of *Posidonia* and high carbonate quantities/high dominance of *Amphibolis*, a relationship also noted by Jernakoff & Nielsen (1998). The greater abundance of epiphytes on *Amphibolis* is attributed to the longevity of the stems; they are biennial (Walker, 1985; Coupland, 1997), whereas *Amphibolis* and *Posidonia* blades have life-spans of 60–100 and 65–130 days respectively. Hence, epiphytes have a longer time to accumulate on stems. Longer immersion times result in higher species abundance and higher species diversity (Turner & Todd, 1993).

Epiphyte carbonate increases significantly, independent of biomass, with increasing water depth to a 8–10 m w.d. maximum associated with *Posidonia* and to a 46 m w.d. maximum associated with *Amphibolis*, and then decreases. This is interpreted to be a function of reduced competition from non-calcareous epiphytes below 10 m w.d. Furthermore, below 10 m w.d. light for coralline algae and food for other epiphytes becomes limited. Reduction in current speed with depth is also a possible control on epiphyte distribution (cf. Eckman, 1983). Calcareous epiphytes appear to be more affected by depth when associated with *Posidonia* than *Amphibolis*, probably because of the longer times available for accumulation on *Amphibolis* stems. This is also because with time calcareous epiphytes out-compete non-calcareous epiphytes (Borowitzka *et al.*, 1990).

There is no apparent correlation between salinity or nutrient concentration at levels prevailing in the local environment, but temperature data indicate optimum temperature as 16–18°C, because maximum variation occurs at 16–18°C, and there is a decrease in abundance over 20°C.

DISCUSSION

Seagrass biomass

Results largely agree with studies elsewhere in western, southern and eastern Australia (West & Larkum, 1979; Silberstein, 1985; Walker, 1985; Walker & McComb, 1988; Hillman *et al.*, 1990; Cambridge & Hocking, 1997; Kendrick *et al.*, 1998b; Sim, 1999; Udy & Dennison, 1999; Seddon, 2000). Variance at the 50–100 m scale at each site is thought to reflect patchiness (cf. Kendrick *et al.*, 1998a,b). Causes of seagrass patchiness include competition between species, lack of equal seagrass growth expansion, limited dispersion of seedlings, non-availability of resources, grazing patchiness, unfavourable sediment conditions, blowouts and anthropogenic influence, e.g. sand harvesting for beach replenishment.

Seventy-four per cent of all seagrass biomass samples lie between 50 and 500 g m^{-2} (Fig. 6) suggesting that conditions enabling higher biomass are rare. Nevertheless, Chinaman Creek, at the head of Spencer Gulf has the highest biomass of *P. australis* recorded in the literature. This is probably because it is located at the head of an inverse estuary, and the seagrass is subject to high temperatures, high salinities and low current velocities suggesting all of these conditions are favourable for the growth of *P. australis*.

The higher biomass of *Amphibolis*-dominated communities is similar to that in Western Australia (Kendrick *et al.*, 1998a,b). This higher biomass is because *Amphibolis* has a greater number of shoots per square metre and the biomass of each individual shoot is greater than *Posidonia* shoots. On the other hand, *Posidonia* has a higher shoot spatial density than *Amphibolis*. The implication is, therefore, that there should be a greater of number of epiphytes per square metre on *Amphibolis*.

Biomass peaks at 2–4 m (Fig. 6) and decreases thereafter. This relationship, tied to light, is well documented (Burkholder & Doheny, 1968; Bulthuis & Woekerling, 1981; Dennison *et al.*, 1993; Walker *et al.*, 1999). The peak at 2–4 m is attributed to photosynthetic inhibition and desiccation in shallower water (Seddon, 2000). In general seagrass does not penetrate below 30 m w.d. In this study the biomass of *Amphibolis* decreases more rapidly than *Posidonia* with depth, suggesting that *Amphibolis* is more sensitive to lower light levels.

Photosynthesis and respiration by seagrasses is affected by temperature (Walker, 1991). Optimum temperatures for *P. sinuosa* have been measured at 13–23°C in comparison with 23°C and above for *P. australis*, *A. griffithii* and *A. antarctica* (Masini & Manning, 1997). Peak biomass for *A. antarctica* is 42.5‰ salinity (Walker, 1985).

Epiphytes

The main parameters controlling epiphyte abundance include seagrass biomass, seagrass genera, water depth and seasonality. *Amphibolis* has 3.5 times more carbonate g kg^{-1} of seagrass than does *Posidonia*. Also, the components are markedly different, with 3.4 times more epiphytes in terms of g kg^{-1} associated with the stems rather than the blades (Table 5).

grass beds are arguably the dominant nearshore carbonate factory throughout this vast temperate water realm.

Finally, the mixed Mg-calcite and aragonite mineralogy further reflects the source partitioning; Mg calcite (corallines, bryozoans, benthic foraminifera) comes largely from the grasses whereas, although some aragonite is produced by annelids, most of the aragonite comes from infaunal bivalves and gastropods.

Accumulation rates

Seagrass epiphytes in temperate environments produce, as in tropical settings, a considerable amount of carbonate sediment. Accumulation rates of that portion of the sediment produced in grass-beds, as measured by sediment cores of Quaternary sediment (Burne & Colwell, 1982; Gostin *et al.*, 1984), ranges between 20 and $270\,cm\,kyr^{-1}$ or overall between 10 and $100\,cm\,kyr^{-1}$ (James, 1997). This sediment comes not only from the epiphytes on the grass but also from the infaunal and epifauanal biota. This study indicates that a large proportion of the sediment comes from the epiphytes.

CONCLUSIONS

1 The predominant control on the abundance of calcareous epiphytes is seagrass biomass. Over 74% of the quadrats sampled during this study had seagrass biomass values between 50 and $500\,g\,m^{-2}$. Seagrass biomass has a peak value at 24 m water depth.

2 Epiphytic carbonate abundance changes significantly with water depth, increasing from 0 to 10 m water depth and then decreasing as depth increases.

3 Each of the 20 sites examined had different abundances of calcareous epiphytes, thus generalizations are dangerous and difficult.

4 Greater abundance of epiphytes is associated with *Amphibolis* compared with *Posidonia* because of the greater accumulation of calcareous epiphytes over time on the relatively long-lived *Amphibolis* stems. *Amphibolis* is more susceptible to epiphyte encrustation than *Posidonia.*

5 Decline in carbonate production during autumn–winter is attributed to simultaneous decrease in seagrass biomass, hence recruitment space.

6 Abundances of coralline algae >> bryozoans = foraminifera >> than all other calcareous epiphytes. Non-geniculate (encrusting) corallines are more common on seagrasses than geniculate (erect articulated) forms.

7 Calculated values of epiphyte production average $210 \pm 26\,g\,m^{-2}\,yr^{-1}$ or $750\,g\,kg^{-1}$ of seagrass per year. The range is $49–661\,g\,m^{-2}\,yr^{-1}$. These amounts are similar to those produced in tropical environments where values average $118–281\,g\,m^{-2}\,yr^{-1}$. Calculated accumulation rates of calcareous epiphyte carbonate is $\sim 7.4\,cm\,kyr^{-1}$.

8 The composition of sediment produced by most epiphytes (corallines, benthic foraminifera, bryozoans) is mostly Mg calcite, with minor aragonite produced by annelids.

ACKNOWLEDGEMENTS

This research was funded principally by the Australian Research Council through a grant to YB and AC and an Australian Postgraduate Research Award to KMB as well as funds from a Natural Sciences and Engineering Research Council of Canada Discovery Grant to NPJ. Assistance in identifying the epiphytes was provided by M. Davies (ostracods), and Q. Li (benthic foraminifera) and R. Schmidt (bryozoans). B. Jones and D. Boscence kindly read and made helpful comments on an early draft of this study while this paper was improved by comments from J. Reijmer and an anonymous reviewer.

REFERENCES

Barnett, E.J., Harvey, N., Belperio, A.P. and **Bourman, R.P.** (1997) Sea-level indicators from a Holocene, tide-dominated coastal succession, Port Pirie, South Australia. *Trans. Roy. Soc. S. Aust*, **121**, 125–135.

Betzler, C., Brachert, T.C. and **Nebelsick, J.** (1997) The warm temperate carbonate province: a review of facies, zonations and delineations. *Courier Forschungsinstitut Senckenberg*, **201**, 83–99.

Bone, Y. and **James, N.P.** (1993) Bryozoans as carbonate sediment producers on the cool-water Lacepede Shelf, southern Australia. *Sed. Geol*, **86**, 247–271.

Borowitzka, M.A. and **Lethbridge, R.C.** (1989) Seagrass epiphytes. In: *Biology of Seagrasses: A Treatise on the Biology of Seagrasses with Special Reference to the Australasian Region* (Eds R.C. Phillips and C.P. McRoy). Amsterdam, Elsevier, pp. 458–499.

Borowitzka, M.A., Lethbridge, R.C. and **Carlton, L.** (1990) Species richness, spatial distribution and colonizatioin pattern of algal and invertebrate epiphytes on the seagrass *Amphibolis griffithii*. *Mar. Ecol. Prog. Ser,* **64**, 281–291.

Boscence, D. (1989) Biogenic carbonate production in Florida Bay. *Bull. Mar. Sci,* **44**, 419–433.

Brasier, M.D. (1975) An outline history of seagrass communities. *Palaeontology,* **18**, 681–702.

Bressan, G. and **Tomini, I.** (1982) Quelques observations sur la croissance des algues rouges calcaires du genre *Fosliella* (Rhodophycophyta, Corallinaceae). *Vie et Milieu,* **31**, 283–291.

Bulthuis, D.A. and **Woelkerling, W.J.** (1981) Effects of *in situ* nitrogen and phosphorous enrichment of the sediments on the seagrass *Heterozostera tasmanica* (Martens ex Ashcer) den Hartog in Western Port, Victoria, Australia. *J. Exp. Mar. Biol. Ecol.,* **53**, 193–207.

Burkholder, J.M. and **Doheny, T.E.** (1968) *The Biology of Eelgrass*. Contribution 3, Department of Conservation and Waterways, Hempstead, Long Island, NY, 120 pp.

Burne, R.V. and **Colwell, I.B.** (1982) Temperate carbonate sediments of northern Spencer Gulf, South Australia; a high-salinity 'foramol' province. *Sedimentology*, **29**, 223–238.

Butler, A.J. (1991) Effect of patch size on communities of sessile invertebrates in Gulf St. Vincent, South Australia. *Biol. Conserv.*, **153**, 255–280.

Bye, J.T. (1972) Ocean circulation south of Australia. In: *Antarctic Oceanology 1, The Australian-New Zealand Sector* (Ed. D.E. Hayes), *Antarctic Research Series*, **19**. American Geophysical Union, Washington, pp. 95–100.

Bye, J.T. (1983) The general circulation in a dissipative basin with longshore wind stress. *J. Phys. Oceanogr,* **13**, 1553–1563.

Cambridge, M.L. and **Hocking, P.J.** (1997) Annual primary production and nutrient dynamics of the seagrasses *Posidonia sinuosa* and *Posidonia australis* in south-western Australia. *Aquat. Bot.*, **59**, 277–295.

Cann, J.H., Belperio, A.P. and **Murray-Wallace, C.V.** (2000) Late Quaternary paleosealevels and paleoenvironments inferred from foraminifera, Northern Spencer Gulf, South Australia. *J. Foramin. Res,* **30**, 29–53.

Cann, J.H., Harvey, N., Barnett, E.J., Belperio, A.P., and **Bourman R.P.** (2002) Foraminiferal biofacies eco-succession and Holocene sealevels, Port Pirie, South Australia. *Mar. Micropaleontol.*, **44**, 31–55.

Carruthers, T. (1994) *Leaf production, canopy structure and light climate of a density manipulated Amphibolis griffithii meadow.* Unpublished MSc. Thesis, University of Western Australia, Perth.

Corlett, H. and **Jones, B.** (2007) Epiphytes communities on *Thalassia testudinum* from Grand Cayman, British West Indies: facies composition, structure, and contribution to lagoonal sediments. *Sed. Geol.* **194**, 245–262.

Coupland, G. (1997) *Rhizome and shoot structure, growth and response to sediment burial in Amphibolis griffithii (Black) den Hartog.* Unpublished Honours thesis, Department of Botany, University of Western Australia, Perth.

Cresswell, G.R. (1991) The Leeuwin Current – observations and recent models. In: *The Leeuwin Current* (Eds A.F. Pearce and D.I. Walker). *Royal Soc. Western Austr. J.*, **74**, 1–15.

Davies, J.L. (1980) *Geographical Variation in Coastal Development*, 2nd ed. Longman, London, 371 pp.

Den Hartog, C. (1970) *The Seagrasses of the World*. North Holland Publishing Co., Amsterdam, 275 pp.

Dennison, W.C., Orth, R.J., Moore, K.A., Stevenson, J.C., Carter, V., Kollar, S., Bergstrom, P.W. and **Batiuk, R.A.** (1993) Assessing water quality with submersed aquatic vegetation: habitat requirements as barometers of Chesapeake Bay health. *BioScience,* **42**, 86–94.

Ducker, S.C. and **Knox, R.B.** (1978) Alleloparasitism between a seagrass and algae. *Naturwissenschaften*, **65**, 391–392.

Ducker, S.C., Foord, N.J. and **Knox, R.B.** (1977) Biology of Australian seagrasses, the genus *Amphibolis* C. Agardh (Cymodoceaceae). *Austr. J. Botany,* **25**, 67–95.

Eckman, J.E. (1983) Hydrodynamic processes affecting benthic recruitment. *Limnol. Oceanogr,* **28**, 241–257.

Edgar, G.J. (2000) *Australian Marine Life*. Reed New Holland, Sydney, pp.101–109.

Edgar, G.J. (2001) *Australian Marine Habitats in Temperate Waters*. Reed New Holland, Sydney, pp. 119–137.

Edyvane, K. (1999a) *Conserving Marine Biodiversity in South Australia, Part 1. Background, Status and Review of Approach to Marine Biodiversity*. Conservation in South Australia. Report 38, Adelaide, PIRSA, SARDI.

Edyvane, K. (1999b) *Conserving Marine Biodiversity in South Australia, Part 2. Identification of Areas of High Conservation Value in South Australia*. Conservation in South Australia, Report 39, PIRSA, SARDI, Adelaide.

English, S., Wilkinson, C. and **Baker, V.** (1997) *Survey Manual for Tropical Marine Resources*. Townsville, Australian Institute of Marine Science, 390 pp.

Frankovich, T.A. and **Zieman, J.C.** (1994) Total epiphyte and epiphytic carbonate production on *Thalassia testudinum* across Florida Bay. *Bull. Mar. Sci,* **54**, 679–695.

Fuller, M.K., Bone, Y., Gostin, V.A. and **von der Borch, C.C.** (1994) Holocene cool-water carbonate and terrigenous sediments from southern Spencer Gulf, South Australia. *Aust. J. Earth Sci.*, **41**, 353–363.

Ginsburg, R.N. (1956) Environmental relationships of grain size and constituent particles in some south Florida carbonate sediments. *AAPG Bull.*, **40**, 2384–2427.

Gostin, V.A., Belperio, A.P. and **Cann, V.A.** (1988) The Holocene non-tropical coastal and shelf carbonate province of southern Australia. *Sed. Geol.*, **60**, 51–70.

Gostin, V.A., Hails, J.R. and **Belperio, A.P.** (1984) The sedimentary framework of northern Spencer Gulf, Southern Australia. *Mar. Geol.*, **61**, 111–138.

Griffin, D.A., Thompson, P.A., Bax, N.J. and **Bradford, R.W.** (1997) The mass mortality of pilchard: no role found for physical or biological oceanographic factors in Australia. *Aust. J. Mar. Freshwat. Res.*, **48**, 27–42.

Hageman, S.J., Bone, Y., McGowran, B. and **James N.P.** (1996) Bryozoan species distributions on the cool-water Lacepede Shelf, southern Australia. In *Bryozoans in Space and Time: Proceedings of the 10th International Bryozoology Conference* (Eds D.P. Gordon, A.M. Smith and J.A. Grant-Mackie). NIWA, Wellington, New Zealand, pp. 109–117.

Hails, J.R., Belperio, A.P. and **Gostin, V.A.** (1984) Quaternary sea levels, Northern Spencer Gulf, Australia. *Mar. Geol.*, **61**, 373–389.

Harlin, M.M. (1980) Seagrass epiphytes. In: *Handbook of Seagrass Biology: an Ecosystem Perspective*

(Eds R.C. Phillips and C.P. McRoy). Garland STPM Press, New York, pp. 117–151.

Harvey, M. and **Bourget, E.** (1997) Recruitment of marine invertebrates onto arborescent epibenthic structures: active and passive processes acting at different spatial scales. *Mar. Ecol. Prog. Ser,* **153**, 203–215.

Hayward, P.J. (1974) Observations on the bryozoan epiphytes of *Posidonia oceanica* from the island of Chios (Aegean Sea). In: *Bryozoa 1974* (Ed. S. Pouyet). Université Claude Bernard, Lyon, pp. 347–356.

Hillman, K., Lukatelich, R.J., Bastyan, G. and **McComb, A.J.** (1990) *Distribution and Biomass of Seagrasses and Algae, and Nutrient Pools in Water, Sediments and Plants in Princess Royal Harbour and Oyster Harbour.* Environmental Protection Authority Technical Series 40, Perth.

Hillman, K., Morrison, P.F., Jernakoff, P. and **Nielsen, J.** (1994) *Perth Coastal Water Study: Determination of Time Series Changes in Marine Communities. Report Project E2*, Water Authority of Western Australia, Perth.

Horner, S.M.J. (1987) Similarity of epiphyte biomass distribution on *Posidonia* and artificial seagrass leaves. *Aquat. Bot,* **27**, 159–163.

Humm, H.J. (1964) Epiphytes of the sea grass *Thalassia testudinum. Bull. Marine Sci.*, **14**, 306–341.

James, N.P. (1997) The cool-water carbonate depositional realm. *In: Cool-water Carbonates* (Eds N.P. James and J.A.D. Clarke). *SEPM Spec. Publ.,* **56**, 1–22.

James, N.P. and **Bone, Y.** (2007) A Late Pliocene to early Pliocene, inner shelf, subtropical, grass-dominated carbonate: Roe Calcarenite, Great Australian Bight, Western Australia. *Palaios,* **22**, 343–359.

James, N.P., Bone, Y., Carter, R.M. and **Murray-Wallace, C.** (2006) Origin of the Roe Plains and their Pliocene calcarenite veneer: implications for sedimentology and tectonics in the Great Australian Bight. *Aust. J. Earth Sci.*, **53**, 407–419.

James, N.P., Bone, Y., Collins, L.B. and **Kyser, T.K.** (2001) Surficial sediments of the Great Australian Bight: facies dynamics and oceanography on a vast cool-water carbonate shelf. *J. Sed. Res,* **71**, 549–567.

James, N.P., Bone, Y., Hageman S., Gostin, V.A. and **Feary, D.A.** (1997) Cool-water carbonate sedimentation during the Terminal Quaternary, high-amplitude, sea-level cycle: Lincoln Shelf, Southern Australia. In: *Cool-Water Carbonates* (Eds N.P. James and J.A.D. Clarke). *SEPM Spec. Publ,* **56**, 53–76.

James, N.P., Bone, Y., von der Borch, C.C. and **Gostin, V.A.** (1992) Modern carbonate and terrigenous clastic sediments on a cool water, high energy, mid-latitude shelf, Lacepede Shelf, southern Australia. *Sedimentology,* **39**, 877–903.

Jernakoff, P. and **Nielsen, J.** (1998) Plant–animal associations in two species of segrasses in Western Australia. *Aquat. Bot.,* **60**, 359–376.

Jernakoff, P., Brearley, A. and **Nielsen, J.** (1996) Factors affecting grazer-epiphyte interactions in temperate seagrass meadows. In: *Oceanography and Marine Biology: An Annual Review* (Eds A.D. Ansell, R.N. Gibson and M.B. Barnes), Vol. **43.** Allen and Unwin, London, pp. 109–162.

Jones, P.L. and **Woekerling, W.J.** (1983) Some effects of light and temperature on growth and conceptacle production in *Fosliella cruciata* Bressan (Corallinaceae, Rhodophyta). *Phycologia,* **22**, 449–452.

Kendrick, G., Campey, M., Lavery, P., Westera, M. and **Wheeler, K.** (1998a) *Shells and dredging environmental management programme. Project S1: Ecological significance of seagrass.* Primary Production Report, Project S1, Task 7. The University of Western Australia and Edith Cowan University, Perth.

Kendrick, G.A., Langtry, S., Fitzpatrick, J., Griffiths, R. and **Jacoby, C.A.** (1998b) Benthic microalgae and nutrient dynamics in wave-disturbed environments in Marmion Lagoon, Western Australia, compared with less disturbed mesocosms. *J. Exp. Mar. Biol. Ecol,* **228**, 83–105.

Keough, M.J. (1983) Patterns of recruitment of sessile invertebrates in two subtidal habitats. *J. Exp. Mar. Biol. Ecol,* **66**, 213–245.

Kirkman, H. (1997) *Seagrasses of Australia.* State of the Environment Technical Paper Series (Estuaries & Seas). Canberra: Department of the Environment. Australia, 12 pp.

Kirkman, H. and **Kuo, J.** (1990) Pattern and process in southern Western Australian seagrasses. *Aquat. Bot.,* **37**, 367–382.

Land, L.S. (1970) Carbonate mud: production by epibiont growth on *Thalassia testudinum. J. Sed. Petrol.,* **40**, 1361–1363.

Lavery, P. and **Vanderklift, M.A.** (2000) Comparison of spatial patterns in seagrass epiphyte assemblages using species and functional group-level data. *Biol. Marina Meditarranea,* **7**, 251–254.

Lavery, P., Vanderklift, M., Brearley, A., Hyndes, G. and **Wells, F.** (1998) *Cockburn Cement Shells and Dredging Environmental Management Programme. The Ecological Significance of Seagrasses on Success Bank; Flora and Fauna Diversity.* Final report of project S-1 Task 6 to Cockburn Cement Pty. Ltd. December 1998.

Lavery, P., Westera, M., Kendrick, G., Brearley, A., Campey, M., Hegge, B. and **Hillman, D.K.** (2000) *Shells and Dredging Environmental Management Programme.* The University of Western Australia and Edith Cowan University, Perth.

Lewis, R., Aust, T., Clarke, S., Edyvane, K., Fisher, R., Fotheringham, D., McLennan, B., McShane, P., Newland, N., Noye, J., Neverauskas, V., Steffensen, D., White, M., Zacharin, W. and **Zeidler, W.** (1997) *Description, Use and Management of South Australia's Marine and Estuarine Environment.* The South Australian Marine and Estuarine Strategy Committee, Adelaide.

Lord, D. Associates (1998) *Estimate of* In Situ *Calcium Carbonate Production on Success Bank, Owen Anchorage (Western Australia)* (Ed. D.A. Lord and Associates Pty Ltd.) Prepared for Cockburn Cement Limited, Perth.

Lowenstam, H.A. (1955) Aragonite needles secreted by algae and some sedimentary implications. *J. Sed. Petrol.*, **25**, 270–272.

Masini, R.J. and **Manning, C.R.** (1997) The photosynthetic responses to irradiance and temperature of four meadow-forming seagrasses. *Aquat. Bot.,* **58**, 21–36.

Middleton, J.F. and **Cirano, M.** (2002) A northern boundary current along Australia's southern shelves: The Flinders Current. *J. Geophys. Res.*, **107**(C9), 3129–3140.

Middleton, J.F. and **Platov, G.** (2003), The mean summertime circulation along Australia's southern shelves: a numerical study. *J. Phys. Oceanogr.*, **33**, 2270–2287.

Nelsen, J.E. and **Ginsburg, R.N.** (1986) Calcium carbonate production by epibionts on *Thalassia* in Florida Bay. *J. Sed. Petrol.*, **56**, 622–628.

Nunes Vaz, R.A., Lennon, G.W. and **Bowers, D.G.** (1990) Physical behaviour of a large inverse estuary. *Cont. Shelf Res.*, **10**, 277–304.

Patriquin, D.G. (1972) Carbonate mud production by epibionts on Thalassia: an estimate based on leaf growth rate data. *J. Sed. Petrol.*, **42**, 687–689.

Perry, C.T. and **Beavington-Penny, S.J.** (2005) Epiphytic calcium carbonate production and facies development within sub-tropical seagrass beds, Ithaca Island, Mozambique, *Sed. Geol.*, **174**, 161–176.

Ridgeway, K.R. and **Condie, S.A.** (2004) The 5500-km-long boundary flow of western and southern Australia. *J. Geophys. Res.*, **109**, CO4017.

Rivers, J., James, N.P., Kyser, T.K. and **Bone, Y.** (2007) Genesis of palimpsest cool-water carbonate sediments on the southern Australian continental margin; Southern Australia. *J. Sed. Res.*, 77, 480–494.

Robertson, E.L. (1984) Seagrasses. In: *The Marine Benthic Flora of Southern Australia, Part 1* (Ed. H.B.S. Womersley), pp. 57–122. Government Printer, Adelaide, South Australia.

Robertson, E.L. (1986) Family Posidoniaceae. In: *Flora of Australia Part IV* (Eds J.P. Jessop and H.B. Toelken), pp. 1724–1729. Government Printer, Adelaide, South Australia.

Rochford, D.J. (1986) Seasonal changes in the distribution of Leeuwin Current waters off Southern Australia. *Aust. J. Mar. Freshwat. Res.*, **37**, 1–10.

Schahinger, R. (1987) Structure of coastal upwelling events observed off the coast of South Australia. *Aust. J. Mar. Freshwat. Res.*, **38**, 439–459.

Searle, D.J. (1984) *Sediment Transport System, Perth Sector Rottnest Shelf, Western Australia.* University of Western Australia, Perth.

Seddon, S. (2000) *Causes and ecological consequences of the Spencer Gulf seagrass dieback.* PhD thesis, University of Adelaide, 171 pp.

Shepherd, S.A. and **Robertson, E.L.** (1989) Regional studies – seagrasses of South Australia, western Victoria and Bass Strait. In: *Biology of Seagrasses: a Treatise on the Biology of Seagrasses with Special Reference to the Australian region* (Eds A.W.D. Larkum, A.J. McComb and S.A. Shepherd). Elsevier, Amsterdam, 841 pp.

Short, A.D. and **Hesp, P.A.** (1982) Wave, beach and dune interactions in southeastern South Australia. *Mar. Geol*, **48**, 259–284.

Silberstein, K. (1985) *The effects of Epiphytes on Seagrasses in Cockburn Sound, Western Australia. Report 63.* Department of Conservation and Environment, Perth.

Sim, C.B. (1991) *Analysis of Calcium Carbonate from Coralline Epiphytes (Corallinaceae Rhodophyta) of the Seagrass Amphibolis antarctica* (Labill.) Sonder et Aschers. Honours thesis, Department of Botany, University of Western Australia, Perth.

Sim, C.B. (1999) *Recruitment Ecology of Crustose Coralline Algae (Corallinales, Rhodophyta) at Rottnest Island, Western Australia.* Unpublished PhD thesis, Botany Department, University of Western Australia, Perth.

Smith, S.V. (1972) Production of calcium carbonate on the mainland shelf of Southern California. *Limnol. Oceanogr.*, **17**, 28–41.

Smith, S.V. and **Atkinson, M.J.** (1983) Mass balance of carbon and phosphorus in Shark Bay, Western Australia. *Limnol. Oceanogr.*, **28**, 625–639.

Taylor, P.D. (1990) Encrusters. In: *Paleobiology; a Synthesis* (Eds D.E.G. Briggs and P.R. Crowther). Blackwell Scientific Publications, Melbourne, pp. 346–351.

Thomas, R.I. and **Clarke, S.M.** (2001) Sedimentology and morphology of clastic seagrass beds in Holdfast Bay, South Australia: a locality of high wave energy for species of the *Posidonia australis* Complex. In: *Proceedings of the Seagrass Restoration Workshop for Gulf St. Vincent*, 15–16 May 2001, pp. 183–204. Adelaide, Australia.

Turner, S.J. and **Todd, C.D.** (1993) The early development of epifaunal assemblages on artificial substrata at two intertidal sites on an exposed rock shore in St. Andrews Bay, N.E. Scotland. *J. Exp. Mar. Biol. Ecol.*, **166**, 252–272.

Udy, J.W. and **Dennison, W.C.** (1999) Seagrass and sediment nutrients: Species comparison and fertilisation responses at Rottnest Island. In: *The Seagrass Flora and Fauna of Rottnest Island, Western Australia* (Eds D.I. Walker and F.E. Wells). Western Australian Museum, Perth, pp. 73–87.

Underwood, A.J. (1981) Techniques of analysis of variance in experimental marine biology and ecology. *Annu. Rev. Oceanogr. Mar. Biol.*, **19**, 513–605.

Walker, D.I. (1985) Correlations between salinity and growth of the seagrass *Amphibolis antarctica* in Shark Bay, Western Australia, using a new method of measuring production rate. *Aquat. Bot*, **23**, 13–26.

Walker, D.I. (1991) The effect of sea temperature on seagrasses and algae on the Western Australian coastline. *J. Royal Soc. Western Australia*, **74**, 71–77.

Walker, D.I. and **McComb, A.J.** (1988) Seasonal variation in the production, biomass and nutrient status of *Amphibolis antarcica* (Labill.) Sonder ex Aschers, and *Posidonia australis* Hook. f. in Shark Bay, Western Australia. *Aquat. Bot.*, **31**, 259–275.

Walker, D.I. and **Woelkerling, W.J.** (1988) Quantitative study of sediment contribution by epiphytic coralline red algae in seagrass meadows in Shark Bay, Western Australia. *Mar. Ecol. Prog. Ser.*, **43**, 71–77.

Walker, D.I., Dennison, W.C. and **Edgar, G.** (1999) Status of Australian seagrass research and knowledge. In: *Seagrass in Australia: Strategic Review and Development of an R&D Plan* (Eds A.P. Butler and P. Jernakoff). CSIRO Publishing, Collingwood, pp. 1–24.

Walker, D.I., Hutchings, P.A. and **Wells, F.E.** (1991) Seagrass sediment and infauna – A comparison of *Posidonia sinuosa* and *Amphibolis antarctica,* Princess Royal Harbour, south Western Australia: I. Seagrass biomass productivity and contribution to sediments. *Proceedings of the Third International Marine Biological Workshop: the Flora and Fauna of Albany*, Perth, Western Australia.

Wanless, H.R. (1981) Fining-upward sedimentary sequences generated in seagrass beds. *J. Sed. Petrol.*, **51**, 445–454.

West, R.J. and **Larkum, A.W.D.** (1979) Leaf productivity of the seagrass, *Posidonia australis*, in eastern Australian waters. *Aquat. Bot.*, **7**, 57–65.

Wilson, C.C. (1991) *Geology of the Quaternary Bridgewater Formation of southwest and central South Australia.*

Unpublished Ph.D. thesis, Flinders University, Adelaide Australia, 341 pp.

Womersley, H.B.S. (1996) *The Marine Benthic Flora of Southern Australia.* Australian Biological Resources Study, Canberra, 392 pp.

Wright, L.D., Guza, R.T. and **Short, A.D.** (1982) Dynamics of a high-energy dissipative surfzone. *Mar. Geol.*, **45**, 41–62.

Zar, J.H. (1996) *Biostatistical Analysis.* Prentice-Hall, Upper Saddle River, New Jersey, 662 pp.

Int. Assoc. Sedimentol. Spec. Publ. (2009) **41**, 149–165

Microbes versus metazoans as dominant reef builders: insights from modern marine environments in the Exuma Cays, Bahamas

MIRIAM S. ANDRES*[1], R. PAMELA REID*, EMILY BOWLIN*, A. PATRICIA GASPAR* and ANTON EISENHAUER†

**Division of Marine Geology and Geophysics, Rosenstiel School of Marine and Atmospheric Science, University of Miami, 4600 Rickenbacker Causeway, Miami, FL 33149, USA (E-mail: miriam.andres@chevron.com)*

†IFM-GEOMAR, Research Center for Marine Geosciences, Department for Marine Environmental Sciences, Wischhofstraße 13, 24118 Kiel, Germany

ABSTRACT

Mixed microbial–metazoan reefs, with distinct lateral zonation of microbialites, coralline algae and coral, fringe the windward-facing margin of two islands in the Exuma Cays, Bahamas: Stocking Island and Highborne Cay. These reefs offer a unique opportunity to identify and characterize controls governing the spatial distribution of microbial versus metazoan reef builders in a modern setting. Distribution patterns characterized by prolific stromatolite development in back-reef lagoons, prominent coralline algal growth along reef crests, and rare coral development on seaward reef edges are a direct response to intensity and frequency of sediment stress, sand abrasion and burial of reef biota. The recent discovery and dating of an outcropping coral reef underlying stromatolites in the southern Highborne Cay reef serves as a basis for interpreting the late Holocene evolution of this reef and addressing the question: what led to the demise of a metazoan and rise of a microbial reef? Comparison of growth histories of Highborne and Stocking Island reefs allows us to explore processes that have governed microbial versus metazoan reef building from mid-Holocene to present, and to speculate on future reef development at these sites. Antecedent topography and late Holocene sea-level transgression determined the timing of reef nucleation at both sites. Initial reef development was relatively unaffected by sedimentation due to a lag in sediment production relative to rising sea level. Once produced, sediment was deposited on the Pleistocene terrace, which due to rapid sea-level rise was below average wave base and sediment suspension. Slowing of sea-level rise combined with increasing sediment production decreased accommodation space. Mobile sediment progressively impinged upon the reef environment and, following the emergence of an algal ridge, sediment was trapped in the back-reef area. Although corals can tolerate some sediment stress, they cannot cope with extended periods of burial. Increasing sediment stress and more importantly increasing amplitude and frequency of sediment burial are interpreted as dominant factors leading to the demise of metazoan reef builders and the prolific growth of microbialites in the back-reef lagoon. Microbial buildups are surprisingly young (<1000 years) and a recent addition to the fringing reef systems along the Exuma margin. The study of modern analogues is important for our process-orientated understanding, which in turn provides concepts and hypothesis to apply and validate in the rock record.

Keywords Microbialite, microbial reef, Holocene, Bahamas, U/Th dating.

INTRODUCTION

Microbialites, i.e. microbial buildups such as stromatolites and thrombolites, are rare in today's metazoan-dominated reef systems. In contrast, microbial deposits dominated the shallows of the Precambrian oceans for over 3 billion years, forming massive reefs, rivalling or exceeding those of the Phanerozoic (Grotzinger & Knoll, 1999). We use the term 'microbialites' following the definition by Burne & Moore (1987): 'Microbialites are organosedimentary deposits that have accreted as a result of a benthic microbial community

[1]Present address: Chevron Energy Technology Company, 6001 Bollinger Canyon Rd, San Ramon, CA 94583, USA.

trapping and binding detrital sediment and/or forming the locus of mineral precipitation'. The term encompasses stromatolites, characterized by a laminated internal structure (Awramik & Margulis, 1974; Walter, 1976b) and thrombolites distinguished by a mesoscopic clotted internal fabric (Aitken, 1967; Kennard & James, 1986).

Although microbial reefs have been in decline since the Late Proterozoic, both stromatolites and thrombolites continued to be important reef-builders during various phases of the Phanerozoic (see Riding (2006) for latest review). The long-term decline of these microbial structures is attributed to eukaryotic competition and interference, competition for space and substrate, changes in the physical environmental conditions (Fischer, 1965; Awramik, 1971, 1990, 1992; Fischer & Arthur, 1977; Vermeij, 1987), and specifically to the rise of grazing and burrowing animals (Garrett, 1970). Based on the sporadic resurgence of microbialites following mass extinctions, they have been referred to as 'disaster forms' (Schubert & Bottjer, 1992). When stromatolite resurgences do not coincide with a mass extinction, authors such as Soja (1994) infer the 'localized eradication of benthic marine communities'. The question of whether metazoan competition is the primary factor governing microbial carbonate abundance was recently addressed in a comprehensive study comparing microbial carbonate abundance to the fluctuation in metazoan diversity through the geological record (Riding, 2006). By comparing the response of microbial carbonates in the aftermath of mass extinctions, i.e. their resurgence or failure to recover, Riding (2006) '... raises doubts that metazoan competition can be invoked as a general explanation for fluctuations in microbial carbonate abundance' and further that '... a variety of factors, each changing through time, must have operated'.

Few studies have addressed the issue of factors controlling microbial versus metazoan reef builders in modern settings, which in part may be explained by the scarcity of suitable localities. Originally thought extinct, modern stromatolites were discovered in Shark Bay, Australia (Logan, 1961) and Bahamas (Dravis, 1983; Dill *et al.*, 1986; Dill, 1991). Since then, Holocene microbialites and in particular stromatolites of all sorts and variety have been reported. Most modern microbialite occurrences are confined to lacustrine (Casanova, 1994; Andrews *et al.*, 1997) or extreme environments with respect to the tolerance limits of most eukaryotes, i.e. temperature (Walter, 1976a), salinity (Dupraz *et al.*, 2004; Vasconcelos *et al.*, 2006) and alkalinity (Kempe *et al.*, 1991).

With respect to true open-marine environments (Shark Bay is restricted marine to hypersaline), microbialites from French Polynesia are considered significant contributors of Holocene reef accretion, i.e. Tahiti (Camoin & Montaggioni, 1994; Camoin *et al.*, 1999) and Tikehau atoll (Sprachta *et al.*, 2001). The Tahitian examples are, however, limited in extent to open cavities within reef framework and lagoonal settings. Bahamian stromatolites, termed 'uncommonly common' after extensive mapping in the Exuma Cays (Reid *et al.*, 1995), are the only known examples of open-marine stromatolites developing as isolated columnar structures, such as the often cited examples from Lee Stocking Island (Dill *et al.*, 1986; Dill, 1991; Feldmann & McKenzie, 1998). In two Bahamian locations, stromatolites '... occur with other reef-building organisms as integral parts of laterally extensive fringing reefs'; on Stocking Island (Reid & Browne, 1991; Macintyre *et al.*, 1996) and Highborne Cay (Reid *et al.*, 1999).

Development of mixed microbialite–metazoan reefs at Stocking Island and Highborne Cay offer a unique opportunity to gain insights into the physical environmental factors controlling microbes versus metazoans as dominant reef builders. The Holocene history of the Stocking Island reef complex has been described (Reid & Browne, 1991; Macintyre *et al.*, 1996) and ecological controls on the stromatolite development in this reef have been assessed (Steneck *et al.*, 1998). Results from Stocking Island, together with the recent discovery and dating of an outcropping basal coral surface underlying stromatolites in the Highborne Cay complex, serve as a basis for interpreting reef history at Highborne Cay: what led to the demise of a metazoan and rise of a microbial reef? Results from these two examples allow discussion on processes and environmental factors determining the evolution and spatial distribution of microbial versus metazoan reef builders in Exuma margin build-ups during the late Holocene.

MIXED MICROBIAL–CORALLINE ALGAE–CORAL REEFS IN THE EXUMAS, BAHAMAS

Fringing reefs composed of mixed microbial–coralline algae–coral reef-building assemblages are reported from two locations in the Exuma

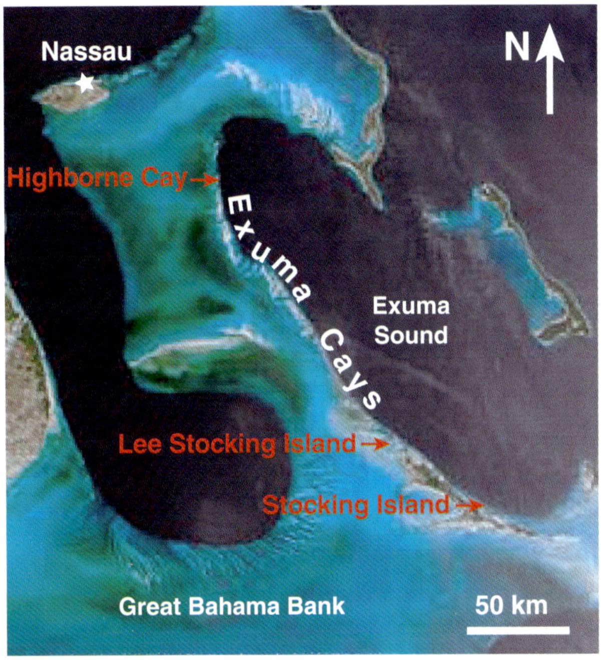

Fig. 1. Overview of the Exuma Cays, Bahamas; a chain of islands on the western rim of the Exuma Platform facing the deep waters of Exuma Sound. Locations of Highborne Cay, Lee Stocking Island and Stocking Island indicated with red arrows.

Cays: Stocking Island (Macintyre *et al.*, 1996; Steneck *et al.*, 1997, 1998) and Highborne Cay (Reid *et al.*, 1999) located in the southern and northern ends of the Exuma chain, respectively (Fig. 1).

The Exuma Cays

The Exuma Cays comprise a northwest–southeast-trending island chain in the central Bahamas. They are bordered to the west by the shallow (<10 m) Exuma Bank and to the east by a narrow terraced shelf dropping off to depths of up to 2000 m into the Exuma Sound (Fig. 1). Surface water in Exuma Sound is characterized by sea-surface temperatures ranging from 20°C to 28°C in winter and summer, respectively and normal marine salinities of ~35 psu (Droxler *et al.*, 1988). Tides are diurnal and range within 0.7–1 m. Surface waters are saturated with respect to both aragonite and calcite (Droxler *et al.*, 1988; Swart & Eberli, 2005). Climatic conditions are dominated by the southeasterly to easterly trade winds with an average wind speed of 10–15 knots. Waves and wave energy are highly dependent on wind direction and strength. Large waves (>1 m) occur mostly in association with strong cold frontal systems and occasional hurricanes. Steady trade winds are constant and strong enough to sustain measured average wave speeds of ~1 m s^{-1} (Eckman and Andres, unpublished data), making these sites moderate- to high-energy environments.

Sediments in the Exuma Cays chain are dominated by shallow-water carbonates, in particular ooids, skeletal and coated grains (Ginsburg *et al.*, 1958; Traverse & Ginsburg, 1966). Skeletal grains and nuclei of coated grains originate from surrounding reefs and shallows. Shallow-water sediment at both Highborne Cay and Stocking Island consists of well-sorted sand-sized sediment (on average 100–270 μm). Large and highly mobile shoals and sand waves are dominantly found in association with strong tidal currents within tidal channels (Gonzalez & Eberli, 1997), and, as observed on Highborne Cay, oceanward of fringing reefs (Andres & Reid, 2006).

Reef development at Stocking Island

On Stocking Island, a small island, 2 km east of Georgetown (Great Exuma Island), a reef complex approximately 50 m wide extends for 3 km along the eastern beach (Figs 1 and 2). The reef facies and internal reef structure are well-documented based on mapping and a suite of shallow cores (Reid & Browne, 1991; Macintyre *et al.*, 1996); these findings are summarized below.

Three zones are described in the Stocking Island complex: back reef, reef flat and fore reef. Stromatolites dominate the back reef area, occurring as metre-sized, sometimes coalescing tabular forms and dome-shaped heads (0.5–1 m diameter) with a maximum relief of ~0.5 m. They form elongated shapes parallel to the outer reef crest/edge, cut by channels reminiscent of the spur and groove structure in coral reefs (Fig. 2). The reef flat is characterized by hard and uneven surfaces of microbial mat in which a number of different cyanobacteria were documented (Macintyre *et al.*, 1996). Various macroalgae and turf communities dominate the seaward part of the reef flat. The fore reef is dominated by the coralline alga *Neogoniolithum strictum* (Steneck *et al.*, 1997). A few metres seaward of the coralline algal platform numerous pinnacles and ridges rise from a sand-swept, smooth rocky bottom. These blade-shaped pinnacles have smooth hard surfaces sometimes covered with macroalgae growth and cyanobacterial mat.

Stratigraphy and evolution of the reef is based on a single shallow-core transect, and dating of discrete samples. The 2.1 m thick Holocene reef

Fig. 2. Stocking Island fringing reef system at low tide. Microbialites coalesce to form metre-sized tabular features. Degraded coralline algal ridge comprised by *Neogoniolithon strictum* below breaking waves.

formed on a Pleistocene limestone terrace, the boundary clearly marked by caliche crusts, leaching and root casts. The initial reef phase consisted of an intertidal vermetid gastropod assemblage; a well-developed build-up by ca. 4000 yr BP. Subsequent late Holocene sea-level transgression (~4000–2000 yr BP) resulted in the growth of coralline algae on the outer edge of the vermetid platform and on more shoreward elevated Pleistocene areas. Stromatolite growth on the reef flat and back-reef area coincided with the formation of a coralline algae-dominated reef crest ridge (~2000–1000 yr BP), thought to be critical for wave dissipation. During this time, an ocean-ward sand bar was partially lithified and differential erosion resulted in the prominent pinnacle features (see Fig. 3e in Macintyre *et al.*, 1996). The reef complex most probably reached its apex ~1000–500 years ago, as documented by prolific inshore stromatolites growth, coalescing to a broad flat reef (Fig. 2) as the stromatolites reach sea level and the coralline algal ridge matured. Degradation has characterized the past 500 years, as a result of extensive bioerosion by echinoderms; this bioerosion has reduced the formerly intertidal algal ridge to a submerged subtidal pavement (Fig. 2). Stromatolites continue to grow in the back-reef area; wave energy is now mostly dissipated by offshore patch reefs.

Macintyre *et al.* (1996) concluded that sediment stress played a key role in the growth history of the Stocking Island reef complex and estimated that 'Sediment-stressed conditions probably first developed in the back-reef area approximately 1500 years ago, when wave energy in this zone was reduced by emergence of the algal ridge'. Steneck *et al.* (1997) attributed the development of stromatolite in the back-reef zone to creation of an ecological refuge as a result of periodic sediment inundation. Sediments reduce eukaryotic species diversity and stromatolite reefs form where ecological pressure is low. Further questions concerning the sediments at Stocking Island and comparisons between the Stocking Island and Highborne Cay reefs are as follows: what is the origin of the sediment that allowed microbial development at Stocking Island, and when and how did sediments impact the reef? How does the Highborne Cay reef compare to the Stocking Island reef and to what extent are factors controlling reef development at Stocking Island relevant to the Highborne Cay system?

The Highborne Cay reef system

Highborne Cay (76°49′W, 24°43′N) is a small island in the northern Exuma island chain (Fig. 1). A fringing reef extends along the windward eastern margin of the island for approximately 2.5 km; for the purposes of field study, the reef zone was subdivided into ten sites, with Sites 1–4 located in the most southern part of the beach (Fig. 3a).

Facies within the Highborne Cay reef were mapped according to the facies identification presented in Reid *et al.* (1999) within GIS-referenced aerial images and transects (Fig. 4). Visual examination of underwater outcropping reef features added additional information regarding reef stratigraphy, and located discrete coral and stromatolite samples for dating.

Present reef zonation

The Highborne Cay reef complex is best developed with respect to size and facies variability in the most southern part of the beach; the reef narrows towards the north (Fig. 3a). Channels up to 4 m deep and 2–3 m wide dissect the reef complex. Reef crest, reef flat and back-reef lagoon compose three distinct zones, varying in width and extent (see Reid *et al.* (1999) for detailed facies zonation). The fore-reef area is dominated by sand, with patchy development of pinnacle structures with smooth eroded surfaces with a veneer of microbial mat (Fig. 3e). Relief on the seaward edge of the reef crest is highly variable as sand waves move along and frequently over the reef crest and reef flat

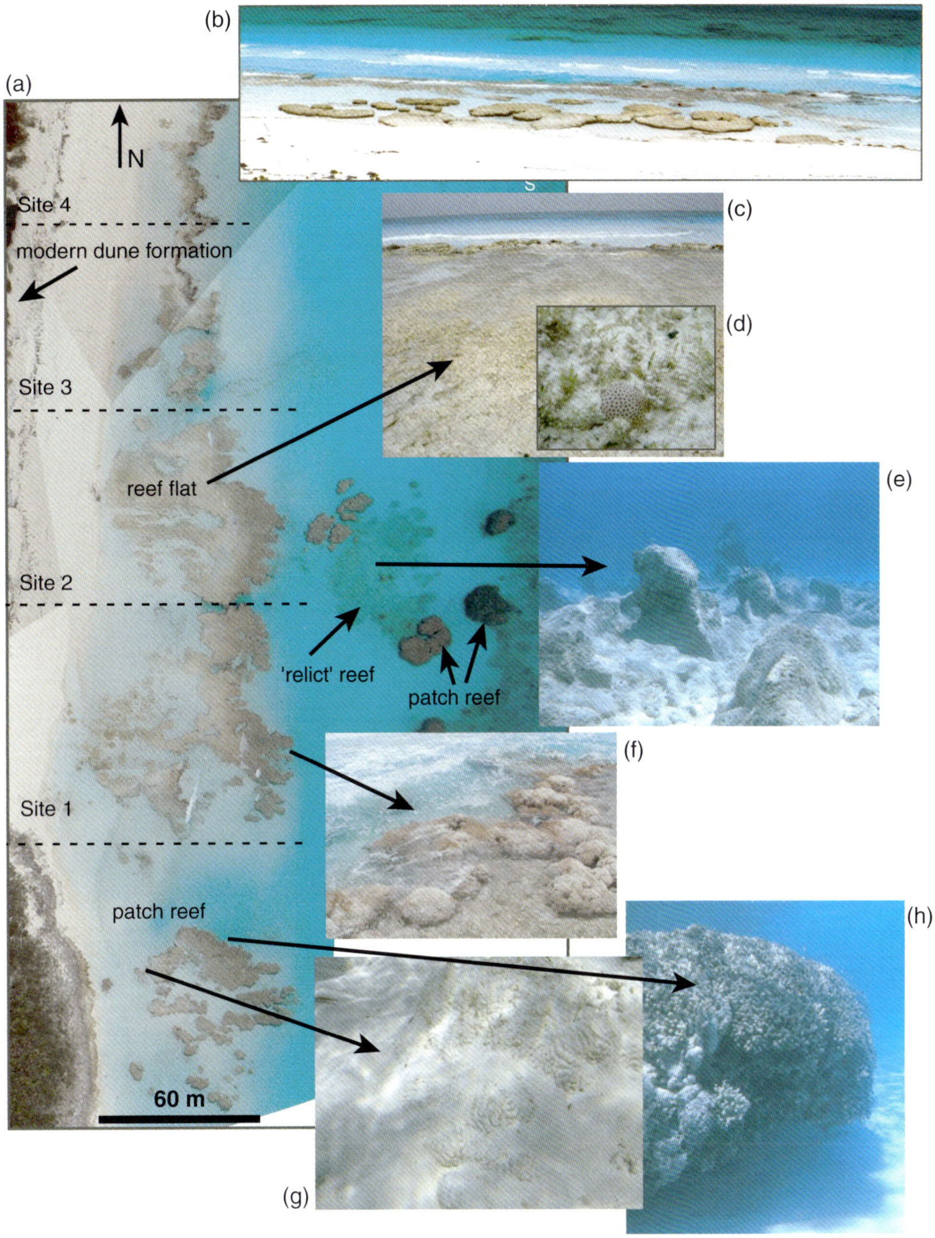

Fig. 3. (a) Aerial overview of the southern part of the Highborne Cay fringing reef system. Mixed microbial–coralline algae–coral system divided into Sites 0–10, from south to north, respectively, but only Sites 1–4 shown here. (b) Reef overview with microbialites such as stromatolites and thrombolites in the back reef lagoon. (c) Reef flat with prominent coralline algal ridge (*Neogoniolithon strictum*) defining the reef crest. (d) Small (5-cm size), sediment-tolerant encrusting brain coral (*Siderastraea radians*). (e) Relict reef seaward of Site 1 and juxtaposed to thriving patch reef (arrow). (f) Coralline algae (*Neogoniolithon strictum*) lining outer reef edge and reef crest thrive in high-energy surf zone. (g) Shoreward side of patch reef south of Site 1: progressive sediment impingement and burial resulting in coral bleaching and abrasion and ultimately demise. We consider this a modern analogue of Site 1 ~1000 years ago. (h) Thriving oceanward side of same patch reef. No sediment stress due to >2 m relief from sediment-covered bottom.

levelling any outer edge relief. Equally variable, due to frequent sand movement, is depth of the back-reef lagoon. Sand frequently fills the back-reef lagoon for weeks to months as illustrated in Fig. 5, whereas the reef flat is covered by sand only a few days. Microbialites develop in the back-reef lagoon; columnar stromatolites dominate the southern part of the beach, whereas formation of stromatolite ridges are restricted to form in the central and northern sites (see detailed discussion in Andres & Reid, 2006). Small patch reefs grow oceanward of Site 1, also to the north and south of the mixed reef complex (Fig. 3a, g and h). The 10–20-m-wide beach is bordered to the west by Holocene and Pleistocene aeolianite dunes. During strong easterly winds, beach sand is blown up against the dunes.

Facies and stratigraphy of southern Highborne Cay reef

The southern part of the reef complex (Site 1) is best developed with respect to size and facies variability and is the most relevant site for the present study due to the outcropping of a basal coral surface. The reef zonation for Site 1 is described

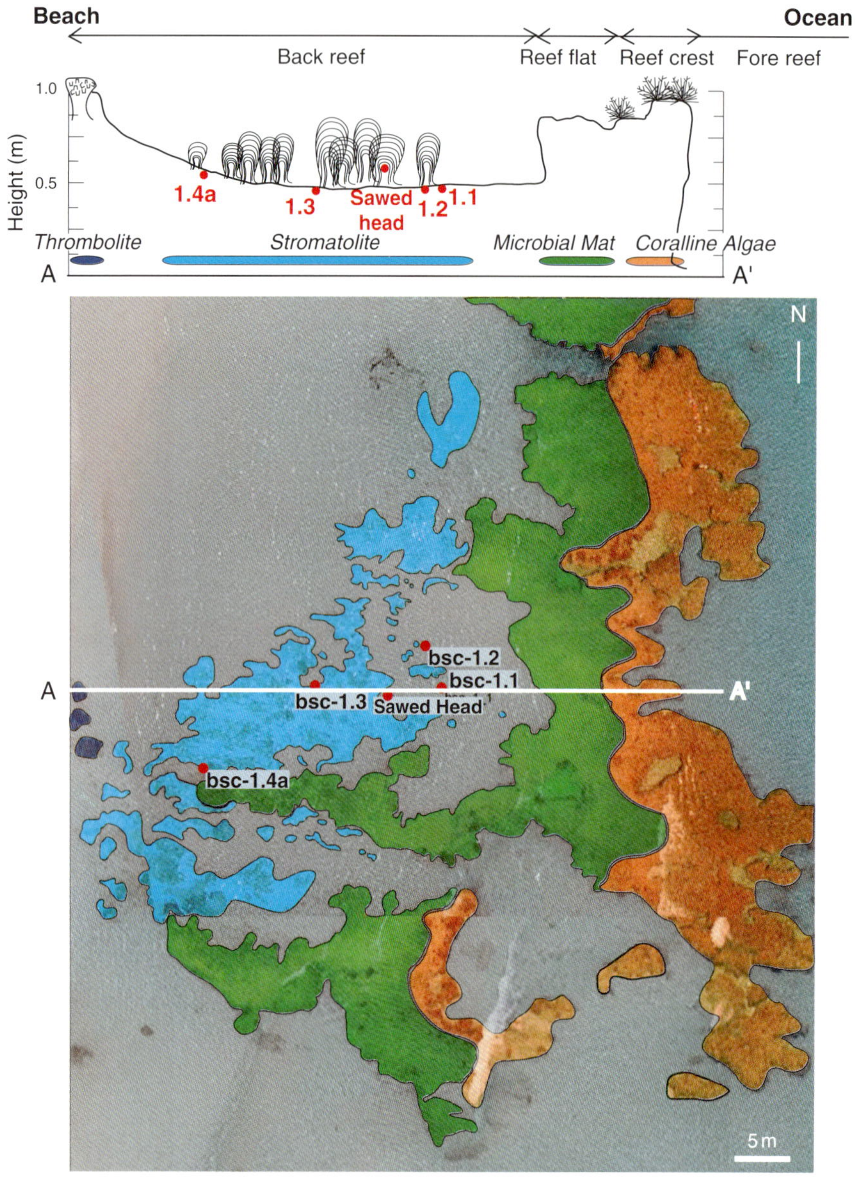

Fig. 4. Transect and aerial image of Site 1. Transect across Site 1 with dated coral and stromatolite samples projected onto transect line. Reef zones (back reef, reef flat, reef crest, and fore reef) characterized by different reef facies (thrombolites, stromatolites, microbial mat and coralline algae). Aerial image overlain by facies to illustrate spatial distribution thereof.

below in a seaward to shoreward direction: fore-reef pinnacle zone, reef crest, reef flat and back-reef lagoon (Fig. 4).

Prominent pinnacle structures, not previously documented in the Highborne Cay reef complex, occur in the fore-reef area, ~25 m east of the coralline algal reef platform (Fig. 3e). Located in ~3 m of water depth, the pinnacles tower up to 2 m in height if their bases are exposed from sand. They are composed of dense coralline algal limestone and sand-sized sediment filling the voids between the branching framework. Living macroalgae and scattered encrusting corals form bulging caps at the top of these pinnacles.

Interestingly, a healthy patch reef grows adjacent to the above-mentioned 'relict' reef (Fig. 3a; arrow indicating patch reef). Its shape is reminiscent of a mushroom: round tabular top with undercut base.

The maximum relief observed at the seaward edge of the reef crest zone is ~2 m; due to a lack of cores, the absolute thickness of the Holocene reef is unknown. Based on visual outcrop observations and samples chiselled from the surface, the wall of the outer reef, like the fore-reef pinnacles, consists of well-indurated, hard, dense algal limestone. Smooth surfaces are the result of constant sand abrasion. Hand samples and observations

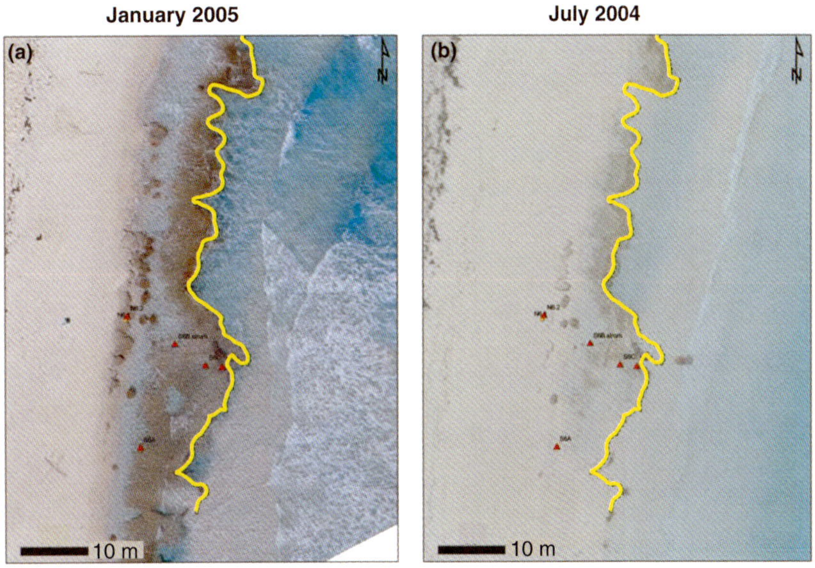

Fig. 5. Aerial image of Site 6 in January 2005 compared to 6 months earlier, July 2004, illustrating the significant change in sedimentation pattern and burial. Geo-referenced yellow circles define the location of the outer reef edge (linked by yellow line).

from holes in the reef flat indicate that at least the upper part of the outer reef flat consists of coralline limestone. The average height of sand burial is clearly discernable in colour and texture on the outer reef face: a thin veneer of microbial mat and turf, brown in colour, covers the upper part, whereas the lower part, if exposed, is a smooth bare light grey limestone.

Prominent red to purple-coloured heads of the branched coralline algae (*N. strictum*) line the intricately shaped reef crest (Fig. 3f) (Steneck *et al.*, 1997). Due to its width, more energy of breaking waves disseminates here before reaching the back-reef lagoon, in comparison to the much narrower reef to the north (Fig. 3a: Sites 1 and 2 in comparison to Sites 3 and 4). Indeed, preliminary data from wave-gauges confirm ~40% less wave energy here (Eckman, personal communication). A few small encrusting brain corals are found on the reef crest and numerous sea urchins inhabit cavities and holes.

A ~25-m-wide reef flat (Fig. 3c) comprises the area between reef crest and lagoon, populated by microbial mats, a variety of macroalgae (Littler *et al.*, 2005), rare small corals (Fig. 3d) and occasional sponges. The most shoreward reef flat and edge facing the back-reef lagoon are covered with centimetre-thick microbial mats.

Stromatolites in the back-reef lagoon develop mainly as columnar structures (Fig. 6a and b), sometimes coalescing together (Andres & Reid, 2006). With up to 40 cm in height and 50 cm diameter, the stromatolite columns at Site 1 are the largest in the Highborne Cay reef complex; height and size decrease shoreward. Water depth in the lagoon is dependent on the height of migrating sand waves, with a maximum of 1 m of water depth at low tide when the lagoon is devoid of sand. There is no apparent preferred growth direction with respect to incoming waves or longshore current. Tops of stromatolites are often colonized by macroalgae, dominantly *Batophora* and *Chondria*, and to a minor extent *Acetabularia* (Andres & Reid, 2006; Littler *et al.*, 2005). Furthermore, clams, bivalves and gastropods are found in holes, which they bored for shelter.

Discovery and dating of corals underlying stromatolites

Detailed mapping and facies descriptions in this and previous studies (Reid *et al.*, 1999; Andres & Reid, 2006) have documented biological surface cover for the Highborne Cay reef, as described above (Fig. 4). To date, however, subsurface information, stratigraphy and chronology have been lacking. A recent discovery of a coral surface underlying the stromatolites at Site 1 (Fig. 6) prompted questions on the relationship between underlying coral and overlying microbial buildups, the age of the reef and controlling factors determining its evolution and development in particular with respect to microbial versus metazoan reef builders (i.e. scleractinian corals and coralline algae). Information on the discovery and dating of the basal coral surface and the dating of a stromatolite head in the back-reef lagoon is presented below.

Fig. 6. Back reef lagoon at Site 1 illustrating the relationship between the underlying basal coral surface and overlying columnar stromatolite development. (a) Large rounded boulders on undulating basal surface; columnar stromatolites ~40 cm high. (b & c) Relationship between gray-blackened basal surface outcropping as sand (white) is washed out of back reef lagoon revealing stromatolite growth directly on this surface; columnar stromatolites ~40 cm high. (d) Small brain coral (under chisel: *Siderastraea radians:* sample H-0410-bsc-1.2) encrusting basal part of stromatolite column dated at 566 ± 11 years. Chisel is 30 cm long for scale. (e) Platy brain coral (*Diploria strigosa*; sample H-0410-bsc-1.3) dated at 667 ± 12 years. Knife is 25 cm long for scale. (f) Platy brain coral (*Diploria clivosa*; sample H-0410-bsc-1.1) dated at 790 ± 11 years. Chisel is 30 cm long for scale.

Basal surface underlying stromatolites

In October 2004 strong longshore currents over an extended period washed out most of the sand from the back-reef lagoon – exposing the basal surface of the back-reef stromatolites (Fig. 6). At the basal surface is a grey to dark grey, partially black, smoothly eroded hard surface. Large well-rounded pebbles to boulders (Fig. 6a) are scattered throughout the surface but tend to accumulate in depressions. Parts of the surface are overgrown by knobbly microbial mat. On closer inspection, numerous centimetre to decimetre-sized platy corals (Fig. 6e and f) and encrusting brain corals (Fig. 6c and d) were found within the grey to blackened basal surface (Fig. 5). Corals were identified as *Diplorian strigosa and clivona* as well as *Siderastraea radians*. Within a few centimetres, corals lie within the same horizontal plane. Most interesting is the relationship between this basal surface and stromatolites: all stromatolites grow off the basal surface (Fig. 6a–d). Small, ~10 cm encrusting *Siderastraea radians* heads grow at the base of some stromatolite build-ups (Fig. 6d).

Stromatolites have been notoriously difficult to date – previously, the age of the stromatolites and the basal corals at Highborne Cay were unknown. These outcrop relationships thus provided the unique opportunity to not only date the reef below the stromatolites but also to determine the maximum age of the stromatolite build-ups.

Material and methods

Coral samples

Ages were determined on four coral samples chiselled from the basal surface in October 2004 (Fig. 6). Two dated corals (H-0410-bsc-1.1, Fig. 6f, and H-0410-bsc-1.3, Fig. 6e) and the sectioned stromatolite head lie on an E–W transect across the northern part, the other dated coral samples (H-0410-bsc-1.2, Fig. 6d, and H-0410-bsc-1.4) are projected onto the transect line (Fig. 4). Coral samples were vertically sectioned and inspected for early diagenetic overprint; they displayed a pristine appearance under the blackened surface.

Stromatolite samples

One columnar stromatolite was vertically sectioned using a hand saw, leaving half the stromatolite

Fig. 7. Columnar stromatolite from Site 1 dated in three horizons. (a) Bottom horizon: laminae overlying coarse grainstone; middle horizon: trapped and bound grains; top horizon: trapped and bound grains. Large cavities due to boring clams; top of stromatolite colonized by macroalgae *Batophora*. (b) Remainder of sawed stromatolite in the back reef lagoon – stromatolite nucleation on coarse grainstone overlying basal surface.

Table 1. U/Th series dating results.

Sample ID	Material	Activity Ratio $^{234}U/^{238}U$ (dpm)	^{238}U conc (μg pm)	^{232}Th conc (μg pm)	Ages (yr)
H-0410-bsc-1.1	Basal brain coral	1.1478 ± 0.0013	3.048 ± 0.002	0.730 ± 0.007	790 ± 11
H-0410-bsc-1.2	Encrusting coral	1.1471 ± 0.0014	2.983 ± 0.002	0.509 ± 0.007	566 ± 11
H-0410-bsc-1.3	Basal brain coral	1.1487 ± 0.0017	3.138 ± 0.003	0.440 ± 0.006	667 ± 12
H-0410-bsc-1.4a	Basal brain coral	1.1477 ± 0.0017	2.769 ± 0.003	0.487 ± 0.005	972 ± 14
H-0310-1-5a	Trapped & bound grains	1.1471 ± 0.0016	2.362 ± 0.003	12.26 ± 0.130	1,590 ± 47
H-0310-1-4a	Trapped & bound grains	1.1437 ± 0.0027	2.439 ± 0.002	14.00 ± 0.150	1,548 ± 55
H-0310-1-6a	Micritic laminae	1.1460 ± 0.0021	2.464 ± 0.003	7.67 ± 0.080	884 ± 29

U/Th series dates for coral and stromatolites samples, Highborne Cay, Bahamas. In coral samples the detrital ^{230}Th was corrected with a $^{230}Th/^{232}Th$ activity ratio of 0.6 ± 0.2.

column standing in the lagoon (Fig. 7). Samples were collected from three horizons; 10 and 20 and 30 cm below the surface, the latter representing an initial lamination (Fig. 6a). Trapped and bound ooids and skeletal grains dominate samples from 10 and 20 cm depths, whereas the sample from 30 cm depth was a micritic lamina comprised mainly of authigenic microcrystalline aragonite.

Uranium–thorium dating

Four coral samples and three stromatolite samples, on average 510 mg, were chemically purified following the published procedures (Cheng *et al.*, 1986, 2000). The U/Th isotopes were measured at the GEOMAR mass spectrometry facilities on a Finnigan MAT 262 RPQ+ multicollector mass spectrometer. Measured $^{234}U/^{238}U$ activity ratios were 1.1478 ± 0.0016 disintegrations per minute (dpm) for corals and 1.1456 ± 0.0021 dpm for stromatolite samples, respectively, indicating a clear seawater signal.

Chronology

U/Th ages of corals and stromatolites are younger than 1000 years, ranging from 970 to 560 calendar years (Table 1). The oldest coral (H-0410-bsc-1.4a) is from the most shoreward location within

Site 1 (Fig. 4), yet the other dates do not suggest progressive oceanward coral demise. Combined ages from the three platy brain corals (H-0410-bsc- 1.1, -1.3, and -1.4a) indicate no pattern, instead random coral death within a patch reef. Noteworthy is the spread of coral ages of roughly 300 years with respect to the lack of vertical growth. The youngest coral sample, H-0410-bsc -1.2 is a small encrusting *Siderastraea radians*, growing at the basal part of a stromatolite column (Fig. 6d).

Ages from the vertically cut stromatolite are seemingly in reverse stratigraphic order: samples from 10 and 20 cm depth date at 1570 years (on average), whereas the underlying initial lamina at 30 cm depth is ~884 year old, or 690 years younger than the overlying material at 10 and 20 cm depth. This can be explained by considering the source of the dated material: the material dated at 30 cm was derived from micritic laminae consisting of authigenically precipitated carbonate; in contrast the older samples from 10 and 20 cm depth were mainly trapped and bound surrounding sediment. Indeed, the 1500-year-old dates for these ooids and skeletal grains agree with previously published ages for oolitic sands elsewhere in the Bahamas (Martin & Ginsburg, 1965) and demonstrate that the trapped and bound sediment in stromatolites represent an old, and well-mixed population.

Demise of a coral reef, rise of a microbial reef

Finding and dating an exposed basal surface in the Highborne Cay reef complex is central to answering one of the fundamental outstanding questions pertaining to the growth history: what is the age of the stromatolites? Previously, it was unclear if the modern stromatolites were growing on an antecedent Holocene or Pleistocene surface. Combining outcrop stratigraphy with the new age dates of the corals below the stromatolites allows this question to be addressed and provides a basis for discussion of Highborne Cay reef development within a temporal domain. Identifying and understanding the factors currently controlling reef development offers an opportunity to explore and speculate on the controls that have been acting in the past and future.

Based on the age of the oldest coral (~970 yr BP), the back-reef lagoon in the southern Highborne Cay reef was dominated by a coral patch reef 1000 years ago. The demise of the most shoreward coral around 970 yr BP raises the possibility of increased sediment input due to a prograding beach. Progressive seaward demise of corals is not, however, supported by the other dates. On the other hand, the first occurrence of stromatolitic micritic laminae over coarse grainstone, dated at 884 yr BP, indicates the presence of coarse-grained sediment and implies a process promoting the stabilization of grains. Coral death around 970 yr BP and initial stromatolitic laminae could be explained by decreasing hydrodynamic energy due to the emerging algal ridge as proposed for Stocking Island (Macintyre *et al.*, 1996) resulting in more sediment, and/or the presence of a stabilizing agent such as microbial mats. Whatever cause or combination thereof, stromatolite laminae overlying this grainstone package are evidence that conditions were right for the growth of microbialites such as stromatolites. Coral dates younger than 884 yr BP (samples H-0410-bsc-1.1, -1.2, and -1.3), in particular the presence of small encrusting corals at the base of some of the stromatolite columns imply the co-habitation of coral and microbial build-ups during this phase of reef growth. All the coral species found in the top of the basal surface (*Siderastraea* and *Diploria*) are tolerant with respect to sediment, wave energy and shallow water; noteworthy that this is an adequate description of the current environment. Today, however, corals such as *Siderastraea* and *Diploria* are predominantly found on the reef flat, reef crest and on the seaward edge of the reef platform as even tolerant coral species cannot cope with months-long burial below 30 cm of sand, which is a common scenario in the back-reef lagoon. Furthermore, the present-day corals on the reef flat are small, only 5–10 cm across, very similar to the 566-year-old *Siderastraea* sampled at the base of the stromatolite (Fig. 6d). We conclude that compared with today, environmental conditions deteriorated with respect to corals but improved with respect to microbialite development.

In summary, up until 1000 years ago, a shallow-water coral community dominated the Highborne Cay reef. The time from 1000 to 500 years ago was characterized by the co-habitation of stromatolites and corals, but with increasingly deteriorating conditions for corals. For the past 500 years, the back-reef lagoon at Highborne Cay has been dominated by microbial build-ups.

CONTROLS ON MICROBIAL VERSUS METAZOAN REEF BUILDERS

Integrating and comparing data and observations from the mixed microbial–coralline algae–coral reef systems at Highborne Cay and Stocking Island provide a basis for exploring controls on modern and Late Holocene reef development. Furthermore, dates in both reef systems provide constraints on the timing of microbial versus metazoan reef development.

Modern reef development, Stocking Island and Highborne Cay

Sediment and sediment dynamics are currently the key factors governing the distribution of microbial versus metazoan reef builders in the Highborne Cay and Stocking Island reef systems. A study investigating the processes that control the distribution of macroscale growth morphologies of stromatolites (i.e. columns vs. ridges) in the present Highborne Cay reef, identified accommodation space, hydrodynamics and sedimentation patterns as the primary controls (Andres & Reid, 2006). Whereas accommodation space is important for all phototropic reef-builders in this reef (stromatolites, coralline algae and corals), the optimal hydrodynamic conditions and intricately linked sedimentary patterns are different for each of these reef builders.

The hydrodynamic setting at both Stocking Island and Highborne Cay is dominated by waves breaking at the reef crest and rolling over the reef flat. Wave height is dependent on wind direction, speed and tide; large waves (>1 m) are generated in association with frontal systems and hurricanes. Maximum wave speeds of up to $2\,m\,s^{-1}$ over the reef flat and lagoon are not unusual, and are accompanied by strong longshore currents.

Thriving along the reef crest today, the branching coralline (*N. strictum)* defines the seaward reef crest, coping with wave energy, hour-long emergence during low tide and sediment abrasion and rare burial (Fig. 3f). We have observed bleaching events, mostly during the winter season, where the normally red to purple coralline algae heads lose their colour, but to date bleached coralline algae have recovered their pink colour within a few months. Sediment is known to negatively affect coral reefs, due to burial, abrasion and increase in turbidity (Hubbard, 1986). The well-sorted sand-sized sediment settles out quickly once stirred up by waves and/or strong currents. Visibility decreases during rough ocean conditions; however, due to a lack of any significant mud, turbidity is not sustained for long after the calming of wind and waves.

Ooid-dominated, highly mobile sand bars frequently move over the reef crest and reef flat, through cuts and channels, filling the back-reef lagoon and occasionally covering the reef flat and crest (Fig. 5). Whereas burial on the reef crest and reef flat is generally limited to days up to a few weeks, month-long burial in the back-reef lagoon can be observed.

Not only burial but also sand abrasion affects the reef habitat. Although sand-sized sediments settle quickly, constant breaking waves and subsequent shoreward surge, as well as swift longshore currents keep sediment in suspension, maintaining an abrading capacity. Evidence for sediment abrasion of the fore-reef area, which, as discussed, is characterized by relict-looking columns that tower up to 2 m high and are composed of hard coralline algae in both Highborne (Fig. 3e) and Stocking Island (see Fig. 3e in Macintyre *et al.*, 1996) reefs. Noteworthy, thriving circular coral-dominated patch reefs grow just adjacent to this pinnacle locale at Highborne Cay (Fig. 3a). Their distinct mushroom-shape and smooth undercut base indicate abrasion today. Furthermore, also at Highborne Cay, the basal surface in the back-reef lagoon today is a smooth, well-abraded surface; all larger pebbles and small boulders found on the basal surface are well-rounded; rounding cannot be explained by currents as these are too weak to move boulders of such size (Fig. 6a). The patch reefs south of Site 1 (Fig. 3a) exhibit a thriving, healthy coral community on the outer, ocean-facing sides; in contrast, the northward and shoreward facing sides of the patch reef (Fig. 3h) are stressed. Frequent sand incursions, burial and abrasion have decreased coral cover; most corals are ghostly white due to the loss of zooxanthellae, some are abraded (Fig. 3g).

It is acknowledged that factors other than sedimentation, such as temperature, salinity, nutrients and carbonate saturation state can be fundamental to reef development and demise (Camoin *et al.*, 1997; Montaggioni & Faure, 1997; Kleypas *et al.*, 1999). High nutrient concentrations, in particular, are invoked for favouring pervasive microbialite development in the lagoon of Tikehau atoll (Sprachta *et al.*, 2001). However, as pointed out by these authors, Bahamian stromatolites are an exception, and form in nutrient-poor water (Sprachta *et al.*, 2001). Nutrient-limited or oligotrophic conditions are reported for both Highborne Cay and Stocking Island (Pinckney *et al.*, 1995a,b; Paerl *et al.*, 2001). Lush patch reefs thrive a few tens of metres to the west of

Sites 1 and 2, as well as to the south and north of the fringing reef system (Fig. 3). Given the active hydrodynamics it is difficult to argue for different nutrient concentrations (or seawater composition) over any sustained period. Undoubtedly, on calm summer days significant temperature differences are measured in the backreef lagoon versus reef edge but are rarely maintained for more than a few days as tides and easterly winds mix the waters. An ambient-water, time-series sampling programme is currently underway to better understand the temporal nutrient characteristics along the fringing reef at Highborne Cay.

In summary, the physical environmental conditions are currently ideal with respect to prolific stromatolite formation in the back-reef lagoon and dominant coralline algae development on the outer reef crest. The limited growth of small encrusting corals on the reef flat imply overall unfavourable conditions with respect to scleractinian reef builders today. Although other factors (i.e. nutrients, temperature, salinity) may play a role, the observations presented here suggest that sediment and sediment dynamics are the key factors governing the distribution of microbial versus metazoan reef builders in the Highborne Cay and Stocking Island reef systems today.

Late Holocene reef development, Highborne Cay and Stocking Island

To what extent were the current controls of reef growth important in the nucleation and initiation of the reefs at Stocking Island and Highborne Cay? Have physical factors been acting with the same intensity and frequency as today? The observation that the back-reef lagoon of the Highborne Cay reef exhibits a basal coral surface, together with the documented Holocene history of the Stocking Island reef imply changes through time in the relative dominance of environmental factors controlling the distribution of reef biota. A major factor to consider when interpreting late Holocene reef history is sea-level rise.

Sea level–sediment interactions

What was the effect of late Holocene sea-level rise with respect to the amount of sediment production, composition and accumulation? For any reef, sea level marks the upper limit on growth, irrespective of the reef-building assemblage. Indeed, sea level in combination with antecedent topography, which provides substrate, are key to the timing of reef initiation and early reef development in general (Macintyre, 1988; Gischler & Hudson, 2004). Although many factors such as temperature, salinity, nutrients and carbonate saturation state may affect reef development (Camoin *et al.*, 1997; Montaggioni & Faure, 1997; Kleypas *et al.*, 1999), the discussion above indicates that sedimentation is the dominant factor controlling present distribution of reef biota at Highborne Cay and Stocking Island. It is therefore of interest to consider how sediment–reef interactions at these sites evolved through time.

Initial flooding of the Pleistocene reef terraces in the Bahamas during the late Holocene sea-level transgression took place around 4600 yr BP, assuming the base of the Pleistocene reef around 2 m below the Holocene reef (Boardman *et al.*, 1989). This time is also marked by an abrupt decrease in the rate of sea-level change from ~15 cm 100 yr^{-1} before 3800 yr BP to ~2 cm 100 yr^{-1} thereafter (Kindler, 1992). This rate of change seemed conducive to overall reef initiation and development in the Caribbean region (Neumann & Macintyre, 1985) and is supported by the basal core date in the Stocking Island reef (3750 ± 70 yr BP).

Sand and ooid shoals started forming in the Bahamas soon after the late Holocene sea-level transgression (Martin & Ginsburg, 1965; Harris, 1979). Significant bioclast production and shoreward transport is proposed for the post-3800 yr BP sea-level rate slowdown (Kindler, 1992). On Stocking Island, the Pleistocene terrace lies roughly 2 m above the sea floor. Thus, mobile sediments seaward of the fore-reef area would have had to overcome this step in order to have a significant impact on initial reef development. Presently, current ripples characterize the sand in more than 2 m of water depth in front of the outer reef edge, indicating that currents dominate at these depths. Occasionally sand in the fore reef accumulates into sand waves, and water depth decreases to such an extent that suspended sediment has an impact on the reef. Based on the rate of sea-level rise up until ~2000 yr BP (Boardman *et al.*, 1989; Kindler, 1992), there was ample accommodation space to fill as reefs had not caught up to sea level, and together with a lag in sediment production, early reef development was not much impacted by sediment. In this context, Steneck *et al.* (1997) concluded that '... the insular shelf off the Exuma Sound, at the time of initial reef and ridge development (*c.* 3000 yr BP) may have had considerably less sediment than it does today'.

Sand bodies stabilize and become cemented in the deepening of the fore-reef area due to sea-level rise – a viable process explaining the fore-reef pinnacle structures. The relict structures in the fore reef at Highborne Cay (Fig. 3e) are of further note: the description, and in particular the photograph of the fore-reef pinnacles at Stocking Island could be the spitting image of each other (see Fig. 3e, Macintyre *et al.*, 1996). According to Macintyre *et al.,* around 2220 yr BP, the offshore sandbars lithified, forming the bases for subsequent coralline algae growth at Stocking Island. No explanation on the processes is given by Macintyre *et al.* (1996). Here it is proposed that as sea level rose, these formerly very agitated bottoms fell below average wave base; sands were shifted around less and less frequently allowing for their stabilization and subsequent cementation.

The slow-down of sea-level rise further enabled sediment production to fill accommodation space and impact adjacent reefs. For approximately 2000 years, reefs began catching up with sea level (Boardman *et al.*, 1989; Kindler, 1992) as evidenced by the formation of an intertidal coralline algal lip at Stocking Island (Macintyre *et al.*, 1996). A slower rate of sea-level rise forced the reef into lateral expansion, but allowed the sand to catch up and fill accommodation space. The differential speed at which sand relative to the reef aggradations filled accommodation space explains most of the variance in the recent reef history as eluded to in the following.

Reef–sediment interactions

Macintyre *et al.* (1996) attributed the start of stromatolite growth behind the algal ridge to the wave-reducing effect of the ridge. We agree that the emerging algal lip dissipated wave energy but suggest that, more importantly, sediment was increasingly trapped behind this barrier, and that the sediment accumulation over increasingly longer time periods negatively affected metazoan reef builders and progressively favoured stromatolite development at Stocking Island. Here, the formation of an emergent algal ridge ~1500 years ago coincided with stromatolite growth in the back-reef lagoon. Similarly, ooids trapped within the vertically cut and dated stromatolite from Highborne Cay yielded ages of ~1500 years. This average age of stromatolite-bound grains suggests that by ~1500 yr BP hydrodynamic conditions were suitable for ooid and sand production; their presence demonstrating the effect on the reef. From 1000 to 500 yr BP the stromatolite and algal ridge at Stocking Island are interpreted to be at the apex of their development. Stromatolites thrived in the back-reef lagoon and once reaching sea level coalesced to form a broad flat reef (Fig. 2). A mature algal ridge characterizes the outer reef crest. In contrast, at Highborne Cay, the coral-dominated basal surfaces indicate that corals were the dominant reef-building organisms ~1000 years ago. Corals of the species *Diploria* and *Siderastraea* are tolerant with respect to sediment, shallow water and elevated salinities suggesting that conditions increasingly shifted to the present-day conditions. Also noteworthy is that although coral ages span roughly 400 years, there is a lack of any significant vertical growth. This could be attributed to limited accommodation space, i.e. the reef had reached sea level within a metre. Alternatively, environmental conditions were not optimal for strong vertical growth, i.e. high energy and/or sediment stress.

Sediment also leaves the reef system. Today, accommodation space at Highborne Cay fills up, at least temporarily. Sand bars, which frequently migrate over the seaward reef edge, cover the reef flat and fill the back reef lagoon; these sand waves bury the reef and obliterate relief (Fig. 5b). Moreover, sand is exported from the reef system today. Emergent sand partially dries during low tides and midday sun. Easterly winds pick up sand grains and blow them onto the beach up against the Pleistocene eolianites, forming modern dune deposits which become cemented (Fig. 3a).

These observations and interpretations highlight the major differences between the two systems: the present day Highborne Cay reef is considered the modern analogue to the Stocking Island reef ~1000 years ago, based on the prominent and intact algal ridge, as well as prolific stromatolite formation (Macintyre *et al.*, 1996). These observations are confirmed with the dating of the underlying coral surface and initial stromatolite growth, yet in turn, beg the question of why stromatolite development at Highborne Cay lags that of Stocking Island by 500–1000 years and further why are there no corals reported from the Stocking Island cores?

With all physical boundary conditions such as rate of sea-level rise, regional climate and oceanographic conditions being equal, the authors speculate that the antecedent topography might have been different at the two island sites. Assuming a one metre thicker Pleistocene buildup at Highborne Cay and the proposed sea-level rise of 15 cm 100^{-1} years (Boardman *et al.*, 1989; Kindler, 1992), it would have taken an additional 660 years before flooding the Pleistocene reef terrace at Highborne Cay and subsequent reef initiation. Incidentally, this coincides with the temporal difference between the developmental stages of Stocking Island versus Highborne Cay.

This study offers no formal explanation for the existence of a coral surface below the stromatolites at Highborne Cay, and apparent lack thereof at Stocking Island. However, it could be speculated that just as the basal coral surface, to date, has only been observed at the southern end of the Highborne Cay reef complex, the single core transect across the Stocking Island reef may not capture the full spatial subsurface variability.

Irrespective of temporal differences in reef growth at Stocking Island and Highborne Cay, both reefs exhibit a dominance in microbial versus reef building organisms for the past 500–1000 years. The extended range of the coral dates (790–566 years) suggests that the demise of the coral patch reef at Highborne Cay was not abrupt, but rather a continuous change from metazoan to microbial reef builders. This study proposes that, in comparison to today, the relative dominance of controls such as sediment stress, sediment patterns, burial amplitude and frequency have changed. The demise of the coral reef and rise of the microbial reef corresponded to increasing sediment input, and as sand filled the accommodation space seaward of the fore reef, it spilled over the reef crest into the back reef lagoon where it buried reef builders for prolonged periods of time. Although corals, coralline algae and vermetid gastropods might have coped with sediment input and abrasion, they could not cope with month-long burial. Stromatolites, however, depend on frequent sand burial, not only for vertical growth, but also to limit eukaryotic surface cover and bioerosion (Andres & Reid, 2006).

In summary, a simple model is proposed: increased sediment stress and corresponding increased frequency and duration of sediment burial led to the demise of a coral reef and the rise of a microbial reef at Highborne Cay and extensive microbialite development in the back reef at Stocking Island. Although proposing a single dominant process might be considered a reduced approach, it explains the late Holocene to present development of the discussed reefs. The lack of microbialite development on all other windward margins of the Exuma Islands, or for that matter in all sediment-dominated reefal settings, indicates that other factors, such as available substrate, act synergistically with sedimentation in controlling reef initiation and determining reef biota. The observed sediment dynamics are necessary to limit or even exclude metazoan reef builders yet possibly not sufficient to explain the presence of microbialites.

FUTURE REEF DEVELOPMENT

Whereas the Stocking Island reef has been degrading for the past ~500 years (Macintyre *et al.*, 1996), the Highborne Cay reef appears to be at or slightly beyond its apex. We predict continued microbial growth in the Highborne Cay reef complex is predicted in the future. With slow sea-level rise or even stagnation and filling of accommodation space, we expect lateral stromatolite expansion, coalescing, and the formation of increasingly tabular structures, such as those at Stocking Island is expected. Alternatively, with predicted increased rates of sea-level rise due to global change of 0.09–0.88 m by 2100 (Houghton *et al.*, 2001), there might be more accommodation space, resulting in predominant vertical rather than lateral growth.

Extensive bioerosion by sea urchins, clams, bivalves and the roots of macroalgae is obvious in both Highborne Cay and Stocking Island reef systems and will continue to occur. The coralline algal ridge at Highborne Cay may degrade to a submerged algal flat as observed at Stocking Island.

The patch reefs south of Site 1 at Highborne Cay (Fig. 3a, g and h) present a good analogue with respect to Highborne Cay roughly 800 years ago. Here, NW to SE sand bars increasingly impinge onto the patch reef resulting in white coral skeletons due to prolonged sand burial and subsequent demise (Fig. 3g). This is in stark contrast to the windward E-facing side, where an over 2 m high relief hinders sediment impingement (Fig. 3h). With continued sediment burial in the future the patch reefs to the south might face the same fate as their northern counterparts: coral demise and microbialite rise.

Geological significance

Modern settings offer the opportunity to characterize the ambient physical environment of a reef and, more importantly to characterize and identify factors controlling reef development: in this case the mixed coral–microbial–coralline algae–coral systems of Highborne Cay and Stocking Island. In the rock record, we are limited to outcrop or core data from which to infer paleoenvironmental information (Feldmann & McKenzie, 1997; Kahle, 1994; Riding *et al.*, 1991). For instance, in the Upper Jurassic Smackover formation, stromatolites were interpreted to form in low-energy peritidal settings characterized by waters with elevated salinity, whereas thrombolites are associated with higher energy conditions in subtidal setting of normal sea water (Mancini *et al.*, 2004). In contrast (Leinfelder *et al.*, 1993) concluded that microbialites are not restricted by water depth, salinity, temperature, light penetration, oxygen content or nutrient supply – but occur where other reef organisms are excluded due to some factor.

How would the Highborne Cay or Stocking Island reef system be interpreted in an outcrop? The high spatial variability might intrigue geologists to interpret the different facies as non time-equivalent. Moreover, the basal coral surface below the stromatolites could be interpreted as an erosional surface formed during sea-level lowstand or representing a non-depositional surface. Alternatively, the demise of the coral reef could be attributed to drowning after a rapid sea-level rise and the microbialite attributed to the next sea-level cycle. From a sequence stratigraphic perspective, stromatolite ridges overlying columnar structures could be interpreted within a shallowing upward sequence – which is correct on a small scale. However, as seen within the Highborne Cay reef system, the high spatial variability is due to the co-habitation and reef development within the same sea-level cycle.

Modern analogue studies offer the unique opportunity of directly characterizing and quantifying the physical environment in three dimensions and time. More importantly, they provide an opportunity to study and understand processes sparking new ideas and alternate scenarios with which to revisit the rock record.

CONCLUSIONS

Based on outcrop stratigraphy and dating, the evolution of the Highborne Cay and Stocking Island reef systems is comparable and events can be correlated within the limits of the dated material and dating itself. Microbial build-ups are surprisingly young (less than 1000 years) and a recent addition to the fringing reef systems along the Exuma margin.

The present distribution of microbial versus metazoan reef builders in the Stocking Island and Highborne cay reef is dominantly controlled by sediment and sediment dynamics, more specifically by the intensity and frequency of sediment stress, sand abrasion and burial of reef biota. Late Holocene reef history appears to reflect variable sedimentation patterns, antecedent topography and late Holocene sea-level transgression. Initial reef development was relatively unaffected by sedimentation, due to a lag in sediment production with respect to sea-level transgression, and more importantly deposition of sediment below the influence of average wave base, and below the influence of current suspension.

Slowing of sea-level rise combined with increasing sediment production decreased accommodation space. Mobile sediment progressively impinged upon the reef environment and, due to an emerging algal ridge, sediment was now trapped in the back reef area. Although corals can tolerate some sediment stress, they cannot cope with extended periods of burial. Increasing sediment stress and more importantly increasing amplitude and frequency of sediment burial are interpreted as dominant factors leading to the demise of metazoan reef builders and the prolific growth of microbialites in the back reef lagoon. The study of modern analogues is important for a process-orientated understanding, which in turn provides concepts and hypothesis to apply and validate in the rock record.

ACKNOWLEDGEMENTS

This paper is a tribute to RNG, a walking encyclopaedia of carbonate sedimentology – whom we admire for his breadth of knowledge, wit and humour. This paper, to us, represents a classic 'Ginsburgian' drama featuring some of RNG's most favourite and long standing research actors: stromatolites, reefs and ooids on one of his favourite geologic stages, the shallow, warm waters of the Bahamas.

We thank reviewers R. Riding and L. Montaggioni for useful suggestions and comments. Funding was provided by the US National Science Foundation. This is RIBS contribution No. 36.

REFERENCES

Aitken, J.D. (1967) Classification and environmental significance of cryptalgal limestones and dolomites, with illustrations from the Cambrian and Ordovician of southwestern Alberta. *J. Sedimentary Petrol.*, **37**, 1163–1178.

Andres, M.S. and **Reid, R.P.** (2006) Growth morphologies of modern marine stromatolites: a case study from Highborne Cay, Bahamas. *Sedimentary Geol.*, **185**, 319–328.

Andrews, J.E., Riding, R. and **Dennis, P.F.** (1997) The stable isotope record of environmental and climatic signals in modern terrestrial microbial carbonates from Europe. *Palaeogeogr., Palaeoclimatol., Palaeoecol.*, **129**, 171–189.

Awramik, S.M. (1971) Precambrian columnar stromatolite diversity: reflection of metazoan appearance. *Science*, **174**, 825–827.

Awramik, S.M. (1990) Stromatolites. In: *Paleobiology: A synthesis* (Eds D.E.G. Briggs and P.R. Crowther), pp. 336–341. Blackwell Scientific Publications, London.

Awramik, S.M. (1992) The history and significance of stromatolites. In: *Early Organic Evolution: Implications for Energy and Mineral Resources* (Eds M. Schidlowski, M.M. Kimberley, D.M. McKirdy and P.A. Trudinger), pp. 435–449. Springer, Berlin.

Awramik, S.M. and **Margulis, L.** (1974) Unpublished manuscript, cited by Walter, M.R. (1976) on page 1.

Boardman, M., Neumann, A.C. and **Rasmussen, K.A.** (1989) Holocene sea level in the Bahamas. In: *Fourth Symposium on the Geology of the Bahamas* (Ed. J.E. Mylroie), pp. 45–52. College of the Finger Lakes Bahamian Field Station, Ft. Lauderdale, Florida.

Burne, R.V. and **Moore, L.S.** (1987) Microbialites: organosedimentary deposits of benthic microbial communities. *Palaios*, **2**, 241–254.

Camoin, G.F. and **Montaggioni, L.F.** (1994) High energy coralgal-stromatolite frameworks from Holocene reefs (Tahiti, French Polynesia). *Sedimentology*, **41**, 655–676.

Camoin, G.F., Colonna, M., Montaggioni, L.F., Casanova, J., Faure, G. and **Thomassin, B.A.** (1997) Holocene sea-level changes and reef development in the southwestern Indian Ocean. *Coral Reefs*, **16**, 247–259.

Camoin, G.F., Gautret, P., Montaggioni, L.F. and **Cabioch, G.** (1999) Nature and environmental significance of microbialites in Quaternary reefs: the Tahiti paradox. *Sedimentary Geol.*, **126**, 271–304.

Casanova, J. (1994) Stromatolites from the east African Rift: A synopsis. In: *Phanerozoic Stromatolites; II* (Eds J. Bertrand-Sarfati and C.L.V. Monty), pp. 193–226. Kluwer Academic Publishers, Dordrecht.

Cheng, H., Edwards, R.L., Hoff, J., Gallup, C.D., Richards, D.A. and **Asmerom, Y.** (2000) The half-lives of uranium-234 and thorium-230. *Chem. Geol.*, **169**, 17–33.

Cheng, J.H., Edwards, R.L. and **Wasserburg, G.J.** (1986) ^{238}U, ^{234}U and ^{232}Th in seawater. *Earth Planetary Sci. Lett.*, **80**, 421–251.

Dill, R.F. (1991) Subtidal stromatolites, ooids and crusted-lime muds at the Great Bahama Bank margin. In: *From Shoreline to Abyss: Contributions to Marine Geology in Honor of Francis Parker Shepard* (Ed. R.H. Osborne), Vol. 46. SEPM Special Publication, Tulsa, Oklahoma.

Dill, R.F., Shinn, E.A., Jones, A.T., Kelly, K. and **Steinen, R.P.** (1986) Giant subtidal stromatolites forming in normal salinity waters. *Nature*, **324**, 55–58.

Dravis, J.J. (1983) Hardened subtidal stromatolites, Bahamas. *Science*, **219**, 385–386.

Droxler, A.W., Morse, D.L. and **Kornicker, W.A.** (1988) Controls on carbonate mineral accumulation in Bahamian basins and adjacent Atlantic Ocean sediments. *J. Sedimentary Petrol.*, **58**, 120–130.

Dupraz, C., Visscher, P.T., Baumgartner, L.K. and **Reid, R.P.** (2004) Microbe–mineral interactions: early carbonate precipitation in a hypersaline lake (Eleuthera Island, Bahamas). *Sedimentology*, **51**, 745–765.

Feldmann, M. and **McKenzie, J.A.** (1997) Messinian stromatolite–thrombolite associations, Santa Pola, SE Spain: an analogue for the Palaeozoic? *Sedimentology*, **44**, 893–914.

Feldmann, M. and **McKenzie, J.A.** (1998) Stromatolite–thrombolite associations in a modern environment, Lee Stocking Island, Bahamas. *Soc. Sedimentary Geol.*, **13**, 201–212.

Fischer, A.G. (1965) Fossils, early life, and atmospheric history. *Proc. Natl. Acad. Sci., USA*, **53**, 1205–1215.

Fischer, A.G. and **Arthur, M.A.** (1977) Secular variations in the pelagic realm. In: *Deep-water Carbonate Environments* (Eds H.E. Cook and P. Enos), Vol. **25**, pp. 19–51. Society for Economic Paleontologists and Mineralogists Special Publication.

Garrett, P. (1970) Phanerozoic stromatolites: Noncompetitive ecologic restriction by grazing and burrowing animals. *Science*, **169**, 171–173.

Ginsburg, R.N., Lloyd, R.M., McCullum, J.S., Stockman, K.W. and **Moody, R.A.** (1958) Surface sediments of Great Bahama Bank. Unpublished EPR report 506, Shell Development Company.

Gischler, E. and **Hudson, H.J.** (2004) Holocene development of the Belize Barrier Reef. *Sedimentary Geol.*, **164**, 223–236.

Gonzalez, R. and **Eberli, G.P.** (1997) Sediment transport and bedforms in a carbonate tidal inlet; Lee Stocking Island, Exumas, Bahamas. *Sedimentology*, **44**, 1015–1030.

Grotzinger, J.P. and **Knoll, A.H.** (1999) Stromatolites in Precambrian Carbonates: Evolutionary Mileposts or Environmental Dipsticks? *Ann. Rev. Earth Planetary Sci.*, **27**, 313–358.

Harris, P.M. (1979) *Facies anatomy and diagenesis of a Bahamian ooid shoal complex.* Sedimenta, **VII**. Comparative Sedimentology Laboratory Publication, Miami, 163 pp.

Houghton, J.T. (2001) Intergovernmental Panel on Climate Change, *Climate Change 2001, Working Group 1: The Scientific Basis.* Cambridge University Press, Cambridge.

Hubbard, D.K. (1986) Sedimentation as a control of reef development: St.Croix, U.S.V.I. *Coral Reefs*, **5**, 117–125.

Kahle, C.F. (1994) Facies and evolution of a Silurian coral-microbialite reef complex, Maumee, Ohio, USA. *J. Sedimentary Res.*, **A64**, 711–725.

Kempe, S., Kazmierczak, J., Landman, G., Konuk, T., Reimer, A. and **Lipp, A.** (1991) Largest known microbialites discovered in Lake Van, Turkey. *Nature*, **349**, 605–608.

Kennard, J.M. and **James, N.P.** (1986) Thrombolites and stromatolites; two distinct types of microbial structures. *Palaios*, **1**, 492–503.

Kindler, P. (1992) Coastal response to the Holocene transgression in the Bahamas: Episodic sedimentation versus continuous sea-level rise. *Sedimentary Geol.*, **80**, 319–329.

Kleypas, J., **McManus, J.** and **Meñez, L.** (1999) Environmental limits to coral reef development: where do we draw the line? *Am. Zool.*, **39**, 146–159.

Leinfelder, R.R., **Nose, M.**, **Schmid, D.U.**, **Werner, S.** and **Werner, W.** (1993) Microbial crusts of the late Jurassic: composition, palaeoecological significance and importance in reef construction. *Facies*, **29**, 26.

Littler, D.S., **Littler, M.M.**, **Macintyre, I.G.**, **Bowlin, E.**, **Andres, M.S.** and **Reid, R.P.** (2005) Guide to the dominant macroalgae of the stromatolite fringing-reef complex, Highborne Cay, Bahamas. *Atoll Res. Bull.*, **531–542**, 67–91.

Logan, B.W. (1961) Cryptozoon and associate stromatolites from the Recent, Shark Bay, Western Australia. *J. Geol.*, **69**, 517–533.

Macintyre, I.G. (1988) Modern coral reefs of western Atlantic: new geological perspective. *AAPG Bull.*, **72**, 1360–1369.

Macintyre, I.G., **Reid, R.P.** and **Steneck, R.S.** (1996) Growth History of Stromatolites in a Holocene fringing reef, Stocking Island, Bahamas. *J. Sedimentary Res.*, **66**, 231–242.

Mancini, E.A., **Llinas, J.C.**, **Parcell, W.C.**, **Aurel, M.**, **Badenas, B.**, **Leinfelder, R.R.** and **Benson, D.J.** (2004) Upper Jurassic thrombolite reservoir play, northeastern Gulf of Mexico. *AAPG Bull.*, **88**, 1573–1602.

Martin, E.L. and **Ginsburg, R.N.** (1965) Radiocarbon ages of oolitic sands on Great Bahama Bank. In: *Proceedings of the 6th International Conference on Radiocarbon Tritium Dating*, pp. 705–719, Pullman, Washington.

Montaggioni, L. and **Faure, G.** (1997) Response of reef coral communities to sea-level rise: a Holocene model from Mauritius (Western Indian Ocean). *Sedimentology*, **44**, 1053–1070.

Neumann, A.C. and **Macintyre, I.G.** (1985) Reef response to sea-level rise: keep-up, catch-up or give-up. In: *5th International Coral Reef Congress*, Vol. 3, pp. 105–110, Tahiti.

Paerl, H.W., **Steppe, T.F.** and **Reid, R.P.** (2001) Bacterially-mediated precipitation in marine stromatolites. *Environ. Microbiol.*, **3**, 123–130.

Pinckney, J.L., **Paerl, H.W.** and **Bebout, B.M.** (1995a) Salinity control of benthic microbial mat community production in a Bahamian hypersaline Lagoon. *J. Exp. Marine Biol. Ecol.*, **187**, 223–237.

Pinckney, J.L., **Paerl, H.W.** and **Fitzpatrick, M.** (1995b) Impacts of seasonality and nutrients on microbial mat community structure and function. *Marine Ecol. Progr. Ser.*, **123**, 207–216.

Reid, R.P. and **Browne, K.M.** (1991) Intertidal stromatolites in a fringing Holocene reef complex in the Bahamas. *Geology*, **19**, 15–18.

Reid, R.P., **MacIntyre, I.G.**, **Browne, K.M.**, **Steneck, R.S.** and **Miller, T.** (1995) Modern marine stromatolites in the Exuma Cays, Bahamas: uncommonly common. *Facies*, **33**, 1–18.

Reid, R.P., **MacIntyre, I.G.** and **Steneck, R.S.** (1999) A microbialite/algal ridge fringing reef complex, Highborne Cay, Bahamas. *Atoll Res. Bull.*, **466**, 1–18.

Riding, R. (2006) Microbial carbonate abundance compared with fluctuations in metazoan diversity over geologic time. *Sedimentary Geol.*, **185**, 229–238.

Riding, R., **Braga, J.C.** and **Martin, J.M.** (1991) Oolite stromatolites and thrombolites, Miocene, Spain: Analogues of recent giant Bahamian examples. *Sedimentary Geol.*, **71**, 121–127.

Schubert, J.K. and **Bottjer, D.J.** (1992) Early Triassic stromatolites as post-mass extinction disaster forms. *Geology*, **20**, 883–886.

Soja, C.M. (1994) Significance of Silurian stromatolite-sphinctozoan reefs. *Geology*, **22**, 355–358.

Sprachta, S., **Camoin, G.F.**, **Golubic, S.** and **Le Campion, T.** (2001) Microbialites in a modern lagoonal environment: nature and distribution, Tikehau atoll (French Polynesia). *Paleogeogr., Paleoclimatol., Paleoecol.*, **175**, 103–124.

Steneck, R.S., **Macintyre, I.G.** and **Reid, R.P.** (1997) A unique algal ridge system in the Exuma Cays, Bahamas. *Coral Reefs*, **16**, 29–37.

Steneck, R.S., **Miller, T.E.**, **Reid, R.P.** and **Macintyre, I.G.** (1998) Ecological controls on stromatolite development in a modern reef environment: a test of the ecological refuge paradigm. *Carbonates Evaporites*, **13**, 48–65.

Swart, P.K. and **Eberli, G.P.** (2005) The nature of the $\delta^{13}C$ of periplatform sediments: implications for stratigraphy and the global carbon cycle. *Sedimentary Geol.*, **175**, 115–129.

Traverse, A. and **Ginsburg, R.N.** (1966) Palynology of the surface sediments of Great Bahama Bank, as related to water movement and sedimentation. *Marine Geol.*, **4**, 417–459.

Vasconcelos, C., **Warthmann, R.**, **McKenzie, J.A.**, **Visscher, P.T.**, **Bittermann, A.G.** and **von Lith, Y.** (2006) Lithifying microbial mats in Lagoa Vermelha, Brazil: Modern Precambrian relics? *Sedimentary Geol.*, **185**, 175–183.

Vermeij, G.J. (1987) *Evolution and Escalation: An Ecological History of Life*. Princeton University Press, Princeton, New Jersey, 527 pp.

Walter, M.R. (1976a) Microbiology and morphogenesis of columnar stromatolites (*Conophyton, Vacerrilla*) from hot springs in Yellowstone National Park. In: *Stromatolites* (Ed. M.R. Walter), pp. 273–310. Elsevier Scientific, Amsterdam.

Walter, M.R. (1976b) *Stromatolites*. Developments in Sedimentology. Elsevier Scientific Publishing Company, Amsterdam.

Int. Assoc. Sedimentol. Spec. Publ. (2009) **41**, 167–178

Microbial dolomite precipitation under aerobic conditions: results from Brejo do Espinho Lagoon (Brazil) and culture experiments

MÓNICA SÁNCHEZ-ROMÁN*, CRISÓGONO VASCONCELOS*, ROLF WARTHMANN*, MARIAN RIVADENEYRA† and JUDITH A. MCKENZIE*

**ETH, Geological Institute, 8092 Zürich, Switzerland (E-mail: monica.sanchez@erdw.ethz.ch)*
†Department of Microbiology, Faculty of Pharmacy, University of Granada, Spain

ABSTRACT

Microbially mediated high-Mg calcite and dolomite precipitation occurs under oxic conditions in Brejo do Espinho lagoon, Brazil, within the upper 5 cm below the sediment–water interface. With burial to < 25 cm in the sediment sequence, early diagenesis associated with sulphate-reducing bacterial activity transforms the mixed carbonate mineralogy to 100% dolomite, as the pore-water becomes undersaturated with respect to calcite, while remaining supersaturated with respect to dolomite. Laboratory culture experiments using moderately halophilic aerobic bacteria (*Virgibacillus marismortui* and *Marinobacter* sp.) isolated from the uppermost part of the microbial mat in Brejo do Espinho demonstrate that microbially mediated dolomite precipitation can occur under ambient Earth's surface conditions in the presence of oxygen. These results add an additional metabolic process, aerobic respiration, to bacterial sulphate reduction and methanogenesis, which have previously been identified with dolomite formation. Furthermore, the formation of carbonate minerals with spherulitic structures in both the natural environment and laboratory culture experiments points to microbial involvement, as recognized in numerous other modern environments and ancient systems. This study suggests that previously recognized modern dolomite-forming environments, such as the supratidal areas of Andros Island, Bahamas, with recent dolomite crusts should be revisited to evaluate the importance of aerobic respiration in dolomite precipitation.

Keywords Moderately halophilic aerobic bacteria, microbial dolomite, hypersaline environment, Brejo do Espinho, Brazil, spherulites.

INTRODUCTION

Dolomite is a common carbonate mineral in sedimentary rocks throughout the geological record, especially in Precambrian carbonate rocks where it is abundant and often found associated with microbial structures, e.g. stromatolites, but it is rarely found precipitating in modern environments. In the 1960s, the discovery of dolomite forming in specific environments, such as beneath the Abu Dhabi sabkhas, UAE and in the ephemeral lakes of the South Australian Coorong Lagoon, provided new insights into the physico-chemical controls on the process. Apropos to the theme of this special publication, Robert N. Ginsburg and his colleagues contributed to this renaissance in dolomite research in the 1960s with their discovery of modern dolomite precipitation in extensive supratidal crusts on Andros Island, Bahamas (Shinn *et al.*, 1965, 1969). Although these modern environments presented the possibility to study the geochemical parameters promoting dolomite precipitation, attempts at laboratory experiments based on these actualistic studies proved to have limited or no success in precipitating dolomite under Earth's surface conditions (Land, 1985, 1998). Thus, the rarity of modern environments relative to the rock record and the lack of success with experimental studies led to a continuing enigma, often called the Dolomite Problem.

In recent years, however, a new approach using microbial experiments has provided fundamentally new data to understand the mechanisms that may be involved in dolomite precipitation under Earth's surface conditions (Vasconcelos *et al.*, 1995; Warthmann *et al.*, 2000) and has inspired the development of the microbial dolomite model based on a study of a modern dolomite-forming hypersaline coastal lagoon, Lagoa Vermehla, Brazil (Vasconcelos & McKenzie, 1997).

In specific modern hypersaline environments, sulphate-reducing bacteria apparently induce dolomite precipitation (Vasconcelos & McKenzie, 1997; van Lith *et al.*, 2002, 2003; Wright & Wacey, 2005). These studies demonstrated the importance of anaerobic processes in carbonate mineral formation, but many modern geological settings, such as the Andros Island, Bahamas, supratidal environment, indicate that aerobic microbial processes may also play an important role in dolomite precipitation. A diverse microbial population is generally involved in carbonate precipitation in natural environments, including photoautotrophic cyanobacteria and bacteria-producing ammonia from the degradation of nitrogen-rich organic matter (Thompson & Ferris, 1990; Rivadeneyra *et al.*, 2000; Sánchez-Román *et al.*, 2007). Rivadeneyra *et al.* (1993) and more recently, Sánchez-Román (2008) have demonstrated that halophilic aerobic bacteria mediate primary precipitation of dolomite at room temperatures. These results indicate that heterotrophic microorganisms, using molecular oxygen as an electron acceptor, may be able to precipitate dolomite at low temperatures and may be important players in modern environments, as well as in the geological past.

Riding (2000) has suggested that morphology (microfabric) studies can provide insight on biomineralization processes, as well as on the environmental conditions where dolomite precipitates. This combined information can be useful for the identification of microbial fossils and/or precipitates, such as biomarkers in carbonate sedimentary rocks (Buczynski & Chafetz, 1991; Allen *et al.*, 2000; Reid *et al.*, 2000; Riding, 2000). Regardless of bacterial type (autotrophic or heterotrophic; marine or non-marine), carbonate morphologies mediated by microorganisms are distinguishable from inorganic precipitation (Krumbein, 1979; Buczynski & Chafetz, 1991; Chafetz & Buczynski, 1992; Knorre & Krumbein, 2000; Braissant *et al.*, 2003).

This study focuses on the role of aerobic bacteria in dolomite formation. A comparative analysis of modern dolomite formation in Brejo do Espinho (Brazil), with generally oxic conditions at the sediment–water interface, and dolomite precipitation in laboratory experiments, using two moderately halophilic aerobic bacteria isolated from Brejo do Espinho, links microbial metabolism with the natural observed conditions, such as pH, salinity, oxygen and carbonate mineralogy. In addition, the role of these two previously undescribed bacterial strains in dolomite precipitation are evaluated and a mechanism for microbially mediated dolomite formation under aerobic conditions is proposed.

GEOLOGICAL SETTING

Brejo do Espinho is a very shallow (<0.5 m) hypersaline coastal lagoon located east of Lagoa Vermelha about 100 km east of Rio de Janeiro city (Brazil) (Fig. 1). Situated within a Pleistocene dune system, the hydrology of Brejo do Espinho is influenced by seepage through the sand dunes from the hypersaline lagoon Lagoa Araruama on the continental side and seawater from the Atlantic Ocean.

Fig. 1. Satellite image showing Brejo do Espinho (22°56″08 S, 42°14′18 W) located between the Atlantic coast and Lagoa Araruama in the State of Rio de Janeiro, Brazil. Image courtesy of Earth Sciences and Image Analysis Laboratory, NASA Johnson Space Center.

Fig. 2. Brejo do Espinho microbial mat with characteristic coloured layers. Green corresponds to cyanobacteria, pink presumably to purple sulphur bacteria, dark brown to zone of anaerobic metabolism and white to carbonate biominerals.

The climatic conditions in this area are peculiar due to the occurrence of an upwelling zone offshore that contributes to the semi-arid climate characteristic for the region. These special conditions lead to a strong interannual variability of the regional climate, whereby evaporation exceeds precipitation (Barbiére, 2002).

Brejo do Espinho lagoon water has a typical seawater Mg/Ca molar ratio of ~5, indicating a seawater origin modified by evaporation and dilution processes (van Lith *et al.*, 2002; Moreira *et al.*, 2004). Brejo do Espinho sediment contains nearly stoichiometric dolomite, which appears to precipitate predominantly in periods of the year with the highest salinities (van Lith *et al.*, 2002). Thick, multicoloured microbial mats develop on the surface of Brejo do Espinho during flooded periods, which are controlled by two factors, surface runoff during the wet season versus influx of seawater through the dunes during the dry season. The microbial mat exhibits a stratification of mat-forming microbial communities, which is visible as different coloured layers (Fig. 2). For this study, microorganisms were isolated from the uppermost layers where oxygen tends to persist.

METHODS

Culture medium

The culture medium (D-1) used in our laboratory experiments has the following composition (%, w/v): 1% yeast extract; 0.5% proteose peptone; 0.1% glucose; and 3.5% NaCl. This medium was supplied with calcium and magnesium acetate to adjust the Mg/Ca molar ratio to 7.4. To obtain a semi-solid medium, 20 g L^{-1} Bacto-Agar (Difco) was added. After the pH was adjusted to 7.0 with 0.1 M KOH, the medium was sterilized at 121°C for 20 minutes.

Microorganisms

The bacterial strains used for this study were isolated from the uppermost part of the microbial mat in Brejo do Espinho lagoon and are designated Strain BE1 and BE2. These strains are heterotrophic, obligate aerobic bacteria. In order to obtain pure cultures of these single strains, dilution series from the isolated samples taken from the microbial mat were inoculated onto Petri dishes containing the culture medium described above and incubated aerobically at 30°C. The Petri dishes were examined periodically to determine if the colonies were able to induce mineral precipitation.

The colonies forming a visible concentric corona of carbonate minerals were isolated. These pure strains were selected for phylogenetic 16S ribosomal deoxyribonucleic acid (rDNA) analysis (Institut für medizinische ünd molekulare Diagnostic AG, Zürich). The sequences obtained (ranging from 794 to 802 base pairs) were compared with the National Center for Biotechnology Information (NCBI) database. The results from the approximate phylogenetic affiliation revealed that the closest relative of strain BE1 is *Virgibacillus marismortui* (NCBI, accession number AJ009793) with 99.8% homology, and strain BE2 has a 100% homology to *Marinobacter* sp. NT N31. Strain BE1 and BE2 are hereafter designated as *V. marismortui* and *Marinobacter* sp., respectively. *V. marismortui* (Heyrman *et al.*, 2003) was first isolated in 1999 from the Dead Sea (Arahal *et al.*, 1999), and was originally described as *Bacillus marismortui. Marinobacter* sp. NT N31 was first isolated from deep-sea sediment below a microbial mat in 3708 m water depth (Okamoto & Naganuma, unpublished).

Study of mineral formation

The ability of *V. marismortui* and *Marinobacter* sp. cells to induce carbonate precipitation was tested using the prepared semi-solid medium D-1 because bacterial carbonates are commonly found in natural gel-like mucilaginous structures (Chafetz & Buzcynsky, 1992), such as microbial mats or biofilms. Biofilms comprise bacteria immersed in a matrix of extracellular polymeric substances (EPS) (Decho, 2000; Riding, 2000). The bacteria studied were inoculated in triplicate onto the surface of semi-solid media. Control experiments, consisting of non-inoculated culture media or media inoculated with autoclaved bacterial cells, were run in parallel.

Recovering of minerals from plates

The Petri dishes were sealed with parafilm to avoid water evaporation and incubated aerobically at 30°C for 30 days. In order to detect the presence of precipitates during incubation, Petri dishes were observed once a day with light microscopy (20×). After the incubation period was completed, precipitates were recovered by scraping the colony from the agar surface. Subsequently, they were washed several times with distilled water to eliminate the nutritive solution, remaining agar and cellular debris, and dried at 37°C. Microscopic observation demonstrated that this treatment does not alter the morphology of the crystals. Measurements of pH were performed at the end of the growth and carbonate formation experiments. The pH-indicator paper (Merck Spezial-Indikatorpapier) was applied directly to the semi-solid surface.

X-ray diffraction analysis

The purified minerals formed in the laboratory culture experiments by the two different bacteria strains and the carbonate sediment sampled at 2.5 and 21 cm depths in the Brejo do Espinho lagoon were powdered and analysed for their mineralogy on a Scintag XDS 2000 X-ray diffractometer (XRD). The samples were scanned by continuous scan at $1°\,min^{-1}$ from 5 to 70° with Cu-Kα radiation. From the d104 of the diffraction spectra, the Mg/Ca ratio of the carbonate minerals was calculated after Lumsden (1979).

Scanning electron microscope analysis

The minerals formed in the laboratory experiments and the Brejo do Espinho sediment were prepared for scanning electron microscopic (SEM) observation. In order to preserve the biological structure as best as possible, the following treatment was applied to the uppermost sediment sample in the laboratory: samples were fixed with 2.5% glutaraldehyde in 0.2 M Na-cacodylate buffer for 90 minutes at 4°C. Subsequently, the samples were washed twice with distilled water, 30% ethanol, dehydrated in acetone, and finally critical-point dried in liquid CO_2. Samples were analysed by a field emission SEM (Leo 1530, 143 eV resolution, LEO Electron Microscopy LTD, Germany).

Stable carbon and oxygen isotopes

The carbon and oxygen stable isotopic compositions of powdered bulk carbonate sediment samples were determined. Samples were dissolved using an on-line common acid bath attached to a VG PRISM mass spectrometer. The reaction time was set at 10 minutes. The isotope data are reported in the standard δ notation relative to the international standard V-PDB (Craig, 1953, 1957). The analytical precision of the mass spectrometer is ±0.1‰ for $\delta^{13}C$, and ±0.2‰ for $\delta^{18}O$. The isotopic composition of the water was determined by equilibration with CO_2 through the use of an automated ISOPREP 18 equilibration device coupled to a FISONS-OPTIMA mass spectrometer. Analytical reproducibility of $\delta^{18}O$ water is ±0.06‰. The oxygen isotopic results are reported in the conventional per mille notation with respect to Vienna standard mean ocean water (V-SMOW). All $\delta^{18}O$ values have been corrected for dolomite – phosphoric acid fractionation at 90°C using the fractionation factor of 1.0093 (Rosenbaum & Sheppard, 1986). Conversions between $\delta^{18}O$ V-PDB and $\delta^{18}O$ V-SMOW were calculated by using the equation in Coplen *et al.* (1983). The sediment and pore-water isotope data are listed in Table 1.

Total organic carbon and sediment pore-water pH

Sediment samples were analysed using UIC, Inc. coulometer system. The amount of total organic carbon (TOC) in the samples was obtained as the difference between the measured inorganic carbon and the total carbon. The percentage TOC data are given in Table 1. Standard pH-microelectrodes (Methrom) were used in the field to measure the pH of the sediment pore water.

RESULTS

Description of the sediment and mineralogy

The 25-cm sediment core, taken from the centre of Brejo do Espinho lagoon, contains fine

Table 1. Mineralogical and geochemical data for Brejo do Espinho sediment and pore water.

Depth (cm)	High-Mg calcite (%)	Dolomite (%)	TOC (%)	$\delta^{13}C_{V\text{-}PDB}$ (‰)	$\delta^{18}O_{V\text{-}PDB}$ (‰)	$\delta^{18}O_{V\text{-}SMOW}$ (‰)
0	84	16	12.50	−3.40	0.59	1.62
2	90	10	11.49	−2.76	0.99	1.78
3	79	21	11.55	−2.23	1.68	2.00
4	80	20	11.35	−2.35	1.64	2.4
5	79	21	9.75	−2.14	1.97	2.63
6	69	31	9.80	−1.99	1.95	2.70
7	75	25	7.75	−2.16	1.86	2.85
8	70	30	7.05	−2.30	2.06	2.85
9	62	38	5.35	−2.71	2.16	2.40
10	58	42	5.30	−2.66	2.06	2.24
11	55	45	4.71	−2.78	2.04	2.55
12	63	37	5.72	−2.70	2.08	2.63
13	67	33	5.73	−2.15	2.04	2.70
14	71	29	4.82	−2.33	2.00	2.89
15	67	33	4.70	−2.09	1.96	3.17
16	69	31	4.15	−2.42	2.06	3.11
17	65	35	3.69	−2.22	2.08	2.60
18	59	41	3.76	−2.67	2.25	2.20
19	44	56	3.43	−4.04	2.67	2.10
20	31	69	2.65	−5.74	2.97	1.78
21	0	100	2.55	−10.16	3.49	1.30
22	0	100	2.45	−9.71	3.62	1.50
23	0	100	2.35	−9.10	3.79	1.62

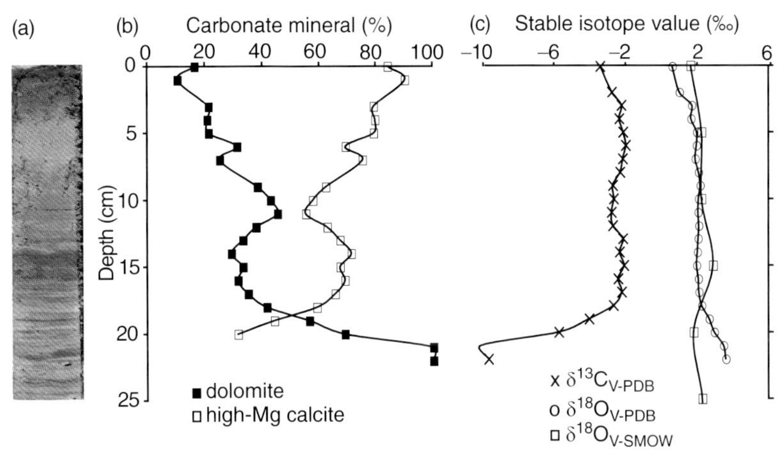

Fig. 3. (a) Photograph of Brejo do Espinho short core. A layer of 100% dolomite is present between 20 and 23 cm, based on the XRD carbonate mineral analysis of samples as shown in (b), which indicates a marked change from predominantly high-Mg calcite above 20 cm to predominantly dolomite below. (c) Stable isotope compositions of bulk carbonate samples and associated pore-water show that the $\delta^{18}O$ value of the 100% dolomite samples below 20 cm becomes more positive, while the $\delta^{13}C$ trends to relatively negative values, indicative of microbial activity.

laminations below 10 cm with less distinct laminations above (Fig. 3a). The carbonate mineral composition of the sediments includes some calcite but predominantly comprises a mixture of high-Mg calcite and dolomite, in varying proportions (Fig. 3b). There is a down-core trend in the relative ratio of high-Mg calcite to dolomite with the latter increasing to 100% at 21 cm. The XRD analysis of the sediment samples from 5 and 21 cm indicates that the dolomite is nearly stoichiometric with approximately 47–52 mol% Mg, respectively, whereas the high-Mg calcite from 5 cm contains 25 mol% Mg (Figs 4a and b). The increase in percentage dolomite with depth is probably related to early diagenetic processes (Vasconcelos & McKenzie, 1997). The primary carbonate precipitate in the uppermost sediments of Brejo do Espinho contains high concentrations of high-Mg calcite, which are susceptible to early diagenetic alteration to dolomite with burial. The sediments are rich in organic carbon, but show a steadily decreasing trend in percentage TOC from 12.5% at the surface to 2.5% in the level of 100% dolomite (Table 1).

Rivadeneyra, M.A., **Delgado, G.**, **Ramos-Cormenzana, A.** and **Delgado, R.** (1998) Biomineralization of carbonates by *Halomonas eurihalina* in solid and liquid media with different salinities: crystal formation sequence. *Res. Microbiol.*, **149**, 277–286.

Rivadeneyra, M.A., **Delgado, G.**, **Soriano, M.**, **Ramos-Cormenzana, A.** and **Delgado, R.** (2000). Precipitation of carbonates by *Nesterenkonia halobia* in liquid media. *Chemosphere*, **41**, 617–624.

Rivadeneyra, M.A., **Paraga J.**, **Delgado, G.**, **Ramos-Coemenzana, A.** and **Delgado, G.** (2004). Biomineralization of carbonates by *Halobacillus trueperi* in solid and liquid media with different salinities. *FEMS Microbiol. Ecol.*, **48**, 39–46.

Roberts, J.A., **Bennett, P.C.**, **Gonzalez, L.A.**, **Macpherson, G.L.** and **Miliken, K.L.** (2004) Microbial precipitation of dolomite in methanogenic groundwater. *Geology*, **32**, 277–280.

Rosen, C. and **Mcnamara, M.R.** (1998). Hydromagnesite replacement of biomineralized aragonite in a new location of Holocene stromatolites, Lake Walyungup, Western Australia. *Sedimentology*, **45**, 1005–1018.

Rosenbaum, J. and **Sheppard, S.M.F.** (1986) An isotopic study of siderites, dolomites and ankerites at high temperatures. *Geochemin. Cosmochi. Acta*, **50**, 1147–1150.

Russell, M.J., **Ingham, J.K.**, **Zedef, V.**, **Maktav, D.**, **Sunar, F.**, **Hall, A.J.** and **Fallick, A.** (1999) Search for signs of ancient life on Mars: expectations from hydromagnesite microbiolites, Salda Lake, Turkey. *J. Geol. Soc. London*, **156**, 869–888.

Sánchez-Román, M., **Rivadeneyra, M.A.** and **McKenzie, J.A.** (2007) Carbonate and phosphate precipitation by halophilic bacteria: influence of Ca^{2+} and Mg^{2+} ions. *FEMS Microbiol. Ecol.*, **61**, 273–284.

Sánchez-Román, M., **Vasconcelos, C.**, **Schmid, T.**, **Dittrich, M.**, **McKenzie, J.A.**, **Zenobi, R.** and **Rivadeneyra. M.A.** (2008) Aerobic microbial dolomite at the nanometer scale: implications for the geologic record. *Geology*, **36**, 879–882.

Shinn, E.A., **Ginsburg, R.N.** and **Lloyd, R.M.** (1965) Recent supratital dolomite from Andros Island, Bahamas. In: *Dolomitization and Limestone Diagenesis* (Eds. L.C. Pray and R.C. Murray). SEPM Spec. Publ. No. **13**, 112–123.

Shinn, E.A., **Lloyd, R.M.** and **Ginsburg, R.N.** (1969) Anatomy of a modern carbonate tidal-flat, Andros Island, Bahamas. *J. Sediment. Petrol.*, **39**, 1202–1228.

Thompson, J.B. and **Ferris, F.G.** (1990) Cyanobacterial precipitation of gypsum, calcite, and magnesite from natural alkaline lake water. *Geology*, **18**, 995–998.

van Lith, Y., **Vasconcelos, C.**, **Warthmann, R.**, **Martins, J.C.F.** and **McKenzie, J.A.** (2002) Bacterial sulfate reduction and salinity: two controls on dolomite precipitation in Lagoa Vermelha and Brejo do Espinho (Brazil). *Hydrobiologia*, **485**, 25–49.

van Lith, Y., **Vasconcelos, C.**, **Warthmann, R.** and **McKenzie, J.A.** (2003) Sulphate-reducing bacteria induce low-temperature Ca-dolomite and high Mg-calcite formation. *Geobiology*, **1**, 71–79.

Vasconcelos, C. and **McKenzie, J.A.** (1997) Microbial mediation of modern dolomite precipitation and diagenesis under anoxic conditions (Lagoa Vermelha, Rio de Janeiro, Brazil). *J. Sediment. Res.*, **67**, 378–390.

Vasconcelos, C., **McKenzie, J.A.**, **Bernasconi, S.**, **Grujic, D.** and **Tien, A.J.** (1995) Microbial mediation as a possible mechanism for natural dolomite formation at low temperatures. *Nature*, **377**, 220–222.

Vasconcelos, C., **McKenzie, J.A.**, **Warthmann, R.** and **Bernasconi, S.** (2005) Calibration of $\delta^{18}O$ paleothermometer for dolomite precipitated in microbial cultures and natural environments. *Geology*, **33**, 317–320.

Vasconcelos, C., **Warthmann, R.**, **McKenzie, J.A.**, **Visscher, P.T.**, **Bittermann, A.G.** and **van Lith, Y.** (2006) Lithifying microbial mats in Lagoa Vermelha, Brazil: modern Precambrian relics? *Sediment. Geol.*, **185**, 175–183.

von der Borch, C.C. and **Jones, J.B.** (1976) Spherular modern dolomite from the Coorong area, South Australia. *Sedimentology*, **23**, 587–591.

Warren, J.K. (1990) Sedimentology and mineralogy of dolomitic Coorong Lakes, South Australia. *J. Sediment. Petrol.*, **60**, 843–858.

Warren, L.A., **Mauricie, P.A.**, **Parmar, N.** and **Ferris, F.G.** (2001) Microbially mediated calcium carbonate precipitation: implications for interpreting calcite precipitation and for solid-phase capture of inorganic contaminants. *Geomicrobiol. J.*, **18**, 93–115.

Warthmann, R., **van Lith, Y.**, **Vasconcelos, C.**, **McKenzie, J.A.** and **Karpoff, A.M.** (2000) Bacterially induced dolomite precipitation in anoxic culture experiments. *Geology*, **28**, 1091–1094.

Wright, D.T. and **Wacey, D.** (2005) Precipitation of dolomite using sulphate-reducing bacteria from the Coorong Region, South Australia: significance and implications. *Sedimentology*, **52**, 987–1008.

Yu, S.H., **Coelfen, H.**, **Hartmann, J.** and **Antonietti, M.** (2002) Biomimetic crystallization of calcium carbonate spherules with controlled surface structures and sizes by double-hydrophilic block copolymers. *Adv. Func. Mater.*, **12**, 541–545.

Int. Assoc. Sedimentol. Spec. Publ. (2009) **41**, 179–197

Karst sub-basins and their relationship to the transport of Tertiary siliciclastic sediments on the Florida Platform

ALBERT C. HINE*, BEAU C. SUTHARD*[1], STANLEY D. LOCKER*, KEVIN J. CUNNINGHAM†, DAVID S. DUNCAN‡, MARK EVANS§ and ROBERT A. MORTON¶

**College of Marine Science, University of South Florida, St Petersburg, FL 33701, USA (E-mail: hine@marine.usf.edu)*

†U.S. Geological Survey, 3110 SW 9th Avenue, Fort Lauderdale, FL 33315, USA

‡Department of Marine Science, Eckerd College, 4200 54th Avenue South, St Petersburg, FL 33711, USA

§Division of Health Assessment and Consultation, NCEH/ATSDR, Mail Stop E-32, 1600 Clifton Road, Atlanta, GA 30333, USA

¶U.S. Geological Survey, 600 4th Street South, St Petersburg, FL 33701, USA

ABSTRACT

Multiple, spatially restricted, partly enclosed karst sub-basins with as much as 100 m of relief occur on a mid-carbonate platform setting beneath the modern estuaries of Tampa Bay and Charlotte Harbor located along the west-central Florida coastline. A relatively high-amplitude seismic basement consists of the mostly carbonate, upper Oligocene to middle Miocene Arcadia Formation, which has been significantly deformed into folds, sags, warps and sinkholes. Presumably, this deformation was caused during a mid-to-late Miocene sea-level lowstand by deep-seated dissolution of carbonates, evaporites or both, resulting in collapse of the overlying stratigraphy, thus creating palaeotopographic depressions.

Seismic sequences containing prograding clinoforms filled approximately 90% of the accommodation space of these western Florida sub-basins. Borehole data indicate that sediment fill is mostly siliciclastic deposited within deltaic depositional systems. The sedimentary fill in the Tampa Bay and Charlotte Harbor sub-basins is mostly assigned to the upper Peace River Formation of late Miocene to early Pliocene age. This fill is part of a >1000 km long, Tertiary siliciclastic deposit that stretches north-to-south down peninsular Florida. Sediment fill of these two sub-basins is linked to erosion and remobilization of pre-existing, middle Miocene quartz-rich sediments via enhanced sediment transport by local, short-length rivers and discharge into coastal-marine depositional environments. Increased sediment discharge possibly resulted from amplified thunderstorm activity and enhanced runoff during a warm period of the Pliocene.

Rather than incised valley fills or reef-margin, backfilled basins, Tampa Bay and Charlotte Harbor represent spatially restricted, sediment-filled karst palaeotopographic lows. The 'dimpling' of a carbonate platform by karst sub-basins provides a previously unrecognized mechanism for the creation of accommodation that can result in the 'drowning' of a carbonate platform by siliciclastics.

Keywords Karst, carbonate platform, siliciclastics, sediment transport, deltas, sea level, deformation, palaeofluvial.

INTRODUCTION

Tampa Bay and Charlotte Harbor, two major estuaries located along Florida's west-central Gulf of Mexico coastline, seem anomalous in that they do not appear to have formed as drowned, incised river valleys typical of other estuarine systems in dominantly siliciclastic settings such as those along coastal plains (Dalrymple *et al.*, 1994). Indeed, Tampa Bay and Charlotte Harbor reside in the centre of the large, dominantly carbonate Florida Platform and are only fed by a few very small, low-sediment and water discharge

[1]Present address: Coastal Planning and Engineering Inc, Tampa Bay Regional Office, 101, 16th Ave So., Suite 4, St. Petersburg, FL 33701, USA (E-mail: bsuthard@coastal-planning.net).

streams that are supported by small, local, upland drainage basins. So, the cause of their estuarine nature appears to be enigmatic.

Seismic data presented in this paper indicate that these shallow (average depth ~2–4 m) estuaries are underlain by karstic, semi-enclosed sub-basins that have as much as 100 m of subsurface relief. The sub-basins are defined as distinct structures that are part of a larger sedimentary basin system – in this case, the entire complex of basins containing siliciclastic fill on the Florida Platform. The sub-basins (tens of kilometres wide) beneath Tampa Bay and Charlotte Harbor are really sub-basin complexes with sub-basins within smaller sub-basins (kilometres wide). Moreover, at even a higher resolution spatial scale these sub-basins reveal significant deformation in the form of folds, warps and sags in the deeper seismic sequences and the seismic basement (hundreds of metres wide). Coring indicates that these basins have been filled with mostly siliciclastic sediments. As a result, these sub-basins present an unusual relationship between carbonates and siliciclastics that has not been previously described.

The new evidence indicates that Tampa Bay and Charlotte Harbor are important siliciclastic repositories and represent a type of mid-carbonate platform accommodation not widely recognized for siliciclastic deposition. This stands in stark contrast to the neighbouring Bahama Banks, which do not reveal any mid-platform basins in the shallow subsurface providing accommodation in this manner (Eberli & Ginsburg, 1987; Ginsburg, 2001). These platforms do contain large, buried linear seaways, are much larger scale than these Florida sub-basins, do not appear to be karst-related and reveal no deformation. The purpose of this paper is to demonstrate the scale, geometry and infilling facies of these sub-basins and to link them to the larger Neogene–Quaternary siliciclastic flux onto the Florida Platform.

GEOLOGICAL BACKGROUND

Ever since the first geologist walked Florida's beaches (Vaughan, 1910), it has been obvious that the Florida Platform had received substantial quantities of quartz-rich sand in its geological past. Some of these notable siliciclastic shorelines that have become classic localities in coastal geology (Davis *et al.*, 1992; Davis, 1997) are a trademark of Florida's tourist-driven economy, and are world renowned as a result. Additionally, the principal geomorphology of central peninsular Florida consists of palaeoshorelines, terraces and scarps composed of a siliciclastic veneer formed on underlying lower Neogene and older carbonates (White, 1970; Winkler & Howard, 1977). Even early sediment distribution studies of the adjacent seafloor indicate that quartz-rich sediments extend some 40 km out onto the west-central Florida shelf and to the upper slope on the east Florida shelf (Gould & Stewart, 1956; Doyle & Sparks, 1970; Hine, 1997). However, the breadth and extent of these siliciclastic sediments comprising the subsurface of peninsular Florida was not fully mapped in these early studies.

Strip mining to acquire phosphate-rich sediments as well as borehole geology driven by the search for groundwater and hydrocarbons show that central peninsular Florida is underlain by a complex array of quartz-rich lithostratigraphic units dominated by the Oligocene-to-Pliocene Hawthorn Group (Riggs, 1979; Scott, 1988, 1997). However, the extent of siliciclastic sediments underlying the Pleistocene and modern carbonate-dominated terrain of southernmost peninsular Florida (Enos & Perkins, 1977) remains more enigmatic. Potentially, as these units are capped by limestone, most remain unseen at the surface and have been largely unstudied for some time as a result. Missimer & Gardner (1976) and Enos & Perkins (1977), for example, began to recognize the broad extent by which quartz-rich sediments were distributed in the subsurface of southernmost Florida.

This work was considerably advanced by the South Florida Drilling Project headed by R.N. Ginsburg. The project was conceptualized in the late 1980s (Ginsburg *et al.*, 1989) and commenced in 1993. Indeed, this scientific effort demonstrated that there had been a significant remobilization of quartz-rich sand and even gravels during the late Miocene through early Pliocene. Missimer & Ginsburg (1998) pointed out that even during the late Oligocene, the Arcadia Formation in south-central Florida was interbedded and mixed with numerous metre-scale units consisting of up to 80% siliciclastic sediments.

The late Miocene to early Pliocene remobilization produced a ~150 m-thick succession of siliciclastics (Cunningham *et al.*, 1998) that extended from the Lake Okeechobee area in south-central peninsular Florida, running beneath the Florida Everglades and Florida Keys, and terminating by downlap onto the approximately 200 m deep Pourtales Terrace (Missimer, 1992; Warseski *et al.*, 1996; Guertin *et al.*, 1999, 2000; Cunningham *et al.*, 1998, 2001a,b, 2003; McNeill *et al.*, 2004). The partial burial of this deep-marine, erosional Miocene terrace marked the southern end of a >1000-km-long siliciclastic transport system that originated with the weathering of crystalline bedrock of the southern Appalachian Mountains and Piedmont (Fig. 1). In general, this siliciclastic transport from the north produced a relatively thin (1–150 m) late Neogene to modern quartz-rich veneer covering a thick (2–6 km) Jurassic-to-Neogene carbonate succession over peninsular Florida (Klitgord *et al.*, 1984).

Results from the South Florida Drilling Project also indicated that there was a late Miocene-to-Pliocene remobilization of siliciclastics in south Florida. Earlier studies had shown that siliciclastics entered the northern Florida peninsula by post mid-Oligocene after the Georgia Channel System (Huddleston, 1993; T. Scott, personal communication) seaway complex had been filled by prograding deltas probably during a major sea-level lowstand that occurred during the early Oligocene (Hull, 1962; Chen, 1965; McKinney, 1984; Popenoe *et al.*, 1987; Popenoe, 1990; Brewster-Wingard *et al.*, 1997). These sediments made their way to central Florida and formed an important part of the Hawthorn Group, possibly transported by extensive longshore sediment transport during sea-level highstands. However, to the south, carbonate sedimentation persisted ultimately depositing the late Oligocene-to-middle Miocene Acadia Formation and lower Peace River Formation. These lithostratigraphic units are unconformably overlain by the siliciclastic sediments of the upper Peace River Formation, which represents renewed siliciclastic transport in the late Miocene to early Pliocene of the Hawthorn Group quartz-rich sediments lying in central peninsular Florida (Cunningham *et al.*, 2003; McNeill *et al.*, 2004).

TAMPA BAY SUB-BASIN

Tampa Bay is a large (~1000 km^2), shallow (average depth ~4 m) estuary located along the west-central Florida Gulf of Mexico coastline. Although there has been considerable geological framework research performed within Tampa Bay in the past, much of this work did not have the benefit of digitally acquired and processed and GPS located, high-resolution seismic-reflection profile data gathered in closely spaced line. Such data enable loop-tying and thus three-dimensional mapping of seismic sequences and bounding surfaces (Stahl, 1970; Willis, 1984; Hebert, 1985; Green *et al.*, 1995; Ferguson & Davis, 2003). Duncan *et al.* (2003) and Suthard (2005) have provided such a data set and correlated their seismic data (~1000 line km) to six neighbouring boreholes on land and numerous short cores within the estuary itself (Fig. 2).

The seismic data of Duncan *et al.* (2003) and Suthard (2005) clearly demonstrate that Tampa Bay is underlain by a sediment-filled sub-basin having multiple smaller-scale sub-basins each separated by bedrock highs (Figs 3 and 4). Recent data have revealed that 'seismic basement' crops out in middle-central Tampa Bay forming hardbottoms supporting appropriate benthic biological communities. The seismic basement consists of the Arcadia Formation – an open-marine limestone–dolostone, with occasional thin beds of phosphatic quartz sands (<1.5 m thick) and clays (<1.5 m thick and of limited areal extent) scattered throughout (Scott, 1988; Suthard, 2005). Suthard (2005) jump-correlated his seismic lines to adjacent onshore boreholes (Green *et al.*, 1995) for chronostratigraphic control. Since these basins have 40–60 m of subsurface relief and the average water depth of Tampa Bay is only 4 m, approximately 5–10% of the remaining accommodation space is unfilled.

The seismic data indicate deformation of the Arcadia Formation and some of the immediately overlying seismic sequences forming the sub-basin fill. Folds and sags dominate the lower Tampa Bay part of the sub-basin (Fig. 5). These deformational features are not anticlines and synclines in the classic structural geology sense in that they have very limited lateral continuity (no axial planes) – rather they are small, broad domes (>1 km across) and narrow (<1 km), circular depressions – some even representing individual sinkholes. This style of deformation is more consistent with deep-seated collapse from below rather than from lateral compression due to tectonic activity.

Finally, seismic reflection and borehole data reveal that the sub-basin underlying Tampa Bay is laterally restricted and does not extend seaward beneath the modern continental shelf more than ~30 km (Fig. 4; Hebert, 1985; Duncan, 1993; Duncan *et al.*, 2003). Rather, the seismic basement underlying Tampa Bay rises seaward where it

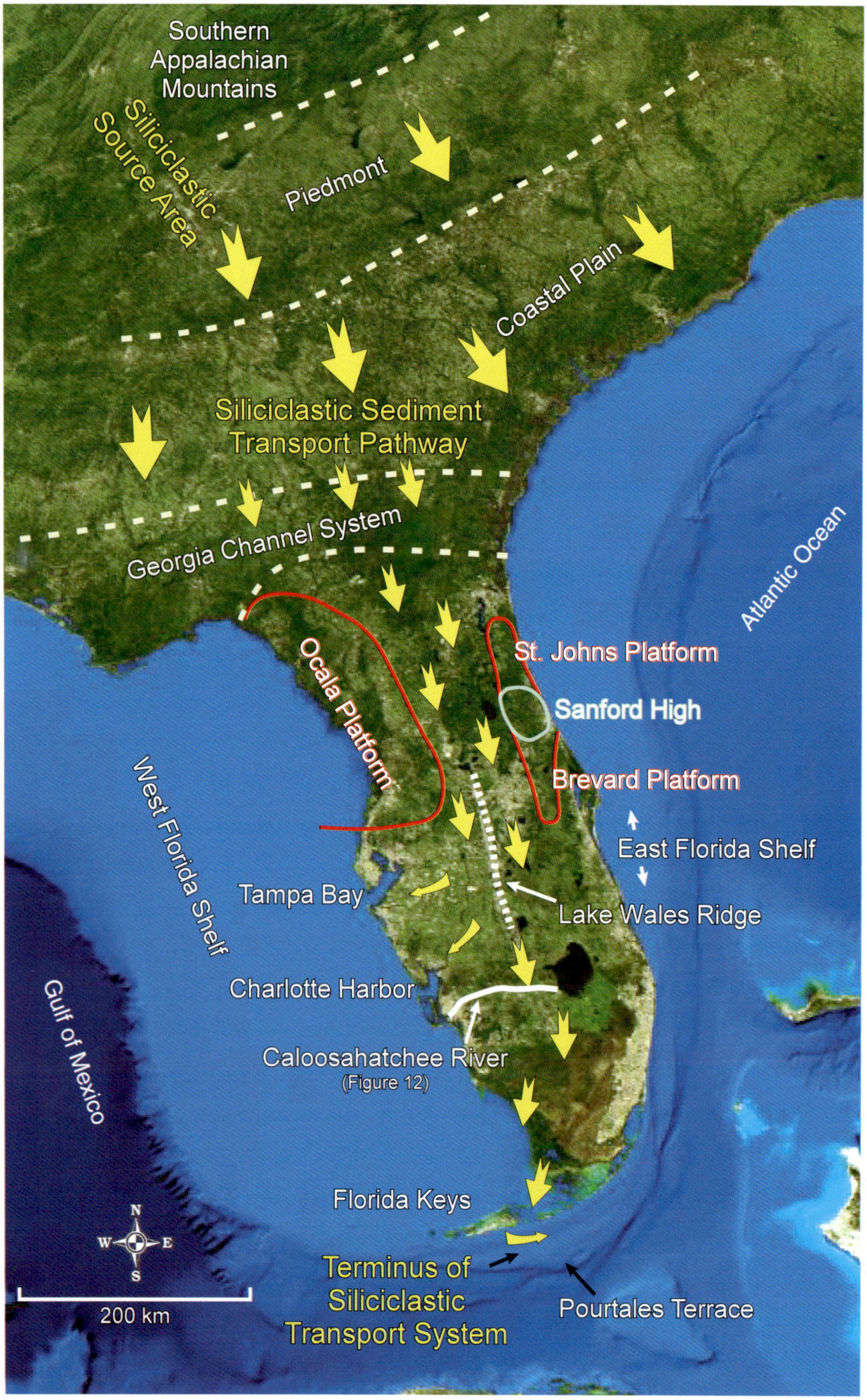

Fig. 1. Satellite image (http://earthobservatory.nasa.gov/Newsroom/BlueMarble/) of the southeast USA showing: (1) a portion of the southern Appalachian Mountains and Piedmont – the siliciclastic source area for the sediments that have been transported to Florida; (2) the transport pathway; and (3) the terminus at the Pourtales Terrace. Key palaeotopographic areas that played a role in guiding these sediments southward, such as the Ocala Platform and Sandford High, are shown as well as other relevant geographical locations. Note the location of the terminus of the Lake Wales Ridge. Note also that Tampa Bay and Charlotte Harbor are located peripherally off to the side of the main north–south sediment transport pathway.

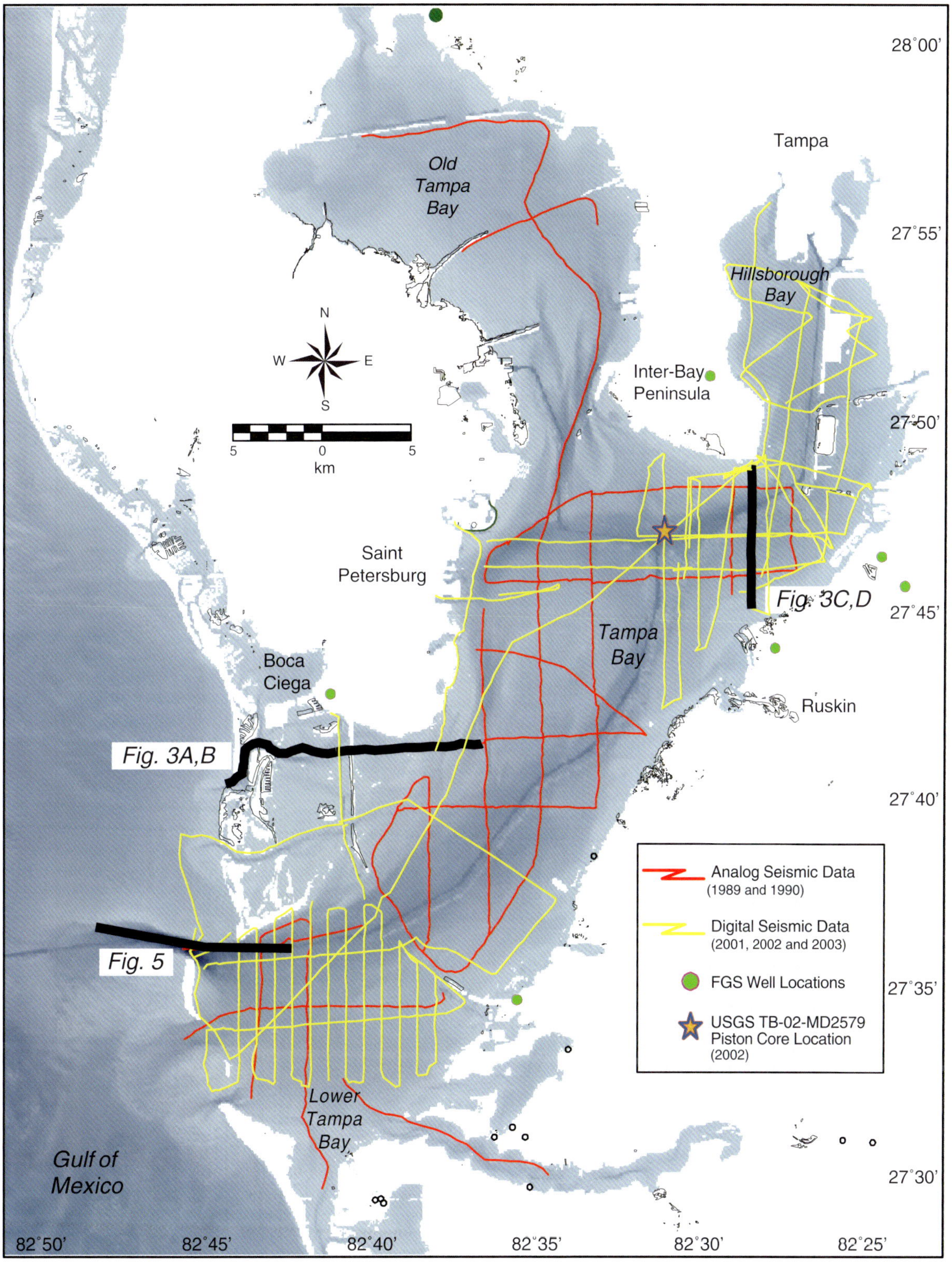

Fig. 2. Seismic data and borehole locations in Tampa Bay. Adjacent boreholes on land provided lithological and chronostratigraphic control for the seismic data. Bathymetry revealed in Tampa Bay roughly coincides with subsurface seismic basement topography (see Fig. 4).

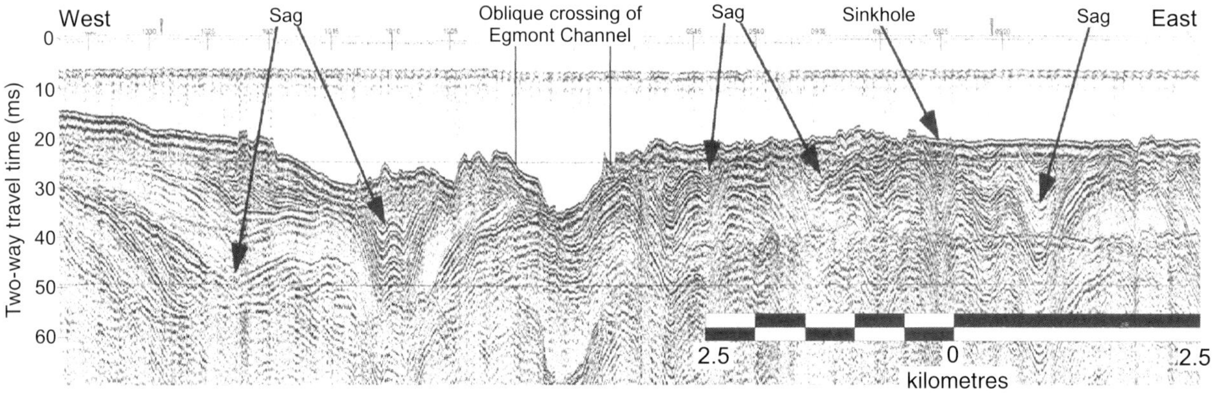

Fig. 5. Detailed seismic line illustrating deformation beneath lower Tampa Bay. Style of deformation indicates collapse from below due to deep-seated dissolution of older carbonates or evaporites (modified fig. 16, Berman *et al.*, 2005).

crops out forming hardbottoms, which dominate portions of the west-central Florida shelf (Locker *et al.*, 2003). This lack of cross-shelf continuity indicates that these sub-basins are not shelf valleys that have been carved by rivers during sea-level lowstands (Dalrymple *et al.*, 1994; Donahue *et al.*, 2003) in contrast to Brooks & Doyle's (1998) contention that the sub-basin beneath Tampa Bay was formed by palaeofluvial incision.

Numerous cores from Tampa Bay (US Army Corps of Engineers, 1969) as well as the surrounding onshore boreholes (Green *et al.*, 1995; Florida Geological Survey, 2005) indicate that overlying the Arcadia Formation is the middle Miocene to lower Pliocene Peace River Formation, which constitutes the basin fill. This lithostratigraphic unit is principally siliciclastic (>66%) with interbedded quartz sands, clays and carbonates. The Peace River Formation is a very complex unit consisting of large amounts of fluvio-deltaic and coastal siliciclastic sediments to minor amounts of restricted and open-marine carbonates (Scott, 1988). The Hawthorn Group, as defined by Scott (1988), is composed of the Arcadia and the Peace River formations.

Suthard (2005) defined 11 different seismic sequences from two major depocentres that constitute the Peace River Formation and the overlying Plio-Quaternary siliciclastic units (Fig. 3). This seismic sequence mapping reveals dominant north-northwest prograding clinoforms (Fig. 3) indicating a source area from the south-southeast – probably from the ancestral Manatee and Little Manatee Rivers (Fig. 6). As mentioned above, the lowermost seismic sequences are deformed as well as the seismic basement indicating syndepositional deformation during the late stage of basin formation and early stage of sub-basin filling.

CHARLOTTE HARBOR SUB-BASIN

Approximately 150 km to the south of Tampa Bay, located along Florida's Gulf of Mexico coastline, lies Charlotte Harbor having very similar dimensions (~725 km^2), shape and average depth (~2.3 m) as Tampa Bay (Fig. 1). Evans (1989) and Evans *et al.* (1989), using a grid of 800 km of analogue, high-resolution seismic-reflection data tied to 22 borehole sites around this estuary, found strong similarities to what was later discovered beneath Tampa Bay (Fig. 7). Additionally, later work by Cunningham *et al.* (2003) in the Caloosahatchee River and southern Charlotte Harbor provided chronostratigraphic control through seismic reflection data and borehole analyses.

Charlotte Harbor is underlain by multiple sub-basins, some having upwards of 100 m of relief as shown by mapping the seismic basement reflection, which also constitutes the top of the Arcadia Formation (Fig. 8). Evans *et al.* (1989) and Evans & Hine (1991) concluded that these underlying Tertiary carbonates had undergone extensive dissolution and collapse creating isolated karst depressions that extend seaward beneath the inner shelf. The same deformational style seen beneath Tampa Bay occurs beneath Charlotte Harbor where broad, high, folded areas are separated by narrow, sinkhole-like sags (Fig. 9). One fold in particular, reveals high-angle faulting, fracturing or both, indicating some degree of lithological induration prior to deformation (Fig. 10).

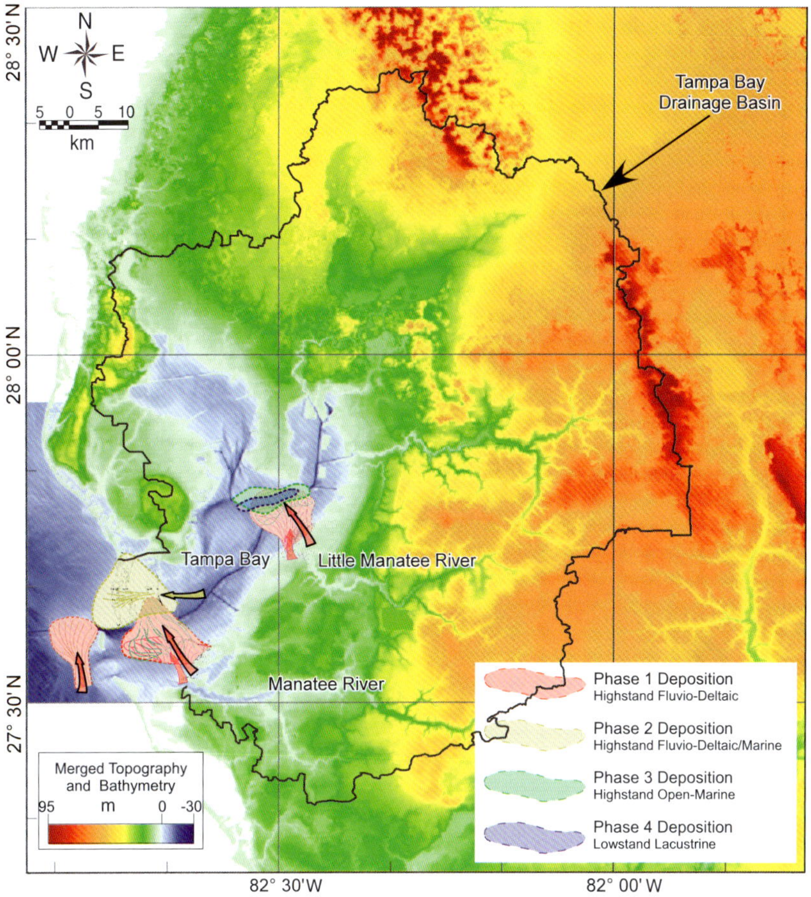

Fig. 6. Terrain model map (USGS) showing rivers and point sources of prograding sequences in Tampa Bay aligned with the modern drainage system. The ancestral counterparts provided the source of siliciclastic sediment that filled in the Tampa Bay sub-basin.

Similarly, these smaller sub-basins have been filled in with at least six sedimentary sequences that are identified and mapped by their bounding unconformable surfaces seen in seismic data (Evans, 1989). These late Neogene to Quaternary sequences are siliciclastic with three Quaternary seismic sequences, based upon the analysis of 40 short (<6 m long) vibracores, consisting of lithological units dominated by mud, shelly sand and quartz gravel. These sediments mostly reflect fluvial to upper estuarine, lagoon and tidal-inlet depositional environments (Evans *et al.*, 1989).

In the southern portion of the Charlotte Harbor estuary near the Caloosahatchee River, lies a prograding deltaic lobe (Fig. 11) interpreted to be part of the Peace River Formation overlying the Arcadia Formation based on seismic data gathered by Missimer & Gardner (1976), Missimer (1999), and Cunningham *et al.* (2001b, 2003). This information combined with seismic data and borehole chronostratigraphy by Cunningham *et al.* (2001b, 2003), indicate that much of the late Neogene and early Quaternary seismic sequences mapped by Evans *et al.* (1989) are the late Miocene to early Pliocene Peace River Formation – same as in the Tampa Bay sub-basin. Seismic data indicate a greater presence of Quaternary cut-and-fill palaeochannels in the seismic sequences beneath Charlotte Harbor than in their counterparts beneath Tampa Bay, suggesting greater palaeofluvial activity.

Finally, an interpreted seismic line (Fig. 12) extending from the Charlotte Harbor sub-basin to Lake Okeechobee run in the Caloosahatchee River illustrates the deformation within the underlying Arcadia Formation that forms a seismic basement high separating two distinct sub-basins. Both basins are filled with prograding clinoforms typical of delta lobes. The clinoforms in the eastern sub-basin have up to 100 m of relief, indicating that a delta lobe prograded into water of at least that depth. Borehole data adjacent to the Caloosahatchee River verify a delta depositional environment (Cunningham *et al.*, 2003). This large delta is the northernmost extent of the palaeofluvial and deltaic depositional

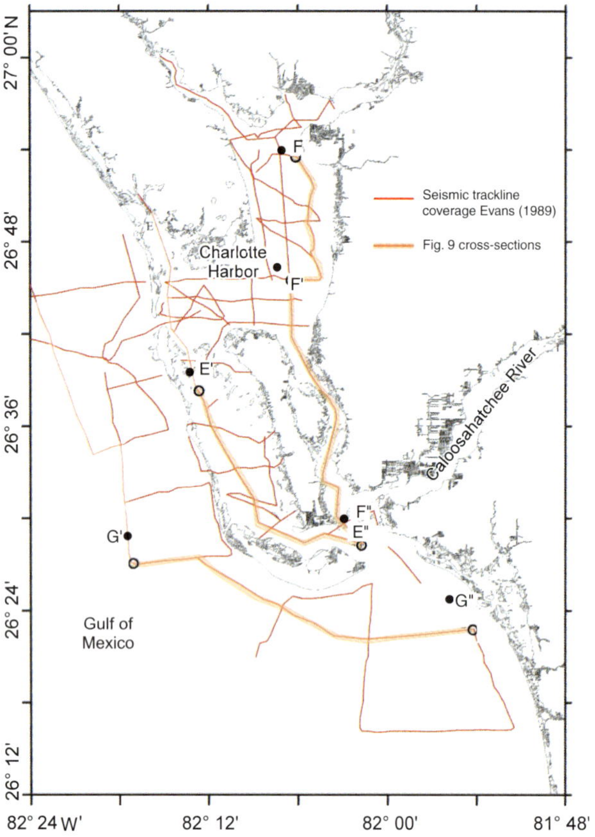

Fig. 7. Seismic data, track-line location in Charlotte Harbor and the Caloosahatchee River. Note location of cross-sections shown in Fig. 9.

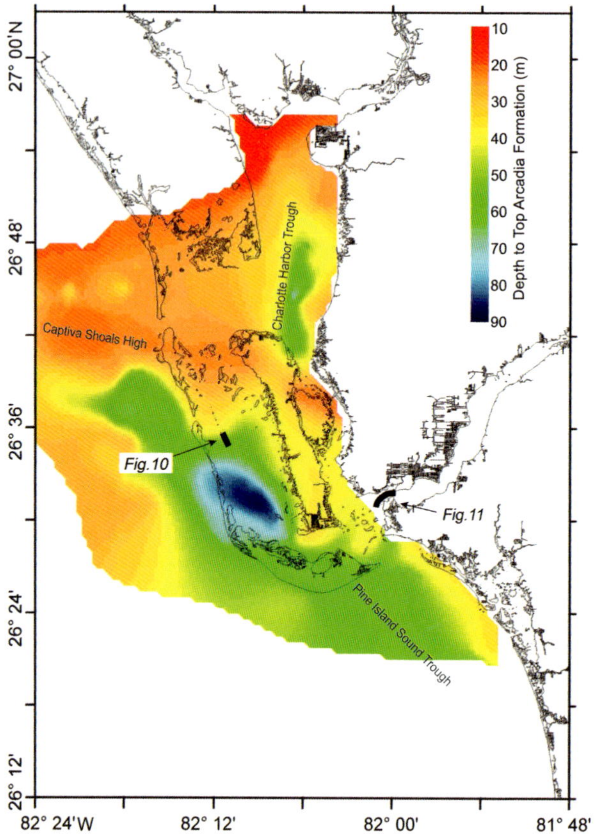

Fig. 8. Depth to basement map of Charlotte Harbor revealing the multiple smaller sub-basins lying beneath this modern estuary. Maximum relief of sub-basins (to Arcadia Formation) is ~100 m.

systems that migrated approximately 200 km south to the Florida Keys (Cunningham *et al.*, 2003).

DISCUSSION

Formation and filling of the Tampa Bay and Charlotte Harbor sub-basins

Since the deformed, underlying limestone is late Oligocene to middle Miocene in age and the overlying siliciclastic sediments are late Miocene to early Pliocene, the sub-basins probably formed during the middle to late Miocene. The style of deformation indicates that deep-seated dissolution caused overlying stratigraphic collapse producing a complex of sags, warps and folds, which combined to form the Tampa Bay and Charlotte Harbor sub-basins. This dissolution and collapse most probably occurred during the extended late Miocene sea-level lowstand shown in Fig. 13 (TB3.1 to TB3.3) although dissolution was probably widespread throughout all Cenozoic lowstands as well (Popenoe *et al.*, 1984; Scott, 1990). During lowstands of sea level, these sub-basins caused the local-to-regional streams and rivers to flow into them forming lakes or swamps (Edgar *et al.*, 2002). During high stands of sea level, these sub-basins became estuaries, or open-marine systems, but were filled primarily by prograding deltas.

It is unclear why the sub-basins are restricted and located where they are. Their location may be related to selective faulting and fracturing of the Mesozoic and Cenozoic carbonate succession overlying the Palaeozoic–PreCambrian basement – perhaps related to reactivation of regional transform faults (Klitgord *et al.*, 1984; Sheridan *et al.*, 1988). Fracturing and faulting may have been stimulated by the early Cenozoic collision with the Cuban arc system (Bralower & Iturralde-Vinent, 1997; Moretti *et al.*, 2003; Pindell *et al.*, 2006). Or, fracturing may have occurred throughout geological time resulting

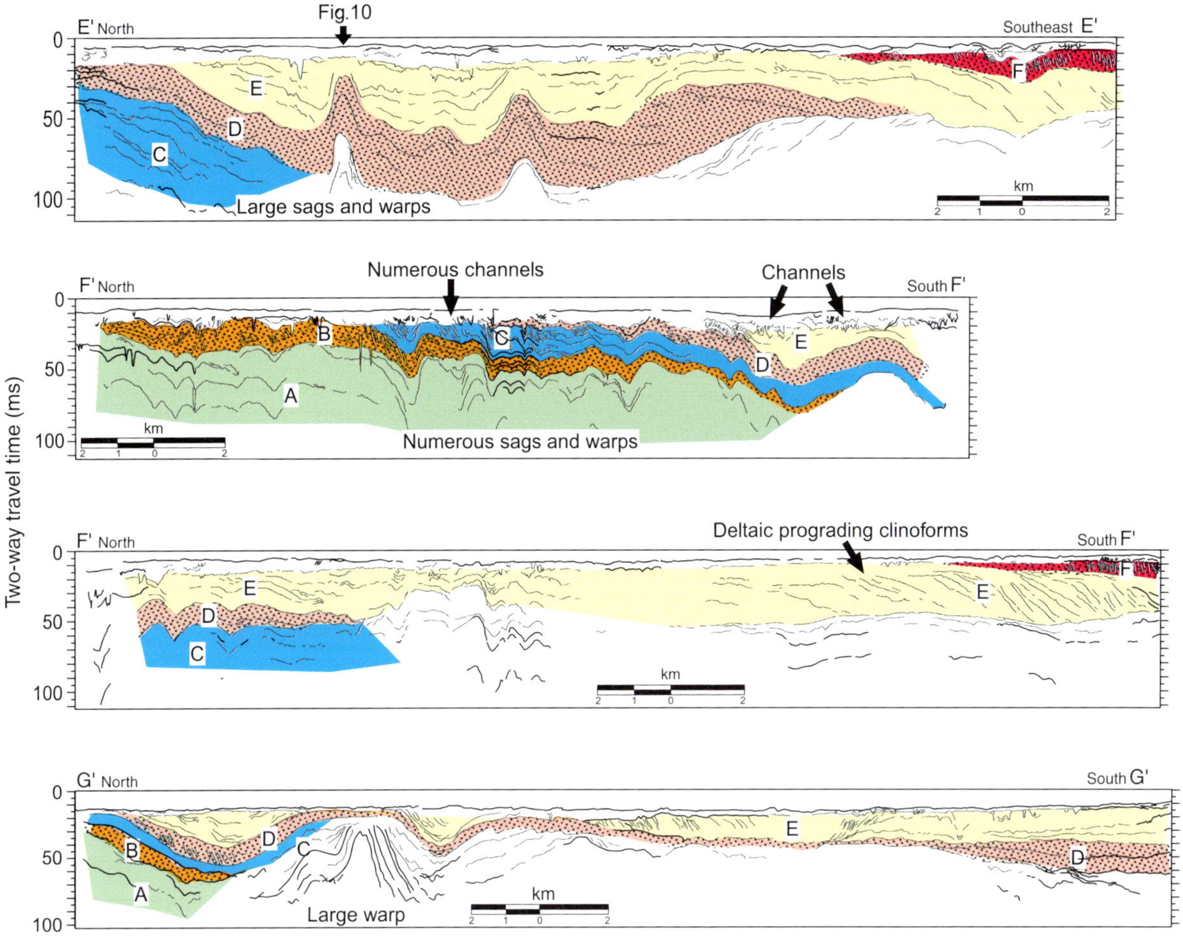

Fig. 9. Interpreted cross-sections from seismic data collected in Charlotte Harbor – see Fig. 7 for location. Cross-sections reveal significant deformation in the form of warps, sags and folds. Also shown are prograding clinoforms, numerous, small palaeofluvial cut-and-fill structures and small, buried sinkholes. Separate seismic sequences are identified as A–F.

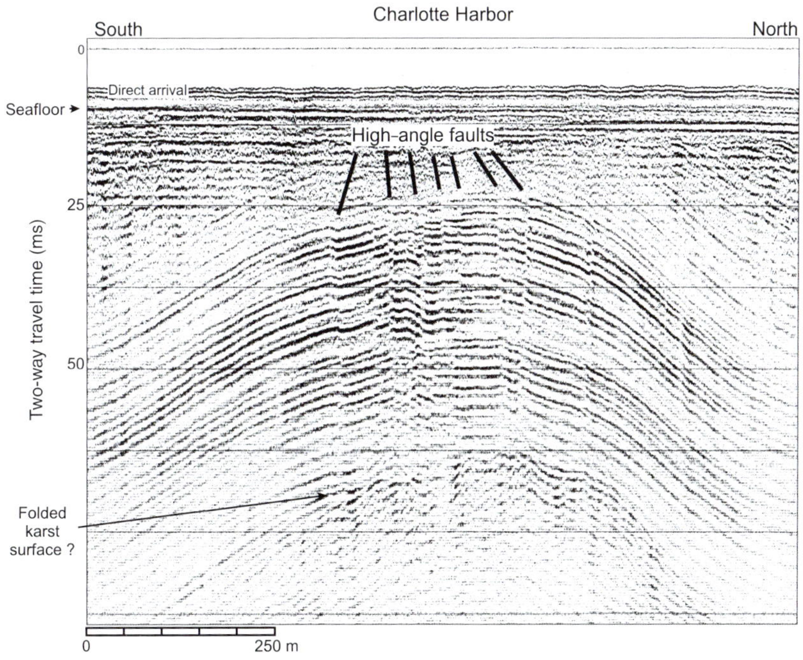

Fig. 10. Detail of fold in seismic data from southern Charlotte Harbor. This fold or warp reveals high-angle faulting indicating lithification prior to deformation. This structure is a dome-like fold exhibiting limited lateral extent and a non anticlinal three-dimensional geometry. See Fig. 8 for location.

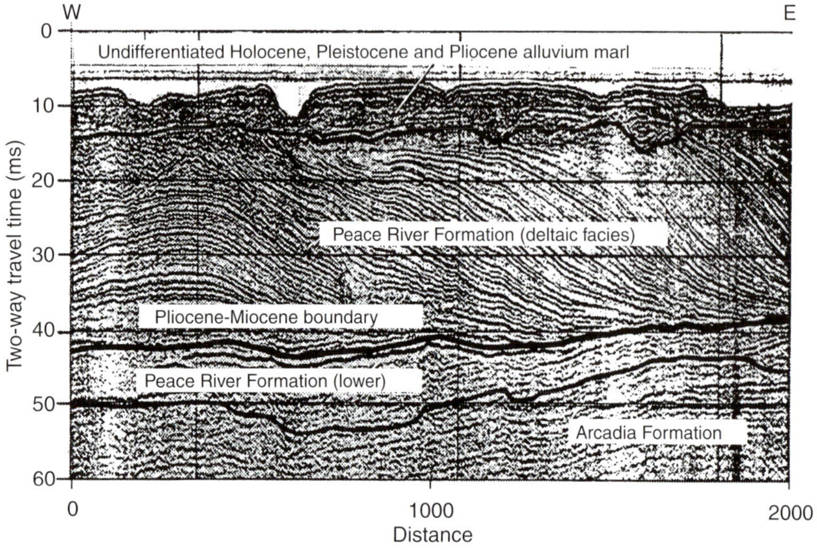

Fig. 11. Seismic line from lower Charlotte Harbor at the west end of the Caloosahatchee River reveals prograding clinoforms form a deltaic lobe as part of the Peace River Formation. This is the lithostratigraphic unit that fills in most of Charlotte Harbor (from Missimer & Gardner, 1976; Missimer, 1999).

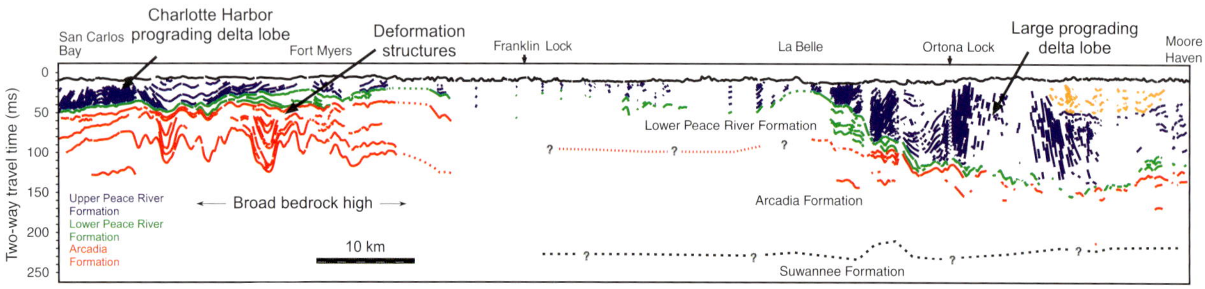

Fig. 12. Interpreted seismic line from Charlotte Harbor to Lake Okeechobee extending west-to-east approximately 50% across the State of Florida. This line reveals two sub-basins separated by an elevated area of the Arcadia Formation. The western, smaller sub-basin is Charlotte Harbor and illustrates the deltaic lobe shown in Fig. 11. The much broader and deeper eastern sub-basin also reveals much higher relief deltaic prograding clinoforms of the upper Peace River Formation. This is the start point of the 200 km long southward delta migration as described by Cunningham *et al.* (2003).

from differential subsidence associated with passive margins facing the Atlantic and the Gulf of Mexico (Klitgord *et al.*, 1988; Sawyer *et al.*, 1991). Indeed, relatively minor seismic activity (earthquake magnitudes <6.0) is ongoing throughout the eastern Gulf of Mexico and Florida Platform region (http://earthquake.usgs.gov/eqcenter/eqinthenews/2006/usslav/). Nevertheless, differential geothermal gradients setting up Kohout-style convection (Tanner, 1976; Fanning *et al.*, 1981; Kohout *et al.*, 1977, 1988; Mitchell-Tapping *et al.*, 1999; Mitchell-Tapping, 2002) mixing ground-waters of different salinities and carbonate saturation states possibly produced selective dissolution of carbonate or evaporite rocks, or both.

Much of the high-resolution seismic reflection profiling on the Little and northern Great Bahama Bank, and Cay Sal Bank presented here reveals no sub-basins or subsurface deformation in the shallow stratigraphy (Hine & Neumann, 1977; Hine, 1977; Hine *et al.*, 1981; Hine & Steinmetz, 1984). Admittedly, there are large gaps in the data coverage and most seismic lines were run across the margins of these huge, modern carbonate platforms. However, there is a complete lack of deformation anywhere, even in the interior seismic lines and even in deeper penetrating seismic data (Eberli & Ginsburg, 1987; Ginsburg, 2001). Perhaps, the fact that the Bahama Banks are isolated and detached as compared to the Florida Platform with its main aquifer system extending into the southeast USA presents a fundamental difference (http://capp.water.usgs.gov/gwa/ch_g/G-text7.html)? During lowstands of sea level, the Florida Platform still received ground-water influx from the northern components of the Floridan Aquifer allowing for continued or

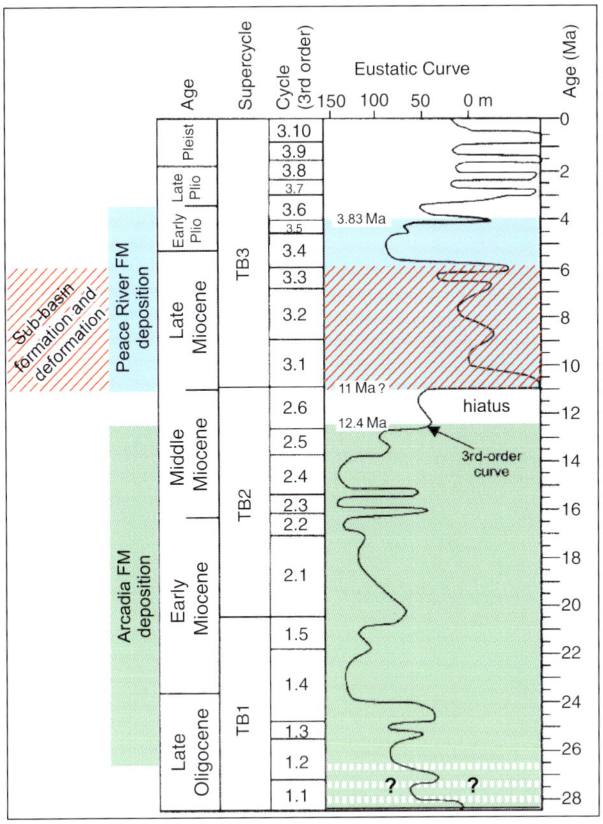

Fig. 13. Relationship between lithostratigraphic units, chronostratigraphy and sea level showing the timing of sub-basin formation, deformation and infilling (modified from Fig. 2, Cunningham *et al.*, 2003; eustatic curve from Haq *et al.*, 1988).

perhaps even stimulated deep-seated dissolution. Whereas, the Bahama Banks, being physically isolated, cannot receive groundwater influx from some lateral source, but only receive their fresh water from local rainfall perhaps limiting subsurface dissolution.

Based upon the lithostratigraphy and chronostratigraphy provided by boreholes adjacent to Tampa Bay, Charlotte Harbor and the Caloosahatchee River, the infilling of the semi-enclosed sub-basins beneath these Florida west-coast estuaries occurred during the very late Miocene and Pliocene sea-level highstand (TB3.4 and TB3.5; Fig. 13). This was part of the major remobilization and southward transport of siliciclastics in south Florida as pointed out by the various authors associated with the South Florida Drilling Project (Guertin *et al.*, 2000), and summarized by Cunningham *et al.* (2003) and McNeill *et al.* (2004). During the late Miocene and Pliocene, a major fluvial-deltaic depositional system migrated southward approximately 200 km to near the southernmost margin of the Florida Platform.

This scenario poses several related critical questions. (1) What was the source of quartz sand and gravel that filled the Tampa Bay and Charlotte Harbor sub-basins and enabled progadation of a deltaic depositional system 200 km south, approaching the margin of the south Florida Platform? (2) What caused the late Miocene-to-Pliocene fluvial-deltaic activity to be significantly enhanced as compared to the present, since today there are no bay-head deltas? (3) What was the nature of the palaeofluvial network – several large long rivers or a complex network of high-discharge local, short-length streams? (4) What shut down this siliciclastic remobilization event and allowed the return to primarily carbonate deposition in the Pleistocene forming the Florida Keys consisting of the Miami Oolite and the Key Largo Limestone (Enos & Perkins, 1977)? Finally (5) what is the geological significance of these karst sub-basins?

The >1000-km long siliciclastic transport pathway

The primarily chemical weathering of exposed silicate-rich bedrock of the southern Appalachian Mountains and Piedmont provided the ultimate source of quartz sands and gravels to the Florida Platform that eventually reached the Pourtales Terrace lying in 200 m water in the southern Straits of Florida. The following scenario (Figs 1 and 14) is suggested in order to partially explain this >1000-km long source-to-sink pathway – a pathway that consisted of multiple sedimentary compartments (coastal plain, river deltas, coastlines, karst entrapment basins and ultimately the open marine shelf and slope) and involved multiple sedimentary transport processes. Thus, the transport and deposition of siliciclastics probably proceeded as a series of steps modulated by Cenozoic sea-level fluctuations, topographic variations and climate changes.

Streams from the Appalachian Mountains and Piedmont brought sediment to the coastal plain where it was deposited along adjacent floodplains or reached deltas discharging into the marine environment. Through time, river deltas prograded across and filled the Georgia Channel System during sea-level lowstands in the late Eocene and early Oligocene (McKinney, 1984). Probably, during the middle Oligocene major

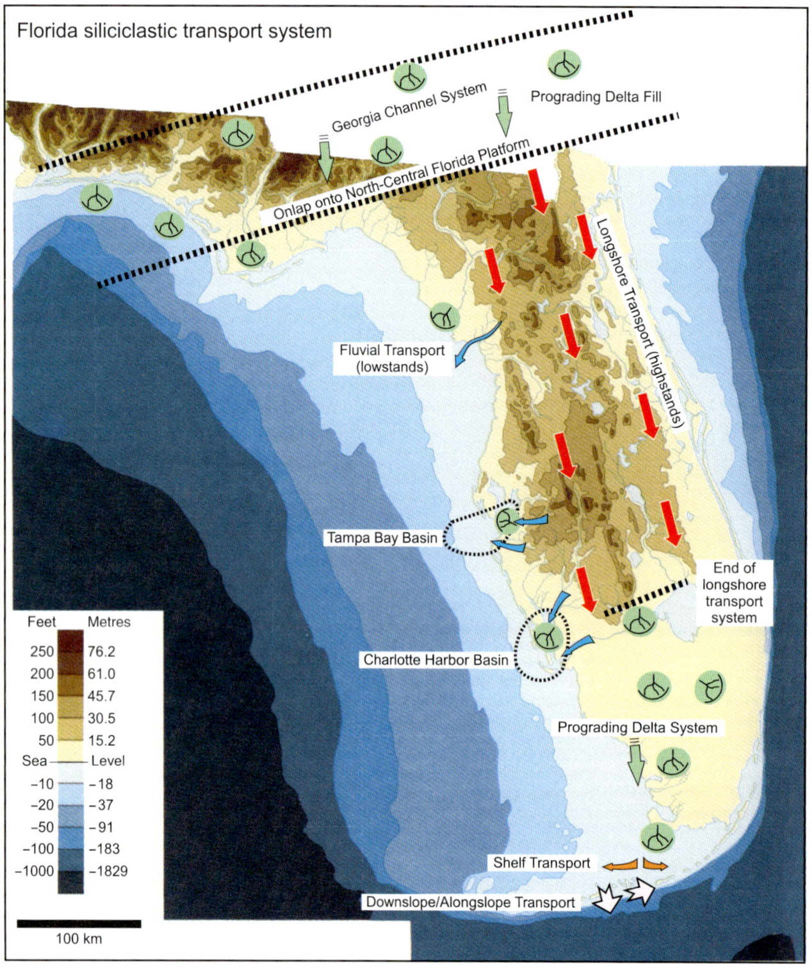

Fig. 14. Map (adapted from Fernald, 1981, p. 16) illustrating the transport pathway and suggested modes of transport of Cenozoic siliciclastic sediment: (1) across the Georgia Channel System from the southeast coastal plain via deltaic progradation infilling this seaway; (2) sediment onlap onto the Florida Platform and transport down north and central peninsular Florida to the southern terminus of the Lake Wales Ridge, primarily by longshore transport during high stands of sea-level; (3) palaeofluvial infilling of mid-platform sub-basins such as Tampa Bay and Charlotte Harbor during lower stands of and/or falling sea level; (4) continued southward palaeofluvial progradation covering an exposed carbonate ramp in south-central peninsular Florida; (5) introduced to the marine environment beneath and seaward of the present Florida Keys forming a shelf–slope system influenced by cross-shelf and downslope currents; and (6) eventually downlapping of sediment onto the 200 m deep Pourtales Terrace.

sea-level lowstand (Haq *et al.*, 1988; Popenoe, 1990; Brewster-Wingard *et al.*, 1997), the Georgia Channel System was filled completely and siliciclastic sediments started to cover north-central peninsular Florida. Peninsular Florida, being elevated (Fig. 1; St. Johns Platform, Sanford High, Brevard Platform and Ocala Platform; Popenoe, 1990; Scott, 1997), could not support a long-distance, north-to-south fluvial system. Consequently, primary sediment movement probably occurred in coastal longshore transport systems during higher sea level.

Secondary sediment movement to the east and west occurred by local rivers during lower sea level. The late Miocene to early Pliocene Bone Valley Member of the Peace River Formation represents a palaeofluvial reworking of older Hawthorn Group siliciclastic and phosphatic-rich units (Riggs, 1979). The surficial geomorphology of peninsular Florida, dominated by numerous palaeoshoreline features, is best illustrated by the Lake Wales Ridge complex. These extensive north–south-trending coastal features have been subsequently incised and eroded by numerous local streams – some of which provided the remobilized siliciclastic sediments that filled in the semi-enclosed, restricted karst sub-basins beneath Tampa Bay and Charlotte Harbor.

The Lake Wales Ridge ends in south-central peninsular Florida (Figs 1 and 14), indicating that extensive north-to-south, coastal longshore transport probably ended there as well. To complete the sediment transport pathway to the southern Straits of Florida, data from Cunningham *et al.* (2003) and other papers associated with the South Florida Drilling Project (McNeill *et al.*, 2004) indicate that a late Miocene-to-Pliocene prograding deltaic depositional system carried quartz sands and gravels on top of a carbonate ramp that had been exposed for 8 million years thus burying the underlying Arcadia Formation. As this delta complex approached the carbonate margin facing south into the Straits of Florida, a new siliciclastic shelf and slope system was formed. Fluvial–deltaic sedimentary processes merged into cross-shelf and downslope sedimentary processes, again all pulsed by variations in sea-level and climate (windiness, storminess, rainfall) as well as oceanographic processes (e.g. Florida Current and Loop Current activity).

High-sediment discharge local rivers?

Since there is no evidence for an extensive long-distance palaeofluvial transport by a single large river flowing down peninsular Florida from the north, the sub-basins beneath Tampa Bay and Charlotte Harbor must have filled in by local, short-length, high-sediment discharge rivers and streams. With no bay-head deltas or sediment-choked areas at the head-of-tides in today's rivers discharging into Tampa Bay or Charlotte Harbor, the rivers of the late Miocene to Pliocene must have had a higher sediment discharge than present. We postulate that the warm period during the Pliocene (Willard *et al*, 1993; Poore & Sloan, 1996; Dowsett *et al.*, 1996) stimulated local thunderstorm activity over peninsular Florida, thus increasing rainfall, runoff and sediment discharge. Increased sea-surface temperatures in the waters surrounding peninsular Florida (~+2–4°C; Dowsett *et al.*, 1996, their Fig. 1) and increased heating of the land mass would have stimulated the local sea-breeze effect, built larger and more vigorous thunderstorms and could have prolonged the thunderstorm season beyond the 4–5 months seen today.

Additionally, the sea-level highstand of the very late Miocene and early–mid-Pliocene lasted several million years (~5.5–3 Ma) and may have been as much as +35 m higher than present sea level (PRISM reconstruction of Dowsett & Poore, 1991). As such, there was sufficient time for the filling of these peripheral sub-basins and for the major fluvial deltaic system to migrate from the southern end of the Lake Wales Ridge complex to the south Florida margin thus covering the pre-existing carbonate ramp. Perhaps, maximum deltaic progradation occurred during multiple, brief intervals or pulses of enhanced sediment discharge during the falling stages of higher-frequency, sea-level events. However, by the late Pliocene, the massive siliciclastic influx seems to have slowed or stopped in south Florida (D.F. McNeill, personal communication, 2006). In south Florida, this ~150 m thick siliciclastic unit is called the Long Key Formation (Guertin *et al.*, 1999) and is the time-transgressive equivalent of the upper Peace River in the Charlotte Harbor–Caloosahatchee River area.

As Warzeski *et al.* (1996) and Cunningham *et al.* (2003) point out, the aggradation of this siliciclastic system provided the substrate for the return of shallow-water carbonates in the Pleistocene. By the late Pliocene–early Pleistocene boundary, a carbonate-dominated depositional environment returned capping the siliciclastics from the Everglades across the Florida Keys to the top of the Pourtales Terrace (Guertin, 1998; Guertin *et al.*, 1999, 2000; Multer *et al.*, 2002; McNeill, personal communication, 2006). These carbonates are overlain by the carbonate Key Largo Formation for which the initiation age is uncertain (D.F. McNeill, personal communication, 2006). The aggradation of the siliciclastic system was essential in creating a widespread shallow-water environment to support a vibrant carbonate factory. But, there must have been a concurrent reduction in the siliciclastic transport as well. We can only speculate that rainfall and runoff (and resulting siliciclastic sediment transport in local rivers) must have been reduced during this time allowing the carbonate factory to flourish.

Geological significance of semi-enclosed karst sub-basins

The siliciclastic-filled sub-basins underlying Tampa Bay and Charlotte Harbor occur in a mid-carbonate platform setting and not along a carbonate margin where siliciclastics have been known to accumulate (e.g. Belize – Ferro *et al.*, 1999; Great Barrier Reef – Dunbar *et al.*, 2000). Rather than incised valley fills or reef-margin, backfilled basins, they represent spatially restricted, semi-enclosed siliciclastic-filled karst features. Kerans (personal communication, 2005) has indicated that there are large-scale, karst collapse systems with up to 320 m of vertical extent and 100–200 m width in the lower Ordovician El Paso Group of west Texas. Kerans (personal communication, 2005) also points out 35 m vertical relief, 100 m diameter cave collapses of mid-Permian age in the Sierra Diablo Range (Texas). Both of these ancient examples are similar scales to the individual warps, folds and sags seen beneath Tampa Bay and Charlotte Harbor.

However, neither of these ancient examples replicates the spatial scale of those seen in the mid-Florida Platform semi-enclosed sub-basins, and neither are filled in with siliciclastic sediments. As a result, the 'dimpling' by coalescing karst basins in the mid Florida Platform setting provides a previously unrecognized mechanism for the creation of accommodation space sometimes resulting in the apparent 'drowning' of a carbonate platform by siliciclastics even though a significant hiatus occurs between the two depositional units.

CONCLUSIONS

1 Multiple, spatially restricted, partially enclosed karst sub-basins with as much as 100 m of relief in a mid-carbonate platform setting lie beneath the modern estuaries of Tampa Bay and Charlotte Harbor located along the Florida Gulf of Mexico coastline.

2 Seismic basement consists of the carbonate, upper Oligocene to middle Miocene Arcadia Formation, which has been significantly deformed into folds, sags, warps and sinkholes. This deformation is interpreted to have been caused by deep-seated dissolution of carbonates at depth allowing the overlying stratigraphy to collapse thus creating a surficial depression. Sub-basin formation occurred during the mid- to late Miocene sea-level lowstand.

3 Approximately 90% of the accommodation space of these surficial sub-basins has been filled by sequences dominated by prograding clinoforms. Adjacent borehole data indicate that the sediment infill is dominantly siliciclastic deposited in a deltaic depositional setting. The infill of the Tampa Bay and Charlotte Harbor basins consists mostly of the Peace River Formation of very late Miocene to Pliocene age. This sediment fill represents a small component of a large and extensive (>1000 km long) siliciclastic Cenozoic invasion of peninsular Florida.

4 The sedimentary fill of the sub-basins was part of a significant remobilization of quartz-rich sediments through enhanced sediment discharge in local, short-length rivers. Enhanced sediment discharge possibly resulted from increased thunderstorm activity during a warm period of the Pliocene particularly during the late stage of basin fill.

5 Rather than incised valley fills or reef-margin, backfilled basins, Tampa Bay and Charlotte Harbor represent spatially restricted, mid-platform, filled-in karst features. The 'dimpling' of a carbonate platform by coalescing karst sub-basins provides a previously unrecognized mechanism for the creation of accommodation that can result in the 'drowning' of a carbonate platform by siliciclastics.

ACKNOWLEDGEMENTS

The Tampa Bay component of the study was funded by the US Geological Survey (USGS) Florida Integrated Science Center – Center for Coastal and Watershed Studies in St Petersburg, FL as part of the Tampa Bay Project; Dr Kim Yates programme manager. We also thank Mark Hansen, and Dana Wiese of the USGS for their advice during this phase of the project. The Charlotte Harbor component of the study was supported by the USGS (Tampa Sub-district), Florida Department of Environmental Regulation and the South Florida Water Management District. The Caloosahatchee River component of the study was supported by the USGS through the South Florida Water Management District. Dr C. Kerans, UT-Austin, provided important insight concerning ancient analogues. We thank Drs Don McNeill, Tom Scott and Len Vacher for reviews and comments. We thank an anonymous reviewer for his/her meticulous comments, which were very helpful. We acknowledge NASA for use of its Blue Marble imagery (used in Fig. 1), which is available at: http://earthobservatory.nasa.gov/Newsroom/BlueMarble/. We thank the USGS for providing the terrain model used in Figure 6.

REFERENCES

Berman, G.A., Naar, D.F., Hine, A.C., Brooks, G.R., Tebbens, S.F., Donahue, B.T. and **Wilson, R.** (2005) Geologic Structure and hydrodynamics of Egmont Channel: an anomalous inlet in the mouth of Tampa Bay, Florida. *J. Coastal Res.*, **21**, 331–357.

Bralower, T.J. and **Iturralde-Vinent, M.A.** (1997) Micropaleontological dating of the collision between North American Plate and the Greater Antilles Arc in western Cuba. *Palaios*, **12**, 133–150.

Brewster-Wingard, G.L., Scott, T.M., Edwards, L.E., Weedman, S.D. and **Simmons, K.R.** (1997) Reinterpretation of the peninsular Florida Oligocene: an integrated stratigraphic approach. *Sed. Geol.*, **108**, 207–228.

Brooks, G.R. and **Doyle, L.J.** (1998) Recent sedimentary development of Tampa Bay, Florida: a microtidal estuary incised into Tertiary carbonate platform carbonates. *Estuaries*, **21**, 391–406.

Chen, C.S. (1965) The regional lithostratigraphic analysis of Paleocene and Eocene rocks of Florida. *Florida Geol. Survey Bull.*, **45**, 105 pp.

Cunningham, K.J., **Bukry, D. Sato, T.**, **Barron, J.A.**, **Guertin, L.A.** and **Reese, R.S.** (2001a) Lithostratigraphy, sequence stratigraphy and biostratigraphy of a carbonate ramp and bounding siliciclastics (late Miocene – Pliocene), southern Florida: *Florida Geol. Survey, Special Paper*, **49**, 35–66.

Cunningham, K.J., **Locker, S.D.**, **Hine, A.C.**, **Bukry, D.**, **Barron, J.A.** and **Guertin, L.A.** (2001b) Surface-geophysical characterization of ground-water systems of the Caloosahatchee River Basin, southern Florida. *U.S. Geological Survey, Water-Resources Investigations Report*, **01–4084**, 76 pp.

Cunningham, K.J., **Locker, S.D.**, **Hine, A.C.**, **Bukry, D.**, **Barron, J.A.** and **Guertin, L.A.** (2003) Interplay of late Cenozoic siliciclastic supply and carbonate response on the southeast Florida platform. *J. Sed. Res.*, **73**, 31–46.

Cunningham, K.J., **McNeill, D.F.**, **Guertin, L.A.**, **Ciesileski, P.F.**, **Scott, T.M.** and **de Verteuil, L.** (1998) New Tertiary stratigraphy for the Florida Keys and southern peninsula of Florida. *Geol. Soc. Am. Bull.*, **110**, 231–258.

Dalrymple, R.W., **Boyd, R.** and **Zaitlin, B.A.** (1994) History of research, types, and internal organization of incised-valley systems: introduction to the volume. In: *Incised-Valley Systems: Origin and Sedimentary Sequences* (Eds R.W. Dalrymple, R. Boyd and B.A. Zaitlin). *SEPM Spec. Publ.*, **51**, 3–10.

Davis, R.A. (1997) Geology of the Florida Coast. In: *The Geology of Florida* (Eds A.F. Randazzo and D.S. Jones). University Press of Florida, Gainesville, FL, pp. 155–168.

Davis, R.A. Jr., **Hine, A.C.** and **Shinn, E.A.** (1992) Holocene coastal development on the Florida Peninsula: In: *Quaternary Coasts of the United States. Marine and Lacustrine Systems. SEPM Spec. Publ.*, **48**, 193–211.

Donahue, B.T., **Hine, A.C.**, **Tebbens, S.**, **Locker, S.D.** and **Twichell, D.C.** (2003) Late Holocene estuarine-inner shelf interactions; is there evidence of an estuarine retreat path for Tampa Bay, Florida? *Mar. Geol.*, **200**, 219–241.

Dowsett, B.T. and **Poore, R.Z.** (1991) Pliocene sea surface temperatures of the North Atlantic Ocean at 3.0 Ma. *Quaternary Sci. Rev.*, **10**, 189–204.

Dowsett, H., **Barron, J.** and **Poore, R.** (1996) Middle Pliocene sea surface temperatures: a global reconstruction. *Marine Micropaleontol.*, **27**, 13–25.

Doyle, L.D. and **Sparks, T.H.** (1970) Sediments of the Mississippi, Alabama, and Florida (MAFLA) continental shelf. *J. Sediment. Petrol.*, **50**, 905–916.

Dunbar, G.B., **Dickens, G.R.** and **Carter, R.M.** (2000) Sediment flux across the Great Barrier Reef shelf to the Queensland Trough over the last 300 ky. *Sediment. Geol.*, **133**, 49–92.

Duncan, D.S. (1993) *Neogene to Recent seismic stratigraphy of the lower Tampa Bay estuary: west-central Florida*. Unpublished M.S. thesis. University of South Florida, FL, USA, 101 pp.

Duncan, D.S., **Locker, S.D.**, **Brooks, G.R.**, **Hine, A.C.** and **Doyle, L.J.** (2003) Mixed carbonate-siliciclastic infilling of a Neogene carbonate shelf-valley system: Tampa Bay, west-central Florida. *Mar. Geol.*, **200**, 125–156.

Eberli, G.P. and **Ginsburg, R.N.** (1987) Segmentation and coalescence of Cenozoic carbonate platforms, northwestern Great Bahama Bank. *Geology*, **15**, 75–79.

Edgar, N.T., **Willard, D.A.**, **Brooks, G.R.**, **Cronin, T.D.**, **Hastings, D.W.**, **Flower, B.P.**, **Swarzenski, P.W.**, **Hollander, D.J.**, **Larson, R.A.**, **Hine, A.C.**, **Suthard, B.C.**, **Locker, S.D.** and **Greenwood, W.J.** (2002) Holocene and Pleistocene marine and non-marine sediment from Tampa Bay, Florida. *Eos Trans. AGU*, **83**(47), Fall Meeting Supplement, Abstract OS61A-0191.

Enos, P. and **Perkins, R.D.** (1977) Quaternary sedimentation in south Florida. *Geol. Soc. Am. Memoir*, **147**, 198 pp.

Evans, M.W. (1989) *Late Miocene to Quaternary seismic and lithologic sequence stratigraphy of the Charlotte Harbor area: southwest Florida*. Unpublished PhD dissertation, University of South Florida, 336 pp.

Evans, M.W. and **Hine, A.C.** (1991) Late Neogene sequence stratigraphy of a carbonate siliciclastic transition. *Geol. Soc. Am. Bull.*, **103**, 679–699.

Evans, M.W., **Hine, A.C.** and **Belknap, D.F.** (1989) Quaternary stratigraphy of the Charlotte Harbor estuarine – lagoon system, southwest Florida: implications of the carbonate-siliciclastic transition. *Mar. Geol.*, **88**, 319–348.

Fanning, K.A., **Byrne, R.H.**, **Breland, J.A.**, **Betzer. P.R.**, **Moore, W.S.**, **Elsinger, R.J.** and **Pyle, T.E.** (1981) Geothermal springs of the west Florida continental shelf: evidence for dolomitization and radionuclide enrichment. *Earth Planetary Lett.*, **52**, 345–354.

Ferguson, T.W. and **Davis, R.A. Jr.** (2003) Post-Miocene stratigraphy and depositional environments of valley-fill sequences at the mouth of Tampa Bay, Florida. *Mar. Geol.*, **200**, 157–170.

Fernald, E.A. (1981) *Atlas of Florida*. Florida State University Foundation, Inc., Florida, 276 pp.

Ferro, C.E., **Droxler, A.W.**, **Anderson, J.B.** and **Mucciarone, D.** (1999) Late Quaternary shift of mixed siliciclastic environments induced by glacial eustatic sea-level fluctuations in Belize. *Advances in Carbonate Sequence Stratigraphy: Applications to Reservoirs, Outcrops, and Models. SEPM Spec. Publ.*, **63**, 385–411.

Ginsburg, R.N. (2001) Subsurface geology of a prograding carbonate platform, Great Bahama Bank: Results of the Bahama Drilling Project. *SEPM Spec. Publ.*, **70**, 271 pp.

Ginsburg, R.N., **Brown, K.M.** and **Chung, G.S.** (1989) Siliciclastic foundations of south Florida's Quaternary carbonates. *Geol. Soc. Am., Abstracts with Programs*, **21**, A-290.

Gould, H.R. and **Stewart, R.H.** (1956) Continental terrace sediments in the northeastern Gulf of Mexico. In: *Finding Ancient Shorelines. SEPM Spec. Publ.*, **3**, 2–19.

Green, R., **Arthur, J.D.** and **Dewitt, D.** (1995) Lithostratigraphic and hydrostratigraphic cross-sections through Pinellas and Hillsborough counties, southwest Florida. *Florida Geological Survey Open File Report*, **61**, 26 pp.

Guertin, L.A. (1998) *A late Cenozoic mixed carbonate/siliciclastic system, south Florida: lithostratigraphy, chronostratigraphy, and sea-level record*. Ph.D. dissertation. University of Miami, FL, USA, 335 pp.

Guertin, L.A., **McNeill, D.F.**, **Lidz, B.H.** and **Cunningham, K.J.** (1999) Chronologic model and transgressive–regressive signatures in the late Neogene siliciclastic foundation (Long Key Formation) of the Florida Keys. *J. Sed. Res.*, **69**, 653–666.

Guertin, L.A., **Missimer, T.M.** and **McNeill, D.F.** (2000) Hiatal duration of correlative sequence boundaries from Oligocene–Pliocene mixed carbonate/siliciclastic sediments of the south Florida Platform. *Sed. Geol.*, **134**, 1–26.

Haq, B.U., **Hardenbol, J.** and **Vail, P.R.** (1988) Mesozoic and Cenozoic chronostratigraphy and cycles of sea-level change. In: *Sea Level Changes – An Integrated Approach* (Eds C.K. Wilgus, B.S. Hastings, C.G.St.C. Kendall, H.W. Posamentier, C.A. Ross and J.C. Van Wagoner). *SEPM Spec. Publ.*, **42**, 72–108.

Hebert, J.A. (1985) *High-resolution seismic stratigraphy of the inner west Florida shelf, west of Tampa Bay.* Unpublished M.S. thesis. University of South Florida, USA, 52 pp.

Hine, A.C. (1977) Lily Bank; Bahamas: response of an active oolite sand shoal to its physical environment and the Holocene sea level rise. *J. Sedimentary Petrol.*, **47**, 1454–1474.

Hine, A.C. (1997) Structural and paleoceanographic evolution of the margins of the Florida Platform. In: *Geology of Florida* (Eds A.F. Randazzo and D.S. Jones). University Press of Florida, Gainesville, FL, pp. 169–194.

Hine, A.C. and **Neumann, A.C.** (1977) Shallow carbonate bank margin growth and structure, Little Bahama Bank, Bahamas. *Am. Assoc. Petroleum Geol. Bull.*, **61**, 376–406.

Hine, A.C. and **Steinmetz, J.C.** (1984) Cay Sal Bank, Bahamas – a partially drowned carbonate platform. *Marine Geol.*, **59**, 135–164.

Hine, A.C., **Wilber, R.J.** and **Neumann, A.C.** (1981) Carbonate sand bodies along contrasting shallow bank margins facing open seaways: northern Bahamas. *Am. Assoc. Petrol. Geol. Bull.*, **65**, 261–290.

Huddleston, P.F. (1993) Revision of the lithostratigraphic units of the coastal plain of Georgia – the Oligocene. *Georgia Geol. Survey Bull.*, **105**, 152.

Hull, J.P.D. (1962) Cretaceous Suwannee Strait, Georgia and Florida. *AAPG Bull.*, **18**, 118–122.

Klitgord, K.D., **Hutchinson, D.R.** and **Schouten, H.** (1988) U.S. Atlantic continental margin; structural and tectonic framework. In: *The Atlantic Continental Margin: U.S.* (Eds R.E. Sheridan and J.A. Grow). *Geol. Soc. Am.*, **I-2**, 1955.

Klitgord, K.D., **Popenoe, P.** and **Schouten, H.** (1984) Florida: a Jurassic transform plate boundary. *J. Geophys. Res.*, **89**, 7753–7772.

Kohout, F.A., **Henry, H.R.** and **Banks, J.E.** (1977) Hydrogeology related to geothermal conditions of the Florida Plateau. In: *The Geothermal Nature of the Floridian Plateau* (Eds D.L. Smith and G.M. Griffin). *Florida Bureau of Geology Spec. Publ.*, **21**, 1–42.

Kohout, F.A., **Meisler, F.**, **Meyer, G.W.** and **Waite, R.F.** (1988) Hydrology of Atlantic continental margins. In: *The Atlantic Continental Margin, U.S., The Geology of North America* (Eds R.E. Sheridan and J.A. Grow). *Geol. Soc. Am.*, **I-2**, 463–480.

Locker, S.D., **Hine, A.C.** and **Brooks, G.R.** (2003) Regional stratigraphic framework linking continental shelf and coastal sedimentary deposits of west-central Florida. *Marine Geol.*, **200**, 351–378.

McKinney, M.L. (1984) Suwannee Channel of the Paleogene coastal plain: support for the carbonate suppression model of basin formation. *Geology*, **12**, 343–345.

McNeill, D.F., **Cunningham, K.J.**, **Guertin, L.A.** and **Anselmetti, F.S.** (2004) Depositional themes of mixed carbonate-siliciclastics in the south Florida Neogene: application to ancient deposits. In: *Integration of Outcrop and Modern Analogs in Reservoir Modeling. AAPG Memoir*, **80**, 23–43.

Missimer, T.M. (1992) Stratigraphic relationships of sediment facies within the Tamiami Formation of southwestern Florida: proposed intraformational correlations. In: *The Plio-Pleistocene Stratigraphy and Paleontology of Southern Florida* (Eds T.M. Scott and W.D. Almon). *Florida Geological Survey Spec. Publ.*, **36**, 63–92.

Missimer, T.M. (1999) Sequence stratigraphy of the Late Miocene – Early Pliocene Peace River Formation, southwestern Florida. *Gulf Coast Assoc. Geol. Soc. Trans.*, **49**, 358–368.

Missimer, T.M. and **Gardner, R.A.** (1976) High-resolution seismic reflection profiling for mapping shallow water aquifers in Lee County, FL. *U.S. Geological Survey, Water Resources Investigation*, **76–45**, 30 pp.

Missimer, T.M. and **Ginsburg, R.N.** (1998), Homogenized carbonates and siliciclastics in the Tertiary of southwest Florida. *Gulf Coast Assoc. Geol. Soc. Trans.*, **48**, 263–274.

Mitchell-Tapping, H.J. (2002) Exploration analysis of basin maturity in the south Florida sub-basin. *Gulf Coast Assoc. Geol. Soc. Trans.*, **52**, 753–764.

Mitchell-Tapping, H.J., **Bellucci, J.R.**, **Woody, G.** and **Lee, T.J.** (1999) Mud Hole: a unique warm-water submarine spring, located offshore southwestern Florida. *Gulf Coast Assoc. Geol. Soc. Trans.*, **49**, 370–383.

Moretti, I., **Tenreyro, R.**, **Linares, E.**, **Lopez, J.G.**, **Letouzey, J.**, **Magnier, C.**, **Gaumet, F.**, **Lecomte, J.C.**, **Lopez, J.O.** and **Zimine, S.** (2003) Petroleum system of the Cuban northwest offshore zone. In: *The Circum-Gulf of Mexico and the Caribbean: Hydrocarbon Habitats, Basin Formation, and Plate Tectonics* (Eds C. Bartolini, R.T. Buffler and J.F. Blickwede). *AAPG Memoir* **79**, 125–128.

Multer, H.G., **Gischler, E.**, **Lundberg, J.**, **Simmons, K.R.** and **Shinn, E.A.** (2002) Key Largo Limestone revisited: Pleistocene shelf-edge facies, Florida Keys, USA. *Facies*, **46**, 229–272.

Pindell, J., **Kennan, L.**, **Stanek, K.P.**, **Maresch, W.V.** and **Draper, G.** (2006) Foundations of Gulf of Mexico and Caribbean evolution: eight controversies resolved. *Geol. Acta*, **4**, 303–341.

Poore, R.Z. and **Sloan, L.C.** (1996) Climates and climate variability of the Pliocene. *Marine Micropaleontol.*, **27**, 326.

Popenoe, P. (1990) Paleoceanography and paleogeography of the Miocene of the southeastern United States. In: *Neogene to Modern Phosphorites* (Eds W.C. Burnett and S.R Riggs) *Phosphate Deposits of the World*, Vol. **3**. Cambridge University Press, New York, NY, pp. 352–380.

Popenoe, P., **Henry, V.J.** and **Idris, F.M.** (1987) Gulf Trough – The Atlantic connection. *Geology*, **15**, 327–332.

Popenoe, P., **Kohout, F.A.** and **Manheim, F.T.** (1984) Seismic-reflection studies of sinkholes and limestone dissolution features on the northeastern Florida shelf. In: *Sinkholes: Their Geology, Engineering and Environmental Impact* (Ed. B.F. Beck). A.A. Balkema, Netherlands, pp. 43–57.

Riggs, S.R. (1979) Phosphate sedimentation in Florida – a model phosphogenic system. *Econom. Geol.*, **74**, 285–314.

Sawyer, D.S., Buffler, R.T. and **Pilger, R.H. Jr.** (1991) The crust under the Gulf of Mexico basin. In: *The Gulf of Mexico Basin* (Ed. A. Salvador). *Geological Society of America*, **J**, 53–72.

Scott, T.M. (1988) The lithostratigraphy of the Hawthorn Group (Miocene) of Florida. *Florida Geol. Survey Bull.*, **59**, 148 pp.

Scott, T.M. (1990) The lithostratigraphy of the Hawthorn Group of peninsular Florida. In: *Neogene to Modern Phosphorites* (Eds W.C. Burnett and S.R Riggs) *Phosphate Deposits of the World*, Vol. **3.** Cambridge University Press, New York, NY, pp. 325–336.

Scott, T.M. (1997) Miocene to Holocene history of Florida. In: *The Geology of Florida* (Eds A.F. Randazzo and D.S. Jones). University Press of Florida, Gainesville, FL, pp. 57–68.

Sheridan, R.E., Mullins, H.T., Austin, J.A. Jr., Ball, M.M. and **Ladd, J.W.** (1988) Geology and geophysics of the Bahamas. In: *The Geology of North America, The Atlantic Continental Margin* (Eds R.E. Sheridan and J.A. Grow). U.S. Geological Society of America, Vol. **I-2**, pp. 329–364.

Stahl, L.E. (1970) *The marine geology of Tampa Bay.* Unpublished M.S. thesis, Florida State University, 69 pp.

Suthard, B.C. (2005) *A siliciclastic-filled sedimentary basin in a mid carbonate platform setting, Tampa Bay, FL.* Unpublished M.S. thesis, University of South Florida, 79 pp.

Tanner, W.F. (1976) Geothermal exploration from deep-well data. *Gulf Coast Assoc. Geol. Soc. Trans.*, **26**, 65–68.

United States Army Corps of Engineers (1969) Survey-review report on Tampa Harbor, Florida. Department of the Army, Corps of Engineers, Jacksonville District, Jacksonville, Florida, 40 pp.

Vaughan, T.W. (1910) A contribution to the geologic history of the Florida Plateau. In: Papers from the Tortugas Laboratory, IV: *Carnegie Institution of Washington Spec. Publ.*, **133**, 99–185.

Warzeski, E.R., Cunningham, K.J., Ginsburg, R.N., Anderson, J.B. and **Ding, Z.D.** (1996) A Neogene mixed siliciclastic and carbonate foundation for Quaternary carbonate shelf, Florida Keys. *J. Sed. Res.*, **66**, 788–800.

White, W.A. (1970) The geomorphology of the Florida peninsula. *FL Department of Natural Resources Bureau of Geol. Bull.*, **51**, 64 pp.

Willard, D.A., Cronin, T.M., Ishman, S.E. and **Litwin, R.J.** (1993) Terrestrial and marine records of climate and environmental changes during the Pliocene in subtropical Florida. *Geology*, **21**, 679–682.

Willis, J.W. (1984) *The shallow structure of Tampa Bay.* Unpublished M.S. thesis, University of South Florida, 84 pp.

Winkler, C.D. and **Howard, J.D.** (1977) Plio-Pleistocene paleogeography of the Florida Gulf coast interpreted from relict shorelines. *Gulf Coast Assoc. Geol. Soc. Trans.*, **27**, 409–420.

Int. Assoc. Sedimentol. Spec. Publ. (2009) **41**, 199–213

Controls on facies mosaics of carbonate platforms: a case study from the Oxfordian of the Swiss Jura

ANDRÉ STRASSER and STÉPHANIE VÉDRINE

Department of Geosciences, University of Fribourg, CH-1700 Fribourg, Switzerland (E-mail: andreas.strasser@unifr.ch)

ABSTRACT

Modern shallow-water carbonate systems commonly display a complex pattern of juxtaposed depositional environments with a patchy facies distribution (facies mosaics). On ancient carbonate platforms, the reconstruction of lateral facies distribution is often hampered not only by discontinuous outcrop but also by lack of sufficiently high time resolution. This case study from the Oxfordian (Late Jurassic) of the Swiss Jura Mountains demonstrates a way to improve the temporal and spatial resolution for the interpretation of carbonate rocks. Sequence-stratigraphic and cyclostratigraphic analyses have been performed that provide a basis for defining depositional sequences, which formed through sea-level changes that were induced by the 400-, 100- and 20-kyr orbital cycles. On the 100-kyr scale, sequence boundaries are well developed and can be correlated between sections. However, identification and correlation of sequences related to the 20-kyr cycle may be difficult if local processes overprinted the record of orbitally controlled sea-level changes. The reconstruction of facies distribution along selected time lines gives a dynamic picture of platform evolution with time steps of a few ten thousand years and helps to interpret the controlling factors such as differential subsidence, low-amplitude eustatic sea-level fluctuations, climate and ecology of the carbonate-producing organisms. Reefs and ooid shoals developed preferentially on topographic highs and thus accentuated platform morphology. Siliciclastics were shed onto the platform during sea-level falls and increased rainfall in the hinterland; their distribution was controlled by platform morphology. Siliciclastics and associated nutrients hindered carbonate production and thus indirectly influenced platform morphology. In addition to these controls, random processes acted on the smaller-scale facies relationships. Sedimentation rates can be estimated for each facies type over time spans of 10–20 kyr. They can thus be compared more easily with Holocene rates than if averaged over millions of years. This study shows that the Oxfordian platform in the Swiss Jura was as complex and dynamic as Holocene carbonate systems.

Keywords Carbonate system, facies, sea-level changes, cyclostratigraphy, Oxfordian, Switzerland.

INTRODUCTION

When flying over modern carbonate platforms, when walking on tidal flats, emergent ooid shoals or coral cays, and when snorkelling in shallow lagoons and over the reef edge it is evident that there is a large array of juxtaposed sedimentary and ecological environments. None of these environments is stable: carbonate mudbanks and sand shoals shift laterally due to tidal and long-shore currents and by doing so modify also the environments adjacent to them. A good example is the spit in Fig. 1a that is about to isolate a lagoon from the open ocean, thus creating a drastic contrast in energy and salinity levels. Storm waves and storm-induced currents lead to abrupt changes by redistributing sediment in large quantities, and by breaching or creating barriers between depositional environments (Fig. 1b). Storms also induce ecological changes by breaking corals, which are later replaced by other species. On longer time-scales (tens, hundreds, or thousands of years), fluctuations in climate and relative sea level become important factors that lead to changes in

Fig. 1. Examples of modern carbonate depositional systems displaying various, juxtaposed environments. (a) Complex spit of carbonate sands migrating northward along the coast of Shark Bay, Western Australia. Note the closing of lagoon to the right, with microbial mats (black in picture) developing in the most protected areas. (b) Reef with spurs and grooves on the eastern (seaward) side of Turneffe Islands, Belize. It is separated from the protected central lagoon (top of picture) by only a thin vegetated land bridge.

composition and distribution of the depositional environments. Increased rainfall in the hinterland carries siliciclastics and nutrients into the carbonate system, whereas aridity increases salinity in shallow lagoons. Water temperature, water transparency, water geochemistry, nutrient availability and sediment input are all factors that control the ecology of carbonate-producing organisms. Sea-level rise leads to flooding of barriers and shallow islands and causes rapid environmental changes once the threshold is passed. Light-dependent ecosystems may lose the photic zone unless they are able to keep up with sea-level rise. Sea-level fall, on the other hand, brings light to deeper environments and can open new habitats for carbonate-producing organisms. In other positions on the platform, sea-level fall leads to subaerial exposure of previously subtidal habitats and shuts off carbonate production.

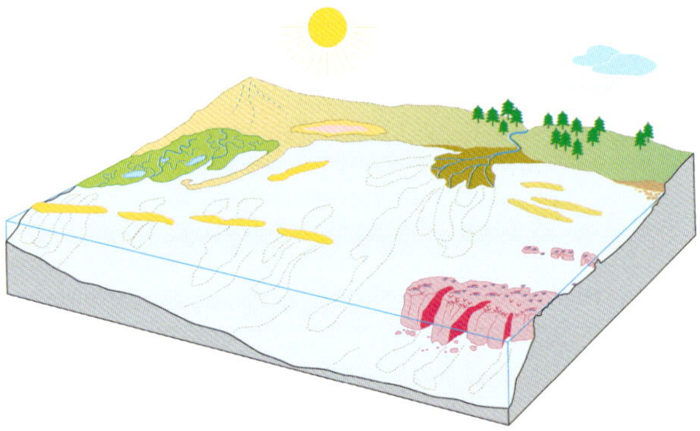

Fig. 2. Sketch illustrating that different depositional environments coexist in shallow carbonate-dominated systems (tidal flats, sabkhas, siliciclastic deltas, beaches, carbonate sand shoals, lagoons, coral reefs). Arid and humid end-members exist but also intermediate cases occur, with seasonal alternations between more humid and more arid phases. Modified from Strasser *et al.* (1999).

Considering the complexity of today's shallow-water carbonate systems (Fig. 2) it must be expected that the sedimentary record of such systems is also complex. There is no reason to believe that Phanerozoic carbonate systems were less complex than modern ones, although plate-tectonic configuration, climatic conditions, amplitude of sea-level changes, seawater geochemistry and the evolution of the participating organisms certainly preclude a direct comparison. Further complexity is added by the fact that only a part of the history is recorded: information is missing due to non-deposition and/or erosion, only part of the organisms participating in the ecosystems are fossilized, and diagenesis may create a preservational bias.

The carbonate facies models of Wilson (1975), Jones & Desrochers (1992) and Flügel (2004) describe facies zones that give a general picture of the potential distribution of sedimentary environments and biota. Read (1985), Burchette & Wright (1992), and Pomar (2001) have emphasized the differences between the geometries of carbonate ramps and platforms and discussed their relation to facies distribution and productivity of carbonate-secreting organisms. Carbonate factories in different palaeoceanographic, palaeoclimatological and palaeoecological settings have been discussed by Schlager (2003) and Mutti & Hallock (2003). Wright & Burgess (2005) acknowledge the high temporal and spatial variability of depositional environments that leads to facies mosaics, which correspond better to reality than the linear arrangement of facies belts as shown in many models. For the Holocene, many examples of facies mosaics have been published: Gischler & Lomando (1999) documented the high complexity of facies distribution of isolated carbonate platforms in Belize; Riegl & Piller (1999) mapped the great lateral variability of coral carpets, reefs, and carbonate sand in Safaga Bay (Egypt); and Rankey (2002) discussed the fractal nature of facies patches on the tidal flats of Andros Island (Bahamas). However, despite the ever more detailed studies of recent and ancient carbonate facies and the undeniable progress in the understanding of shallow-water carbonate systems, there is still a discrepancy in the space and time resolution between the interpretations of ancient and modern depositional environments. This is a result of the difficulty in obtaining continuous, three-dimensional information on facies and sedimentary structures in ancient carbonates, and from the difficulty of defining a precise time-frame that allows reconstructing isochronous depositional surfaces in the rock record.

In this paper a case study is presented where the evolution of carbonate facies through space and time is monitored with a time resolution of 10–20 kyr. This allows for a realistic evaluation of the controlling parameters such as sea-level and climate changes.

STUDIED SECTIONS, PALAEOGEOGRAPHY AND STRATIGRAPHY

The Oxfordian mixed carbonate–siliciclastic sedimentary succession in the Swiss Jura Mountains has been studied extensively over the past few decades (Ziegler, 1956; Bolliger & Burri, 1970; Gygi, 1986, 1995; Gygi & Persoz, 1986; Pittet, 1996; Dupraz, 1999; Hug, 2003; Védrine, 2007). For the present work, five sections were selected (Fig. 3) focusing on the uppermost part of the Günsberg and Röschenz members and the lower part of the Hauptmumienbank and Steinebach members (Fig. 4). The sections were logged along road cuts in the cases of Pertuis, Savagnières, Gorges de Court and Vorbourg, and along an abandoned forestry path in the case of Hautes-Roches. Unfortunately, outcrop conditions did not allow the strata to be followed laterally over more than a few metres, and the sections are widely spaced. The facies analysis is based on continuous outcrop observation with a hand lens and on 433 thin-sections and marl washings (Védrine, 2007). Particular attention was paid to bedding surfaces in order to distinguish between hardgrounds, firmgrounds and erosion surfaces.

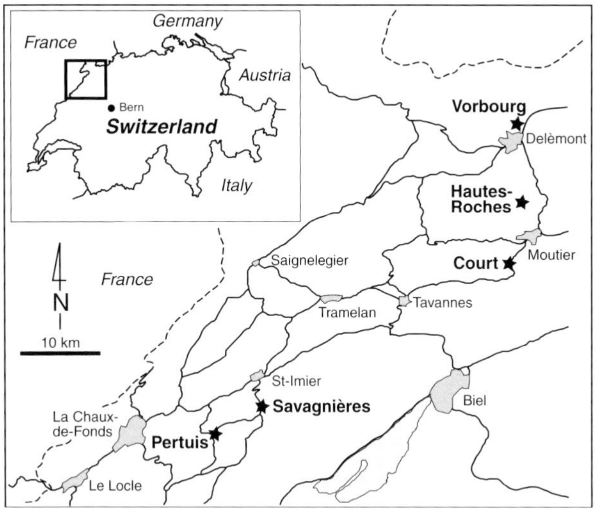

Fig. 3. Location of the five studied sections in the Swiss Jura Mountains. Solid lines are roads, dashed line is the Swiss–French border.

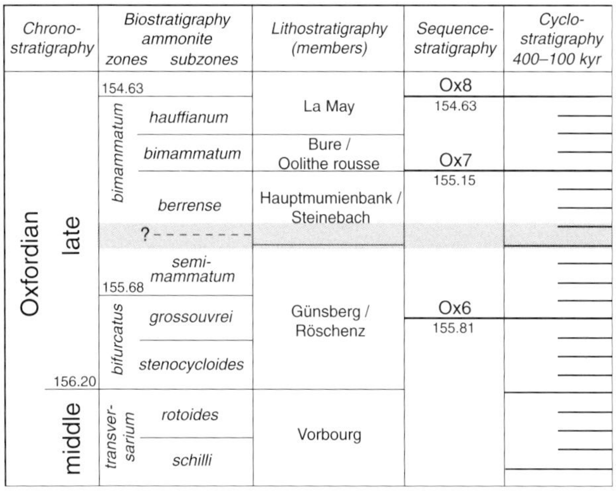

Fig. 4. Stratigraphic framework of the sections studied. A high-resolution analysis is presented for the interval marked in grey (Fig. 9). Chrono-, bio- and sequence stratigraphy based on Hardenbol *et al.* (1998), lithostratigraphy according to Gygi (1995), cyclostratigraphy according to Pittet (1996), Pittet & Strasser (1998) and Hug (2003).

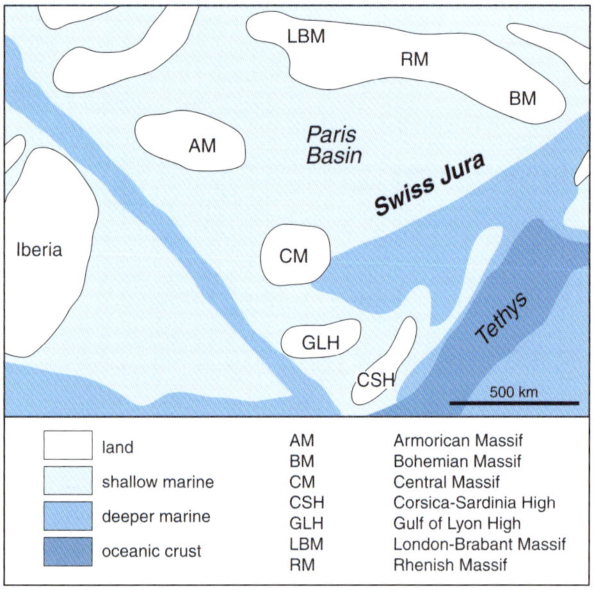

Fig. 5. Late Jurassic palaeogeography of the northern margin of the Tethys Ocean, the Swiss Jura and the Paris Basin (modified from Ziegler, 1988).

The upper part of the Günsberg Member and its lateral equivalent, the Röschenz Member, is dominated by marls and isolated limestone beds with shallow-lagoonal flora and fauna. Freshwater algae, root traces and plant debris occur locally. The Hauptmumienbank Member is composed of massive limestone beds rich in oncoids. It is laterally replaced by the Steinebach Member, which is characterized by oolites and coral patch reefs. These sediments were deposited on the shallow platform between the Paris Basin to the north and the Tethys Ocean to the south, in the position of today's Swiss Jura (Fig. 5). Siliciclastics were shed from the Bohemian and the Central massifs. Palaeolatitude was about 26–27°N (Dercourt *et al.*, 1993).

The interval studied has been dated by mineral-stratigraphic correlations with ammonite-bearing horizons (Gygi & Persoz, 1986; Gygi, 1995): it lies in the lower part of the bimammatum zone (semimammatum and berrense subzones) of the late Oxfordian (Fig. 4). Gygi *et al.* (1998) and Hug (2003) have established the sequence-stratigraphic framework, which implies that the interval studied is situated between sequence boundaries Ox6 and Ox7 of Hardenbol *et al.* (1998). The sequence-stratigraphic terminology follows Vail *et al.* (1991).

Pittet (1996), Pittet & Strasser (1998), Strasser *et al.* (2000) and Hug (2003) have performed a cyclostratigraphic analysis of the middle and late Oxfordian of the Swiss Jura, applying the methodology presented in Strasser *et al.* (1999). Detailed analyses of thin-sections and sedimentary structures allow identification of deepening and shallowing trends of facies evolution. Accordingly, elementary, small-scale and medium-scale depositional sequences have been defined in each section studied, which then were correlated over the entire study area. It is seen that four small-scale sequences regularly build up a medium-scale sequence, and that there are two to six (in many cases five) elementary sequences comprised in a small-scale sequence. The bifurcatus and bimammatum zones contain 16 small-scale sequences (Fig. 4). According to Hardenbol *et al.* (1998), based on the absolute dating by Gradstein *et al.* (1995), this interval lasted about 1.6 Myr. This implies that one small-scale sequence lasted 100 kyr, which corresponds to the duration of the short eccentricity cycle of the Earth's orbit (Berger *et al.*, 1989). A medium-scale sequence would then represent the long eccentricity cycle of 400 kyr, and an elementary sequence (where it can be identified) the precession cycle of 20 kyr. From the observed facies trends it appears that sea level was a major controlling factor in sequence formation. It is therefore suggested that orbital cycles translated into climate changes (Milankovitch cycles), which in turn controlled sea-level fluctuations. Between sequence boundaries Ox6 and Ox7 lie eight small-scale sequences but only 660 kyr are allocated for this interval by Hardenbol *et al.* (1998). On the other hand, four small-scale sequences are contained between Ox7 and Ox8, an interval which Hardenbol *et al.* (1998) estimate to have lasted 520 kyr (Fig. 4). These discrepancies are due to the error margins in absolute dating of stage boundaries (Gradstein *et al.*, 1995) from which Hardenbol *et al.* (1998) have interpolated their age estimations, and/or to the possibility that a sequence boundary dated in the basin by Hardenbol *et al.* (1998) may have its strongest expression on the platform not necessarily at the same time (Strasser *et al.*, 2000). Also, some orbitally induced sea-level changes may not have been recorded in the sedimentary system (see below).

DEFINITION AND CORRELATION OF DEPOSITIONAL SEQUENCES

Small- and medium-scale sequences

Within the general sequence-stratigraphic and cyclostratigraphic framework discussed above, the sections studied are first correlated on the level of small-scale (100-kyr) and medium-scale (400-kyr) sequences (Fig. 6). Sequence boundaries are placed where facies indicate relatively shallow water or intertidal conditions with birdseyes and microbial mats, where very thin elementary sequences suggest that accommodation was low, and/or where channels and overlying lithoclasts imply erosion. Maximum-flooding surfaces or intervals are indicated by relatively deep facies, by shifts from carbonate-dominated deposits to marls that imply highstand progradation, and/or by strong bioturbation due to low sedimentation rates. In one section alone it is often difficult to place these boundaries unequivocally and to estimate their magnitude (i.e. if they belong to elementary, small-scale, or medium-scale sequences). Through correlation with other sections, however, a best-fit solution can be found (Fig. 6).

The model in Fig. 7 illustrates how superimposed high-frequency sea-level changes lead to the formation of complexly stacked depositional sequences in one spot on the shallow platform. Adding to this platform morphology and the fact that different depositional environments react differently to the same sea-level change, it is to be expected that sequence definition and correlation is often not straightforward. For example, a reef may create a shallowing-up succession if it catches up with rising sea level (Kendall & Schlager, 1981), whereas during the same time a lagoon will produce a deepening-up succession. Also, if sea-level fall is minor, it will lead to emersion and a well-developed sequence boundary in previously very shallow environments, whereas in a deeper lagoon the same sea-level fall will not be recorded at all, or only indirectly through input of material reworked from the emergent areas ('hidden SB' in Fig. 7).

Overlying the coral reefs, ooid shoals and oncoid lagoons shown in the lowermost parts of the sections in Fig. 6, there is rapid change towards low-energy, plant-bearing strata (at 4 m in Gorges de Court and at 6.3 m in Hautes-Roches) and local erosion surfaces (at 5 m in Pertuis and at 6.3 m in Hautes-Roches). This is interpreted as representing a sequence boundary on the 400-kyr scale. In Savagnières, this boundary has been placed according to the stacking pattern of the elementary sequences at 4.8 m: a thick elementary sequence following several thin ones implies a rapid increase in accommodation, which is attributed to transgression after the sequence boundary. The maximum flooding of the first medium-scale sequence coincides with the maximum-flooding surface of small-scale sequence 3, which is developed as a strongly bioturbated surface in Pertuis, Savagnières and Court, and as a hardground at Hautes-Roches (Figs 6 and 8). According to the biostratigraphic position, the following medium-scale sequence boundary corresponds to Ox6 of Hardenbol *et al.* (1998). It displays plant fragments at Pertuis and tidal-flat facies at Gorges de Court but is not very conspicuous in Savagnières or Hautes-Roches.

The overlying medium-scale sequence begins with locally well-developed ooid shoals in small-scale sequence 5. Maximum-flooding conditions are implied by a rapid change to a marl-dominated succession in Pertuis and Hautes-Roches, and by a change from high-energy shoals to low-energy lagoonal deposits in Savagnières. However, there is no feature in Gorges de Court that would allow placing this maximum flooding. The following medium-scale sequence boundary marks the end of the marly facies and the beginning of the massive beds of the Hauptmumienbank and Steinebach members. Small-scale sequence 9 generally is very thick and thus suggests rapid gain in accommodation. This sequence will be the subject of a high-resolution analysis in the next chapter. Maximum flooding of the corresponding medium-scale sequence is inferred by the appearance of open-marine fauna in all sections and by a change from ooid dunes and coral reefs to marly and low-energy facies in Court, Hautes-Roches and Vorbourg (Fig. 6).

When comparing the facies of time-equivalent small-scale sequences between the sections it becomes evident that there are significant lateral facies changes: for example, ooid shoals are laterally replaced by oncoid lagoons, or tidal flats are juxtaposed to marls and palaeosols (Figs 6 and 8). Ooid shoals do not always occur at the same time in the sections studied, which suggests that they migrated across the platform through time or were stranded in one position and later reinitiated in another one. Futhermore, the distribution of siliciclastics is heterogeneous, which may be due to preferential deposition in troughs. This and the varying thicknesses of the sequences point to a highly structured platform.

Elementary sequences

An elementary sequence is the smallest recognizable unit of the sedimentary record and is defined as having formed through one cycle of environmental change (Strasser *et al.*, 1999). However, if this cycle is not strong enough to create a

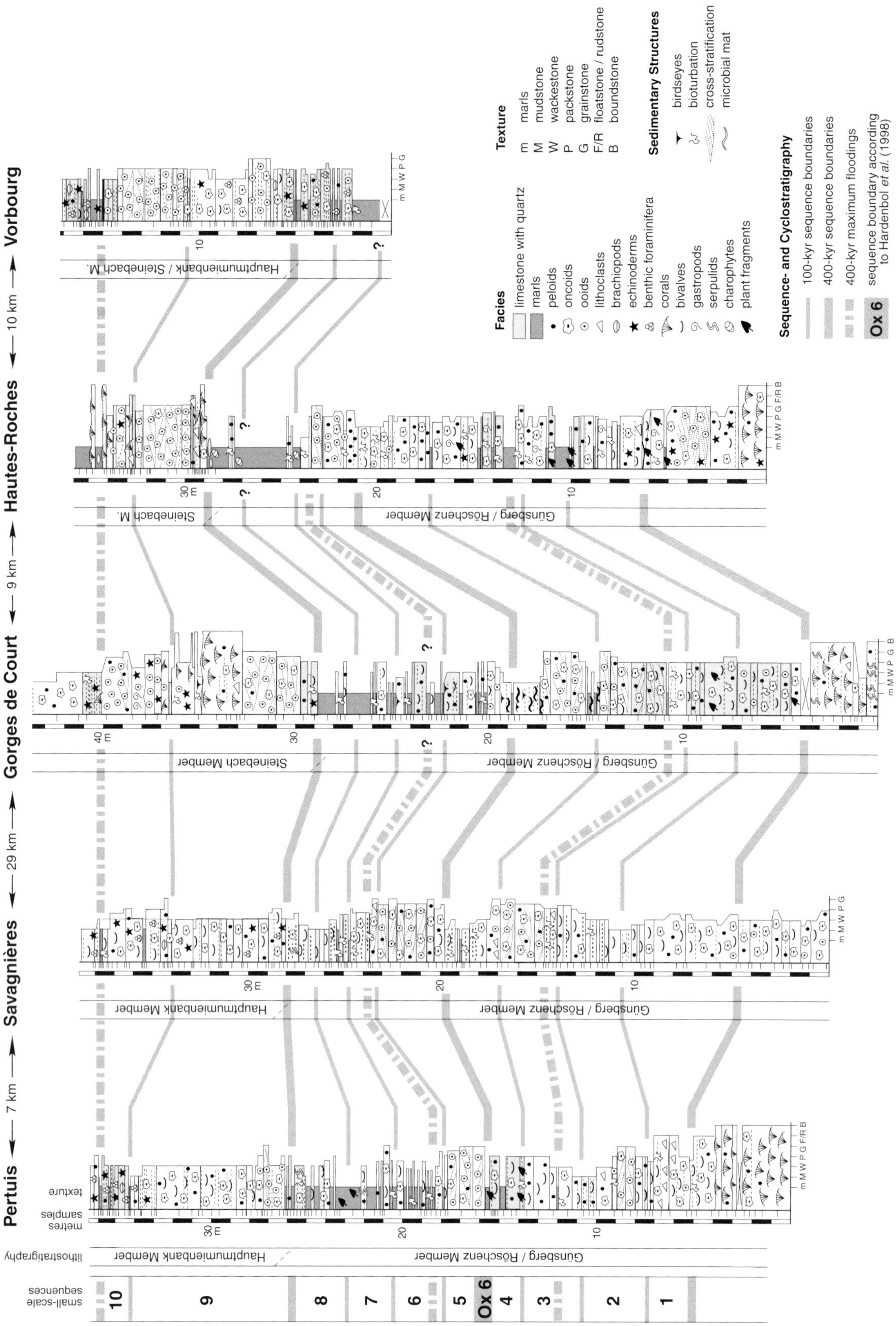

Fig. 6. Correlation of 100- and 400-kyr sequences between the sections studied. Only major facies elements and sedimentary structures are indicated in the lithological columns. For discussion refer to text.

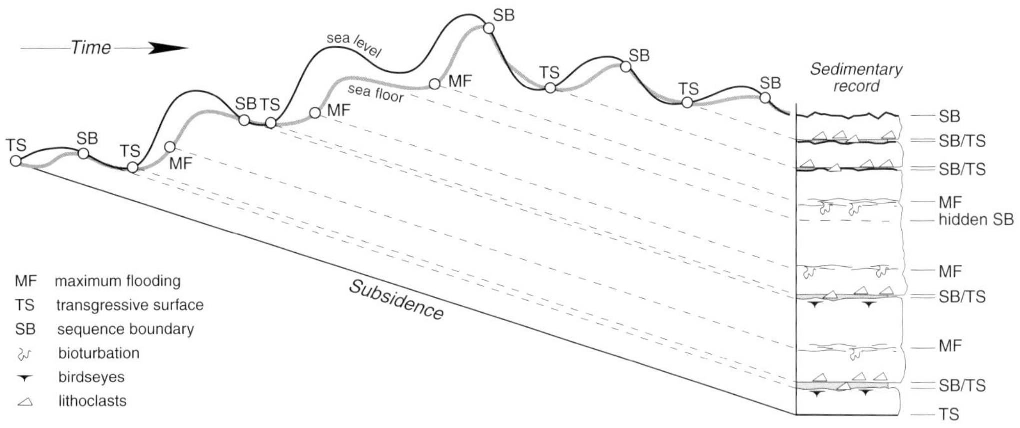

Fig. 7. Superimposed frequencies of sea-level fluctuations lead to a complex sedimentary record. Note that sequence boundaries may be hidden if sea-level fall is not sufficient to create a facies contrast. When accommodation is low due to a long-term sea-level fall, elementary sequences become very thin and have eroded tops, or they are not deposited at all. For more discussion refer to text. Modified from Strasser *et al.* (2004).

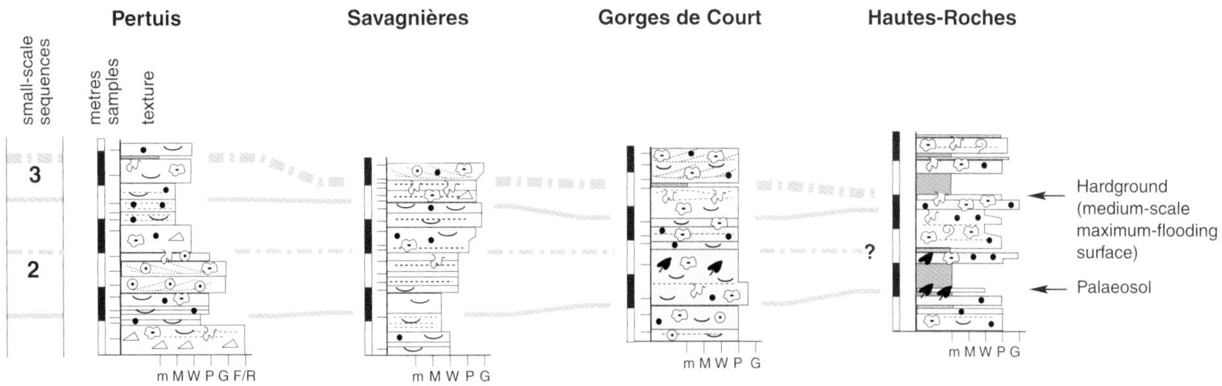

Fig. 8. Correlation of small-scale sequence 2 (compare with Fig. 6). While at Pertuis there is a well developed maximum-flooding surface separating the transgressive ooid shoals from the muddy lagoonal highstand, the transgressive deposits overlying a palaeosol at Hautes-Roches are rich in siliciclastics and the maximum flooding of the small-scale sequence is difficult to place. The maximum flooding of the medium-scale sequence, however, is expressed by a hardground within small-scale sequence 3. Symbols are as in Fig. 6.

facies contrast, it will not be expressed in the record. When accommodation is low, elementary sequences may not be deposited or eroded ('missed beats' of Goldhammer *et al.*, 1990). Consequently, the best chance of recording elementary sequences is during long-term transgression when accommodation space is created while carbonate production keeps up and the depositional environments stay shallow. For a detailed analysis we have therefore chosen small-scale sequence 9 (Figs 6 and 9), which formed at the beginning of a medium-scale (400-kyr) transgression (Hug, 2003).

Figure 9 presents a correlation between the five sections studied. Small-scale sequence 9 is composed of individual beds, which are separated by thin layers of marls. The number of beds varies from one section to the other. The identification of elementary sequences is most straightforward at Vorbourg, where small channels indicate shallowest water and thus sequence boundaries. The maximum-flooding surface of small-scale sequence 9 is implied by the change from high-energy to low-energy facies and by bioturbation. In the second and third elementary sequence, their maximum flooding can be identified by thin marly layers and subtle facies changes. The Hautes-Roches section displays coral carpets at the base. The top of a conspicuous ooid sand wave is interpreted as the maximum-flooding surface of small-scale sequence 9. In the following ooid dunes it is

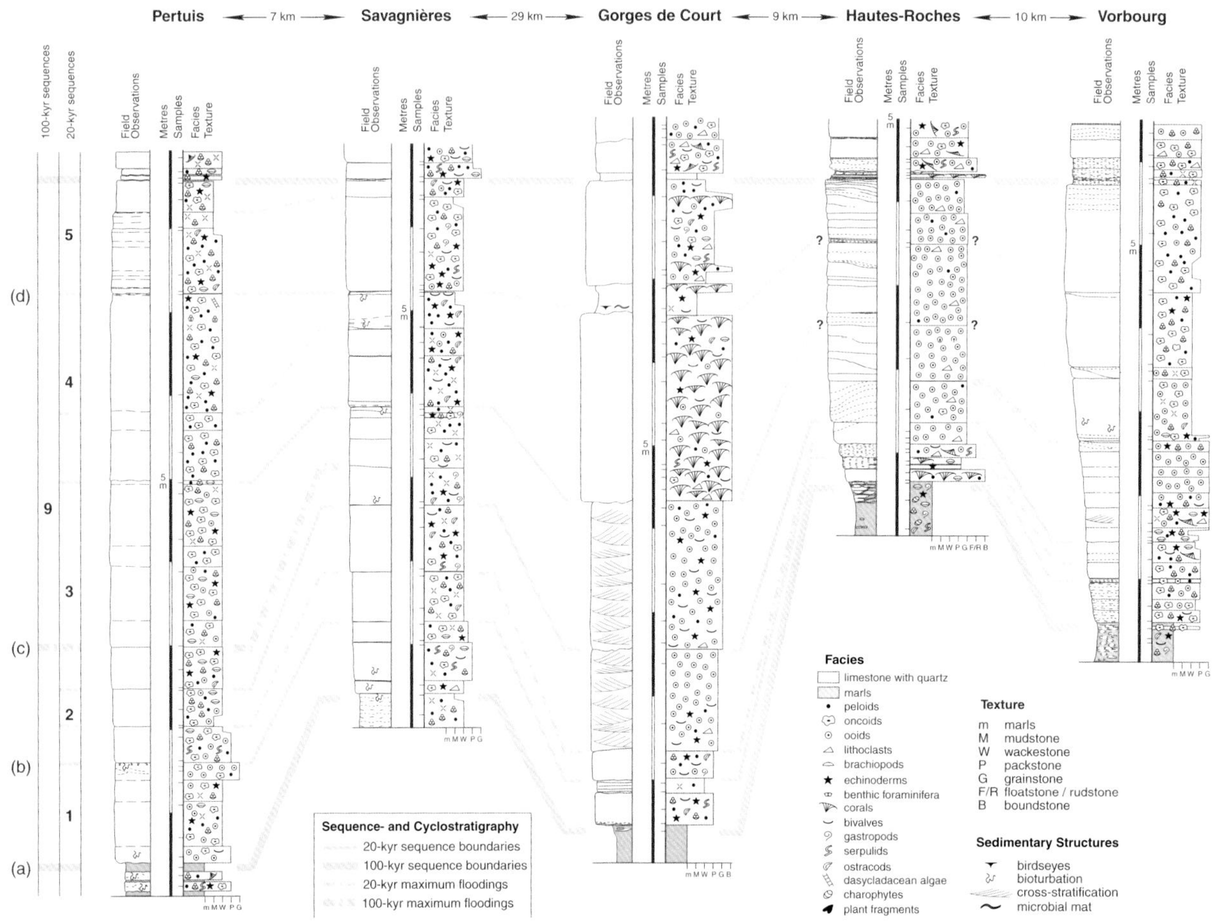

Fig. 9. Correlation of elementary (20-kyr) sequences that compose small-scale sequence 9 (Fig. 6). Only major facies elements and sedimentary structures are indicated in the lithological columns. a, b, c and d indicate time lines along which the reconstructions in Fig. 10 have been made. For discussion see text.

not possible to distinguish between reactivation surfaces and elementary sequence boundaries, and the correlation is speculative. In the Gorges de Court section, a thin interval with birdseyes and microbial mats covers a coral reef. This is attributed to the top of elementary sequence 4. The sections of Pertuis and Savagnières display only slight facies contrasts. The correlation of the elementary sequence boundaries is therefore guided by the most conspicuous bedding surfaces. However, maximum-flooding surfaces may also be expressed as bedding surfaces. This is explained by the ponding of clays during fastest relative sea-level rise, while the clays at sequences boundaries are washed onto the platform as a result of rapid sea-level fall (Strasser & Hillgärtner, 1998). As for the small-scale sequences (Fig. 6), the correlation at the level of the elementary sequences represents a best-fit solution.

CONTROLS ON FACIES DISTRIBUTION

The complex interplay of many factors controls carbonate-producing ecosystems, sediment distribution and the final sedimentary record; however, the most important parameters will be discussed separately.

Sea-level change

Facies evolution in many small-scale sequences clearly indicates that water depth, water energy and/or marine influence have increased then decreased during the deposition of one sequence (Figs 6 and 8). In other small-scale sequences, facies show a rather aggradational pattern, then a rapid shallowing at the sequence boundary. The fact that the boundaries and in some cases also the maximum-flooding surfaces of these sequences can be followed over tens of kilometres and that they are independent of facies is best explained by sea-level changes that affected the entire platform. Depending on the position on the platform and on the type of environment (Fig. 2), a sea-level cycle produced a deepening and shallowing or an aggradational pattern. Thicknesses of the small-scale sequences vary between 2 and 8 m (Fig. 6). If facies-dependent decompaction is applied (procedure described in Strasser *et al.*, 2004) this would correspond to about 3 to 15 m. The amplitude of a sea-level change that created a specific sequence must have been at least as high as this decompacted sequence if it ends in tidal facies, or somewhat higher if the sequence stayed subtidal throughout.

In many elementary sequences it is difficult to discern deepening and shallowing facies evolutions. However, bed limits may display features (e.g. hardgrounds, bioturbation, channelling) that indicate changes in the depositional environment (Fig. 9). Bed limits are caused by thin marly layers. The input of clays can be related to sea-level changes (erosion of soils during sea-level lowstands, ponding in deeper environments during maximum flooding), to more rainfall in the hinterland increasing fluvial transport, and/or to currents that redistribute the clays independently of sea level or climate (Strasser & Hillgärtner, 1998). In ooid shoals, bedding limits form through bedform migration and channelling. These limits are first of all controlled by tidal currents, and only the intensity of the latter may be related to sea level. Consequently, a bed of ooid grainstone seen in outcrop does not necessarily represent a cycle of sea-level change.

The hierarchical stacking of elementary, small-scale and medium-scale sequences and the biochronostratigraphic time control suggest that the formation of these sequences was in tune with the orbital cycles of precession, short eccentricity and long eccentricity, respectively (see above). The quasi-periodic perturbations of the Earth's orbit produce changes in the insolation received at the top of the atmosphere, which then translate into climate changes (Schwarzacher, 1993; De Boer & Smith, 1994; Matthews & Perlmutter, 1994). In the Late Jurassic, ice in high latitudes probably was present, but land-bound ice-volumes were too small to induce important glacio-eustatic fluctuations (Frakes *et al.*, 1992; Eyles, 1993; Valdes *et al.*, 1995). However, volume changes of alpine glaciers could have made a small contribution (Fairbridge, 1976; Valdes *et al.*, 1995). An important component in low-amplitude sea-level changes is thermal expansion and contraction of the uppermost layer of ocean water (Gornitz *et al.*, 1982; Church & Gregory, 2001). Contributions may also have been made by thermally induced volume changes in deep-water circulation (Schulz & Schäfer-Neth, 1998), and/or by water retention and release in lakes and aquifers (Jacobs & Sahagian, 1993). Consequently, it is assumed that Oxfordian high-frequency sea-level changes were coupled to climate changes, which themselves were linked to the Earth's orbital parameters. The above-mentioned mechanisms could easily produce the few metres of amplitude necessary for the accommodation of the observed depositional sequences.

On a shallow platform, even very low-amplitude sea-level changes can have dramatic effects. For example, a sea-level fall of a few tens of centimetres can lead to emersion and erosion of a reef that has previously built up to sea level, and ooid shoals can be forced to migrate laterally or will emerge to form islands. On the other hand, a small sea-level rise can flood a barrier that has hitherto protected a lagoon and produce a rapid change in water energy, salinity and ecology. Facies changes and depositional sequences can also be induced independently of sea level changes by lateral migration of sediment bodies controlled by currents (Pratt & James, 1986) or by progradation (Ginsburg, 1971). Such 'autocyclic' processes may lead to stacked sequences but will never be able to create a hierarchical stacking pattern as observed in the sections studied (Strasser, 1991). In the definition of individual elementary sequences, however, such processes must be considered.

Differential subsidence

The Swiss Jura in Late Jurassic times was part of the passive northern margin of the Tethys Ocean (Fig. 5). The tectonic regime was extensional, and a pattern of structural highs and depressions formed (Allenbach, 2001). Average subsidence rates are estimated at 20–40 m per million years (Wildi *et al.*, 1989). The thickness variations of the sections studied (Fig. 6) can therefore be attributed to differential subsidence of individual tectonic blocks. The general facies distribution in the sections studied suggests that the blocks were slightly tilted: coral reefs and ooid shoals developed on the higher margins towards the

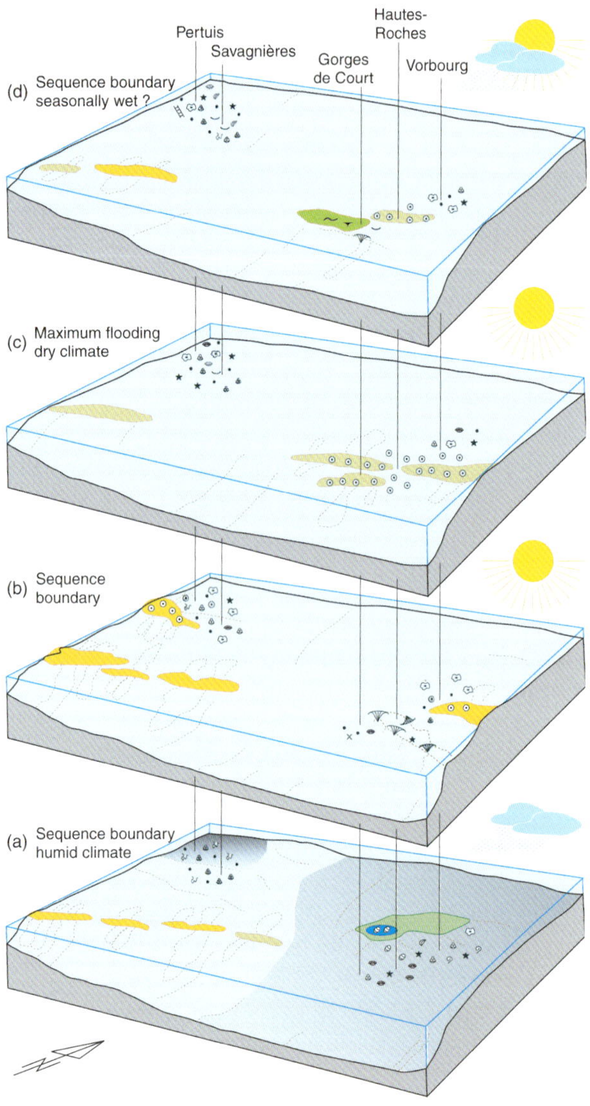

Fig. 10. Sketches of depositional environments along the time lines indicated in Fig. 9. Platform morphology has been reconstructed to be compatible with the evolution of facies and sequences observed in the sections studied. For discussion refer to text. No palinspastic correction has been applied: the distances between the sections may therefore be somewhat different from Late Jurassic times but their relative positions have not changed.

south, whereas lagoons formed in the depressions towards the north (Fig. 10). The relief created by tectonics therefore contributed significantly to the general facies distribution. Furthermore, the irregular distribution of siliciclastics (Fig. 6) can be explained by depressions, which were created by differential subsidence and served as conduits (Pittet, 1996; Hug, 2003). The preferential growth of reefs and the initiation of ooid shoals on tectonic highs further enhanced the relief.

Climate

General climate models for the Late Jurassic have been developed by Oschmann (1990), Frakes *et al.* (1992), Sellwood & Valdez (1997), or Abbink *et al.* (2001). These models, however, do not consider short-term climate variations. Orbital cyclicity causes changes in the total amount of insolation and in its seasonal distribution, which then result in latitudinal shifting of atmospheric circulation cells and concomitant changes in seasonality, rainfall pattern, oceanic circulation and sea-surface temperatures (Matthews & Perlmutter, 1994).

In the sections studied, siliciclastics are in many cases concentrated around small-scale sequence boundaries and associated with plant material (Fig. 6). This implies that during sea-level fall climate in the hinterland was rather humid, increasing the terrigenous runoff and allowing vegetation growth. Coral reefs and ooid shoals are situated preferentially in the trangressive parts of small-scale and medium-scale sequences in the absence of siliciclastics, suggesting a rather dry and warm climate. However, there are many exceptions to these trends, so that humid-dry changes can be considered as only one controlling factor among several others. There is no evidence for arid conditions as no evaporites have been observed in the sections studied. It is interesting to note that in Upper Oxfordian sections in Spain, in a more southern palaeolatitudinal position, the siliciclastics are preferentially associated with maximum flooding in small-scale sequences (Pittet & Strasser, 1998). This implies that rainfall there occurred during rising sea-level and places the Spanish sections in a different palaeoclimatic zone than those in the Swiss Jura.

The estimation of water temperature is difficult. Based on $\delta^{18}O$ analyses on modified whole rock, Plunkett (1997) derived an average palaeotemperature of 26–27°C for samples in the Vorbourg and Gorges de Court sections. The isotopic variations within individual depositional sequences, however, are too weak to be interpreted confidently as high-frequency temperature changes.

Ecology

The most conspicuous carbonate-producing ecosystems in the sections studied are coral patch reefs. They occur in different stratigraphic positions and the individual reef bodies never lived longer than the duration of an elementary sequence, i.e. 20 kyr (Figs 6 and 9). The Oxfordian reefs in the Swiss Jura have been studied extensively by Dupraz (1999) and Dupraz & Strasser (2002). Coral diversity is low to moderate. The corals are commonly bioeroded and covered by micro-encrusters (cyanobacteria, foraminifera, bryozoans, serpulids, sponges) and by microbialites. Conditions varied from oligotrophic allowing for healthy coral growth to meso- and eutrophic when nutrient levels were high and microbialite growth was favoured. Nutrient input was coupled with terrigenous input, which in addition diminished water transparency. This situation occurred preferentially when sea level was low (see above). Consequently, the trophic and photic conditions that controlled coral growth were directly linked to climatic and sea-level changes.

Micro-encrusters and microbialites also occur in oncoids, which are important components in the sections studied (Fig. 6). Other carbonate producers include echinoderms, foraminifera, brachiopods, bivalves, gastropods, ostracods and serpulids. Dasycladacean algae are rarely found, possibly due to early diagenetic aragonite dissolution. Charophytes occur only in the uppermost part of the Günsberg/Röschenz Member at Hautes-Roches and point to local development of freshwater ponds. Oncoids and the heterotrophic fauna are found in pure carbonates as well as in the facies containing siliciclastics, while the coral reefs flourished in the absence of siliciclastics.

Facies mosaics

Reconstructing the facies distribution on an entire platform and its evolution through time based on only five sections is of course a highly speculative enterprise. Nevertheless, there are certain clues that are given through the time lines established by the cyclostratigraphical analysis, and by the logical interpretation of the facies occurring along these time lines. For example, a wackestone with oncoids and ooids implies a low-energy lagoon, which was protected from the open ocean by an ooid shoal from where the ooids were washed in. If the ooids are corroded and/or encrusted, a time lag between ooid formation and deposition in the lagoon has to be postulated. For the reconstruction presented in Fig. 10, four time lines have been chosen: three sequence boundaries and a maximum-flooding surface (Fig. 9). According to our cyclostratigraphic analysis, the intervals between these lines represent 20, 30 and again 30 kyr, assuming in a simplified manner that the sea-level cycles were symmetrical and that the maximum-flooding surfaces formed 10 kyr after the sequence boundaries.

In order to best explain the observed features along the chosen time lines, an initial tectonic control has to be postulated, which determined the position of the reefs and ooid shoals at Hautes-Roches and Gorges de Court (Fig. 10). Periodically, there was also emersion in this area, as testified by charophytes at Hautes-Roches and tidal flats at Court. This topographic high was very close to the open-marine realm because stenohaline brachiopods and echinoderms are found together with charophytes. A barrier system of unknown nature must be assumed south of Pertuis and Savagnières, in order to produce the protected lagoonal facies there. Occasionally, ooid shoals developed close to these sites. At the time when sequence boundary (a) formed, climate must have been humid, and siliciclastics were shed onto the platform. Savagnières was less affected by this input, and a slight topographic high is assumed in that area. This high could be the result of differential subsidence but may have been enhanced by the accumulation of thick ooid shoals 300 kyr before (Fig. 6). Siliciclastic input then diminished rapidly, partly due to a drier climate, partly to sea-level rise at the beginning of a 400-kyr eccentricity cycle. Biogenic carbonate production was high because the ecological conditions were favourable, and ooids formed because the water was warm and well saturated. Sediment accumulation kept pace with rising sea level. Sea-level falls related to the 20-kyr precession cycle were attenuated by the general rising trend and inflicted only minor channel formation at Vorbourg. Hautes-Roches was dominated by ooid shoals, with a sand wave forming in a tidal pass. The development of a tidal flat with microbial mats on top of the coral reef at Gorges de Court possibly implies a seasonally wet climate at sequence boundary (d) (Wright & Burchette, 1996).

These reconstructions, albeit interpretative, clearly show that at any time the platform was structured and hosted several, juxtaposed depositional environments. Seafloor topography was created by differential subsidence but also by reefal constructions and accumulation of ooid shoals. The input of siliciclastics, triggered by falling sea level and by increased rainfall in the hinterland, slowed down carbonate production. It is therefore not a contradiction to postulate relatively deep lagoons at sequence boundary (a) and shallower environments at the time of maximum flooding (c) (Fig. 10). The source areas for the siliciclastics

Int. Assoc. Sedimentol. Spec. Publ. (2009) **41**, 215–238

The allocyclic interpretation of the 'Latemar Cycles' (Middle Triassic, the Dolomites, Italy) and implications for high-frequency cyclostratigraphic forcing

ROB M. FORKNER[*1], LINDA A. HINNOV[‡], ROBERT K. GOLDHAMMER[*†] and LAURIE A. HARDIE[‡]

[*]University of Texas at Austin, Jackson School of Geosciences, Department of Geological Sciences 1 University Station C1100, Austin, Texas, 78712, USA (E-mail: rob@metloef.com)

[‡]Department of Earth and Planetary Sciences, Olin Hall, 3400 North Charles Street, Johns Hopkins University, Baltimore, Maryland 21218, USA

ABSTRACT

The Middle Triassic Latemar Platform is an approximately 5 km wide, 700 m thick isolated carbonate platform succession containing over 500 cycles (<1 m average thickness), which have been attributed to allocyclic forcing by Milankovitch to sub-Milankovitch cycle driven composite eustasy. These interpretations are based on the facies composition of the cycles (thicker subtidal units overlain abruptly by thin, centimetre-scale, subaerial caps), the 5:1 bundling of the 'fundamental' metre-scale cycles into lower-frequency 'megacycles', and spectral analyses of thickness and rank series showing conspicuous matches to predicted Milankovitch periodicities. The influence of periodic Milankovitch cycle-driven composite eustasy on the development of the Latemar cyclic succession has been questioned based on the dating of ashfall tuffs, palaeomagnetic analysis of the platform, and correlation to biostratigraphic markers in nearby basinal deposits. In order to test the interpretation of allocyclic forcing for Latemar cycles, the cyclic succession preserved at Mendola Pass (located 30 km northwest of Latemar Platform) was investigated where Latemar-equivalent cyclic platform interior strata are well exposed. The Mendola cycles (average 0.70 m per cycle) are also bundled into upward-thinning packages with an approximate 5:1 ratio. However, unlike the subtidal deposits immediately overlain by vadose diagenetic caps found in the Latemar succession, Mendola cycles consist of a mud-rich subtidal unit gradationally overlain by a cryptomicrobial (peritidal) laminite cap. A measured section from Mendola Pass of 36 cycles correlates biostratigraphically and statistically to a unique interval within the Latemar succession. Although laminite-capped cycles are often attributed to autocyclicity, the similarity of the stacking patterns of the Mendola cycles to those of Latemar support an allocyclic interpretation. In addition, depositional rates calculated from dated Holocene shallow-water carbonate facies equivalent to those at Latemar and Mendola Pass are shown to be consistent with Milankovitch or multimillennial periodicities, rather than millennial (1 kyr) cycle periods. Finally, the question of whether comparative sedimentology can be relied upon for the interpretation of relative cycle duration is considered, or if comparative sedimentology has reached its useful limit for identifying facies and depositional environments with respect to the Middle Triassic. It is concluded that statistical and biostratigraphic correlations, in addition to comparative sedimentology, indicate that the Latemar and Mendola cycles were deposited under the control of an allogenic forcing mechanism, and that this mechanism generated depositional cycles with multimillennial individual periodicities.

Keywords Middle Triassic carbonates, cyclostratigraphy, high-frequency eustasy.

INTRODUCTION: THE PROBLEM

Investigation of the cyclic succession of platform carbonate deposits at Latemar began over 20 years

[†]Deceased.
[1]Present address: Shell International Exploration and Production B.V. Kessler Park 1, 2288 GS Rijswijk, The Netherlands.

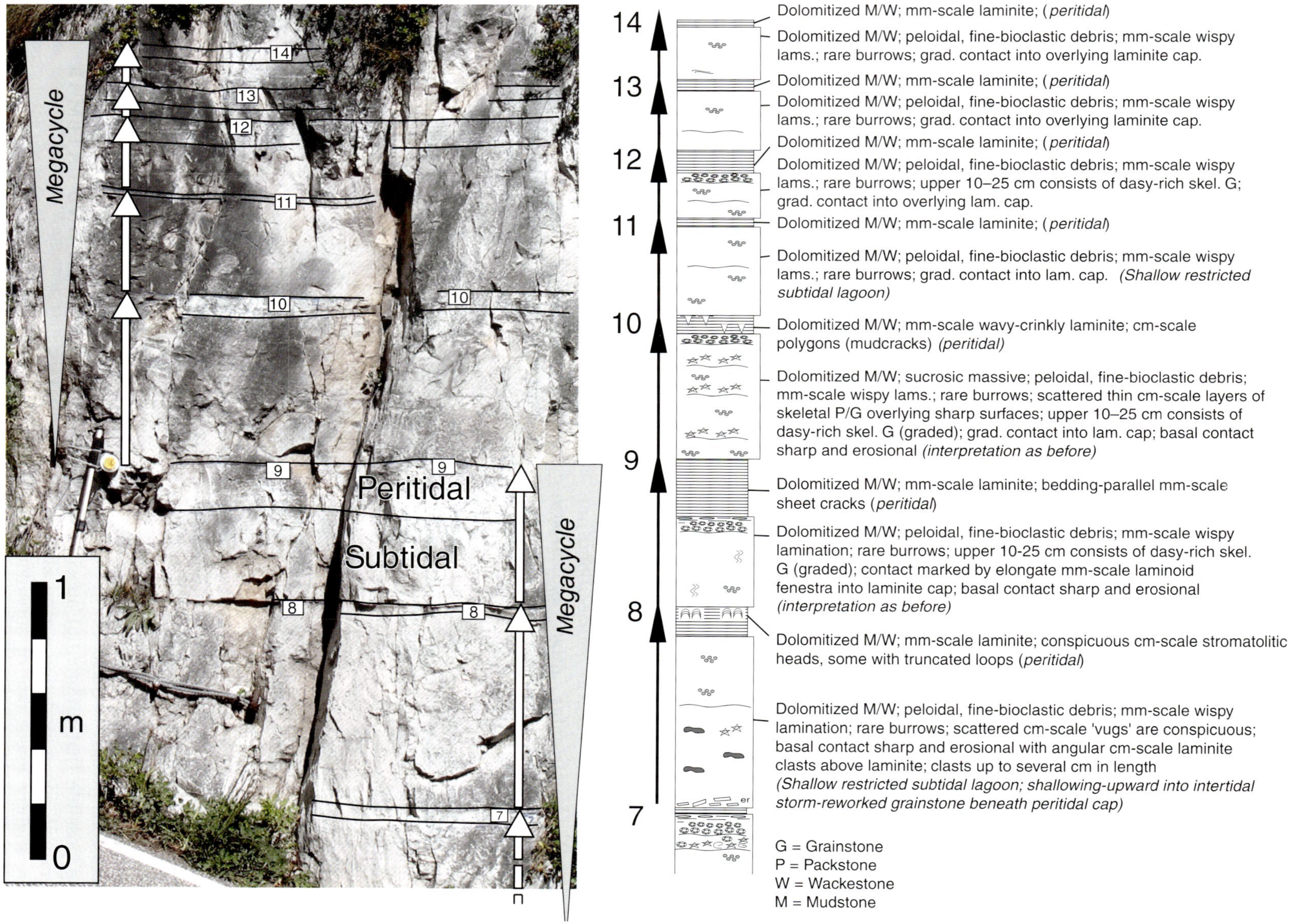

Fig. 5. Photograph of the Mendola road cut section showing cycles 7–14, as well as facies descriptions and megacycle bundling interpretation. Cycles are metre-scale, dolomitized, laminite-capped peritidal deposits probably formed in environments similar to modern carbonate tidal flats.

time little published data have become available with reference to its age. However, a recently published geological map includes clues to the age of the Mendola Dolomite (Avanzini, 2002). Notes from the map indicate that the 'Mendola Formation' is a platform carbonate body of late Anisian–early Ladinian age (Avanzini, 2002). Portions of the map's notes are paraphrased below (translated by N. Preto, pers. comm. 2006).

> "Within the Mendel Dolomit, Ogilvie Gordon (1927) recognized a lower member with dasycladacean algae *Diplopora anullatissima* and *Physioporella paucifoliata* corresponding to the Sarl-Dolomit, and an upper member with *Diplopora annulata* corresponding to the Schlern-Dolomit. These conclusions were later corroborated by von Klebelsberg (1935) and Van Hilten (1960). Further stratigraphic studies carried out for the purpose of geological mapping of the area distinguished two informal units within the Mendola Formation. The first is the subtidal member, corresponding to the Upper Serla Dolomite and Contrin Dolomite of the eastern Dolomites. The second is the peritidal member. The lower boundary of the peritidal member is marked by a palaeosol and sometimes associated with tepee horizons. The upper boundary is more variable. At Mendola Pass, strongly karstified portions are infilled with andesitic and basaltic volcanics that are laterally equivalent with the basinal Roen Limestones. Additionally, a few ammonoids of the *avisianum* subzone occur at the base of the peritidal member (Preto, pers. comm.). The sedimentary environment is that of an aggrading, periodically exposed carbonate platform."

The section measured in this study belongs to the 'peritidal member' of the Mendola Formation, which overlaps with the Latemar Platform interior biostratigraphy (Fig. 6). The assemblage of dasycladacean algae extracted from the Mendola section includes *Physoporella leptotheca* (Cycle 7), *Diplopora annulata annulata* (Cycles 7, 8 and 20) and *Gyroporella ladinica* (Cycle 20). Dasycladacean algae of the subtidal member of the Mendola Formation, including

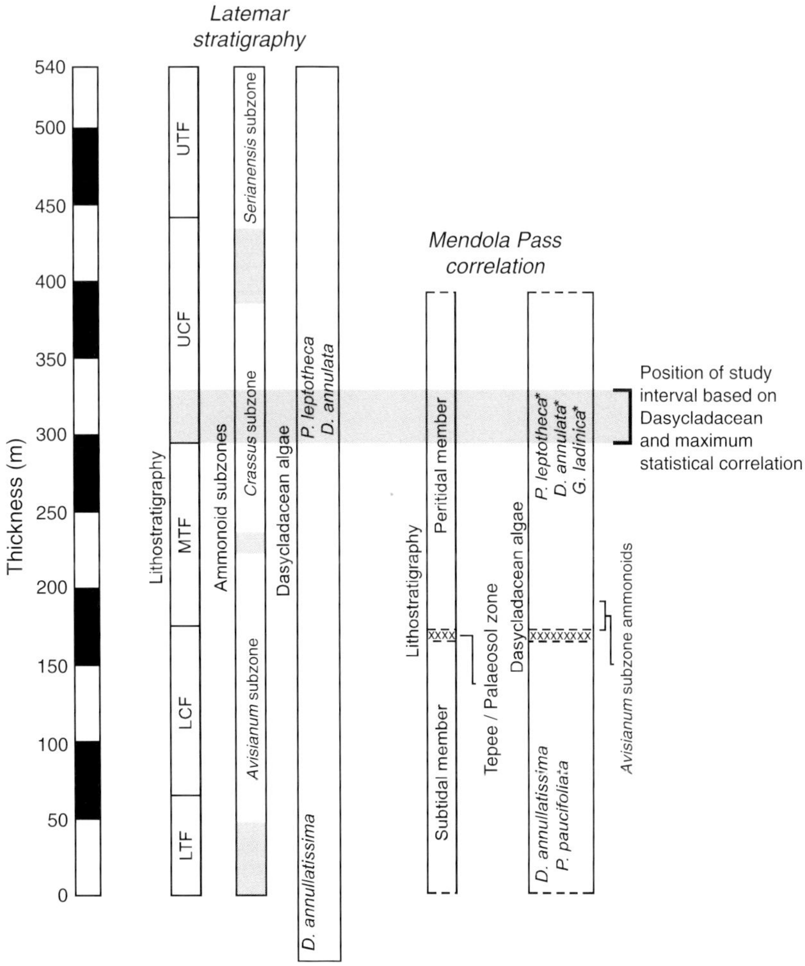

Fig. 6. Stratigraphic column of the Latemar Platform modified from Egenhoff *et al.* (1999), Preto *et al.* (2001) and Manfrin *et al.* (2005) on the left and proposed correlative stratigraphic chart from Mendola Pass to the right. Depth scale applies to both charts. Dasycladacean algae from Latemar section are reported in Gaetani *et al.* (1981). LTF, Lower Tepee Facies; LCF, Lower Cyclic Facies; MTF, Middle Facies; UCF, Upper Cyclic Facies; UTF, Upper Tepee Facies. Dasycladacean algae and lithostratigraphy from Mendola Pass are reported in Ogilvie Gordon (1927) and Avanzini (2002). Dasycladacean algae marked with an asterisk (*P. leptotheca, D. annulata,* and *G. ladinica*) were collected at the Mendola study interval, indicating placement in the 'Peritidal Member' of the Mendel Dolomit and providing a zone of possible correlation to the Latemar Platform based on the occurrence of *P. leptotheca, D. annulata* there. Correlation of Mendola study interval to Latemar section (marked by horizontal grey bar) is based on dasycladacean biostratigraphy and maximum statistical correlation of Mendola and Latemar cycle thickness series.

D. annullatissima and *P. paucifoliata* have not been found within the study interval (N. Preto and O. Piros, pers. comm. 2006; Piros and Preto, 2008). In terms of correlation to the Latemar, *D. annullatissima* occurs only in the lowermost part of the Latemar succession (Gaetani *et al.*, 1981) and *P. paucifoliata* is even older. However, *D. annulata* was found within the study interval and is consistent with the biostratigraphy of Ogilvie Gordon (1927) for the upper peritidal member of the Mendola Formation. The lowermost portion of the peritidal member is reported to contain ammonoids of the *avisianum* subzone indicating a late Anisian age, which is consistent with a 175-m-thick succession of cycles in the Latemar that includes the upper portion of the Lower Tepee Facies, the entirety of the lower cyclic facies, and the lower half of the Middle Tepee Facies (see Manfrin *et al.*, 2005, p. 480, Fig. 3). The upper portion of the peritidal member of the Mendola Formation may correlate to portions of the Latemar stratigraphy above this interval, including the upper portion of the Middle Tepee Facies and/or the upper cyclic facies. In sum, while biostratigraphy does not tie the Mendola succession to a specific interval within the Latemar, it does indicate that there is a biostratigraphic overlap of the Mendola and Latemar cyclic successions.

STATISTICAL CORRELATION OF THE MENDOLA AND LATEMAR SECTIONS

Statistical time-series analysis of the Latemar and Mendola cycles sheds light on the stratigraphic patterns in the sections and their interrelationships. *Running cross-correlation analysis* pinpoints the most likely position of the Mendola section within the Latemar extended cyclic stratigraphy. Importantly, this sets up a common time origin for the two sections and an opportunity to investigate specific cyclic patterns shared by the two sections.

The assessment of stratal frequencies (i.e. bundling frequencies) common to the two sections is extremely challenging due to the short length of the Mendola measured section (36 cycles). Multiple graphical and statistical signal analysis techniques are needed to focus on different aspects of the Mendola section's stratal patterns and its corresponding Latemar counterpart. To present a full range of views, Fischer plots, harmonic analysis, spectral coherency and cross-phase analysis were used as follows.

Running cross-correlation analysis

To study the similarity of stacking between cyclic sections, a running cross-correlation procedure may be applied. This technique was developed by Anderson & Kirkland w(1966) to correlate intrabasinal sequences of varves in the Permian Castile Formation. Subsequently, the technique was formalized in Dean & Anderson (1974). Running cross-correlation was later used by Holland *et al.* (2000) as an independent means to correlate regional stratal patterns and to corroborate biostratigraphic correlations.

Preto *et al.* (2004) applied this 'semi-automated' technique to the Latemar Platform, successfully correlating their Cimon del Latemar cycle thickness series to the Cima Forcellone cycle thickness series. The results are replicated here in Fig. 7a and give a perfect match for the 'marker horizon M' that Egenhoff *et al.* (1999) observed in their field measurements of both localities (Fig. 7c).

Fig. 7. Running cross-correlation analysis of cycle thickness series within the Latemar Platform, and between Mendola Pass and the Latemar Platform. The 95% confidence limits were estimated by interpolating the '95% Confidence Belt' of Taylor (1990, Table 8.6, p. 139) to sample sizes (i.e. degrees of freedom) of 180 (for intra-Latemar correlation) and 36 (for Mendola–Latemar correlation). (a) Pearson correlation coefficients obtained by progressively shifting the Cimon del Latemar cycle thickness series of Preto *et al.* (2001, 2004) relative to the cycle thickness series at Cima Forcellone of Goldhammer *et al.* (1987), with 57 additional cycle measurements within the Tepee Facies from Dunn (1991) (see Appendix 1 for details). A singular peak correlation occurs between the Cimon del Latemar series and Cima Forcellone Cycles 356 to 504. (b) Pearson correlation coefficients obtained by progressively shifting the Mendola Pass cycle thickness series (this study) against the Cima Forcellone cycle thickness series. Peak correlation occurs at a unique interval of 36 cycles in the Cima Forcellone series, Cycles 351 to 386, which corresponds to the lowermost 30 cycles of the Cimon del Latemar series. In both (a) and (b) the 95% confidence limits widen at the top of the section as the shifting shorter series (Cimon del Latemar or Mendola) rolls off the upper end of the longer (Cima Forcellone) series, resulting in ever-decreasing degrees of freedom in the calculated coefficients. (c) Detail highlighting the relative positions of the correlated intervals with respect to Cima Forcellone.

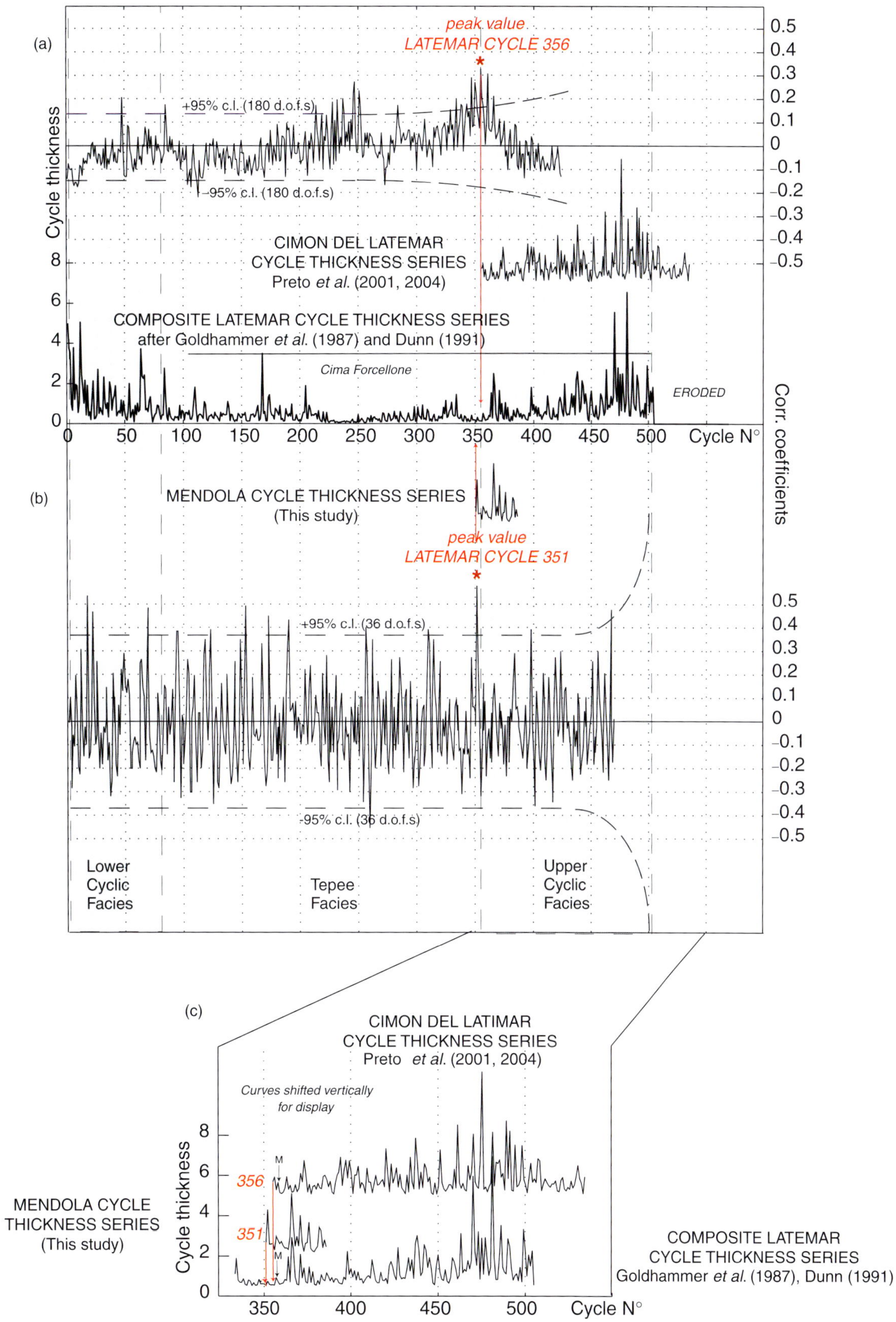

Fig. 7. See opposite page for caption.

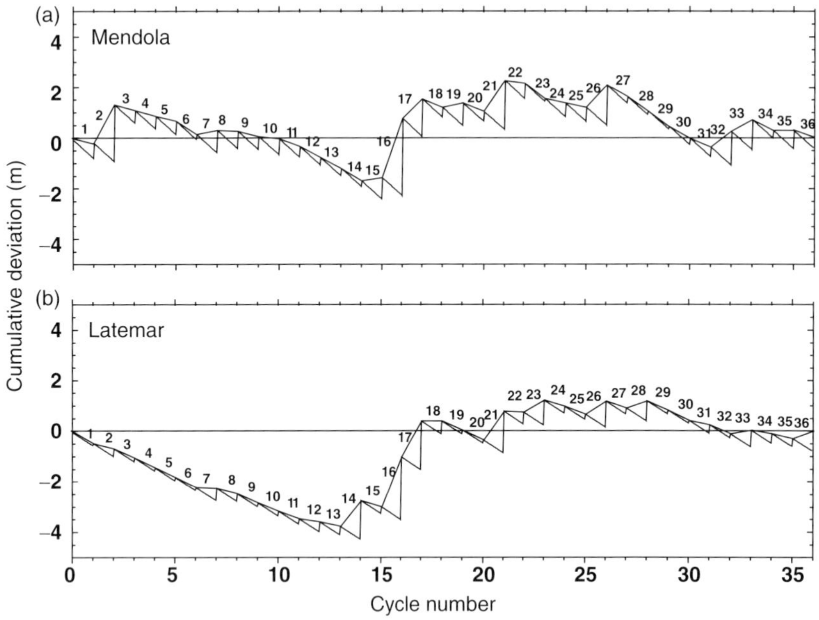

Fig. 8. Accommodation (Fischer) plot scripts were calculated using the Fischer plot script (see Appendix 2 for MATLAB script). Accommodation plots presented here are for (a) the Mendola Pass cycle thickness series (Cycles 1–36) and (b) the correlative Latemar cycle thickness series (Cycles 351–386 from the Cima Forcellone section, see Fig. 7).

The same technique was performed between the Mendola and Cima Forcellone cycle thickness series, shown in Fig. 7b. With the Mendola series being only 36 points long, the running cross-correlation values are more variable ('noisier') than the comparatively highly organized cross-correlation curve between the two Latemar localities (compare Fig. 7a and 7b). In addition, the Mendola Pass locality is around 30 km distant from the Latemar Platform, and may be associated with a separate tectonic block, if not a unique subsidence history, as several of the carbonate platforms of the Mid-Triassic of the Dolomites are reported to be (Doglioni, 1987, 1988; Brack & Muttoni, 2000). This alone would be expected to minimize the chances for a positive Mendola–Latemar correlation at the metre-scale level of this analysis. Nonetheless, an extremely high correlation coefficient value of +0.6, greatly exceeding the 95% confidence limit of 0.375 for 36 degrees of freedom, occurs between the Mendola and the Cima Forcellone series from Cycles 351 to 386. This strongly supports the idea that this is the stratigraphic placement of the Mendola section within the Latemar Platform.

Fischer plots

Charts plotting the cumulative deviation from mean cycle thickness (i.e. 'Fischer plots', after Fischer, 1964) highlight relative accommodation trends in platform cycle successions. The Fischer plot of the Mendola section (Fig. 8a) shows clear bundling of cycles into megacycles with an average grouping of five fundamental cycles into one megacycle (5:1). The Fischer plot of the correlative Latemar section (Fig. 8b) is remarkably similar, with both bundling and overall stacking trends synchronized with those at Mendola Pass. Both Fischer plots suggest a regressive interval over the first 15 cycles (cycles thinning upward). This trend is particularly pronounced in the Latemar section and is followed by the deposition of a 2.5-m-thick tepee-capped cycle (i.e. Cycle 366; see Goldhammer, 1987), which is the thickest cycle in the study interval. The equivalent cycle at Mendola Pass is also the thickest in the study interval (3.06-m-thick), but is laminite-capped. The remaining, upper 21 cycles are the most strongly bundled in both sections and both have a long-term trend suggestive of transgressive to highstand conditions.

Harmonic analysis, coherency and cross-phase spectra

The full suite of stratal patterns in the sections is revealed through statistical time-series analysis. The multitaper method, in particular, is an exceptional tool for frequency assessment of short, highly noisy time-series (Thomson, 1982). The multitaper method harmonic analysis procedure evaluates signal-to-noise ratios in spectra, and searches for statistical outliers indicative of the presence of lines, at ultra-high-frequency resolution. The method also provides averaged spectral coherency and cross-phase estimators that outperform others, harvesting more information (degrees of freedom) from the data, and with less bias, than other methods (see Hinnov & Goldhammer (1991) for an example pertaining to the Latemar cycles).

The multitaper method harmonic analysis of the Mendola and Latemar sections (Fig. 9a,b) shows that some of the bundling frequencies are shared by the two sections, although their relative amplitudes are quite different. This may be due to different, very low-frequency tectonic influences on the development of the Mendola versus Latemar cycles or the presence of differential cycle thicknesses resulting from normal sedimentary processes (e.g. lateral sediment erosion and transport) on Mendola tidal flats. The lowest frequency bundling term, Mendola at 1/18.9 and Latemar at 1/27, while measured with a high *F*-test in both sections, is based on only two repetitions of the term in the 36-cycle-long Mendola section and less than two repetitions for the Latemar section, thus may not represent a true, sustained bundling at this frequency. The next to lowest bundling term, Mendola at 1/8.5 and Latemar at 1/9.8, is only 1/8.5–1/9.8 = 1/64 apart, i.e. less than the elementary bandwidth $\Delta f = 1/36$. Thus, this line may represent the same process in both sections.

Examination of the Fischer plots (Fig. 8) indicates that one source of this peak in both of the sections appears to involve a single bundle of 8–9 cycles (cycles 7–15). Surprisingly, the 5:1 bundling frequency has a very low *F*-test (60–75%) in both sections. In both, the line is slightly broader than Δf, and in the Latemar data it has a substantially lower amplitude (Latemar 0.2 m, Mendola 0.35 m). This suggests that this bundling component, while quite visible in the Fischer plots of both series, is somewhat variable about 5:1, causing the *F*-test to fail. Finally, both sections show a bundling at 1/2.4. Again, these bundles can be observed in the Fischer plots (Fig. 8, e.g. Mendola cycles 19–20, or Latemar cycles 14–15) and some of the power at this frequency could also be a spectral artefact from asymmetry in the 5:1 bundling.

Magnitude-squared coherency and cross-phase spectral analysis identify significant frequency components shared by two time series. For example, frequencies present in both series that are consistently coordinated over the duration of the series. The magnitude-squared coherency spectra measure correlation coefficients between time-series as a function of frequency. To maximize the degrees of freedom of magnitude-squared coherency estimates, averaging over two or more sampled frequencies is necessary. Cross-phase spectra indicate the phasing of frequency components between the two series. For example, when cross-phase registers 0° at a particular frequency, the two series are said to be 'in-phase' at that frequency.

In Fig. 9c, again 2π multitapers have been applied, which preserves the narrowest possible averaging bandwidth ($\beta = 3\Delta f$), while gathering a substantial 5–6 degrees of freedom across most of the frequency range (see degrees of freedom curves, Fig. 9a,b). To be on the conservative side, the significance levels for zero coherence for 5 degrees of freedom. Three bands of non-zero coherence are present, centred at 1/10, 1/5 and 1/2.4. The double-peak nature of the magnitude-squared coherency at 1/10 could be artefacts from excessive weighting imposed by one of the higher order 2π multitapers on the outer edges of the averaging bandwidth β and magnified by the frequency interpolation (zero-padding). The other striking magnitude-squared coherency peak is centred at $f = 1/5$, which indicates that 5:1 bundling is present and coordinated between the series at a statistically significant level.

The cross-phase spectrum (Fig. 9d) can be interpreted at frequencies where magnitude-squared coherency values indicate significant non-zero coherence. The cross-phase of 20° at 12:1 bundling may indicate that the two series are phase-shifted relative to each other at this frequency. In the 'time' domain this would correspond to a shift of $(20°/360°) \times 12 = 0.67$ cycles, i.e. less than a one cycle lead of the Latemar series over the Mendola series. Similarly, at 9:1, the Latemar series leads the Mendola series by around 1 cycle; at 5:1, Mendola leads the Latemar series by around 0.5 cycles; and at 2.4:1, the two series are statistically in-phase. Taken together, the Mendola and Latemar series show close coordination to within +1 cycle at all significant bundling frequencies, constituting powerful evidence of a broadband statistical fit and phase lock between them.

RESULTS

The following results were derived from the analysis of the cyclic successions.

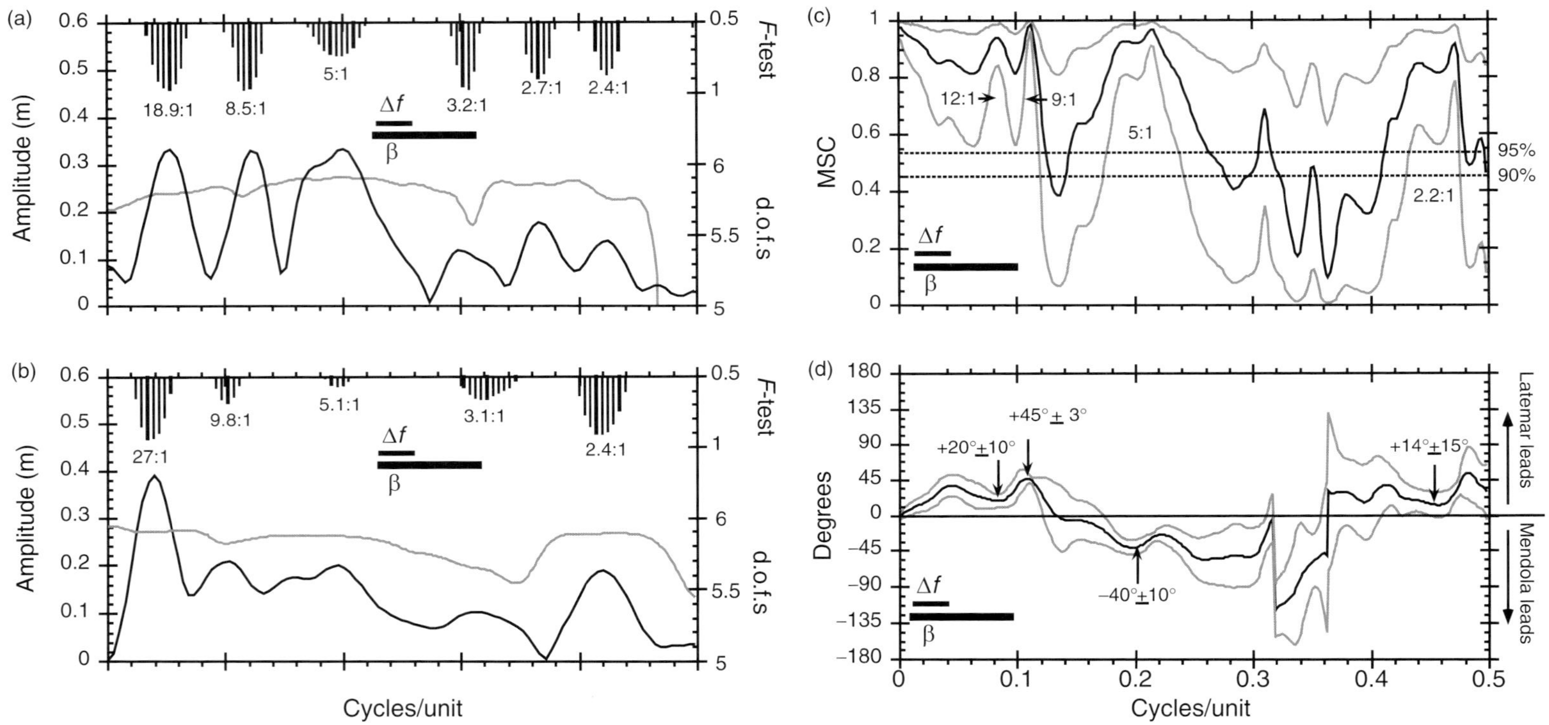

Fig. 9. The Multitaper Method spectral analysis of the Mendola and Latemar cycle thickness series using three 2π multitapers (orders 0, 1 and 2) with adaptive weighting (Thomson, 1982). Both series have been zero-padded from their initial length of 36 to 500, resulting in interpolated spectra with 250 estimated dimensionless frequencies. However, the elementary sampling frequency remains at Δf=1/36=0.028, and the 2π multitapering has the effect of averaging over $\beta=3\Delta f$=0.083 (see horizontal lines). Harmonic analysis of (a) Mendola series and (b) the Latemar series. For both (a) and (b), the black curves are the line (or amplitude, or first moment) spectra, and refer to the *y*-axis to the left; the grey lines are effective degrees of freedom (or adaptive weighted d.o.f.), and refer to the *y*-axis on the right (bottom). The bars descending from the top are *F*-test results for the lines, reported as significance (50–100%), and refer to the *y*-axis on the right (top). (c) Magnitude-squared coherency spectrum of the Mendola versus Latemar series. The horizontal dashed lines labelled 90% and 95% indicate significance levels for zero coherence for Magnitude-squared coherency estimates of normally distributed variables with 5 d.o.f. (Carter *et al.*, 1973). The black curve gives the estimated Magnitude-squared coherency values, corrected for the statistical bias imposed by 5 d.o.f. (Carter *et al.*, 1973); the grey curves indicate the 95% confidence limits. Labels indicate bundling components at frequencies with lower than 95% confidence limits exceeding the 95% level of zero coherence. (d) Cross-phase spectrum. The black curve indicates estimated cross-phase; the grey curves delineate the 95% confidence limits after Bendat & Piersol (1986). For cross-phase values less than 0°, Mendola leads Latemar; for values greater than 0°, Mendola lags Latemar. Labels identify the phase relations associated with statistically significant (>95%) peaks in the Magnitude-squared coherency spectrum depicted in (c).

Shallowing-upward theme of the cycles

The Mendola and Latemar cycles are late Anisian–early Ladinian-aged shallowing upward, metre-scale peritidal platform carbonate cycles that formed as a result of repetitive environmental changes over time. The presence of the shallowing-upward cycles allows for physical measurement and statistical analysis of their stacking trends (i.e. trends in the vertical arrangement of cycle thicknesses). The cyclic successions have stacking patterns that not only appear similar to the eye (Fig. 8), but cross-correlate with a correlation coefficient exceeding 95% (Fig. 7), suggesting a shared external driver. Site-specific controls (e.g. autogenic sedimentary processes, local subsidence) exerted comparably less influence on the development of the cyclic successions, but could explain the differing average cycle thicknesses between the successions (the average cycle thickness of the Latemar interval is 0.52 m; the average cycle thickness of the Mendola interval is 0.73 m).

Recording of cyclic processes

Both Mendola and Latemar successions formed in environments conducive to recording rhythmic, cyclic sea-level oscillations. The shallow-marine realm (0–5 m water depth) is almost certainly the depositional environment in which these deposits formed, where repeated flooding and exposure of the sediment–water interface was recorded within each depositional facies succession. This is especially true at Mendola Pass where facies within shallowing-upward cycles are consistent with those found on modern carbonate tidal flats (Hardie & Ginsburg, 1977). Strong and consistent sedimentological evidence for early subaerial exposure of Latemar cycle caps provides evidence for a similar facies–water-depth relationship at Latemar (Preto *et al.*, 2004).

The significant statistical correlation between the Mendola and Latemar stacking patterns provides additional support for the same sea-level oscillations affecting both depositional environments. This observation is important because constantly deep-submerged environments, as favoured by Blendinger (2004) for the Latemar cycles, would 'miss beats', not have a non-random bundling pattern, and not be readily correlatable to a clearly peritidal counterpart at Mendola (Preto *et al.*, 2005). The presence of Mendola cycle cap laminites and Latemar cycle cap dolomite crusts indicating subaerial exposure during sea-level fall means that Latemar deposition was not only able to 'keep up' with subsidence, but also experienced extended periods of time with the platform top in peritidal and supratidal conditions; environments not known for exceptionally fast accumulation rates.

The case for allocyclicity

The fact that correlated cycle stacking patterns are recognized at separate locations that are constrained by biostratigraphy is strong evidence for the existence of a common allocyclic mechanism driving the development of both successions. The presence of allocycles at Latemar and Mendola Pass sections requires that mechanisms existed during the Mid-Triassic that acted to generate both sea-level oscillations and resultant carbonate depositional cycles at both localities. In the case of the Latemar controversy, it can be established to a first order that shallowing-upward depositional cycles have been identified and cross-correlated to the Mendola Pass section, implying that the driving mechanism is at the very least regional in its stratigraphic influence.

DISCUSSION

Comparative sedimentology

Comparison of Holocene and Pleistocene dated allocycles to the more distant past record is useful for several reasons. First, composite sea-level oscillations regarded as drivers for shallow-marine carbonate sedimentation in the Holocene and Pleistocene are grouped into five approximately 20 kyr oscillations per 100 kyr oscillation (Broecker *et al.*, 1968; Bloom *et al.*, 1974), a bundling pattern that conspicuously matches the cycle stacking at Latemar and Mendola Pass successions. The fact that sea-level oscillations within the Milankovitch band form dateable carbonate depositional cycles demands consideration. Second, if the composite allocyclic forcing mechanisms that formed depositional cycles in the Mid–Late Triassic are still active today, then those same mechanisms ought to be recorded as depositional cycles today. Discovery of modern or recent depositional cycles with millennial periodicities and bundling patterns similar to that of the Latemar pattern would better point the cyclostratigraphic community towards identifying the correct forcing mechanism. Third, shallowing-upward facies successions similar to those found at Mendola Pass can be found in Florida Bay, the Bahamas and, to an extent, in the Persian Gulf. By comparing the rates at which these deposits form in modern environments, one may better understand average cycle durations for similar deposits in the geological past.

Recent publications offer new insights into sea-level change during the past 20 kyr. Two compilations are provided here. Table 1 summarises Pleistocene and Holocene cyclic climatic processes operating at sub-Milankovitch periodicities. While many types of variations are listed, not every process is tied (observationally or otherwise) to metre-scale sea-level change at these periodicities, and none are tied to the formation of shallowing-upward depositional cycles in modern marine carbonate sedimentary settings. Table 2 compiles literature involving the dating of Pleistocene and Holocene carbonate shallowing-upward successions. This suggests that in most cases, the sediments making up the Holocene carbonate successions, most of which are not yet complete shallowing-upward cycles, formed over the past 5–7 kyr as the result of a longer-term sea-level rise over the past 10–20 kyr. While a few examples of dated shallowing-upward facies successions from modern environments appear to have formed within the past 1000 years (see Strasser & Samankassou, 2003), dating and analyses of Holocene and Pleistocene carbonate successions from around the world have not yet produced a correlatable stack of metre-scale shallowing-upward carbonate depositional cycles with millennial (1–2 kyr) periodicity.

Holocene and Pleistocene 'icehouse' conditions, so-called because of the presence of major land-locked ice sheets, store and release large amounts of seawater, and allow for high-frequency

Table 1. Compilation of dated modern and recent cyclic climatic processes and behaviours with millennial periodicities.

Study	Cycle type	Cycle period
Roth and Reijmer (2005)	Oxygen isotope excursions in a 30 m-long core from the leeward margin of the Great Bahama Bank related to aragonite content	Multi-millenial 1.725 kyr 0.54 kyr
Niggemann *et al.* (2003)	Oxygen isotope excursions in 61-cm stalagmite	*c.* 1.450 kyr
Bond *et al.* (1993, 1997, 2001)	Oxygen isotope excursions and ice-rafted sediment cycles	1.5 kyr DO cycle 5–6 kyr Bond cycle
Munk *et al.* (2001)	Tidal forcing – Millennial Scale	1.795 kyr
van de Plassche *et al.* (1998)	Mean high water marks from Hammock River marsh, Clinton, Connecticut, USA	Century scale

Table 2. Compilation of dates of modern carbonate shallowing-upward (SU) facies successions. Most have not yet filled accommodation and none of these examples exhibit stacking of shallowing-upward facies successions at millennial periodicities.

Study	Locality	Depositional cycle type	Age of most recent SU succession
Strasser and Samankassou (2003)	Florida Bay, Bahamas, Bermuda	Variable, Shoaling, tidal flat, and subtidal facies successions	Variable oldest 5630 BP youngest 680 BP
Gischler (2003)	Belize platforms	Variable, Reefal, Shoaling and subtidal facies successions	4.5 kyr
Parkinson (1989)	Southwest coast of Florida	T–R cycle, capped by red mangrove peat	*c.* 7 kyr
Tudhope (1989)	Davies Reef, central Great Barrier Reef complex	Shallowing-up from: gravel lag; bioturbated muddy sand; shoal or exposure cap	Mid-cycle sediments dated at *c.* 3 kyr. Actual cycle must be older
Logan *et al.* (1969)	Shark Bay, Western Australia	Shallowing-up from: grainstone; skel. M/W; laminite cap	*c.* 5 kyr
Taft *et al.* (1968)	New Providence Platform, Bahamas	Coarsening-up from mud; skel. P/G; grapestone GS	6.7 kyr

Note: SU, shallowing upward; T–R, transgressive–regressive; M/W, mudstone/wackestone; P/G, packstone/grainstone; GS, grainstone.

sea-level oscillations with amplitudes of several tens of metres. As an example, dating of coral reef terraces in Barbados by Bloom *et al.* (1974) suggests that six major sea-level oscillations have occurred over the past 130 kyr, with amplitudes of 20–100 m (Bloom *et al.*, 1974; Chappell & Shackleton, 1986). Shallow-marine carbonate cycles produced by sea-level oscillations of these magnitudes are also comparably thicker than those found in the Alpine Triassic (Goldhammer *et al.*, 1990). Pleistocene depositional cycles identified in Florida and the Bahamas are of the order of 5–15 m thick, and consist of subtidal carbonate capped by laterally correlative red soil crusts (Goldhammer & Kaufman, 1995). Both modern carbonate shallowing-upward successions as well as the record of Pleistocene and Holocene sea-level oscillations operate at cyclic frequencies that are in tune with known orbital periodicities (Table 2; Logan *et al.*, 1969; Bloom *et al.*, 1974; Tudhope, 1989; Parkinson, 1989; Goldhammer & Kaufman, 1995; Strasser & Samankassou, 2003).

The Early to Mid-Triassic is generally considered to have been a period of greenhouse (or possibly transitional icehouse–greenhouse) climate, lacking major continental ice sheets, and is therefore thought to have been characterized by comparably low-amplitude (10 m or less), high-frequency sea-level oscillations (Vail *et al.*, 1977; Goldhammer *et al.*, 1990; Wright, 1992; Read, 1995). Consequently, shallowing-upward depositional cycles are relatively thin (1–2 m), but in many cases still form stacks that seem to reflect orbitally driven composite eustasy. Many workers have suggested that Triassic depositional cycles formed as the result of Milankovitch-band high-frequency–low-amplitude eustatic oscillations related to rhythmic orbital perturbations that affected the spatial and temporal distribution of solar energy at the Earth's surface (Fischer, 1964; Schwarzacher, 1975, 1993, 2005; Goldhammer *et al.*, 1990; Goldhammer & Kaufman, 1995; Yang & Lehrmann, 2003; Maurer *et al.*, 2004; Zühlke, 2004; Cozzi *et al.*, 2005).

The geo-climatic conditions of the Holocene and Pleistocene stand in contrast to those of the Early to Mid-Triassic. However, observations from modern sediments suggest that cyclic climatic perturbations with millennial periodicities are not being recorded as shallowing-upward carbonate depositional allocycles, even in areas where carbonate sediments have infilled available accommodation space (e.g. tidal flats of Andros Island). Instead, most shallowing-upward facies successions dated from the Holocene and Pleistocene seem to have formed with periodicities that are commensurate with Milankovitch and/or other multimillennial processes. It is certainly possible that the processes driving Triassic depositional cycles were unique to that period of geological time and that comparison between modern and ancient records is not warranted. However, the presence of Triassic allocycles that have an apparent statistical profile that is consistent with Milankovitch orbital parameters, a link that is temporally justified through comparative rate studies of Holocene facies successions, leaves two main options regarding their origin discussed in the following sections.

The successions of carbonate depositional cycles at Latemar and Mendola Pass sections record periodic, allogenic processes directly linked to Milankovitch cycle driven composite eustasy

The case for Milankovitch cycle driven composite eustasy forcing the deposition of either fundamental cycles and/or megacycles at Latemar and Mendola Pass sections is substantial, and is based on extensive sedimentological and statistical analyses. While the two main models supporting the presence of Milankovitch composite eustasy involve different timescales (see Fig. 1), both studies strongly argue for the presence of multimillennial astroclimatic forcing in the development of carbonate depositional cycles. Here, it is recognized that both the models of Goldhammer *et al.* (1987) and Zühlke *et al.* (2003) argue that the development of Latemar cyclicity is linked to orbital forcing.

From a comparative sedimentological standpoint, Latemar and Mendola cyclic stratigraphy agrees with known parameters of Milankovitch band driven sedimentation. Dating of shallowing-upward successions from modern carbonate settings indicates that most have formed (most of which are not yet complete 'cycles') over the past 5–7 kyr in response to Holocene sea-level rise (Table 2). The fact that many platforms have not yet aggraded to sea level and/or formed exposure caps suggests that the periodicity associated with the cyclic driver is longer than the time in which the depositional cycle actually forms. Additionally, sea-level oscillations inferred from coral reefs and deep-sea sediments confirm that five oscillations with amplitudes of 20–100 m occurred over the past 100 kyr (Broecker *et al.*, 1968; Bloom *et al.*, 1974; Chappell & Shackleton, 1986). The frequencies of these oscillations match those of the precession and eccentricity cycles, and they provide a driver for the bundling of depositional cycles that is fundamentally linked to carbonate sedimentation.

Additional sedimentological evidence for Milankovitch band forcing of depositional cycles relates to the rates at which complete carbonate cycles form, particularly vadose diagenetic facies. While the cycles measured at Mendola Pass do not have diagenetic caps, many of the cycles in the equivalent Latemar succession do have well-developed diagenetic caps. The presence of these diagenetic caps must be accounted for in a temporal sense for the Latemar cycles and in a time-correlative sense for Mendola cycles. If Holocene examples are used as a temporal benchmark, rapid, millennial-scale sea-level oscillations do not provide adequate time for observed vadose diagenetic features to form. Radiocarbon-calibrated rates of Holocene caliche formation average 2–4 cm kyr^{-1} (James, 1972; Robbin & Stipp, 1974; Handford *et al.*, 1984; Demicco & Hardie, 1994), and several studies suggest even lower rates, at 0.4–1 cm kyr^{-1} (Lucia, 1968; Davies, 1970; Evamy, 1973). The caliche that caps Latemar cycles ranges in thickness from 1 to 30 cm, averaging 6 cm in thickness (Goldhammer *et al.*, 1987). In order for diagenetic caliche caps to form within a millennial cyclic framework (entire cycle in 0.9–1.97 kyr), rates of caliche formation must operate at least an order of magnitude faster (20–40 cm kyr^{-1}), if not more, than those in the Holocene. Additionally, studies of Holocene tepees at Lake MacLeod, Western Australia by Handford *et al.* (1984) show that sediment buckles and is tilted into tepee antiforms at an average rate of 4 cm kyr^{-1}, and void-filling aragonite cements within tepees grow at extremely slow rates of 0.2–0.4 cm kyr^{-1}. Goldhammer *et al.* (1987) and Hardie *et al.* (1991) reported tepee zones in the Latemar successions with thicknesses varying from 1 to 13 m, and individual cement bands up to 20 cm thick. Individual tepee antiforms are typically onlapped by sediments from subsequent cycles, suggesting that tepees were completely formed prior to the deposition of the next cycle(s). If Holocene rates of tepee formation from Lake Macleod are even close approximations of rates of tepee development in the Latemar, the multimetre tepee structures at Latemar section alone represent several hundred thousand years of time. In order for these same tepee structures to have formed caps on millennial cycles, both tepee antiforms and cements would have to form at rates that are at least two orders of magnitude faster (40–400 cm kyr^{-1}antiform; 20–40 cm kyr^{-1} cement) than those documented at Lake MacLeod.

In addition to sedimentological evidence, spectral analysis of Latemar cycle thickness and rank series indicates a close match between bundling frequencies of Latemar cycles and Milankovitch orbital cycles. The clarity of the record and the non-trivial nature of the match is especially apparent in studies using rank series analysis, where plots of depth-ranked units appear to have formed in lock-step synchroneity with Milankovitch cycle insolation changes (see Fig. 8 of Preto *et al.*, 2004). However, tuning the Latemar cycles to a sub-Milankovitch component (i.e. 4.2 kyr; Zühlke *et al.*, 2003) also preserves Milankovitch frequency components, but at a different order (see Figs 11 and 12 and Table 2 of Zühlke, 2004), pointing to a 'fractal-like' aspect of the different orders of the Milankovitch rhythms (first noted by Schwarzacher, 1998). The difference lies in the assignment of the precession: the pure Milankovitch interpretation suggests that the precession has driven fundamental cycle formation (i.e. 20 kyr Latemar cycles and 100 kyr megacycles), while the mixed Milankovitch scale and sub-Milankovitch scale interpretation suggests that precession drove the formation of megacycles (i.e. 4 kyr Latemar cycles and 20 kyr megacycles).

In light of this 'fractal' equivalence, the mixed Milankovitch and sub-Milankovitch model deserves closer attention. Thus far, no specific proposal has been offered for the required eustatic driver operating at a millennial scale of around 4 kyr. However, it is notable that recent work on Pleistocene eustasy may lead to the identification of such a driver. Chappell (2002) reported 6 kyr recurrent coral terraces uplifted in the Late Pleistocene of Papua New Guinea suggestive of sea-level oscillations with amplitudes as large as 15 m. Possibly this is the analogue to the 4.2 kyr Latemar driver. There is also evidence of similar eustatic behaviour in the uplifted Pleistocene coral terraces of Barbados (Thompson & Goldstein, 2005). However, both locations are situated in tectonically active areas, involving rapid uplift of the islands related to the geodynamics of adjacent convergence zones. In particular, the rather large (>10 m) amplitudes that have been suggested for these millennial eustatic changes depend on the accuracy of the uplift rates that have been estimated for these islands. It is also important to note that these measurements were taken from fast-growing corals, not from stacked peritidal depositional cycles.

In either case (pure Milankovitch or mixed Milankovitch and sub-Milankovitch), both models rely on an astroclimatic forcing mechanism as a cyclic driver for carbonate sedimentation. The signs of this forcing are recognized via statistical analysis of Latemar cycles and cross-applied to the Mendola succession. Comparative sedimentology indicates that Holocene carbonate successions formed at rates commensurate with multimillennial processes, providing further evidence for a link between an allocyclic, Milankovitch cycle driven process and the sedimentary record.

The successions of carbonate depositional cycles at Latemar and Mendola Pass sections record periodic, allogenic processes operating at millennial periodicities

In reference to the question of cyclicity in sedimentation in Latemar, Emmerich *et al.* (2005) wrote 'radiometric age dating on detrital zircons in air-borne tuff layers intercalated within the cyclic succession of the Latemar solved the controversy' (Emmerich *et al.*, 2005, p. 11). However, this statement is by no means universally accepted. Indeed, the mean ages in their correlation scheme (see Fig. 7b in Emmerich *et al.*, 2005) indicates two age-reversals (i.e. older-upward rather than younger-upward), the first occurring from 'Tc' (241.2 +0.8/–0.6 Ma) to LAT-31 (242.6 ± 0.7 Ma) and the second occurring between the platform-interior ash beds LAT-30 (241.2 +0.7/–0.6 Ma) and LAT-32 (241.7 +1.5/–0.7 Ma). Moreover, the ±0.6 kyr and larger uncertainties may preclude the direct use at these ages to assign any specific millennial timing to the Latemar cycle. Nonetheless, the uncertainties point towards a millennial rhythm, and here existing proposals for specific millennial timings from experiments are simply chosen.

Emmerich *et al.* (2005), and independently, Kent *et al.* (2004), suggested that both cycles and megacycles in the Latemar succession operate at sub-Milankovitch cycle frequencies, i.e. 0.9–1.97 kyr per cycle and 3.5–10 kyr per megacycle. As a result, the accumulation of the entire Latemar buildup was calculated to have occurred over approximately 1 Myr. Indeed, the prevalent 5:1 megacycle bundling originally identified by Goldhammer *et al.* (1987) was not recognized by Kent *et al.* (2004).

Periodic drivers for Triassic millennial shallowing-upward cycles are difficult to identify because no such drivers have yet been linked to the deposition of stacked shallowing-upward cycles in the Holocene or Pleistocene (see previous discussion). Kent *et al.* (2004) argued for tidal forcing operating at around 1.8 kyr, citing a theoretical tide at the periodicity identified by Munk *et al.* (2002). However, Munk *et al.* (2002, p. 382) stated that 'the equivalent tidal amplitude of the millennial term is estimated at 0.04 mm', which is clearly not enough to generate the metre-scale Latemar cycles. Similarly, there is no known connection between millennial-period Dansgaard–Oeschger climate cycles and Heinrich-type warming events cited by Emmerich *et al.* (2005) as possible cycle drivers. No link has yet been established between these cycles and formational periodicities of shallowing-upward carbonate successions of the Holocene and Pleistocene.

Recently, Roth & Reijmer (2005) identified oxygen isotope excursions in a 30-m-long core from the leeward margin of the Great Bahama Bank that have both centennial and millennial (1.7 kyr) periodicities. While this discovery establishes a link between centennial and millennial cyclic processes and carbonate sedimentation, the cycling has yet to be linked to the depositional periods of platform-interior shallowing-upward depositional cycles. The identification of periodic megacycle drivers within the millennial model is likewise problematic. Studies from the Holocene and Pleistocene have yet to identify a lower-frequency (5–10 kyr) cyclic process by which millennial tides, Bond cycles, or other climatic processes will produce bundled shallowing-upward carbonate megacycles.

While the rates of eustatic rise and fall required for millennial cyclic carbonate deposition are rapid, rates for subsidence and sediment accumulation required for the development of depositional cycles within this context are likewise very high. The subsidence required to generate 700 m of section in 1 Myr is approximately 0.7 m kyr^{-1}. As both the Latemar and Mendola sections preserve cyclic successions of similar stacking and bundling trends, subsidence would have operated at near-equal rates at both localities during the study interval in question. In addition, accumulation rates (including time for sedimentation and subaerial exposure) of platform interior and tidal-flat sediments would need to be extraordinarily high in the millennial framework, requiring a sustained accumulation rate in excess of 0.7 m kyr^{-1} approaching the growth rates of Holocene corals, to generate both successions (see Schlager (1981) for rates). If one considers low-energy carbonate tidal flats as sediment sinks rather than sediment sources (e.g. Ginsburg *et al.*, 1977), source sediments would need to be produced at rates that are even higher than 0.7 m kyr^{-1} in order to generate enough sediment to fill the tidal-flat sink.

However, the paucity of evidence for millennial cyclicity of shallowing-upward carbonate successions in the Holocene and Pleistocene must be considered with the fact that biotic, climatic and oceanic conditions have changed throughout geological time. It is possible that the cyclic drivers in the Triassic were unique to that period and cannot be compared with those operating in the Holocene and Pleistocene. Indeed, it may be that in this case comparative sedimentology has met its useful limit as a temporal reference, and that its only appropriate use in Mid-Triassic carbonates is as a tool for identifying facies and depositional environments.

The conclusion that processes with millennial periodicities formed the depositional cycles at the Latemar section comes from palaeomagnetic analysis, the dating of platform interior ash beds, and the correlation of these dates to basinal deposits proximal to the Latemar Platform. The decision to rely on these techniques to a greater degree than comparative sedimentology and/or statistical analyses of the cyclic succession is one that must be made in order to accept the millennial model.

CONCLUSIONS

This work has undertaken a new examination of the issue of allocyclicity in the origin of the Latemar Platform cycles of the Middle Triassic of the Dolomites, northern Italy. This involved the measurement and analysis of a coeval cyclic platform carbonate formation external to the Latemar locality, 30 km to the northwest at Mendola Pass. The objective was to determine if the Mendola Pass succession contained a stratigraphic signal comparable to the one found in the Latemar succession. Both successions have metre-scale cyclic bedding with a shallowing-upward depositional theme, with each cycle exhibiting subtidal (<10 m water depth) to supratidal successions of subfacies. Biostratigraphic ties were established between the two successions and statistical correlation analysis of measured cycle thickness series from the two localities allowed for stratigraphic placement of the shorter Mendola Pass section within the Latemar succession. Harmonic analysis identified the full suite of stacking pattern frequencies in both sections, and spectral coherency and cross-phase analysis confirmed significant coherency and phase-locking between the cyclic signals of both localities.

These results have the following implications for high-frequency cyclostratigraphic forcing of sedimentation on the Latemar Platform.

1. The current state of knowledge regarding Holocene and Pleistocene composite eustasy, the formation of shallowing-upward cycles, and the rates at which these cycles and their component facies form indicate that carbonate allocycles in modern environments form as the result of multimillennial processes (Tables 1 and 2). If the drivers and depositional rates that have been identified for these cycles can serve as a comparative model, it is then likely that the depositional cycles at Mendola Pass and in the Latemar section, as well as their conspicuous bundled stacking patterns, formed as the result of composite eustatic forcing.
2. Resolving the issue of millennial forcing versus composite Milankovitch forcing by comparing the ancient to the modern record is difficult because metre-scale carbonate allocycles stacked with millennial periodicities remain undiscovered in modern carbonate sedimentary settings. In addition, a purely sub-Milankovitch scale cyclic driver that would act to form bundled carbonate megacycles remains undiscovered in modern settings.
3. At the present time, eustatic oscillations involving Milankovitch cycle driven composite eustasy are still the most likely explanation for the record of cyclic stratigraphy observed at Mendola Pass and in the Latemar section either in the form of pure Milankovitch forcing (Goldhammer *et al.*, 1987) or mixed Milankovitch and sub-Milankovitch forcing (Zühlke *et al.*, 2003).
4. If one accepts the conclusion that Latemar and Mendola allocycles and allocycle bundles were formed at millennial periodicities, one must also accept one of the following conclusions:
   (a) That the controls on the sedimentary system, including amplitudes and periodicities of composite eustatic drivers and sustained, very high (≥0.7 m kyr^{-1}) sedimentation rates in areas of rapid subsidence are unique to Mid-Triassic carbonate systems.

(b) That drivers for Triassic cycles also operate in the Holocene and Pleistocene but do not generate carbonate depositional cycles.

ACKNOWLEDGEMENTS

This manuscript represents the last body of work done by the late Professor Robert K. Goldhammer in the Dolomite Alps of northern Italy. Bob had a great love of fieldwork in the Alps due to the excellent preservation of cyclic carbonate stratigraphy at various localities there. Bob also had deep respect for his scientific forebears. Bob would commonly refer to publications and ideas put forth by 'the Gins' when teaching to remind students of the many ideas within the scientific community regarding the origin of carbonate rock fabrics and depositional cycles. What we present here is a written attempt to steward Bob's latest ideas to the geological and broader scientific community.

The authors gratefully acknowledge Dr Nereo Preto and Dr Olga Piros for identification of dasycladacean algae collected from the Mendola cycles, and for translating important references and ammonite records relating to the Mendola stratigraphic framework. We also thank Dr Florian Maurer, Dr David Osleger and Dr Thomas Stevens for reviews of the manuscript and their constructive commentary. R.M. Forkner was supported by The R.K. Goldhammer Mesozoic Margins Consortium, ConocoPhillips, and The John A. and Katherine G. Jackson School of Geosciences. L.A. Hinnov and L.A. Hardie were supported by National Science Foundation Grant EAR-9909528.

APPENDIX 1

Datasets with the original Goldhammer thickness series from Goldhammer (1987) and the Goldhammer + Dunn thickness series published by Preto *et al.* (2004) (used in this study) may be downloaded from **WDC for Paleoclimatology Data Access and Data Contribution**: http://www.ncdc.noaa.gov/paleo/data.html.

APPENDIX 2

A MATLAB™ script for Fischer plots, adapted from Read and Sriram (1990), may be downloaded from **WDC for Paleoclimatology Free Software**: http://www.ncdc.noaa.gov/paleo/softlib/softlib.html.

REFERENCES

Anderson, R.Y. and **Kirkland, D.W.** (1966) Intrabasin varve correlation, *Geol. Soc. Am. Bull.*, **77**, 241–256.

Avanzini, M. (2002) Tavola 26 III – Fondo, Note Illustrative. *Provincia Autonoma Trento, Servizio Geol.*, pp. 159, Trento, Italy.

Bendat, J.S. and **Piersol, A.G.** (1986) *Random Data*, 2nd Edn. John Wiley and Sons, New York, 566 pp.

Black, M. (1933) The algal sediments of Andros Island, Bahamas. *Phil. Trans. Roy. Soc. London, Ser. B*, **122**, 165–191.

Blendinger, W. (2004) Sea level changes versus hydrothermal diagenesis: origin of Triassic carbonate platform cycles in the Dolomites, Italy, *Sed. Geol.*, **169**, 21–28.

Blendinger, W. (2005a) Sea level changes versus hydrothermal diagenesis: origin of Triassic carbonate platform cycles in the Dolomites, Italy. *Reply. Sed. Geol.*, **178**, 141–144.

Blendinger, W. (2005b) Sea level changes versus hydrothermal diagenesis: origin of Triassic carbonate platform cycles in the Dolomites, Italy. *Reply. Sed. Geol.*, **178**, 151–153.

Bloom, A.L., **Broecker, W.S.**, **Chappell, J.M.A.**, **Matthews, R.K.** and **Mesolella, K.J.** (1974) Quaternary sea level fluctuations on a tectonic coast. *Quat. Res.*, **4**, 185–205.

Bond, G., **Broecker, W.S. Johnsen, S.J. McManus, J.**, **Labeyrie, L.D.**, **Jouzel, J.** and **Bonani, G.** (1993) Correlations between climate records from North Atlantic sediments and Greenland ice. *Nature*, **365**, 143–147.

Bond, G., **Kromer, B.**, **Beer, J.**, **Muscheler, R.**, **Evans, M.N.**, **Showers, W.**, **Hoffman, S.**, **Lotti-Bond, R.**, **Hajdas, I.** and **Bonani, G.** (2001) Persistent solar influence on North Atlantic climate during the Holocene. *Science*, **294**, 2130–2136.

Bond, G., **Showers, W.**, **Cheseby, M.**, **Lotti, R.**, **Almasi, P.**, **deMenocal, P.**, **Priore, P.**, **Cullen, H.**, **Hajdas, I.** and **Bonani, G.** (1997) A Pervasive millennial-scale cycle in North Atlantic Holocene and glacial climates. *Science*, **278**, 1257–1266.

Bosellini, A. (1991) Geology of the Dolomites: An introduction. *Dolomieu Conf. Carbonate Platforms and Dolomitization*. Ortisei, Italy 43 pp.

Bosellini, A. and **Hardie, L.A.** (1985) Facies e cicli della Dolomia Principale delle Alpi Venete. *Mem. Soc. Geol. Ital.*, **30**, 245–266.

Brack, P. and **Muttoni, G.** (2000) High-resolution magnetostratigraphic and lithostratigraphic correlations in Middle-Triassic pelagic carbonates from the Dolomites (Southern Alps, Italy). *Palaeogeogr., Paleaoclimatol. Palaeoceanogr.*, **161**, 361–380.

Brack, P. and **Rieber, H.** (1993) Towards a better definition of the Anisian/Ladinian boundary: new biostratigraphic data and correlations of boundary sections from the Southern Alps. *Eclogae Geol. Helv.*, **86**, 415–527.

Steinernes Meer (Northern Calcareous Alps, Austria). *Sed. Geol.*, **181**(1–2), 93–106.

Shinn, E.A. (1986) Modern carbonate tidal flats: their diagnostic features. In: *Carbonate Depositional Environments, Modern and Ancient* (Eds L.A. Hardie and E.A. Shinn). *Colorado School Mines Quart.*, **81**(1), 7–33.

Strasser, A. and **Samankassou, E.** (2003) Carbonate sedimentation rates today and in the past: Holocene of Florida Bay, Bahamas, and Bermuda vs. Upper Jurassic and Lower Cretaceous of the Jura Mountains (Switzerland and France). *Geol. Croat.*, **56**(1), 1–18.

Taft, W.H. (1968) Lithification of modern marine carbonate sediments at Yellow Bank, Bahamas. *Bull. Mar. Sci.*, **18**(4), 762–828.

Taylor, J.K. (1990) *Statistical Techniques for Data Analysis*. Lewis Publishers, Inc., Chelsea, Michigan, 200 pp.

Thompson, D.J. (1982) Spectrum estimation and harmonic analysis. *Proc. IEEE*, **70**, 1055–1096.

Thompson, W.G. and **Goldstein, S.L.** (2005) Open-system coral ages reveal persistent suborbital sea-level changes. *Science*, **308**, 401–404.

Tudhope, A.W. (1989) Shallowing-upwards sedimentation in a coral reef lagoon, Great Barrier Reef of Australia. *J. Sed. Petrol.*, **59**(6), 1036–1051.

Vail, P.R., Mitchum jr., R.M. and **Thompson III, S.** (1977) Seismic stratigraphy and global changes of sea level, part 4: Global cycles of relative changes of sea level. In: *Seismic Stratigraphy – Applications to Hydrocarbon Exploration* (Ed. C.E. Payton). *AAPG Mem.*, **36**, 83–97.

Van de Plassche, O., van der Borg, K. and **de Jong, A.F.M.** (1998) Sea-level-climate correlation during the past 1400 yr. *Geology*, **26**(4), 319–322.

Van Hilten, D. (1960) Geology and Permian paleomagnetism of the Val di Non area. *Geol. Ultra.*, **5**, 1–95.

Venzo, G.A. and **Fuganti, A.** (1965) Il Trias della Mendola (Trentino – Alto Adige). *Studi. Trent. Sci. Natur., Sez. A*, **XLII**(1), 55–86.

Wright, V.P. (1992) Speculations on the controls on cyclic peritidal carbonates: ice-house versus greenhouse eustatic controls. *Sed. Geol.*, **76**, 1–5.

Yang, W. and **Lehrmann, D.** (2003) Milankovitch climatic signals in Lower Triassic (Olenekian) peritidal carbonate successions, Nanpanjiang Basin, South China. *Palaeogeogr., Palaeoclimatol., Palaeoecol.*, **201**(3–4), 283–306.

Zühlke, R. (2004) Integrated cyclostratigraphy of a model Mesozoic carbonate platform – the Latemar (Middle Triassic, Italy). In: *Multidisciplinary Approach to Cyclostratigraphy* (Eds B. D´Argenio, A. Fischer, I. Premoli Silva and H. Weissert). *SEPM Spec. Publ.*, **81**, 183–212.

Zühlke, R., Bechstädt, T. and **Mundil, R.** (2003) Sub-Milankovitch and Milankovitch forcing on a model Mesozoic carbonate platform – the Latemar (Middle Triassic, Italy). *Terra Nova*, **15**(2), 69–80.

Int. Assoc. Sedimentol. Spec. Publ. (2009) **41**, 239–254

Phylloid algal mounds in the Paradox Basin, southwestern USA: an alternative to the *in situ* constructional growth model?

G. MICHAEL GRAMMER and AUDREY L. RITTER

Michigan Geological Repository for Research and Education, Department of Geosciences, Western Michigan University, 1903 W. Michigan Avenue, Kalamazoo, MI 49008, USA (E-mail: michael.grammer@wmich.edu)

ABSTRACT

Phylloid algae are recognized as being common reef builders during the late Palaeozoic and, in many parts of the world, are known to form important hydrocarbon reservoirs. Most genetic models suggest that phylloid mounds develop primarily through *in situ*, constructional processes. The dominant processes in these models revolve around the baffling of sediment by the algae themselves, most often in dense meadows, combined with the binding of sediment by various organisms and pervasive early submarine cementation. Most models suggest that phylloid mound deposition occurred in widely distributed belts along open carbonate platforms or ramps, at or slightly above wave base in water depths of 15–20 m. In the Paradox Basin, most of the individual algal mounds and mound complexes seem to fit previous interpretations, not only for depth, but also for deposition in a relatively shallow open platform or ramp setting.

There are, however, local accumulations of individual phylloid algal mounds exposed in outcrop whose distribution and geometry suggest deposition in an environment more closely associated with strong currents. These algal mounds, which are exposed along the San Juan River in southwestern Utah, have 10–12 m amplitudes and are distributed in a near sinusoidal pattern suggesting that they may have formed under the influence of wave- or tide-generated ocean currents. In modern environments, tide-generated currents may pile similar-sized plates of the modern algae *Halimeda* into dunes or sand waves along the sea floor with a similar near-sinusoidal distribution. Ascertaining the conditions under which these mounds initiated and grew may provide an alternative explanation for the genesis of phylloid algal mounds at least locally, and may have important implications for exploration concepts devised for algal mound reservoirs. High energy, current-emplaced mounds, for instance, may be aerially limited in the subsurface and may trend preferentially along strike (e.g. wave currents) or, conversely, along dip (e.g. tidal currents). In contrast, constructional mounds may have formed in much broader areas across much of an entire shelf or platform.

Keywords Phylloid algae, algal mounds, Halimeda.

INTRODUCTION

Late Palaeozoic phylloid algal mounds have been described from many localities in Laurasia, including the USA (Konishi & Wray, 1961; Pray & Wray, 1963; Heckel & Cocke, 1969; Toomey *et al.*, 1977; Mazzullo & Cys, 1979; Toomey, 1980; Dawson & Carozzi, 1986; Roylance, 1990; Baars, 1992; Kirkland *et al.*, 1993; Grammer *et al.*, 1996, 2000; Soreghan & Giles, 1999; Forsythe, 2003; Samankassou & West, 2002, 2003), Canada (Reid, 1986; Beauchamp *et al.*, 1989; Breuniger *et al.*, 1989) and Russia (Antoshkina, 1998). The term phylloid algae was first used by Pray & Wray (1963) to describe a growth form of leaf-like algae that includes both calcareous red algae and codiacean green algae. The genus *Ivanovia*, a codiacean green algae, is the most abundant type of phylloid algae in the Paradox Basin (Pray & Wray, 1963). *Ivanovia* is generally thought to have grown in warm, shallow marine environments, at depths of <20 m (Wray, 1968; Wilson, 1975; Toomey, 1980). Deposition of phylloid-dominated mounds has been attributed to a variety of mechanisms, with most authors describing some combination of sediment baffling by the algae, along with binding and encrusting by various organisms, such as gastropods and foraminiferans coupled with the precipitation of early marine cements (Forsythe *et al.*, 2002; Samankassou & West, 2003).

Phylloid algal mounds exposed in the Paradox Basin region of Utah, southwestern USA, locally exhibit a sinusoidal pattern of 10–12 m high mounds with 'wavelengths' that average some 35–50 m. Whereas most phylloid mounds are interpreted as having been deposited in relatively shallow open-marine environments on carbonate shelves and platforms, the unique geometry and distribution of these local Paradox examples suggests a possible alternative genetic history. Here the distribution of phylloid algal mounds in the Paradox Basin in general are reviewed. Localized mounds exposed along the San Juan River are then compared to a modern environment in the Florida Keys where similar algal plates (i.e. *Halimeda)* make up sand waves with amplitudes of several metres, formed in response to tidal currents.

METHODS

Paradox Basin

Seventy-five polished slabs from outcrop and subsurface core were analysed for composition, fabric and texture. Rock colours were determined using the GSA Rock Color Chart (1991). A total of 180 thin-sections were analysed from outcrop and subsurface examples to identify and evaluate the phylloid algal mounds and associated facies. General composition, allochem type, and abundance of cement and carbonate mud were determined through a combination of point counting (200–400 counts per slide) according to the guidelines established by Plas & Tobi (1965) to ensure statistical relevance, and visual estimation using the charts provided in Flugel (1982). Algal mounds in outcrop were measured for height, width (in two dimensions) and spacing where exposed along the San Juan River in southwestern Utah.

Florida Quicksands

Visual reconnaissance was done by snorkelling at five sites within the sand belt. Seven samples were collected by hand and grab sampler at the same locations. Samples were air dried and sieved and relative abundance of allochems was established using the visual estimation charts in Flugel (1982).

GEOLOGICAL SETTING

The Paradox Basin is an elongate, northwest to southeast trending intracratonic evaporite basin that extends from the northwestern part of New Mexico into east-central Utah (Fig. 1). The basin covers an area of approximately 27,000 km^2 that also encompasses southwestern Colorado and southeastern Utah (Stevenson & Baars, 1988). Although the Paradox Basin was most actively subsiding and accumulating sediments during the Late Carboniferous (Pennsylvanian), the basin rests on a structural framework that was developed by late Precambrian time (Baars & Stevenson, 1981). Continued tectonic movement throughout the Palaeozoic resulted in repeated vertical movements along these basement fractures (Baars & Stevenson, 1982). In the basin centre, shale, anhydrite, salt and potash were deposited, while along the northern Uncompahgre Uplift, thick coarser-grained siliciclastic wedges accumulated. Deposition on the Paradox shelf, further to the south and west, was characterized primarily by cyclical deposition of carbonates with subordinate siliciclastic sandstones.

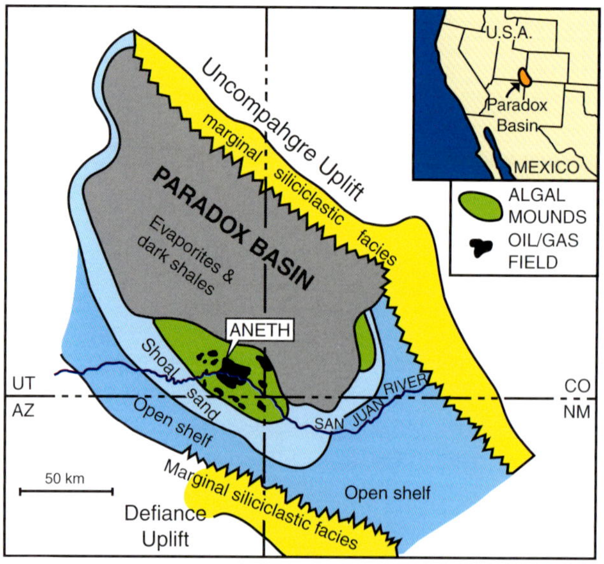

Fig. 1. Map showing location of the Paradox Basin in the southwestern portion of the USA. Modified from Choquette (1983) and Grammer *et al.* (1996).

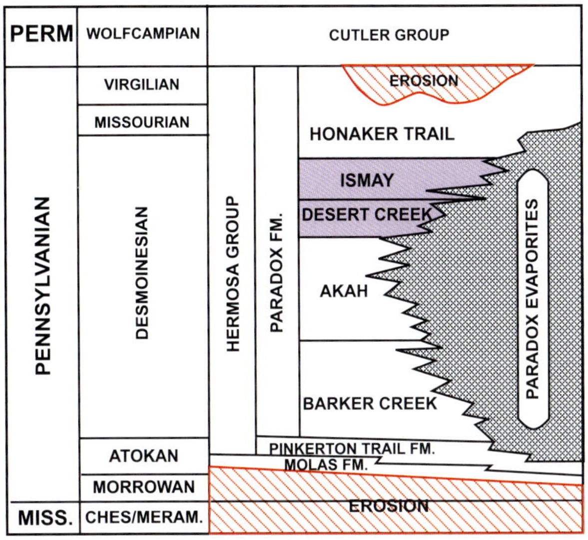

Fig. 2. General Pennsylvanian stratigraphy of the Paradox Basin area showing the Desmoinesian age of the studied Desert Creek and Ismay intervals. Modified from Goldhammer *et al.* (1991) and Baars & Stevenson (1982).

The Paradox Formation is defined as an evaporite facies in the basin centre, but there was coeval deposition of carbonates and siliciclastics on the shelf that are included in the formation. On the shelf, the mixed carbonate–siliciclastic cycles have been grouped into four major intervals or stages, which in ascending order are the Barker Creek, Akah, Desert Creek and Ismay (Fig. 2). The Honaker Trail Formation overlies these intervals and is composed of cherty limestone and calcareous sandy siltstones in the west. The Honaker Trail Formation grades into more massive, arkosic sandstones and gravels toward the Uncompahgre Uplift in the east. Towards the end of Pennsylvanian time, the Hermosa sea withdrew from the Paradox Basin due to regional tectonics, creating a major unconformity between the Carboniferous and Permian strata (Grammer *et al.*, 1996).

The Desert Creek and Ismay intervals form the top of the shelfal equivalent of the Paradox Formation. These shelfal cycles are correlated to the evaporite cycles in the subsurface through the use of black shales and dark carbonate mudstones, which appear to be chronostratigraphic units (Choquette & Traut, 1963). The cycle-bounding shales include the Chimney Rock Shale (base of Desert Creek), the Gothic Shale (base of Lower Ismay) and the Hovenweep Shale (base of the Upper Ismay). The carbonate cycles of the Desert Creek are correlated with evaporite cycles 4 and 5 of Peterson & Hite (1969). The Lower and Upper Ismay are separated by a regional marker bed on outcrop, the Horn Point Limestone, which is overlain by a siliceous dolostone that weathers a characteristic yellow colour and is known locally as the 'Old Yeller'. The Lower Ismay is correlated with evaporite cycle 3, and the Upper Ismay with evaporite cycle 2. The Desert Creek, Lower and Upper Ismay intervals are characterized by extensive development of phylloid algal buildups in the subsurface and are important hydrocarbon reservoirs in different regions throughout the Paradox Basin.

PHYLLOID ALGAL MOUND AND ASSOCIATED FACIES IN THE PARADOX BASIN

The following facies descriptions are summarized from Grammer *et al.* (1996, 2000), and describe the phylloid algal mound and associated facies observed both in outcrop and in the subsurface of the Paradox Basin (Fig. 3). In an attempt to maintain some consistency between research projects, and to facilitate comparisons between different studies and study areas, we have utilized the same basic facies terminology first established by Pray & Wray (1963) and more recently by Goldhammer *et al.* (1991), with minor modification and augmentation.

Intermediate facies-diverse biota

The intermediate facies (diverse) is a bioturbated skeletal wackestone to packstone that, in outcrop, has thin- to medium-scale undulatory bedding, rare cross-bedding and a diverse normal marine fauna consisting of crinoids, brachiopods (atrypids, athyrids, spiriferids, terebratulids and orthids), tubular and encrusting foraminifera, fusulinids, molluscs, branching and fenestrate bryozoans, and minor phylloid algae. Most of the skeletal grains are broken but not very abraded. This facies is bioturbated to burrow-mottled indicating a healthy infaunal element. Much of the matrix material consists of peloidal mud. The facies contains 5–15% admixed quartz-dominated sand and coarse silt and 3–5% dolosilt. Bioturbation may result in a dismicrite fabric with blocky calcite spar filling voids in the matrix mud. Intergranular blocky calcite spar as well as intragranular calcite spar is common. Some fossil fragments are

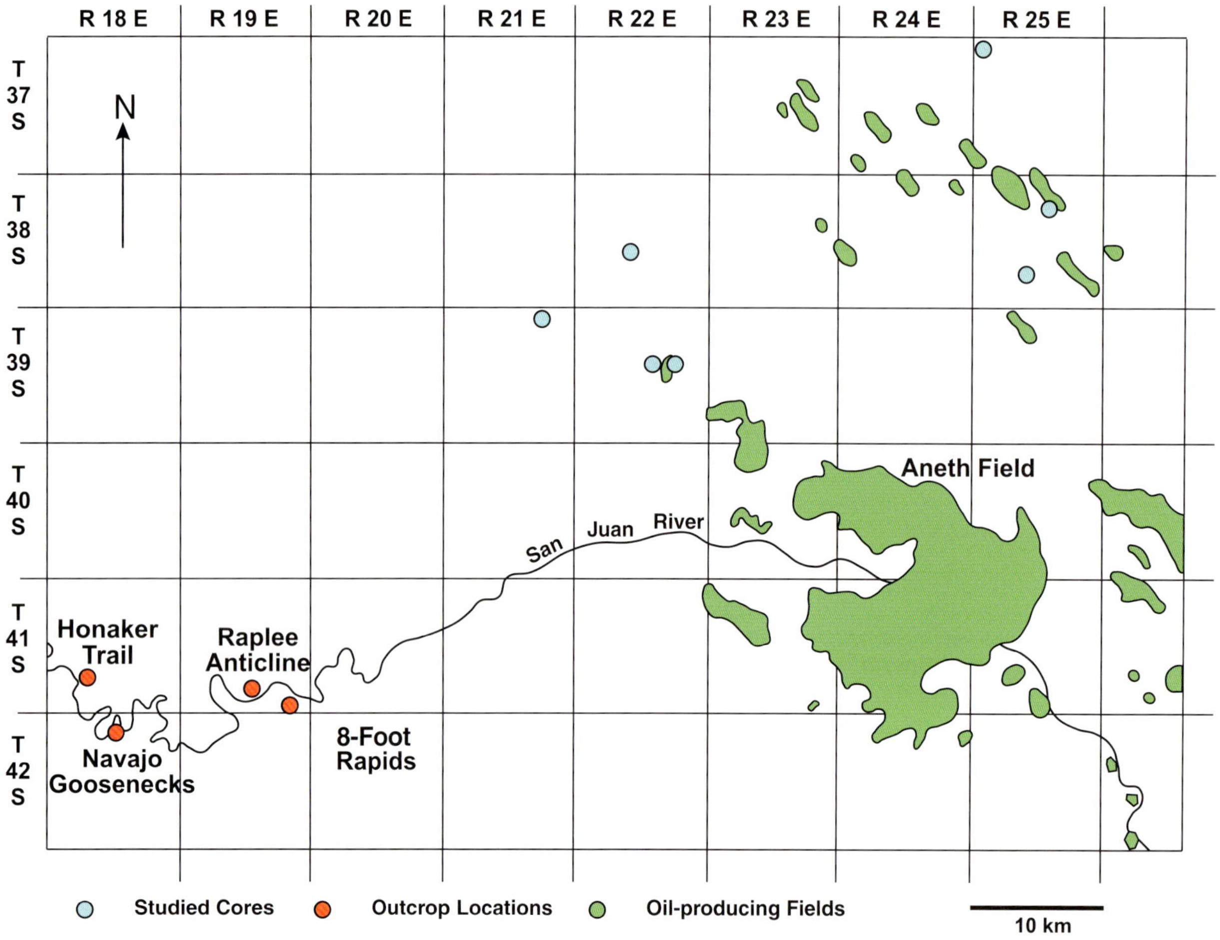

Fig. 3. Map showing location of the four primary outcrop locations on the Paradox shelf and the cored wells studied through the same interval in the basin. The focus of this paper is the exposed algal mounds around the 8-Foot Rapids location in the southwest part of the map. Note also how the individual mound fields (except for the giant Aneth field), located in the basin to the northeast, are generally of limited aerial extent and trend in a NW–SE direction. Strike of the basin is also in a NW–SE orientation.

preserved only by micrite envelopes around the perimeter of the grains. Wispy and low-amplitude (subcentimetre scale) sutured-seam stylolites are typically present.

The intermediate facies (diverse) is interpreted as a low to moderate energy, open-marine platform facies with good circulation, deposited when flooding of the platform had proceeded to a point that was conducive to the establishment of a normal subtidal carbonate factory. Lower contacts are gradational. The upper contact is gradational with the incipient mound facies.

Incipient mound facies

The incipient mound facies is a phylloid algal mudstone to wackestone with a matrix of peloidal mud and scattered diverse skeletal grains (Fig. 4a). This low-energy facies is found as a regionally pervasive, muddy phylloid algal unit present at the base of algal mounds in both outcrop and subsurface. The incipient mound facies forms a laterally persistent, flat bed at the base of algal mounds in outcrop that is approximately 1.5–2.0 m thick, with a sharp basal contact and a gradational transition into the overlying phylloid algal mound. The facies is interpreted as a low-energy accumulation of phylloid algal plates and carbonate mud on an open platform. The mechanism(s) responsible for the transition from muddy algal facies to well-developed phylloid algal mounds is unknown, but probably results from changes in environmental conditions such as wave and/or current energy, and therefore circulation, which would then affect temperature, salinity and water quality.

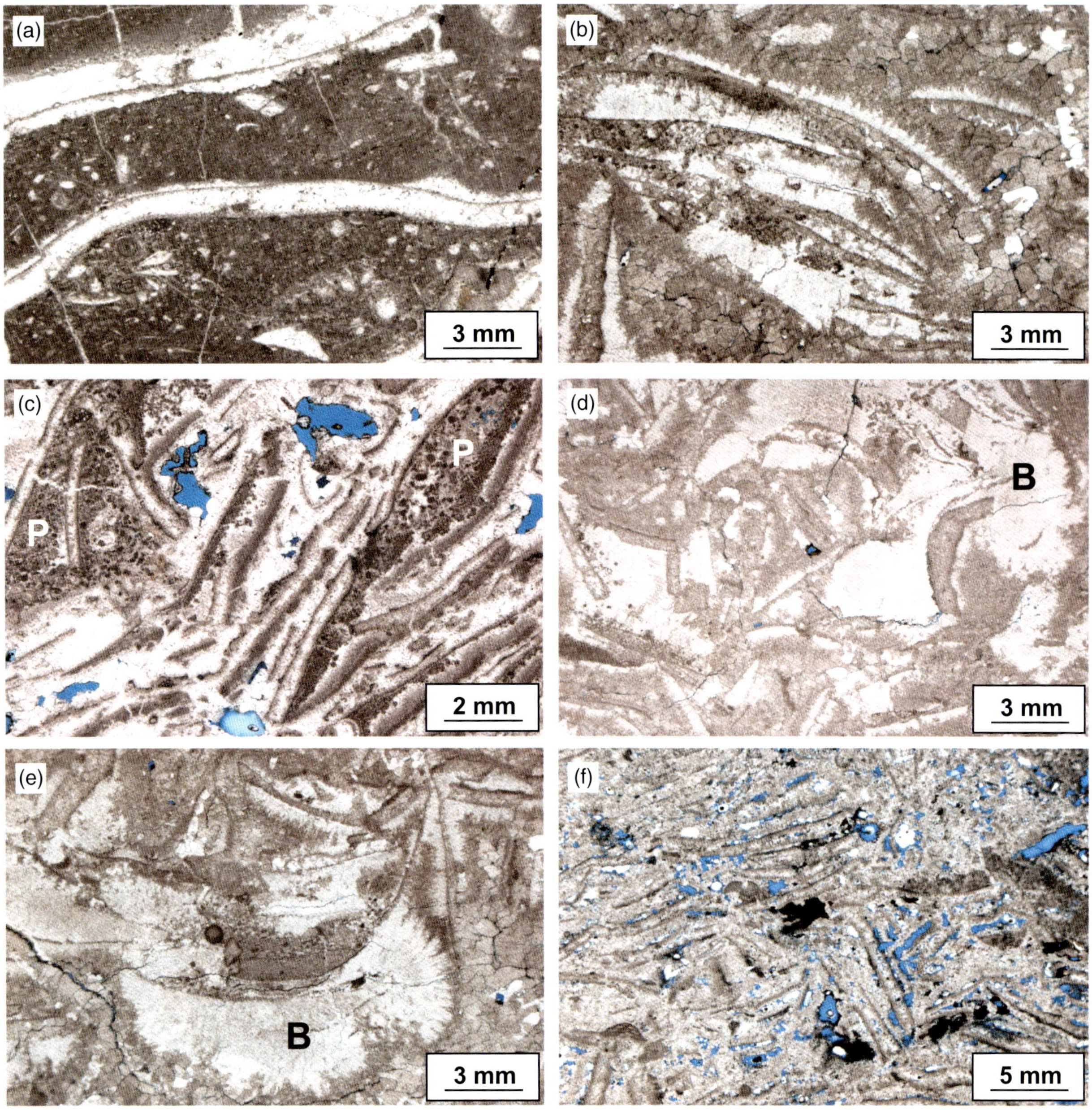

Fig. 4. Thin-section photomicrographs, plane polarized light: (a) incipient mound facies, showing local large phylloid algal plates in mud matrix; (b) interior of phylloid algal mound with abundant early marine cement; (c) abundant phylloid algal plates with primary pores partially filled with early cement and peloidal sediment and/or cement (P), blue is pore space; (d) fragmented algal plates and abundant early marine cement, some botryoidal (B); (e) botryoidal (B) and other early marine cements between broken phylloid algal plates; (f) low power photomicrograph showing classic distribution of algal plates and abundant early cement. Blue is pore space.

Phylloid algal mound facies

Throughout the Paradox Basin, in both outcrop and in the subsurface, the phylloid algal facies consists of phylloid algal bioherms that vary in geometry from biostromal (flat and elongate) to a mound-like morphology with flat bases. The internal fabric of the algal facies varies from grain-supported algal bafflestones to algal wackestones and packstones with a highly variable biotic component ranging from near monospecific to a diverse normal marine faunal assemblage (Fig. 4). Abundant primary porosity is partially filled with syndepositional marine cements (fibrous and botryoidal) as well as peloidal sediment or cement (MacIntyre, 1985). The algal facies

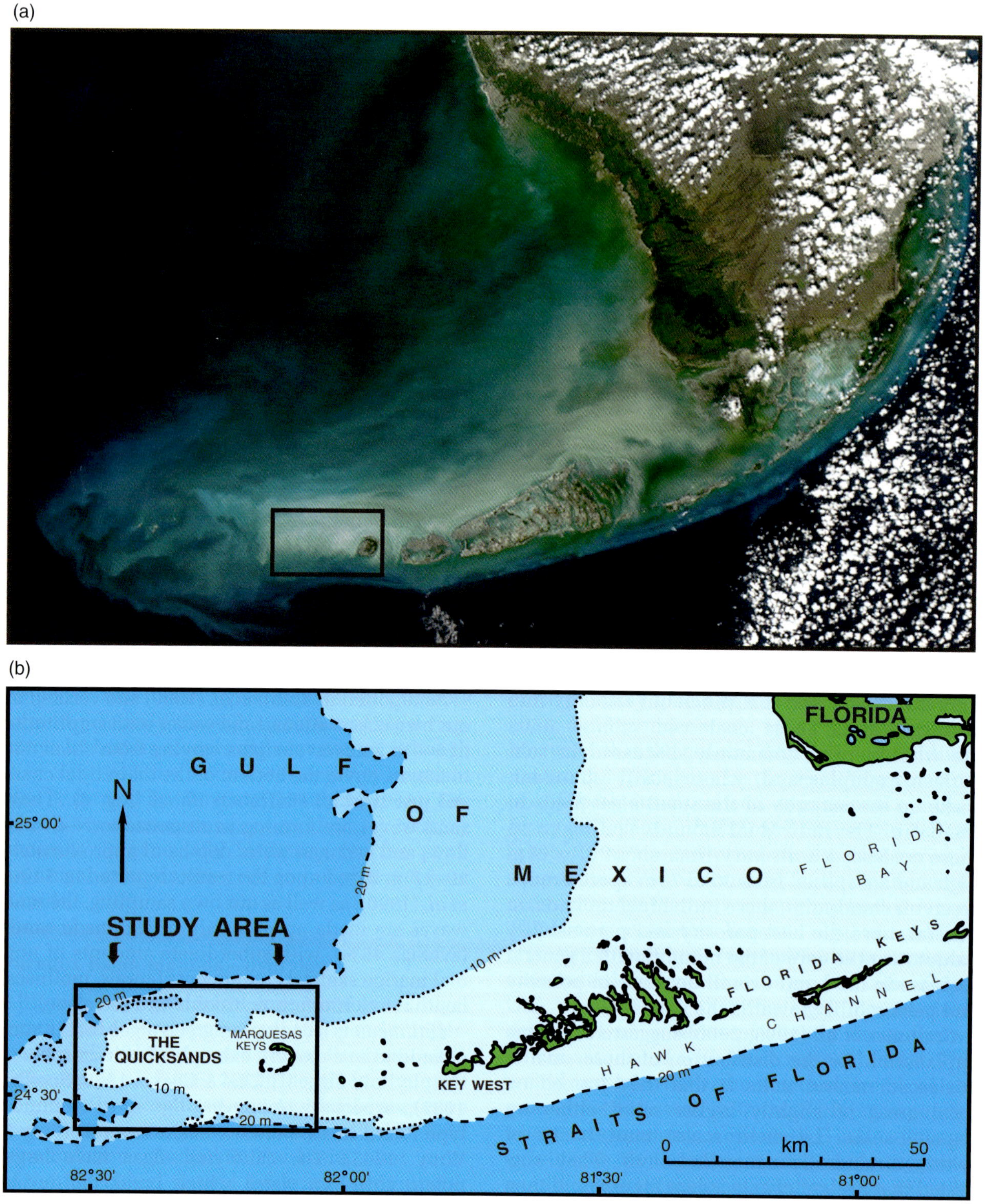

Fig. 6. (a) Space shuttle photograph and (b) map showing lower Florida peninsula and Florida Keys. Box in lower left of photograph outlines the Marquesas Keys and the Quicksands area where large amplitude sand waves comprised mostly of *Halimeda* plates are actively forming due to tidal currents. Image courtesy of the Image Science and Analysis Laboratory, NASA, Johnson Space Centre. Map modified from Shinn *et al.* (1990).

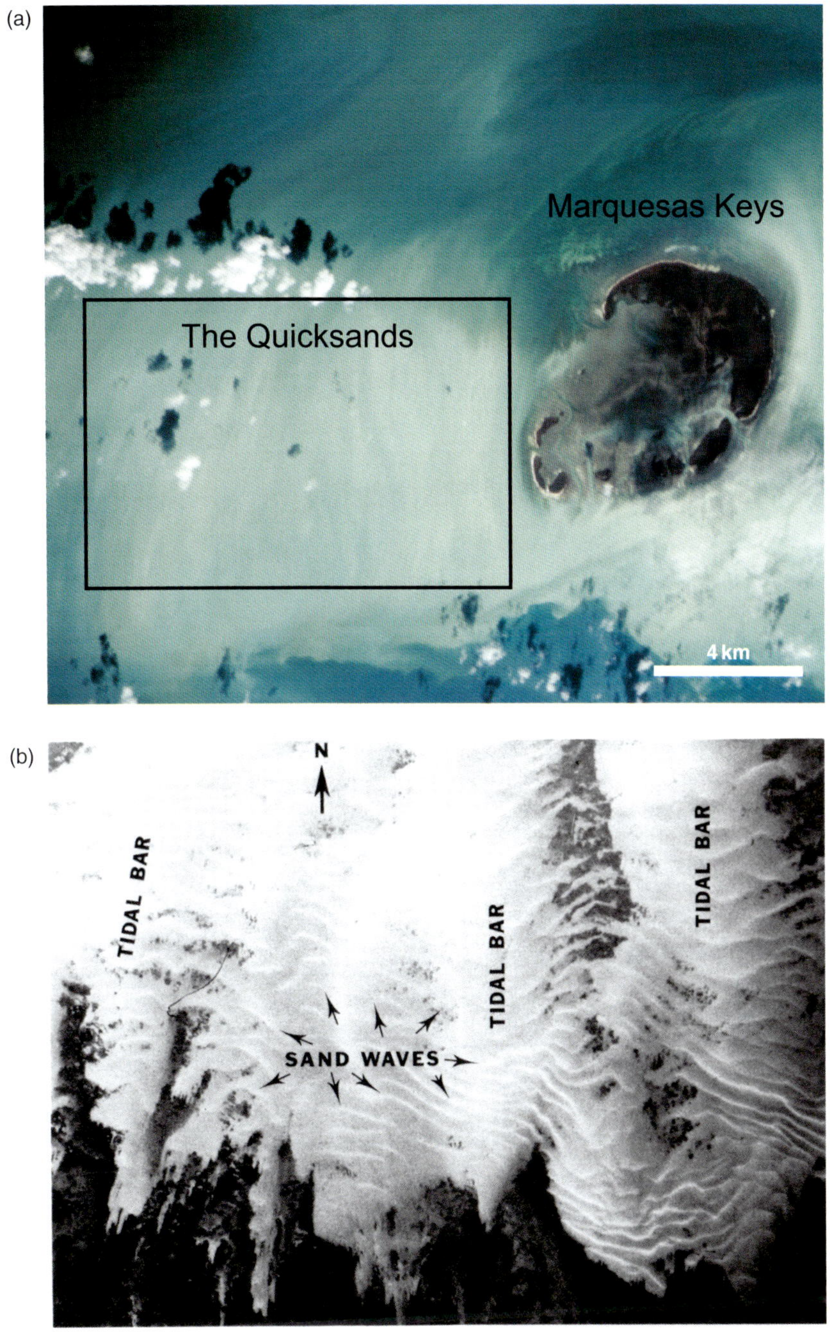

Fig. 7. (a) Space shuttle photograph showing location of The Quicksands west of the Marquesas Keys. (b) Low-altitude aerial photograph of Quicksands showing well-developed tidal bar belts of carbonate sand trending N–S, and large (several metres) sand waves with crests oriented in a roughly E–W orientation. Image courtesy of the Image Science and Analysis Laboratory, NASA, Johnson Space Centre. Aerial photograph modified from Shinn *et al.* (1990).

Grain-size analysis of these samples shows that the individual *Halimeda* plates making up the sand bars in this region average about 3–5 mm (longest dimension), with maximum lengths of 6–8 mm (Fig. 9), making them similar in both size and shape to the Phylloid algal plates found in the Paradox Basin mounds (average 4–5 mm, maximum approximately 8–10 mm). In both the Paradox Basin mounds and the modern sand waves, platy algae make up the predominant grain type (8-Foot Rapids: range 9–55%, average 34%; Quicksands: range 10–66%, average 45%) with subordinate amounts of normal marine biota contributing to the sediment.

PHYLLOID ALGAL MOUNDS DEVELOPED IN TIDAL PASSES?

The question of whether the Paradox mounds exposed at 8-Foot Rapids might have formed as a result of tidal currents piling algal plates into large sand waves is significant for a number of reasons.

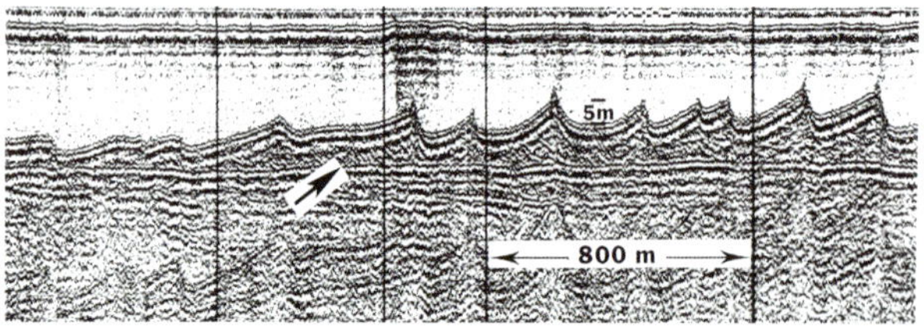

Fig. 8. High-resolution single-channel seismic profile from Shinn *et al.* (1990) showing large (5+m high) sand waves that are migrating across the bottom of a tidal channel just west of Halfmoon shoal. Sand waves are forming in depths of about 12 m. Arrow indicates Pleistocene bedrock surface as originally illustrated by Shinn *et al.* (1990).

Fig. 9. (a) Photograph showing living codiacean (green) alga *Halimeda* sp. (b) The white sands are from the Quicksand area and are comprised mainly of disarticulated plates from *Halimeda* sp. The plates are released from the living organism upon death, or due to storms, etc. that break off part of the alga.

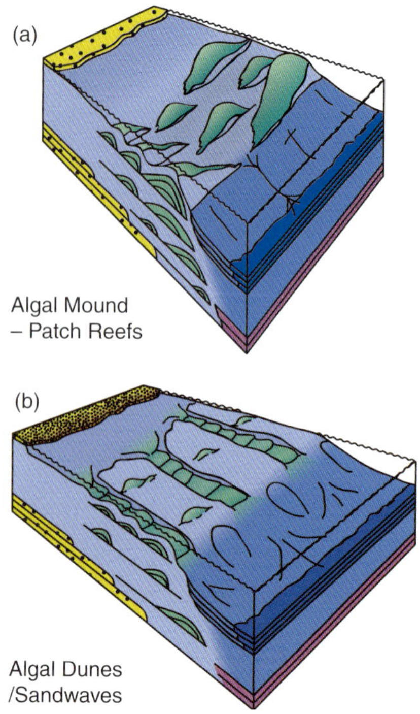

Fig. 10. Schematic diagrams (a,b) illustrating variations in algal buildups that form through commonly accepted baffling and binding (a) and as a result of tidal currents building sand waves or dunes. In the classic model, phylloid algal mounds are formed by a combination of baffling, binding and syndepositional marine cementation. These mounds are distributed much like other patch reefs, and would extend preferentially along strike. In contrast, phylloid algal sand waves forming in tidal channels or passes, would be much more limited aerially, and found in belts that are dip-elongated with widths along strike varying from hundreds of metres to a few kilometres based upon modern examples.

Of particular interest is that such a mechanism would provide a new model for the formation of some of the Paradox algal mounds, while also providing support for the few studies to date which have suggested a similar genetic process in phylloid mounds in other parts of the world (Hamblin, 1969; Heckel & Cocke, 1969; Giles & Soreghan, 1999; Samankassou & West, 2003). In addition, if some algal mounds form within tidal channels or passes, it could change the way that exploration models are developed for some fields. Instead of following regional structural trends, or mapping mounds as a series of 'patch reefs' that are widely distributed across a shallow carbonate platform, mound fields that develop in tidal channels or passes will be located within a relatively narrow belt that is oriented roughly perpendicular to the strike of a platform (Fig. 10).

There are a number of similarities between the modern sand waves forming in the Quicksands region and the Upper Carboniferous phylloid algal mounds exposed along the San Juan River. In both locations, large mounds/sand waves are made up primarily of coarse, platy carbonate grains that range from coarse grained sand to gravel size. Based on the common interpretation that *Ivanovia*, the most common phylloid in the Paradox Basin, is related to the modern codiacean alga *Halimeda*, it is likely that the hydrodynamic behaviour of these two types of platy carbonate grains would be very similar under similar current conditions. Whereas plate-like algae is the dominant grain type in both locations, the subordinate grains are also similar and characterized by a diverse, shallow-marine biota. In the Paradox region, the mounds/dunes are interpreted to have formed in water depths of 12–15 m, while the larger Holocene sand waves in the Quicksand region are being deposited in depths of about 10–12 m.

In addition, Grammer *et al.* (1996, 2000) suggested that the algal mounds near 8-Foot Rapids were probably distributed in an elongate, roughly dip-oriented trend that was about 1 km wide by 3–4 km long, similar in scale and geometry to the main tidal bar belts near Halfmoon Shoal. In the Paradox example, the present-day San Juan River, and therefore the outcrop of exposed mounds, runs approximately E–W, or approximately along dip during Pennsylvanian deposition. Modern feeder streams have carved secondary canyons that intersect the main river canyon at near right angles, which locally provides a means to observe the distribution of mounds in the strike dimension. Along such a secondary stream channel at 8-Foot Rapids (Wildhorse Canyon), it is possible to walk-out and follow the rapid change from the 10–12 m high mounds/dunes exposed along the main river as they pinch out within just a couple hundred metres to the south. This type of rapid reduction in large sand waves/dunes would be expected near the edges of the tidal channels and passes, and is similar to what is observed near Halfmoon Shoal. Unfortunately, the distribution of the mounds to the north of 8-Foot Rapids, which would be towards the other side of a hypothesized tidal pass, is unknown as the section is not exposed by the river or subordinate streams.

The many similarities between the large sand waves forming in the Quicksands area and the Paradox mounds at 8-Foot Rapids strongly suggest that the phylloid 'mounds' might have been deposited as sand waves or dunes within a tidal channel or pass (Fig. 11). Such an interpretation could have a significant effect on the exploration strategies employed within the basin for algal mound reservoirs. The distribution of relatively small satellite fields within the Paradox Basin, which are typically 1–2 km wide and 3–5 km long, have previously been interpreted as being associated with deep-seated structural trends, and are mapped and explored for accordingly. Whereas there is most likely a structural component to these near-dip oriented fields, based upon the abundant subsurface data that has been studied to date, the possibility of tidal pass concentration of algal mounds should not be discounted.

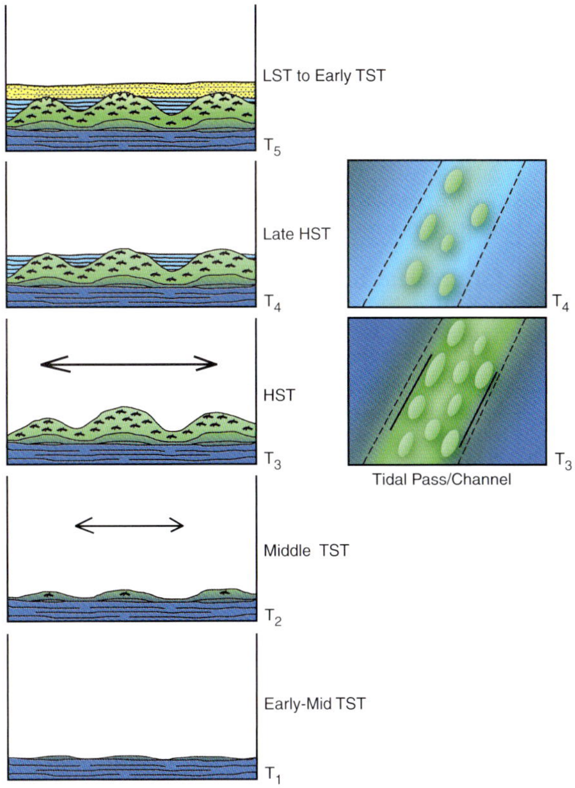

Fig. 11. Diagram showing model for development of phylloid algal mounds/dunes in a tidal channel or tidal pass. Mound distribution would be elongated in dip direction and relatively narrow in strike dimension (hundreds of metres to few kilometres). The diagram shows the sand waves/dunes initiating during the early to middle TST as the platform becomes flooded to a point where tidal currents start to develop (T_1 and T_2). The bases of the dunes in this model are muddier than later when the tidal currents reach higher velocities. The algal dunes, consisting primarily of phylloid algal plates, form during maximum tidal flow during highstand (T_3). As available accommodation space becomes more limited during the latest highstand (T_4), the topographic lows between the dunes would become filled with more diverse shallow-marine skeletal grains from nearby areas. In the Paradox Basin example discussed here, lowstand results in exposure of the mounds, followed by a capping of the cycle with siliciclastic sands

CONCLUSIONS

The striking similarity of the phylloid algal mounds exposed along the San Juan River to the sand waves reported by Shinn *et al.* (1990) in southwesternmost Florida (i.e. in grain size, shape, allochem distribution, platy component abundance and near sinusoidal distribution, as well as amplitude and wavelengths of the individual mounds), suggests that the mounds exposed along the river near 8-Foot Rapids may have also been deposited in response to tidal currents. The formation of phylloid algal mounds due to hydrodynamic emplacement of algal plates moved by high-energy tidal currents would provide not only another possible mechanism to explain the formation of these Late Palaeozoic mounds, but could be significant to exploration geologists searching for mounds that are hydrocarbon reservoirs (see Table 1). In contrast to the standard model of broad shelfal distribution of algal mounds, those emplaced through the work of tidal currents are most likely going to be in relatively narrower belts that are oriented primarily in a dip direction. Realization that these important hydrocarbon reservoirs may have multiple orientations may prove to be a valuable tool in many hydrocarbon basins throughout the world.

Exploration and Production Scales: Case Studies from the Pennsylvanian of the Paradox Basin and the Upper Devonian of Alberta (Eds P.W. Homewood and G.P. Eberli). *Elf Aquitaine Memoir*, **24**, 29–69.

Grammer, G.M., Ginsburg, R.N., Swart, P.K., McNeill, D.F., Jull, A.J.T. and **Prezbindowski, D.R.** (1993) Rapid growth rate of syndepositional marine aragonite cements in steep marginal slope deposits, Bahamas and Belize. *J. Sed. Petrol.*, **63**, 983–989.

Hamblin, W.K. (1969) Marine paleocurrent directions in limestones of the Kansas City Group (Upper Pennsylvanian) in eastern Kansas. *Kansas Geol. Survey Bull.*, **134**, 24 pp.

Heckel, P.H. and **Cocke, J.M.** (1969) Phylloid alga-mound complexes in outcropping upper Pennsylvanian rocks of mid-continent. *AAPG Bull.*, **53**(5), 1058–1074.

Herrod, W.H. and **Gardner, P.S.** (1988) Upper Ismay reservoir at Tin Cup Mesa field. *Rocky Mountain Association of Geologists, 1988 Carbonate Symposium*, pp. 175–192.

Hine, A.C., Hallock, P., Harris, M.W., Mullins, H.T., Belknap, D.F. and **Jaap, W.C.** (1988) Halimeda bioherms along an open seaway: Bawihka Channel, Nicaraguan Rise, SW Caribbean Sea. *Coral Reefs*, **6**, 173–178.

Hudson, J.H. (1985) Growth rate and carbonate production in *Halimeda opuntia*, Marquesas Keys, Florida. In: *Paleoalgology: Contemporary Research and Applications* (D.F. Toomey and M.H. Nitecki). Springer-Verlag, Berlin, pp. 257–263.

Kirkland, B.L., Moore, C.H. Jr. and **Dickson, J.A.D.** (1993) Well-preserved, aragonitic, phylloid algae (Euogonphyllum, Udoteaceae) from the Pennsylvanian Holder Formation, Sacramento Mountains, New Mexico. *Palaios*, **8**, 111–120.

Konishi, K. and **Wray, J.L.** (1961) *Eugonophyllum*, a New Pennsylvanian and Permian algal genus. *J. Paleontol.*, **35**(4), 659–667.

James, N.P. (1983) Reef environment. In: *Carbonate Depositional Environments* (Eds P.A. Scholle, D.G. Bebout and C.H. Moore). *AAPG Mem.*, **33**, 346–453.

MacIntyre, I.G. (1985) Submarine cements-the peloidal question. In: *Carbonate Cements* (N. Schneidermann and P.M. Harris). *SEPM Special Publication*, **25**, 113–124.

Mazzullo, S.J. and **Cys, J.M.** (1979) Marine aragonite sea-floor growths and cements in Permian phylloid algal mounds, Sacramento Mountains, New Mexico. *J. Sed. Petrol.*, **49**(3), 917–936.

Peterson, J.A. and **Hite, R.J.** (1969) Pennsylvanian evaporite-carbonate cycles and their relation to petroleum occurrence, southern Rocky Mountains. *AAPG Bull.*, **53**, 884–908.

Petroleum Frontiers (1984) Paradox Basin: Unravelling the mystery. Petroleum Information Incorporated, summer 1984, 54 pp.

Plas, L. van der and **Tobi, A.C.** (1965) A chart for judging the reliability of point counting results. *Am. J. Sci.*, **263**, 87–90.

Pray, L.C. and **Wray, J.L.** (1963) Porous algal facies (Pennsylvanian) Honaker Trail, San Juan Canyon, Utah. In: *Shelf Carbonates of the Paradox Basin* (Ed. R.O. Bass). Four Corners Geological Society 4th Field Conference Guidebook, pp. 204–234.

Reid, R.P. (1986) Discovery of Triassic phylloid algae: possible links with the Paleozoic. *Can. J. Earth Sci.*, **23**, 2068–2071.

Rine, J.M. and **Ginsburg, R.N.** (1985) Depositional facies of a mud shoreface in Suriname, South America – a mud analogue to sandy, shallow-marine deposits. *J. Sed. Petrol.*, **55**, 633–652.

Roylance, M.H. (1990) Depositional and diagenetic history of a Pennsylvanian algal-mound complex: Bug and Papoose Canyon fields, Paradox Basin, Utah and Colorado. *AAPG Bull.*, **74**, 1087–1099.

Samankassou, E. and **West, R.R.** (2002) Construction versus accumulation in phylloid algal mounds: an example of a small constructed mound in the Pennsylvanian of Kansas, USA. *Palaeogeogr. Palaeoclimatol. Paleoecol.*, **185**, 379–389.

Samankassou, E. and **West, R.R.** (2003) Constructional and accumulation modes of fabrics in selected Pennsylvanian algal-dominated buildups in Eastern Kansas, Midcontinent, U.S.A. In: *SEPM Spec. Pub.*, **78** and *AAPG Mem.*, **83**, pp. 219–237.

Schlager, W. (1981) The paradox of drowned reefs and carbonate platforms. *Geol. Soc. Am. Bull.*, **92**, 197–211.

Scoffin, T.P. (1987) *An Introduction to Carbonate Sediments and Rocks.* Blackie/Chapman & Hall, New York, 274 pp.

Shinn, E.A., Lidz, B.H. and **Holmes, C.W.** (1990), High-energy carbonate-sand accumulation, the Quicksands, southwest Florida Keys. *J. Sed. Pet.*, **60**, 952–967.

Soreghan, G.S. and **Giles, K.A.** (1999) Facies character and stratal responses to accommodation in Pennsylvanian bioherms, Western Orogrande Basin, New Mexico. *J. Sed. Res.*, **69**, 893–908.

Stevenson, G.M. and **Baars, D.L.** (1986) The Paradox: A pull-apart basin of Pennsylvanian age. In: *Paleotectonics and Sedimentation in the Rocky Mountain Region* (Ed. J.A. Peterson). *AAPG Mem.*, **41**, 513–539.

Stevenson, G.M. and **Baars, D.L.** (1988) Overview: Carbonate reservoirs of the Paradox Basin. *Rocky Mountain Association of Geologists, 1988 Carbonate Symposium*, pp. 149–162.

Toomey, D.F. (1980) History of a Late Carboniferous Phylloid Algal Bank Complex in Northeastern New Mexico. *Lethaia*, **13**, 249–267.

Toomey, D.F., Wilson, J.L. and **Rezak, R.** (1977) Evolution of Yucca mound complex, Late Pennsylvanian phylloid algal buildup, Sacramento Mountains, New Mexico. *AAPG Bull.*, **61**, 2115–2133.

Weber, L.J., Wright, F.M., Sarg, J.F., Shaw, E., Harman, L.P., Vanderhill, J.B. and **Best, D.A.** (1995) Reservoir delineation and performance: application of sequence stratigraphy and integration of petrophysics and engineering data, Aneth field, southeast Utah, U.S.A. In: *Hydrocarbon Reservoir Characterization: Geologic Framework and Flow Unit Modeling* (Eds E.L. Stoudt and P.M. Harris). *SEPM Short Course*, No. 34, pp. 1–30.

Wilson, J.L. (1975) *Carbonate Facies in Geologic History.* Springer-Verlag, New York, 471 pp.

Wray, J.L. (1968) Late Paleozoic phylloid algal limestones in the United States. *Proceedings of the 23rd International Geological Congress*, Prague, **8**, pp. 113–119.

Int. Assoc. Sedimentol. Spec. Publ. (2009) **41**, 255–275

The Cincinnati Arch: a stationary peripheral bulge during the Late Ordovician

MICHAEL C. POPE*, STEVEN M. HOLLAND† and MARK E. PATZKOWSKY‡

**School of Earth and Environmental Sciences, Washington State University, Pullman, WA 99164-2812, USA (E-mail: mcpope@wsu.edu)*

†Department of Geology, University of Georgia, Athens, GA 30602-2501, USA

‡Department of Geosciences, Pennsylvania State University, University Park, PA 16802-2714, USA

ABSTRACT

Shallow-water features, including peritidal carbonate complexes, tidally influenced shoal-water facies, biostromes, and coincident subaerial unconformities on the peritidal complexes were restricted to the Cincinnati Arch during the Late Ordovician Taconic Orogeny. Persistence of these features along the Cincinnati Arch indicate that this area experienced decreased subsidence relative to surrounding areas, forming the peripheral bulge within the Taconic foreland basin. These shallow-water features are surrounded on all sides by deeper water, shallow subtidal to deep subtidal facies. The Jessamine and Nashville Domes were prominent topographic highs on the arch and were the locus of shallow-water deposition and subaerial erosion, but many of these shallow-water features also occur in the subsurface between the two domes. The shallow-water features and unconformities on the Cincinnati Arch occur in a narrow band repeatedly throughout the Late Ordovician. The width of these features is much too narrow to have formed solely by flexure of the lithosphere and their persistent formation in the same location indicates they did not form as a viscoelastic response to orogenesis. Rather the boundary between the Grenville Front and the Eastern Granite–Rhyolite Province acted as a focusing mechanism for the orogenic load stresses to the east, producing a relatively stationary peripheral bulge. A temporal and spatial change in the stratigraphy and subsidence on the Nashville and Jessamine Domes during the Edenian probably records a northward shift of loading, possibly sedimentary loading, during the Taconic Orogeny.

Keywords Peripheral bulge, foreland basin, Taconic Orogeny, flexure, stratigraphy.

INTRODUCTION

Stratigraphic modelling of foreland basins demonstrates that the regional stratigraphy in these basins reflects the interplay of sediment supply, eustasy, flexural loading from thrust sheet emplacement and possibly long-wavelength dynamic loading (Mitrovica *et al.*, 1989; Gurnis, 1992a,b). The stratigraphic models are useful both in the forward mode, for understanding the principles underlying the stratigraphic architecture of foreland basins in general (Flemings & Jordan, 1989, 1990), and in the inverse mode, for reconstructing the shifting geographical location of thrust loads and changing eustatic sea level during the history of particular foreland basins (Quinlan & Beaumont, 1984; Beaumont *et al.*, 1988). Studies of foreland basins have frequently emphasized the behaviour of the peripheral bulge (or forebulge) over time, in particular, whether it is a static or mobile feature, and what such behaviour implies about the rheology of the lithosphere and the movement of the loads that generate the foreland basin (Sinclair *et al.*, 1991; Sinclair & Allen, 1992).

As such, the peripheral bulge and its sedimentary record may provide insights into large-scale tectonism. For instance, the backstepping of carbonate reefs may record advancement of a thrust load where there is little record of this loading recorded in the basin itself (Dorobek, 1995; Galewsky *et al.*, 1996; Galewsky, 1998). Additionally, the geographical position of the peripheral bulge with respect to the orogenic belt can be calculated if the peripheral bulge is modelled as being solely generated by flexure; commonly its position is dependent on the age of the continental crust being deformed (Crampton &

Allen, 1995). Conversely, the location of the peripheral bulge can become relatively 'fixed' along lateral lithospheric variability in the foreland (Waschbusch & Royden, 1992a,b) or when pre-existing structural weaknesses in the crust are reactivated (Dorobek *et al.*, 1991).

Whether the Cincinnati Arch was a positive structural feature that influenced deposition in Kentucky and Tennessee during the Late Ordovician Taconic Orogeny is contentious. Interpretations range from the arch having little or no influence on sedimentation (Gustadt, 1958; Weiss *et al.*, 1965; Cressman, 1973; Weir *et al.*, 1984), to limited uplift that affected sedimentation primarily during the Cincinnatian (Ettensohn, 1992, 1994), to geographical uplift only on the domes with an intervening saddle (Borella & Osborne, 1978). However, integration of outcrop and core data (Wilson, 1962; Hrabar *et al.*, 1971; Mackey, 1972; Pope & Read, 1997a,b, 1998), favours uplift or decreased rates of subsidence along the Cincinnati Arch in Kentucky and Tennessee during much of the Late Ordovician and that this affected sedimentation, a view predicted by stratigraphic modelling (Quinlan & Beaumont, 1984). During this time a shallow-water carbonate platform developed on the Cincinnati Arch and was surrounded by deeper water, storm-influenced carbonate and shale (Keith, 1988a,b; Holland, 1993; Holland & Patzkowsky, 1996, 1997; Pope & Read, 1997a,b; McLaughlin *et al.*, 2004). This paper combines the detailed sequence stratigraphic framework of the Upper Ordovician strata of the Cincinnati Arch (Holland, 1993; Holland & Patzkowsky, 1996, 1998; Pope & Read, 1997a,b, 1998) to decipher the geographical position and depositional history of the peripheral bulge in this basin during the Taconic Orogeny. This history is compared with that predicted by several types of flexural models to gain insight into the mechanical behaviour of the Appalachian foreland basin system during its initial formation.

PERIPHERAL BULGE BEHAVIOUR

Models of foreland basin stratigraphy indicate that peripheral bulge behaviour is controlled by several factors, including whether the lithosphere displays an elastic or a viscoelastic response, whether the lithosphere is mechanically continuous or broken, and whether the locations of loads are relatively static or mobile. The lithosphere may deform under a load in two distinct modes, depending on whether the lithosphere is modelled as a thin sheet of elastic or viscoelastic material overlying an asthenosphere modelled as an inviscid fluid. In elastic models (Fig. 1), the foreland basin adjacent to the thrust load undergoes immediate subsidence proportional to the mass of the load, with the amount of subsidence decreasing exponentially with distance (Nádai, 1963). Distal areas experience broad, but minor uplift on the peripheral bulge. For an elastic plate, the amount of uplift or subsidence at any location will change over time only if the distribution or magnitude of the load changes. In contrast, the uplift or subsidence at any location will change for a viscoelastic plate even if the distribution or magnitude of the load does not change. For a viscoelastic rheology (Fig. 1), the plate will deform initially as if it were elastic, but over time it will behave as if it were a weaker elastic plate. In effect, this will cause the basin to deepen and narrow, with the peripheral bulge becoming higher and narrower, while moving towards the thrust load.

In both of these cases, the location of the peripheral bulge will shift if the location of loading shifts. For example, in the case of a peripheral foreland basin, in which the edge of a continent undergoes attempted subduction, such as the modern-day Persian Gulf, the location of loading shifts considerably as the continent approaches the subduction zone (Jacobi, 1981). Relative to a fixed point on the continental interior, the location of loading initially lies over the adjacent, attached oceanic lithosphere. As the continent approaches the subduction zone, the location of loading moves progressively toward this fixed point. Ultimately, the loading takes place at the edge of the continental lithosphere, before subduction is aborted, owing to the buoyancy of continental lithosphere. As a result, the forebulge migrates from the oceanic lithosphere across the edge of the continent to the interior of the continent. In the case of a retro-arc foreland basin, located on the craton side of a volcanic arc, such as in the Precordillera of Argentina or the Western Interior Basin of North America, thrust-sheet motion is limited to a relatively defined area of shortening (Allen & Allen, 2005). As a result, the peripheral bulge experiences less overall motion towards the craton. Additionally, if the lithosphere beneath the foreland basin varies laterally in its rigidity, the position of the peripheral bulge may become 'fixed' for extended periods of

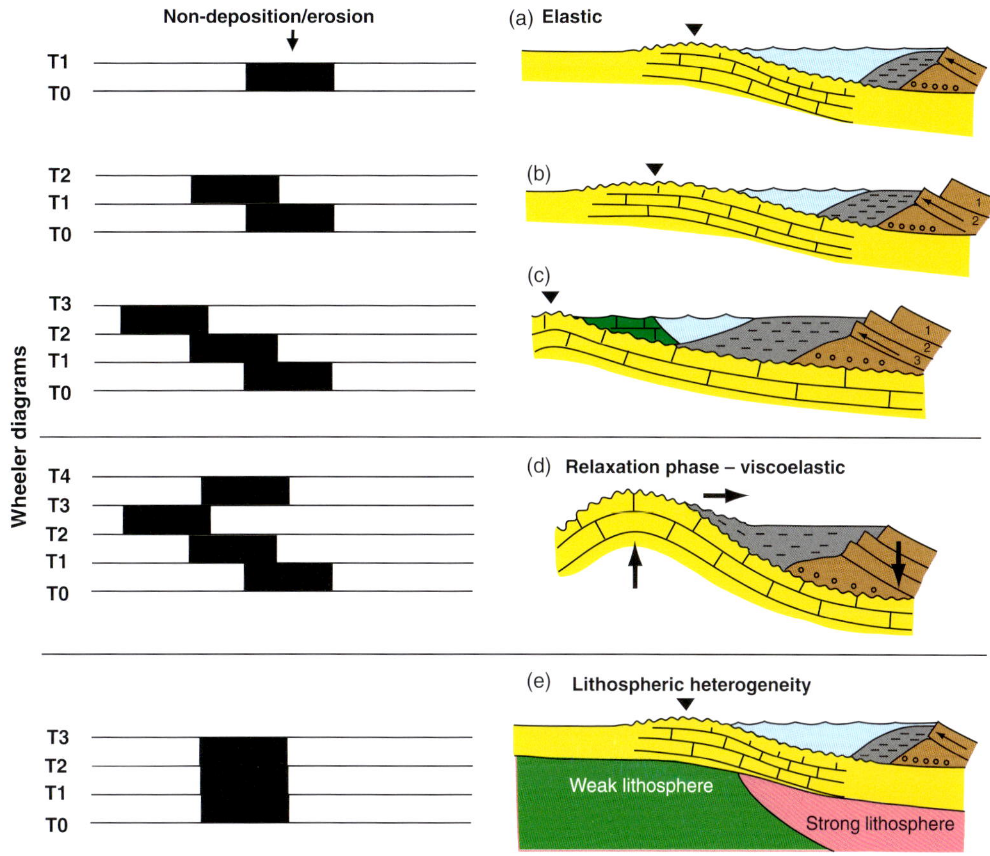

Fig. 1. Cross-sections of foreland basin evolution for a continuous elastic plate (a–c), a continuous viscoelastic plate (a–d), and an elastic plate with lithospheric heterogeneities (e) plotted next to their respective Wheeler diagrams. Areas of non-deposition or erosion over a subaerially exposed peripheral bulge indicated in black on the Wheeler diagrams.

time (Grotzinger & Royden, 1990; Waschbusch & Royden, 1992a,b). In such cases, the peripheral bulge may appear as a stationary feature within the foreland as the basin continues to deepen both cratonward of the peripheral bulge and in the foredeep (Fig. 1).

SCALING CONSIDERATIONS

Although peripheral bulges frequently are depicted in schematic figures as topographically prominent features, flexural modelling indicates that they are actually rather subtle. In particular, peripheral bulges are commonly depicted as too tall relative to the depth of the basin, as too narrow relative to the width of the basin, and consequently, with slopes that are too steep. The theory of bending of flat, thin plates offers predictions about the scaling of peripheral bulges. For an infinite elastic slab deformed by a line load (Nádai, 1963), the vertical deflection (w) at any distance (x) from the load is

$$w = w_0 e^{-Lx[\cos(Lx) + \sin(Lx)]} \tag{1}$$

where w_0 is the vertical displacement at the load, that is, the maximum vertical displacement and L is the flexural parameter, equal to

$$L = \left(\frac{(\rho_m - \rho_f)g}{4D} \right)^{1/4} \tag{2}$$

where ρ_m is the density of the asthenosphere, ρ_f is the density of the basin fill, g is gravitational acceleration and D is flexural rigidity.

Based on this equation, the width of the foreland basin ($width_{\text{basin}}$), that is, the distance from the load to the leading edge of the peripheral bulge, is

$$width_{\text{basin}} = \frac{3\pi}{4L} \tag{3}$$

Table 1. Calculated foreland basin parameters using equations from Allen and Allen (2005).

Flexural rigidity, D (dyne cm)	Air-filled $\rho_f = 0.0$	Water-filled $\rho_f = 1.0$	Sediment-filled $\rho_f = 2.5$
(a) Basin width – distance from load to edge of peripheral bulge (in km)			
10^{28}	44	48	61
10^{29}	78	85	109
10^{30}	139	151	193
10^{31}	247	269	344
10^{32}	439	478	611
10^{33}	780	851	1087
(b) Distance from load to top of peripheral bulge (in km)			
10^{28}	58	64	82
10^{29}	104	113	145
10^{30}	185	202	258
10^{31}	329	359	458
10^{32}	585	638	815
10^{33}	1040	1135	1450
(c) Width of peripheral bulge (in km)			
10^{28}	58	64	82
10^{29}	104	113	145
10^{30}	185	202	258
10^{31}	329	359	458
10^{32}	585	638	815
10^{33}	1040	1135	1450
(d) Width of secondary basin (back-bulge basin; in km)			
10^{28}	58	64	82
10^{29}	104	113	145
10^{30}	185	202	258
10^{31}	329	359	458
10^{32}	585	638	815
10^{33}	1040	1135	1450

Depending on flexural rigidity and whether the basin is filled with air, water or sediment, the width of the basin may range from 44 to 1087 km (Table 1). Given the values of flexural rigidity most commonly reported (Allen & Allen, 2005), basin widths of 100–300 km are typical for most sediment-filled foreland basins.

For elastic plates, peripheral bulges should be approximately one-third wider ($Width_{bulge}$) than their associated foreland basin:

$$width_{bulge} = \frac{\pi}{L} \quad (4)$$

Given typical values of flexural rigidity, peripheral bulges associated with sediment-filled basins should be 130–390 km wide (Table 1). The height of a peripheral bulge should scale linearly with the maximum amount of subsidence in the basin (that is, subsidence at the load), regardless of the type of fill in the basin:

$$\frac{w}{w_0} = -e^{-\pi} = 0.0432 \quad (5)$$

In other words, uplift on the bulge should be approximately 4% of the subsidence in the basin. For a 1000-m deep flexure, this translates into only 40 m of uplift. Because the width of the peripheral bulge is likely to be greater than a hundred kilometres, slopes on the flanks of the peripheral bulge will be negligible (less than 0.1°). Distal to the peripheral bulge is an area of minor subsidence generated by flexure, sometimes called the back-bulge basin (Decelles & Giles, 1996). The amount of subsidence in this basin is related to the amount of subsidence at the load by the relationship:

$$\frac{w}{w_0} = e^{-2\pi} = 0.00187 \quad (6)$$

which is less than 0.2% of the subsidence at the load. For a 1000-m deep flexure, this translates into a mere 2 m of subsidence. So, although sedimentary basins commonly are developed on the distal side of the peripheral bulge, it is unlikely that their origin is solely or even primarily driven by flexure related to thrust loading.

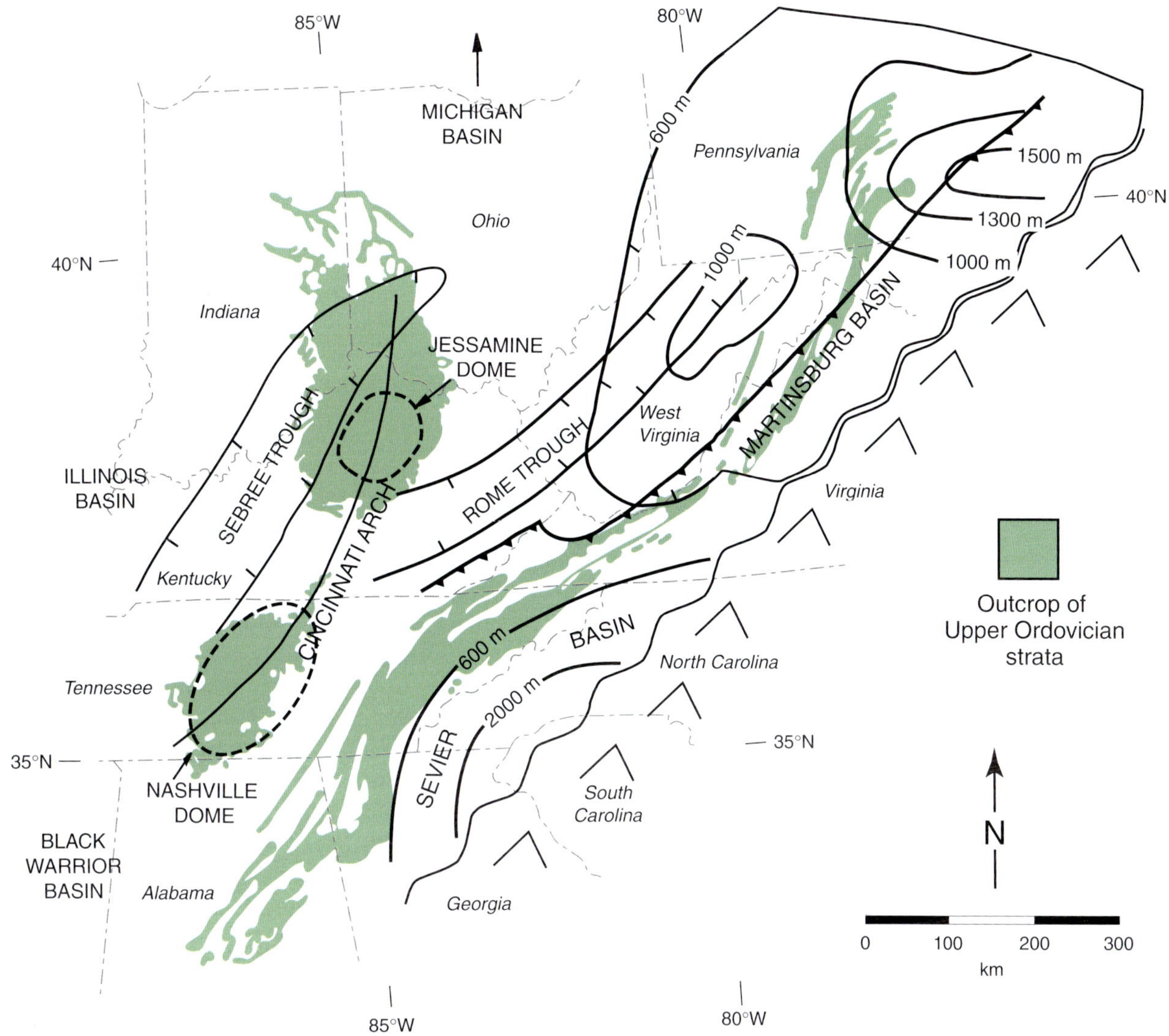

Fig. 2. Map of study area, showing outcrop belts of Upper Ordovician rocks in green, as well as the locations of principal structural features in the region. Contours of Upper Ordovician sediment thickness in the Martinsburg and Sevier basins are indicated.

THE CINCINNATI ARCH AND APPALACHIAN FORELAND BASIN

The Cincinnati Arch is interpreted to be the peripheral bulge of the Appalachian foreland basin, which formed by collision of Laurentia with a magmatic arc (Quinlan & Beaumont, 1984; Ettensohn, 1992). The Jessamine and Nashville Domes on the Cincinnati Arch are further interpreted to have formed from overlapping peripheral bulges, generated by orogenic and sedimentary loading in the Appalachian foreland basin and by sedimentary loading in the intracratonic Michigan and Illinois Basins (Quinlan & Beaumont, 1984; Fig. 2). The Middle to Upper Ordovician records the foundering of the Cambrian–Early Ordovician passive margin (Knox Group) of eastern North America (Read, 1989). This attempted subduction of the eastern margin of Laurentia under a volcanic arc/accretionary wedge complex (Rodgers, 1971; Drake *et al.*, 1989) created a westward-migrating peripheral bulge that produced the regional Knox unconformity and a foreland basin extending from Alabama to Newfoundland (Jacobi, 1981; Mussman & Read, 1986; Lash, 1988). The global correlation of the Knox unconformity overlapping with the Sauk–Tippecanoe boundary (Sloss, 1963) also points to a eustatic role in its generation.

The proximal foredeep was bordered on the east by tectonic highlands that served as the principal siliciclastic source to the region for the remainder of the Ordovician. The western side of this foredeep passed into a shallow carbonate ramp developed over the Cincinnati Arch (Fig. 2). Major structural elements associated with this arch were the Nashville and Jessamine Domes on the crest, the Sebree trough on the western side and border faults of the Rome trough on the eastern side (Borella & Osborne, 1978; Cable & Beardsley, 1984; Weir *et al.*, 1984; Diecchio, 1993; Kolata *et al.*, 2001).

The Taconic foreland basin in the central and southern Appalachians experienced its earliest development during the Late Ordovician, and its history commonly is subdivided into two distinct phases (Fig. 3). The first phase, known as the Blountian (W-b to M4 sequences), began in the latest Llanvirn to earliest Caradoc and its depocentre was in eastern Tennessee (Shanmugam & Walker, 1980, 1982; Shanmugam & Lash, 1982; Ettensohn, 1994). The second phase, known as the Taconic (M5–C6 sequences), began in the middle to late Caradoc and its depocentre was primarily in eastern Pennsylvania and northern Virginia (Ettensohn, 1994). Each of these phases produced a distinctive second-order (10–15 Myr duration) depositional sequence, characterized by pronounced rapid deepening in the foredeep followed by prolonged shallowing and progradation on the adjacent shallow-water carbonate ramp (Rodgers, 1971).

The Blountian phase (Fig. 3) records the development of the rapidly subsiding Sevier foredeep in Alabama, Georgia, Tennessee and Virginia, which was subsequently filled with a thick turbidite and deltaic package (Shanmugam & Walker, 1980, 1982; Shanmugam & Lash, 1982; Ettensohn, 1991, 1994). These deep-water facies pass westwards and up-ramp into mid-ramp facies of the Chickamauga Group in Tennessee and coeval limestone units in Virginia (Read, 1980; Ruppel & Walker, 1984; Pope & Read, 1998). Farther to the west along the Cincinnati Arch, peritidal carbonates of the High Bridge and Stones River Groups accumulated under humid conditions on a nearly flat, muddy carbonate ramp in Kentucky, Virginia and Tennessee (Cressman & Noger, 1976; Holland & Patzkowsky, 1996, 1997; Pope & Read, 1998).

The Taconic phase (Fig. 3) is characterized by the Martinsburg foredeep, best developed in northern Virginia, eastern Maryland and eastern Pennsylvania (McBride, 1962; Read, 1989; Lash, 1988; Ettensohn, 1991). Similar to the Sevier foredeep, the Martinsburg foredeep was filled with a thick turbidite to deltaic package. These coarse siliciclastic sediments prograded westward, culminating in the Queenston Delta (Dennison, 1976; Kreisa, 1981; Diecchio, 1993). To the west, the foredeep passed up-dip onto a mixed carbonate–siliciclastic ramp developed over the Cincinnati Arch.

These Taconic phase rocks along the Cincinnati Arch record a fundamental change in sedimentation style from relatively pure peritidal carbonates deposited in warm waters (Stones River and High Bridge Groups), to cool, temperate water subtidal carbonates interbedded with varying amounts of shale (Brookfield, 1988; Patzkowsky & Holland, 1993; Lavoie, 1995; Pope & Read, 1998). Relatively pure subtidal and peritidal carbonates were deposited along the crest of the Cincinnati Arch and are surrounded by interbedded carbonate and shale of the Sebree trough and Appalachian basin foredeep (Keith, 1988a). The cleaner carbonate facies of the Lexington Limestone in Kentucky and Nashville Group in Tennessee (Fig. 3) are laterally equivalent to the well-described Trenton facies of the northern Appalachian Basin (Keith, 1988a,b; Brett *et al.*, 2004). Biostratigraphic evidence suggests that uppermost Ordovician (Gamachian) rocks are not preserved on the Cincinnati Arch or in the Appalachian foreland basin in Kentucky (Sweet, 1984). Such strata were probably not deposited or were eroded during the eustatic fall associated with short-lived terminal Ordovician glaciation (Dennison, 1976; Brenchley *et al.*, 1994). Uppermost Ordovician siliciclastics do occur in Virginia and Tennessee (Dorsch *et al.*, 1993) and in the Queenston delta farther north (Dennison, 1976).

EVIDENCE OF PERIPHERAL BULGE POSITION

Several lines of evidence suggest that the Cincinnati Arch was a persistent and relatively fixed site of shallow-water deposition throughout the Late Ordovician. In particular, the distribution of peritidal facies, shoal-water facies, buildups and unconformities provide evidence that the Cincinnati Arch was a shallow-water to emergent feature throughout most of the Late Ordovician.

Peritidal facies

Peritidal facies are consistently associated with the crest of the Cincinnati Arch, particularly the

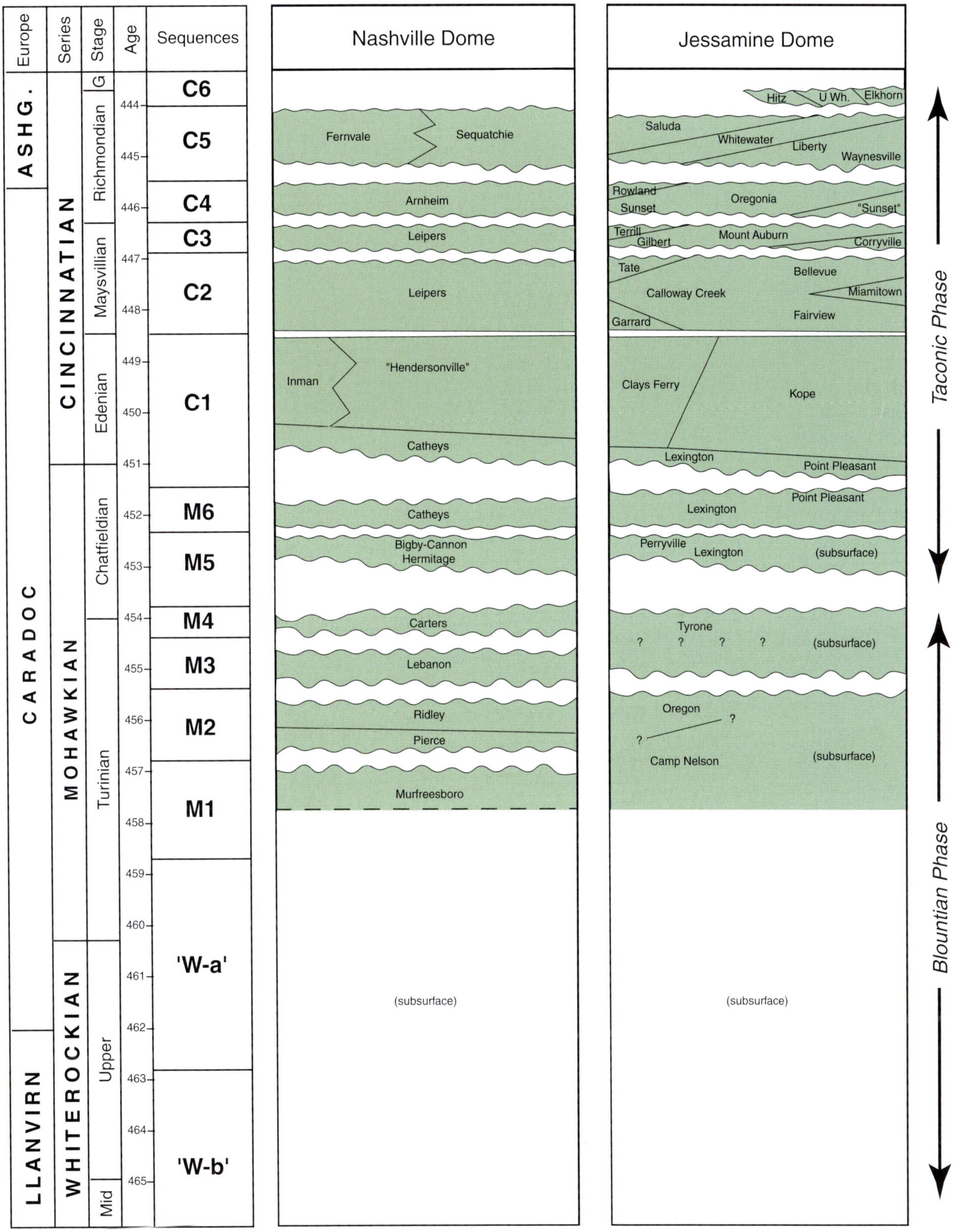

Fig. 3. Lithostratigraphic units and sequence stratigraphic architecture of Upper Ordovician strata on the Cincinnati Arch, following nomenclature of Holland & Patzkowsky (1996, 1998). Age is based on a calibration of the conodont-based graphic correlation of Sweet (1984); see Holland & Patzkowsky (1998) for details of age calibration.

Jessamine and Nashville Domes in the Upper Ordovician.

Blountian phase

On the Jessamine Dome in Kentucky, the basal unit of the Blountian phase is the Wells Creek Formation, a locally sandy dolostone deposited in a peritidal setting. The bulk of the overlying High Bridge Group consists predominantly of intensely bioturbated shallow subtidal to intertidal limestone (Camp Nelson and Tyrone Limestones) and lesser amounts of dolostone (Oregon Formation; Cressman & Noger, 1976; Kuhnhenn & Haney, 1986; Pope & Read, 1998). Parasequences in the High Bridge Group commonly consist of thin, intraclastic packstone bases that grade upward into pellet packstone. This unit is in turn overlain by skeletal/pelletal wackestone/mudstone capped by fenestral mudstone (Pope & Read, 1998). Most of these parasequences are difficult to trace laterally and probably formed by autocyclic processes (Ginsburg, 1971; Pratt & James, 1986).

The Oregon Formation, a thin, diagenetic unit of peritidal dolostone occurs in a roughly semicircular region (~1000 km^2) over the Jessamine Dome (Fig. 4) on the Cincinnati Arch within the otherwise limestone-rich High Bridge Group (Wolcott *et al.*, 1972; Cressman & Noger, 1976; Kuhnhenn *et al.*, 1981; MacQuown *et al.*, 1984; Kuhnhenn & Haney, 1986). The Oregon Formation contains abundant desiccation and prism cracks, which suggest increasing aridity and emergence of the platform during this time (MacQuown *et al.*, 1984). The dolostone is fine-grained, fabric-preservative, and probably formed by evaporative pumping on a topographically high area (Horrell, 1981; Kuhnhenn *et al.*, 1981; Kuhnhenn & Haney, 1986), indicating that this area of the arch was emergent and probably surrounded by marine waters when the dolostone formed.

On the Nashville Dome in Tennessee, peritidally capped parasequences occur in the Blountian phase in the Murfreesboro, Ridley, Lebanon and Carters Limestones of the Stones River Group on the Nashville Dome (Fig. 3; Holland & Patzkowsky, 1998). Most of these peritidal caps are bioturbated fenestral lime mudstones, suggesting humid depositional conditions, but some, such as those at the top of the Upper Carters, consist of desiccation-cracked, laminated, dolomitic mudstones, indicative of more arid conditions. The distribution of all of these tidal-flat deposits shows no clear pattern over the Nashville Dome, suggesting

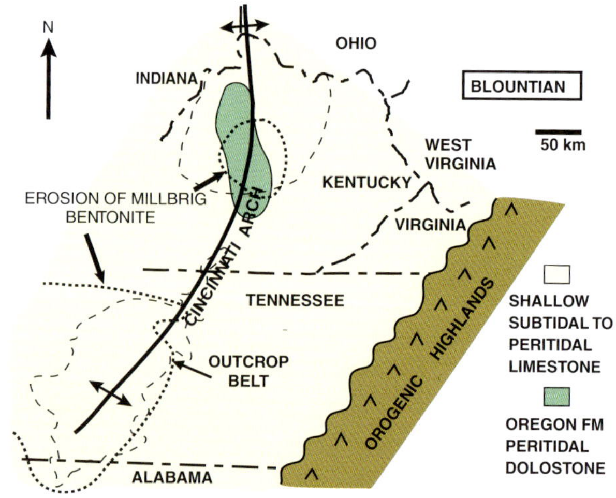

Fig. 4. Map of Blountian features along the Cincinnati Arch. The distribution of the Oregon Formation indicates that it formed along a topographic high (Wolcott *et al.*, 1972; Kuhnhenn *et al.*, 1981). The Millbrig K-bentonite is eroded over much of the Jessamine and Nashville Domes (adapted from Scharpf, 1990; Haynes, 1994; Kolata *et al.*, 1998) suggesting that these areas were topographically high or exposed during deposition of this widespread ash bed.

that the tidal flats developed as a series of isolated islands (cf. Pratt & James, 1986) rather than from a persistent localized high. However, the lack of similar coeval peritidal facies in the Sequatchie Valley to the immediate southeast of the Nashville Dome indicates that the Nashville Dome was a broad, shallow-water feature during the Blountian phase of the Taconic Orogeny.

Taconic Phase

The Lexington Limestone and Nashville Group (Fig. 5) record shallow to deep, predominantly subtidal, cool-water carbonate and shale deposition over the Cincinnati Arch (Cressman, 1973; Patzkowsky & Holland, 1993; Holland & Patzkowsky, 1996; 1997; Pope & Read, 1997b). Peritidal facies in these units are rare and consist of bioturbated fenestral limestone. They are localized over the arch (Fig. 6) and pass in all directions into deeper-water facies (Hrabar *et al.*, 1971; Mackey, 1972; Cressman, 1973; Ettensohn, 1992; Holland & Patzkowsky, 1996, 1997, 1998; Pope & Read, 1997a,b). The peritidal facies of the humid Perryville tidal-flat complex (Mackey, 1972) were best developed along a north–south line along the Cincinnati Arch, beneath a third-order sequence boundary in the Lexington Limestone (Pope & Read, 1997a, 1998). This tidal-flat succession is correlative with similar facies in the

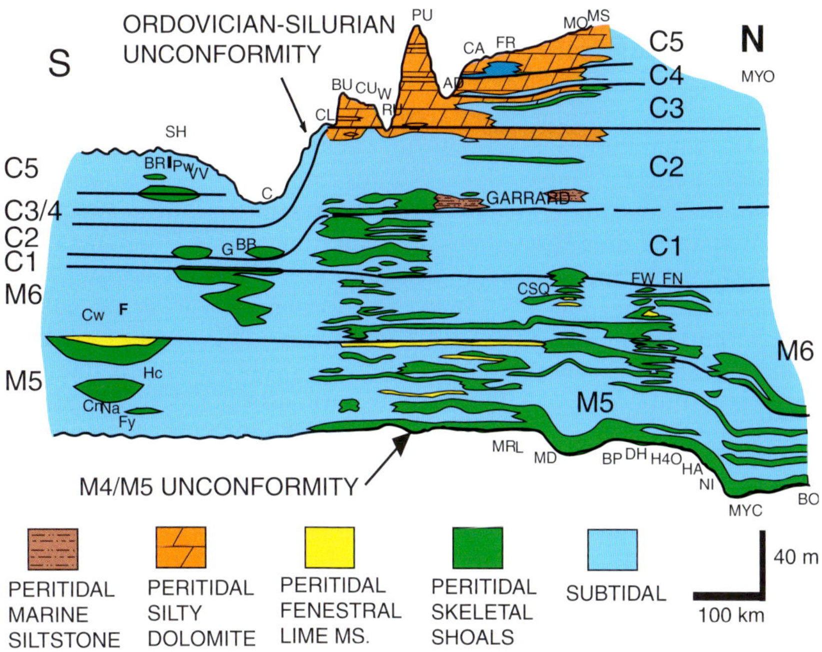

Fig. 5. Regional north–south cross-section of M5–C5 strata on the Cincinnati Arch. Initials are locations of measured sections, see Pope & Read (1997b) and Holland & Patzkowsky (1996, 1997, 1998) for individual sections. Note how sedimentation and stratigraphy in M5 and M6 are similar but there is a significant change in sedimentation and stratigraphy during C1–C5.

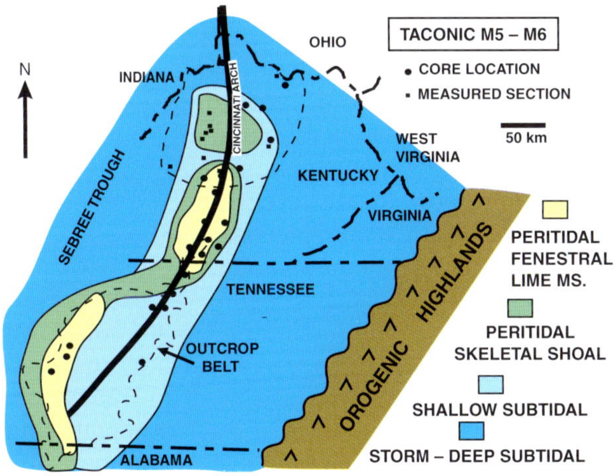

Fig. 6. Map showing distribution of facies during the M5 and M6 sequences along the Cincinnati Arch. Note restriction of peritidal and shoal facies to a stable belt during this interval.

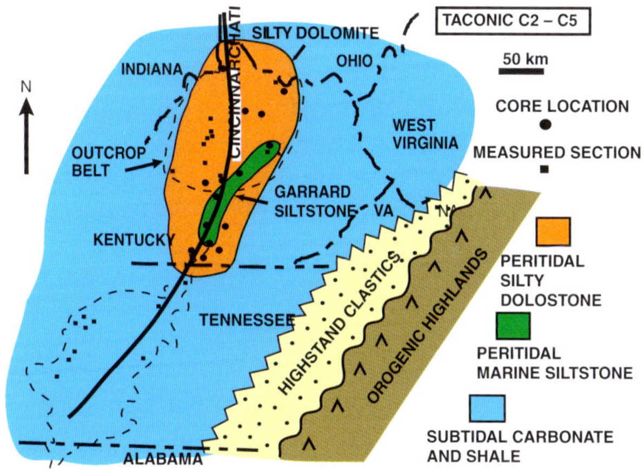

Fig. 7. Map of shallow-water features of the C2–C5 sequences, including silty peritidal dolomite and the tidally influenced Garrard Siltstone, both of which were deposited along the Cincinnati Arch.

Bigby-Cannon limestone (Holland & Patzkowsky, 1997). Peritidal caps in the Bigby-Cannon and Catheys are thicker and more numerous along the western side of the Nashville Dome (Wilson, 1949), suggesting that the western flank of this dome was a persistently shallow region during the M5 and M6 sequences. As waters continued to deepen during the M6 and C1 sequences, peritidal facies became exceedingly rare, only occurring as isolated islands within grainstone shoals on the Jessamine Dome (e.g. Devils Hollow Member of the Lexington Limestone, Cressman, 1973; Pope & Read, 1997a; Ettensohn *et al.*, 2004) and restricted to the southern margin of the Nashville Dome in the Catheys and Inman formations (Wilson, 1949; Holland & Patzkowsky, 1998). Peritidal facies in the Inman Formation are green dolomitic mudstones with desiccation cracks.

Peritidal facies (Tate, Terrill, Rowland, Saluda and Preachersville Members) occur in the Cincinnatian (C2–C5) over the Jessamine Dome (Fig. 7) and nearly everywhere consist of silty laminated dolostone with abundant desiccation cracks and sparse, short vertical burrows. These peritidal facies consistently pass northwards into coeval shallow subtidal facies (Weir *et al.*, 1984; Holland, 1993; Pope & Read, 1997b). Localization of these peritidal facies demonstrates that the Jessamine Dome in central Kentucky was a persistent shallow area throughout the Cincinnatian Series. Only in the upper Cincinnatian (C5–C6) do these peritidal facies expand northward and down-ramp on the Cincinnati Arch into Ohio and Indiana (Hatfield, 1968).

Shoal-water facies

Peloidal–skeletal grainstone shoals were commonly affected by tidal currents along the Cincinnati Arch during the Late Ordovician. These shoals separated open-marine facies from more restricted facies deposited in semi-restricted lagoons, subtidal pools or tidal flats. The facies behind the grainstone shoals are mud-rich wackestone/packstone containing abundant stromatoporoids, molluscs, ostracods and rare corals (Wilson, 1949; Walker & Laporte, 1970; Mackey, 1972; Cressman, 1973; Holland, 1993). Shoal facies consistently developed in relatively fixed positions on the Jessamine and Nashville Domes in the Late Ordovician.

Blountian phase

On the Jessamine Dome, cross-bedded peloidal packstone/grainstone is rare, and only locally developed in the High Bridge Group. Cross-bedded carbonate sand-shoal facies developed sporadically in the Pierce, Ridley, Lebanon and Carters limestones of the Nashville Dome, and they are nearly always local in their occurrence, suggesting the presence of isolated, ephemeral skeletal to peloidal sand shoals (Holland & Patzkowsky, 1998). One exception to this pattern is the massive member of the Lebanon Limestone, which is a widespread ooid–skeletal cross-bedded grainstone, that occurs over the entire Nashville Dome without systematic variations in thickness (Wilson, 1949).

Taconic phase

On the Jessamine Dome, skeletal packstone and grainstone of the transgressive Curdsville Member of the Lexington Limestone are widespread and form a series of hardground-bounded parasequences (Pope & Read, 1997a). These grainstones are rich in phosphate, and indicate that strong waves and tidal currents commonly affected the Jessamine Dome during this transgression.

Tidally influenced skeletal grainstone facies of the Lexington Limestone (Fig. 6) are best developed along the axis of the Cincinnati Arch and over the Jessamine Dome (Hrabar *et al.*, 1971; Cressman, 1973; Borella & Osborne, 1978; Pope & Read, 1997, 1998). These shoal-water facies are rich in phosphate, comprised predominantly of brachiopods, bryozoans and crinoids and commonly have herringbone cross bedding oriented to NW–SE (Hrabar *et al.*, 1971; Pope & Read, 1997a). These grainstone shoals developed on local topographic highs over the Jessamine Dome (Hrabar *et al.*, 1971; Cressman, 1973; Ettensohn, 1992).

The Garrard Siltstone, the coarsest siliciclastic unit along the Cincinnati Arch in the Upper Ordovician, was deposited in shallow to deep waters (Ettensohn *et al.*, 1986a; Jacobs, 1986) and is laterally equivalent to high-energy skeletal grainstones in the subsurface to the south (Pope & Read, 1997b). Bimodal palaeocurrent measurements in the Garrard Siltstone suggest at least part of the Garrard Siltstone was deposited in a tidal setting. Outcrops of the Garrard Siltstone are restricted to a linear trend along the southeastern side of the Cincinnati Arch, suggesting its distribution was structurally controlled (Weir *et al.*, 1984; Ettensohn, 1986a; Jacobs, 1986).

A belt of grainstone shoals is sporadically present during the C3 to C5 sequences on the northern flank of the Jessamine Dome, but is oriented east–west (Weir *et al.*, 1984; Holland, 1993). These grainstones are restricted to the transgressive systems tracts of these sequences and separate restricted lagoonal mudstones and wackestones to the south (e.g. Gilbert, Sunset Members) from open-marine shallow subtidal packstones and fossiliferous mudstones to the north (e.g. Mount Auburn, Oregonia formations). These grainstone shoals are typically approximately 1 m thick and display large-scale trough cross-bedding. They are composed primarily of crinoid ossicles and highly abraded skeletal material, but are also rich in phosphate and are locally enriched in thick ramose, frondose and massive bryozoans, all highly abraded and bored.

On the Nashville Dome, transgressive grainstones of the Curdsville Member lying above the M4–M5 unconformity are restricted to the flanks of the Nashville Dome and correlate to a well-developed hardground/unconformity developed on the Stones River Group (i.e. Curdsville Island of

Wilson, 1962; Holland & Patzkowsky, 1997). Like the peritidal facies, shoal facies on the Nashville Dome display a more organized pattern of distribution in the M5–C5 sequences. Cross-bedded, crinoidal to skeletal, phosphatic grainstones are a prominent feature of the Bigby-Cannon, Catheys, Leipers and Fernvale formations. Although such grainstones are present locally at the base of several parasequences, they are thickest and best developed along a persistent north–south belt (Fig. 6) on the western flank of the Nashville Dome (Wilson, 1949; Holland & Patzkowsky, 1998). There, the cross-beds display a strong bimodal pattern, suggesting tidal influence and the development of a tidal-bar belt (Ball, 1967) along the edge of the Sebree trough, just west of the Nashville Dome. This belt of carbonate sands passes westward into a thick package of deep subtidal storm beds and passes eastward into bioturbated shallow subtidal packstones.

Buildups

Organic buildups (generally biostromes, less commonly low-relief bioherms) are not common on the Jessamine and Nashville Domes and are rarely more than a metre thick. Commonly, the organisms capable of producing organic biostromes are associated with shoal and peritidal facies and have a relatively fixed distribution in the Late Ordovician.

Blountian phase

On the Jessamine and Nashville Domes, numerous organisms capable of constructing biostromes are present in the High Bridge and Stones River Groups. Most commonly, these include stromatoporoids and tabulate corals (favositids and tetradiids; Wilson, 1949; Patzkowsky & Holland, 1999). However, in only a few cases do they occur in concentrations dense enough to constitute biostromes. For example, small, isolated buildups developed in the Carters Limestone near the southern margin of the Nashville Dome (Alberstadt *et al.*, 1974). These biostromes consist of tabulates, stromatoporoids and bryozoans and are up to a few metres thick. More commonly, biostrome-producing organisms are present but only sparsely distributed through shallow subtidal bioturbated wackestone and packstone facies (Cressman & Noger, 1976; Patzkowsky & Holland, 1999).

Taconic phase

In the M5 through C6 sequences of the Jessamine and Nashville Domes, a wider array of potential biostrome-producing organisms is present than during the Blountian phase and includes chaetetids (*Solenopora*; see Riding, 2004), stromatoporoids, tabulates, colonial rugosans and bryozoans. Several distinct biostromes are present, but all are restricted to a few square kilometres in area or less and do not repeatedly occupy the same region. Shallow subtidal, shoal and peritidal facies also all commonly contain these potential buildup-constructors (Wilson, 1949; Weir *et al.*, 1984; Holland & Patzkowsky, 1998; Pope & Read, 1998; Patzkowsky & Holland, 1999).

On the Jessamine Dome, the Sulphur Well Member of the Lexington Limestone is a bryozoan biostrome up to 5 m thick, surrounded by a grainstone shoal along the axis of the Cincinnati Arch (Cressman, 1973; Ettensohn *et al.*, 1986b). Deposition of the Sulphur Well Member is hypothesized to be structurally controlled (Ettensohn *et al.*, 2004). Stromatoporoids and chaetetids locally form biostromes in the Lexington Limestone (McLaughlin *et al.*, 2004).

Bryozoans also form small, localized biostromes in shallow subtidal facies throughout the Cincinnatian (Cuffey, 1998), but these are typically less than a metre thick and less than a few dozen square metres in area. A small patch biostrome consisting of chaetetids is present in one outcrop in the Cincinnatian near Winchester, Kentucky (Weir *et al.*, 1984). The C5 sequence in northern Kentucky and southeastern Indiana contains a series of biostromes composed of colonial rugosans and tabulates, including the Bardstown, Kentucky 'reef' (Browne, 1964), the Otter Reef Coral Bed near Richmond, Kentucky (Simmons & Oliver, 1967), and various biostromes associated with Saluda peritidal facies (Hatfield, 1968). In most places, these biostromes are approximately 1 m thick.

On the Nashville Dome, few biostromes are known from the Taconic phase. One low (50 cm), small (100 m in length) biostrome composed entirely of the chaetetid *Solenopora* is present in the C1 sequence along the western flank of the Nashville Dome near Franklin, Tennessee (Holland & Patzkowsky, 1998, their Fig. 11). Large domal stromatoporoids are common in grainstone-rich facies of the M5 and M6 sequences, as they are on the Jessamine Dome, and form only localized biostromes up to a metre thick. Rugosan and tabulate biostromes are absent in the Cincinnatian on the Nashville Dome, unlike the Jessamine Dome and northern Cincinnati Arch.

Unconformities

Unconformities separating third-order depositional sequences record their greatest erosion or period of non-deposition on the Jessamine and Nashville Domes along the crest of the Cincinnati Arch in the Late Ordovician, and commonly pass off the arch into downdip conformities.

Blountian phase

Regional truncation has not been documented at the M2–M4 sequence boundaries over the Jessamine Dome or the M1–M4 boundaries on the Nashville Dome, at least in part owing to a lack of marker horizons such as K-bentonites. Nonetheless, these erosional surfaces provided the material for multiple intraclastic beds at the bases of these depositional sequences (Kuhnhenn *et al.*, 1981; Holland & Patzkowsky, 1996). Additionally, the occurrence of a laterally continuous dolostone (Oregon Formation) in this otherwise limestone dominated unit, located on the Jessamine Dome suggests that this area was subaerially exposed for a protracted interval following deposition.

K-bentonite ash beds are common in the upper part of the Blountian succession and two of these beds (Deicke and Millbrig) are regionally widespread, facilitating correlations throughout the eastern mid-continent and Appalachian Basin (Haynes, 1994; Kolata *et al.*, 1996). The Millbrig (locally known as Mud Cave or T-4) K-bentonite, which occurs immediately below the High Bridge Group–Lexington Limestone contact, is eroded over much of the Jessamine Dome and is preserved only in structural lows (Fig. 4; Haynes, 1994; Kolata *et al.*, 1996, 1998). The erosional relief of the post-High Bridge Group unconformity (M4–M5) is localized over the Jessamine Dome on the Cincinnati Arch where up to 5 m of the Tyrone Limestone is missing (Cressman, 1973). Over the Jessamine Dome, the basal Lexington Limestone commonly contains clasts up to 10 cm in diameter of Tyrone Limestone lime mudstone. This unconformity passes down-ramp into a correlative conformity (Pope & Read, 1997b).

Over the Nashville Dome, up to 10 m of the Carters Formation, including both the Deicke and Millbrig K-bentonites are locally eroded, with the amount of erosion increasing towards the western flank of the Nashville Dome (Wilson, 1962; Wilson & Stearns, 1963; Holland & Patzkowsky, 1997). However, the K-bentonites are present on the eastern flank of the dome and in the subsurface surrounding the dome (Haynes, 1994), indicating that erosion was limited to the topographically higher dome. Erosional fragments of the Carters Formation occur locally in the basal Hermitage Formation (Wilson, 1962; Holland & Patzkowsky, 1998).

Taconic phase

Widespread subaerial unconformities in the Lexington Limestone developed only above the regionally developed tidal-flat facies (Perryville Member – Lexington Limestone; Dove Beds – Nashville) on the Cincinnati Arch (Mackey, 1972; Holland & Patzkowsky, 1996, 1997; Pope & Read, 1997a,b) and pass down-ramp into conformable subtidal surfaces. Along the arch the unconformity above the Perryville tidal-flat facies has small karstic cavities up to 1.5 m deep that are filled with phosphatic sand produced during the subsequent transgression (Pope & Read, 1997b). Locally developed tidal flat facies (cf. Devils Hollow Member – Lexington) show less evidence for subaerial exposure and pass laterally into phosphatic hardgrounds often capped by phosphatic grainstone.

Subaerial unconformities are more easily recognized over the Nashville Dome due in large part to the greater abundance of peritidal lime mudstones (Dove Beds) in this area (Wilson, 1962). However, the most widespread unconformity occurs at the M6 boundary and this surface is directly correlative with the unconformity above the Perryville Member in Kentucky (Holland & Patzkowsky, 1996, 1997).

Third-order sequence boundaries in the Cincinnatian Series (C1–C6) on the Jessamine Dome and northern Cincinnati Arch rarely display obvious evidence of differential truncation of underlying strata, in part because of an absence of marker horizons such as bentonites. The C6 sequence, however, is preserved only on the northern half of the Cincinnati Arch and shows a depositional pinchout to the south. This restriction to the northern arch is consistent with the distribution of peritidal facies in the Cincinnatian (Figs 5 and 7), as well as the east–west orientation of the grainstone shoals in the C2–C4. Collectively, these features suggest a persistent northward dip of Jessamine Dome and northern Cincinnati Arch in the C2–C6 sequences, with persistently shallower conditions to the south.

On the Nashville Dome, truncation at the M4–M5 boundary, made visible by K-bentonites as in central Kentucky, reveals a strong east–west pattern of truncation, with at least 7 m more strata eroded at this sequence boundary in the west than the east (Holland & Patzkowsky, 1998, their Fig. 9). A major truncation surface is also present at the C5 sequence boundary, which displays the progressive loss of the underlying C4, C3 and partially the C2 sequences from north to south (Holland & Patzkowsky, 1998, their Fig. 11). This pattern, strikingly different from that seen in the M5 through C4, suggests a tectonic reconfiguration on the Nashville Dome prior to deposition of C5.

EVIDENCE OF TECTONIC ACTIVITY

Several lines of evidence, including seismites, syntectonic faulting and differential subsidence, all suggest that the Cincinnati Arch was subjected to seismic activity during the Late Ordovician.

Seismites

Several horizons of seismically induced, synsedimentary ball-and-pillow beds (seismites) occur within Upper Ordovician strata of the Jessamine Dome (Pope *et al.*, 1997; Ettensohn *et al.*, 2002; Jewell & Ettensohn, 2004; MacLaughlin & Brett, 2004). These seismites indicate that occasional large earthquakes (Moment Magnitude 5–8) occurred in this region of the Cincinnati Arch during the Late Ordovician (Pope *et al.*, 1997). However, whether these faults formed as a response to local stress or by stresses generated much farther afield in the foreland basin (Jewell & Ettensohn, 2004; McLauglin & Brett, 2004) is unclear.

Similar seismite beds occur in M5–C5 strata on the Nashville Dome (Holland & Patzkowsky, 1997). Most of these are visible in the southern half of the Nashville Dome, where individual horizons can be correlated for tens of miles from east to west. Some of these seismites are closely associated with sequence boundaries, such as at the C5 sequence boundary and at the Ordovician–Silurian unconformity; it is unclear if this association reflects enhanced preservation or increased seismicity (cf. McLaughlin & Brett, 2004).

Faults and structural features

The Jessamine Dome is a fault-bounded structure that localized the development of tidally influenced skeletal grainstones and stromatoporoid buildups in the Lexington Limestone (Hrabar *et al.*, 1971; Cressman, 1973; Ettensohn, 1992; Pope & Read, 1995, 1997a). The Kentucky River Fault system and other unnamed faults that delineate this structure are basement features that were probably reactivated by Taconic orogenesis (Ettensohn, 1992; Ettensohn *et al.*, 2002, and references therein). Synsedimentary growth faults are documented from the Lexington Limestone in Kentucky (Black & Haney, 1975; Pope *et al.*, 1997). It is not known whether fault structures are responsible for the persistent position of shoals and tidal flats on the Nashville Dome, but such consistency in their occurrence hints at the role of prior structures.

Peritidal facies of the Perryville Member of the Lexington Limestone commonly are laterally equivalent to deeper water shale that occurs in east–west trending zones, which may reflect differential subsidence perpendicular to the regional trend of the Cincinnati Arch (Mackey, 1972; Kulp, 1995; Kulp & Ettensohn, 1995; Ettensohn *et al.*, 2002). Structurally controlled east–west trending troughs produce alternating topographic highs and lows along the arch with tidal flats restricted to highs and slightly deeper water channels (Mackey, 1972) occupying lows. The close association of these channels with exposed fault segments suggests that these were produced by reactivation of Precambrian or earlier Palaeozoic faults (Cressman, 1973; Black, 1986; Kulp, 1995; Kulp & Ettensohn, 1995; Rast & Goodman, 1995; Pope & Read, 1997a; Ettensohn *et al.*, 2002, 2004).

The Sebree Trough is an enigmatic narrow (25–150 km wide) and elongate (600 km long) deep-water basin that formed west of the Cincinnati Arch and filled with black shale during the Chatfieldian–Richmondian (Kolata *et al.*, 2001). The Sebree Trough is interpreted to have formed over the Precambrian–Cambrian Reelfoot rift (Kolata *et al.*, 2001) or through the reactivation of basement structures by far-field stresses from foreland basin loading (Ettensohn *et al.*, 2002).

Subsidence history

Backstripping was used to reconstruct the accommodation history of the Jessamine and Nashville Domes (Fig. 8; see Holland & Patzkowsky (1998) for details on methods). Although the eustatic and tectonically driven subsidence components cannot be separated from these limited data, it is assumed that the longer-term features of this accommodation history reflect tectonic subsidence, whereas the short-term features (<1–3 Myr) are thought to reflect eustasy (Holland & Patzkowsky, 1998).

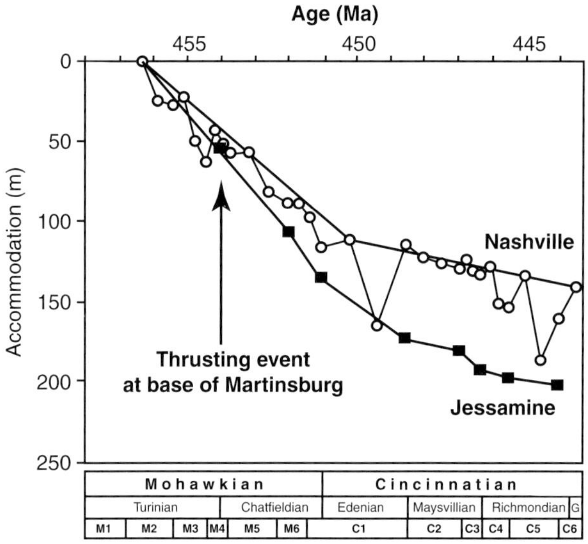

Fig. 8. Accommodation histories generated by backstripping analysis of sections from the northern Nashville Dome and from the central Jessamine Dome. Long-term patterns are thought to reflect primarily subsidence histories, whereas short-term patterns (<1 Myr) are thought to reflect eustasy.

Through the M6 sequence, the Jessamine and northern Nashville Domes have comparable subsidence rates, suggesting that these regions were undergoing moderate subsidence, although it is possible that this history actually reflects slow long-term eustatic rise. Beginning in the C1 sequence, however, the subsidence histories of these two regions began to diverge. Although subsidence rates on the Jessamine Dome are only slightly lower than before, those on the northern Nashville Dome underwent a fourfold decrease. This pattern of relatively faster subsidence rates to the north along the Cincinnati Arch is true even north of the Jessamine Dome, where Cincinnatian strata in southwestern Ohio and southeastern Indiana are 40% thicker than those on the Jessamine Dome. Likewise, the pattern of slower subsidence rates to the south is consistent with the erosional loss of the C2–C4 sequences before C5 deposition on the southern Nashville Dome. This change in subsidence rates is notable in that it does not coincide with the onset of the Taconic phase early in the M5, but is initiated 2–3 Myr later in the Edenian Stage (C1).

These subsidence histories, combined with the lack of any long-duration unconformities, suggest that the Jessamine and Nashville Domes were slowly subsiding features throughout the Late Ordovician, rather than actively uplifting features. The difficulty of disentangling eustasy and tectonic subsidence in single locations, however, does open the possibility that the Cincinnati Arch may have experienced some uplift, particularly in the northern Nashville Dome region, but that this uplift was combined with a somewhat larger rate of eustatic rise to produce a net slow positive rate of accommodation. The truncated record of the C2–C4 sequences on the southern Nashville Dome more strongly suggests uplift.

DISCUSSION

Comparison to flexural models

Based on viscoelastic and elastic modelling of foreland basin stratigraphy, a peripheral bulge should migrate towards the craton as the orogenic load advances onto the continent (Quinlan & Beaumont, 1984; Flemings & Jordan, 1989, 1990; Jordan & Flemings, 1991; Dorobek, 1995). Furthermore, viscoelastic models predict that the peripheral bulge should move toward the thrust load and the foredeep should deepen as loading is 'relaxed'. The Taconic foreland basin has been interpreted to record such a viscoelastic response to loading (Quinlan & Beaumont, 1984; Tankard, 1986; Beaumont *et al.*, 1988; Ettensohn, 1991, 1992, 1994). However, if the Cincinnati Arch was a peripheral bulge, it appears that its position did not migrate during the Late Ordovician. Instead, it appears to have been a relatively fixed, or static, feature throughout the Taconic Orogeny.

Additionally, the boundaries for many of these features (e.g. the Jessamine Dome, Sebree Trough) are fixed and clearly associated with faults (Ettensohn, 1992; Kolata *et al.*, 2001; Ettensohn *et al.*, 2002).

Scale issues

The width of the preserved Taconic foreland basin is ~200–300 km, from the western edge of its thrust front to the eastern edge of the peripheral bulge (Fig. 2). Undoubtedly, the width of the Taconic foreland basin was probably larger during deposition due to subsequent shortening during thrusting, however, according to equation (4) the peripheral bulge for a basin of this width should be ~250–400 km wide. However, all shallow-water features in the Upper Ordovician along the Cincinnati Arch occur within an area less than 150 km in width (Figs 4, 6 and 7).

Similarly, wireline-log correlations across the Taconic foreland basin from Kentucky to Virginia indicate that the zone of shallow-water sediments across the Cincinnati Arch is less than 180 km wide (Hohman, 1998; Pope & Read, 1997a). This suggests the Cincinnati Arch is far too narrow a feature to be a peripheral bulge formed by flexural loading of a continuous elastic or viscoelastic plate. We suggest the Cincinnati Arch is so narrow because it was forming as a response to the reactivation of basement features, where the flexural load stresses were transmitted to the same location repeatedly over a prolonged period of time (10–12 Myr).

The Sebree Trough also could be interpreted as a back-bulge basin to the Cincinnati Arch. However, the Sebree Trough is also too narrow a feature to be a back-bulge basin, since the width of the back-bulge basin should be comparable to the width of the forebulge (Turcotte & Schubert, 2002). Likewise, the extremely small amount of subsidence that could be generated from lithospheric flexure in a back-bulge basin (<0.2% of the subsidence at the load) also argues against such an origin for the deep-water Sebree Trough.

Persistence of Cincinnati Arch

The localization of peritidal and shoal-water facies, as well as buildups and unconformities in the Upper Ordovician along the Cincinnati Arch all indicate that the crest of the Cincinnati Arch was a relatively high topographic feature throughout much of the Taconic Orogeny. Such topographic persistence does not necessarily imply uplift and could reflect lower subsidence rates on the domes relative to surrounding regions. The Jessamine and Nashville Domes were the most common sites of shallow-water sedimentation on the arch, which was surrounded at all times by deeper water. Additionally, while the record of sedimentation on these domes is similar, it is not identical and these variations in sedimentation and stratigraphy probably provide information about coeval tectonism and its sedimentary response.

Erosion of the High Bridge and Nashville Groups as well as the Millbrig and Deicke K-bentonites indicates that the Jessamine and Nashville Domes were topographically positive features subjected to subaerial exposure and erosion prior to Taconic phase deposition on these domes. Erosion of the Deicke K-bentonite over much of the Nashville Dome area and greater amount of missing Blountian sediments indicate that this structure experienced more widespread erosion than the Jessamine Dome prior to deposition of the overlying M5 strata.

Basement control on location of the Cincinnati Arch

Evidence for the persistence of the Cincinnati Arch indicates that it formed in the same location repeatedly. Peripheral bulges may not migrate, or can appear to reoccur in the same position repeatedly (Waschbusch & Royden, 1992a,b), owing to inherited structural weaknesses or heterogeneities in the underlying lithosphere (Fig. 1). In such a scenario, the weaker regions of the lithosphere bend more easily than stronger regions, so plate bending is concentrated within the weak zone and the peripheral bulge becomes 'fixed'. In this case, the width of the peripheral bulge should reflect the width of the weak zone, not the flexural wavelength of the plate (Waschbusch & Royden, 1992a).

The most prominent heterogeneity in the basement along the Cincinnati Arch (Fig. 9) is the boundary between the Grenville front and the Eastern Granite–Rhyolite Province (EGRP) and the overlying East Continent Rift Basin (ECRB) that parallels the eastern limit of the Cincinnati Arch (Drazhoval *et al.*, 1992; Van Schmus *et al.*, 1993). The Grenville front consists of high-grade metamorphic rocks that were thrust over the calc-alkaline rocks of the EGRP and rift-related basalts and sediments of the ECRB (Drazhoval *et al.*, 1992; Hauser, 1993). In Kentucky the ECRB occurs in a restricted zone that narrows to the south and is probably completely buried in Tennessee (Fig. 9). We suggest that the ECRB was a zone of lithospheric weakness and that stresses imposed on the eastern edge of the Grenville lithosphere were transmitted westward and concentrated in this area, producing increased uplift or decreased subsidence along the overlying Cincinnati Arch, adjacent to the boundary between these two zones. This hypothesis explains the narrow width of this peripheral bulge (as manifested in the north–south alignment of shallow-water features and unconformities), and its occurrence in the same location for such a prolonged period during the Late Ordovician. A similar scenario, involving isostatic uplift of low-density Proterozoic sediments to the west of the Grenville basement was proposed for uplift along the Cincinnati Arch associated with Acadian (Silurian–Devonian) deformation (Rast & Goodman, 1995).

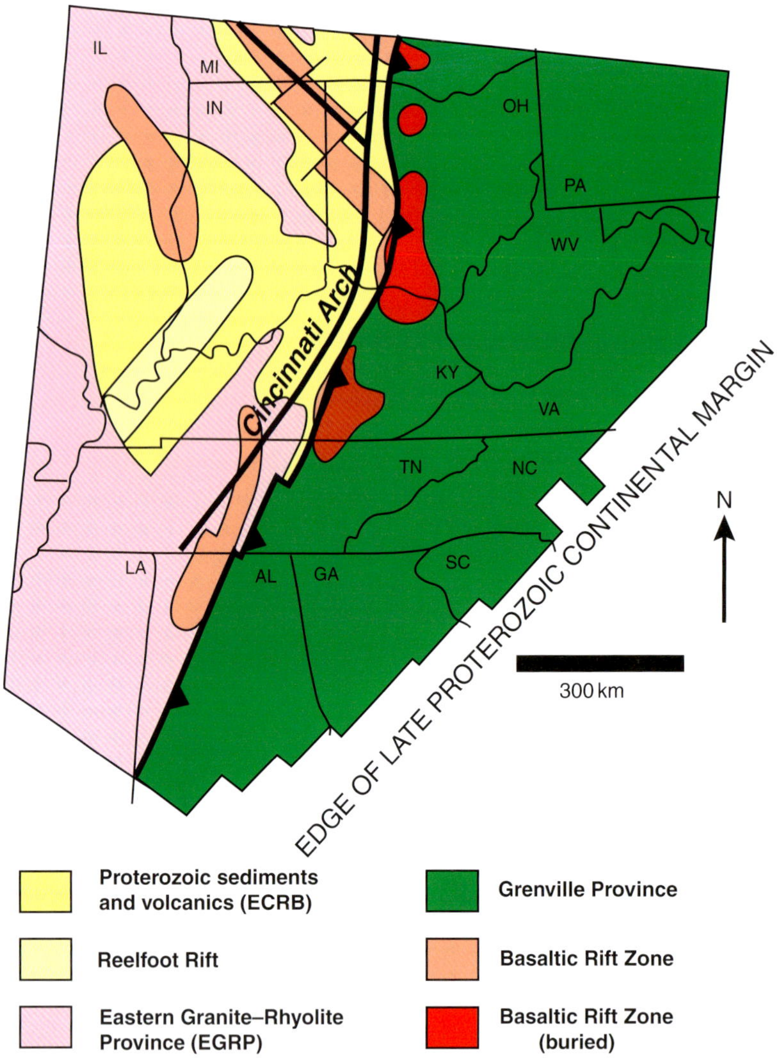

Fig. 9. Map showing the relation between the Cincinnati Arch and the boundary between Grenville crust and the Eastern Granite–Rhyolite Province (EGRP) and the East Continent Rift Basin (ECRB). Adapted from Drazhoval *et al.* (1992). The Cincinnati Arch lies just west of this boundary and it is suggested that its relatively fixed position and narrow width are due to uplift or decreased subsidence within this zone of weak lithosphere. In this scenario, orogenic stresses generated to the east were repeatedly transmitted to this zone of weak lithosphere, which responded by syndepositional faulting and resisting subsidence.

The stresses that produced the Cincinnati Arch in the Late Ordovician were probably responsible for the reactivation of structural features and formation of syndepositional faults and seismites in this area (Ettensohn *et al.*, 2002). Similar features in the Upper Ordovician and Middle Silurian farther north along the Cincinnati Arch also are attributed to reactivation of basement faults along the boundary between the Grenville Front and EGRP/ECRB (Onasch & Kahle, 1991, 2002). Additionally, the long-distance transport of these Taconic orogenic stresses may have reactivated extensional faults of the underlying Reelfoot rift to produce the Sebree Trough (Kolata *et al.*, 2001).

Lateral sedimentary variation along the peripheral bulge

Inherited structural features can be reactivated and can affect sedimentological and stratigraphic patterns in ways not predicted by stratigraphic models, such as alternating highs and lows along the axis of an arch (Bradley & Kidd, 1991; Diecchio, 1993; Holland & Patzkowsky, 1997). North-to-south variations in sedimentology and stratigraphy of the Upper Ordovician on the Cincinnati Arch, specifically from the Jessamine and Nashville domes, indicate that this peripheral bulge did not behave uniformly throughout this period, instead recording spatial and temporal variations in its response to coeval tectonism.

For instance, there is a change in depositional style from sequences M1–M4 to sequences M5–C6. Sequences M1–M4 are a thick succession of shallow subtidal to peritidal carbonates that were deposited on a homogeneous, flat to extremely gently dipping ramp where facies changes across this platform were subtle. Peritidal carbonates rarely amalgamated into regional markers because there was probably a series of peritidal islands that evolved separately, rarely forming extensive regional units (Pratt & James, 1986).

After the M4 sequence, the ramp was more clearly differentiated, with grainstone shoals, peritidal facies and skeletal buildups arrayed as linear features along the Cincinnati Arch (Mackey, 1972; Holland & Patzkowsky, 1996, 1997; Pope & Read, 1997a; Ettensohn *et al.*, 2002, 2004). Beginning with the C1 sequence, there was a significant change in sedimentation and stratigraphy from the Nashville and Jessamine Domes. The Jessamine Dome continued to be the locus of peritidal and shoal-water facies whereas the Nashville Dome has a dearth of these facies. There also is a change in the orientation of shoals from persistent north–south orientations to subtle east–west orientations. This C1 (Edenian) change in sedimentation and stratigraphy coincides with a structural reorganization on the Nashville Dome as evidenced by abrupt change in subsidence rates, followed by the erosional bevelling of the C2–C4 sequences on the southern portion of the Nashville Dome. The abrupt change in subsidence rates on the Nashville Dome, but not on the Jessamine Dome, is most likely a reflection of the coeval northward migration of the orogenic load (i.e. Rodgers, 1971; Read, 1989). However, it is quite possible that the differential subsidence along this arch also may reflect sedimentary loading onto the continent from erosion of the coeval orogenic highland, which is more pronounced to the north, or the reactivation of basement faults transverse to the arch controlling lateral subsidence patterns (Burgess & Gayer, 2000).

CONCLUSIONS

1 The Cincinnati Arch was the peripheral bulge to the Blountian and Taconic phases of the Taconic Orogeny. The Cincinnati Arch, with the Nashville and Jessamine Domes in particular, were the consistent locus of peritidal, shoal and build-up facies, as well as third-order unconformities throughout the Late Ordovician, and collectively indicate that this peripheral bulge was a stationary feature for 10–12 Myr during this period. Subsidence analyses suggest that the Cincinnati Arch was either a slowly subsiding feature or that it was a slowly uplifting feature during a period of even greater net eustatic rise.
2 The static position of the Cincinnati Arch and Sebree Trough, as well as their narrow widths relative to the width of the coeval foreland basin are best explained by a lithospheric slab that contains lateral heterogeneities. The crustal boundary between the Grenville front and continental crust (EGRP) weakened by a subsequent rift event (ECRB) probably controls the flexural rigidity of the lithosphere in this area, serving as the locus of basement fault reactivation, syndepositional seismites and control of facies patterns along the Cincinnati Arch during the Taconic Orogeny.
3 Large-scale lateral variations in the facies along the Cincinnati Arch were probably caused by regional variations in subsidence produced by shifting centres of orogenic loading and syndepositional reactivation of basement faults.

ACKNOWLEDGEMENTS

Funding for Mike Pope's portion of this research was provided by NSF EAR-9316057 (P.I. Fred Read). Acknowledgment is made to the Donors of the Petroleum Research Fund, administered by the American Chemical Society, for support of this research to Steven Holland. Mike Pope thanks Fred Read, Aus Al-Tawil, Taury Smith and Anna Balog for instructive discussions in forming many of the ideas put forth in this paper. Insightful reviews by Frank Ettensohn and Maya Elrick on an earlier version of this manuscript are greatly appreciated.

REFERENCES

Alberstadt, L.P., **Walker, K.R.** and **Zurawski, R.P.** (1974) Patch reefs in the Carters Limestone (Middle Ordovician) in Tennessee, and vertical zonation in Ordovician reefs. *Geol. Soc. Am. Bull.*, **85**, 1171–1182.

Allen, P.A. and **Allen, J.R.** (2005) *Basin Analysis – Principles and Applications,* 2nd edition. Blackwell Publishing, Malden, MA, 549 pp.

Ball, M.M. (1967) Carbonate sand bodies of Florida and the Bahamas. *J. Sed. Petrol.*, **37**, 556–591.

Beaumont, C., **Quinlan, G.** and **Hamilton, J.** (1988) Orogeny and stratigraphy: numerical models of the Paleozoic in the eastern interior of North America. *Tectonics*, 7, 389–416.

Black, D.F.B. (1986) Basement faulting in Kentucky. In *Proceedings of the 6*th *International Conference on Basement Tectonics* (Eds M.J. Aldrich and A.W. Laughlin). International Basement Tectonics Association, Salt Lake City, Utah, Vol. 6, pp. 125–139.

Black, D.F.B. and **Haney, D.C.** (1975) Selected structural features and associated dolostone occurrences in the vicinity of the Kentucky River Fault System. *Annual Field Conference, Geological Society of Kentucky*, Kentucky Geological Survey, 27 pp.

Borella P.E. and **Osborne, R.H.** (1978) Late Middle and early Late Ordovician history of the Cincinnati Arch Province, central Kentucky to central Tennessee. *Geol. Soc. Am. Bull.*, **89**, 1559–1573.

Bradley, D.C. and **Kidd, W.S.F.** (1991) Flexural extensions of thee upper continental crust in collisional foredeeps. *Geol. Soc. Am. Bull.*, **103**, 1416–1438.

Brenchley, P.J., Marshall, J.D., Carden, G.A.F., Robertson, D.B.R., Long, D.G.F., Meidla, T., Hints, L. and **Anderson, T.F.** (1994) Bathymetric and isotopic evidence for a short-lived Late Ordovician glaciation in a greenhouse period. *Geology*, **22**, 295–298.

Brett, C.E., McLaughlin, P.I., Cornell, S.R. and **Baird, G.C.** (2004) Comparative sequence stratigraphy of two classic Upper Ordovician successions, Trenton Shelf (New York-Ontario) and Lexington Platform (Kentucky-Ohio): implications for eustasy and local tectonism in eastern Laurentia. *Palaeogeogr., Palaeoclimatol., Palaeoecol.*, **210**, 295–329.

Brookfield, M.E. (1988) A mid-Ordovician temperate carbonate shelf-The Black River and Trenton Limestone Groups of southern Ontario, Canada. *Sed. Geol.*, **60**, 137–154.

Browne, R.G. (1964) The coral horizons and stratigraphy of the Upper Richmond group in Kentucky west of the Cincinnati Arch. *J. Paleontol.*, **38**, 385–392.

Burgess, P.M. and **Gayer, R.A.** (2000) Late Carboniferous tectonic subsidence in South Wales: implications for Variscan basin evolution and tectonic history in SW Britain. *J. Geol. Soc. London*, **157**, 93–104.

Cable, M.S. and **Beardsley, R.W.** (1984) Structural control on Late Cambrian and Early Ordovician carbonate sedimentation in eastern Kentucky. *Am. J. Sci.*, **284**, 797–824.

Crampton, S.L. and **Allen, P.A.** (1995) Recognition of forebulge unconformities associated with early stage foreland basin development: example from the North Alpine foreland basin. *AAPG Bull.*, **79**, 1495–1514.

Cressman, E.R. (1973) Lithostratigraphy and depositional environments of the Lexington Limestone (Ordovician) of central Kentucky. *US Geol. Surv. Prof. Pap.* 768, 61 pp.

Cressman, E.R. and **Noger, M.C.** (1976) Tidal-flat carbonate environments in the High Bridge Group (Middle Ordovician) of central Kentucky. *Kentucky Geol. Surv. Report of Investigations* **18**, 15 pp.

Cuffey, R.J. (1998) The Maysville bryozoan reef mounds in the Grant Lake Limestone (Upper Ordovician) of north-central Kentucky. In: *Sampling the Layer Cake That Isn't: The Stratigraphy and Paleontology of the Type-Cincinnatian* (Eds. R.A. Davis and R.J. Cuffey). Ohio Division of Geological Survey Guidebook **13**, 38–44.

DeCelles, P.G. and **Giles, K.A.** (1996) Foreland Basin Systems. *Basin Res.*, **8**, 105–123.

Dennison, J.M. (1976) Appalachian Queenston Delta related to eustatic sea-level drop accompanying Late Ordovician glaciation centered in Africa. In: *The Ordovician System: Proceedings of a Palaeontological Association Symposium* (Ed. M.G. Bassett), September, 1975, Cardiff, Wales, University of Wales and National Museum of Wales, pp. 107–120.

Diecchio, R.J. (1993) Stratigraphic interpretation of the Ordovician of the Appalachian Basin and implications for Taconian flexural modeling. *Tectonics*, **12**, 1410–1419.

Dorobek, S.M. (1995) Synorogenic carbonate Platforms and reefs in foreland basins: controls on stratigraphic evolution and platform/reef morphology. In: *Stratigraphic evolution of foreland basins* (Eds. S.M. Dorobek and G. Ross), *SEPM Spec. Publ.*, **52**, Tulsa, pp. 127–146.

Dorobek, S.M., Reid, S.K., Elrick, M., Bond, G.C. and **Kominz, M.A.** (1991) Subsidence across the Antler foreland of Montana and Idaho: tectonic versus eustatic effects. In: *Sedimentary Modeling, Computer Simulations and Methods for Improved Parameter Definition* (Eds. E.K. Franseen, W.L. Watney and C.G.St.C. Kendall). *Kansas Geol. Surv. Bull.* 233, Lawrence, Kansas, pp. 231–251.

Dorsch, J., Bambach, R.K. and **Driese, S.K.** (1993) Basin-rebound origin for the 'Tuscarora Unconformity' in southwestern Virginia and its bearing on the nature of the Taconic orogeny. *Am. J. Sci.*, **294**, 237–255.

Drahovzal, J.A., Harris, D.C., Wickstrom, L.H., Walker, D., Keith, B. and **Furer, L.C.** (1992) The east continental rift basin: a new discovery. *Kentucky Geol. Surv. Spec. Publ.* **18**, Kentucky Geological Survey, Lexington, Kentucky, pp. 25.

Drake, A.A., Jr., Sinha, A.K., Laird, J. and **Guy, R.E.** (1989) The Taconic orogen. In: *The Appalachian-Ouachita Orogen in the United States* (Eds R.D. Hatcher, Jr., W.A. Thomas and G.W. Viele). Boulder, Colorado, *Geol. Soc. Am., Geology of North America*, F-2, 101–177.

Ettensohn, F.R. (1991) Flexural interpretation of relationships between Ordovician tectonism and stratigraphic sequences, central and southern Appalachians, USA. In: *Advances in Ordovician Geology* (Eds C.R. Barnes and S.H. Williams). *Geol. Surv. Can. Pap.*, 90–9, 213–224.

Ettensohn, F.R. (Ed.) (1992) Changing interpretations of Kentucky geology – layer-cake, facies, flexure, and eustacy. *State of Ohio, Department of Natural Resources, Miscellaneous Report* 5, pp. 184.

Ettensohn, F.R. (1994) Tectonic control on formation and cyclicity of major appalachian unconformities and associated stratigraphic sequences. In: *Tectonic and Eustatic Controls on Sedimentary Cycles* (Eds J.M. Dennison and F.R. Ettensohn). *SEPM Concepts Sedimentol. Paleontol.* **4**, 217–242.

Ettensohn, F.R., Pashin, J.C. and **Jacobs, G.W.** (1986a) Characteristics of shallow-water, marine shelf silts and sands: two Paleozoic examples from eastern Kentucky. *Spring Program, Appalachian Basin Industrial Associates*, **10**, pp. 197–207.

Ettensohn, F.R., Amig, B.C., Pashin, J.C., Greb, S.F. and **others.** (1986b) Paleoecology and paleoenvironments of the bryozoan-rich Sulphur Well Member, Lexington Limestone (Middle Ordovician), central Kentucky. *Southeastern Geol.*, **26**, 199–210.

Ettensohn, F.R., Hohman, J.C., Kulp, M.A. and **Rast, N.A.** (2002) Evidence and implications of possible far-field responses to Taconic Orogeny: Middle-Late Ordovician Lexington Platform and Sebree Trough, east-central United States. *Southeastern. Geol.*, **41**, 1–36.

Ettensohn, F.R., Kasl, J.M. and **Stewart, A.K.** (2004) Structural inversion and origin of a Late Ordovician (Trenton) carbonate buildup: evidence from the

Tanglewood and Devils Hollow members, Lexington Limestone, central Kentucky (USA). *Palaeogeogr., Palaeoclimatol., Palaeogeogr.*, **210**, 249–266.

Flemings, P.B. and **Jordan, T.E.** (1989) A synthetic stratigraphic model of foreland basin development. *J. Geophys. Res.*, **94**, 3851–3866.

Flemings, P.B. and **Jordan, T.E.** (1990) Stratigraphic modeling of foreland basins: interpreting thrust deformation and lithosphere rheology. *Geology*, **18**, 430–434.

Galewsky, J. (1998) The dynamics of foreland basin carbonate platforms: tectonic and eustatic controls. *Basin Res.*, **10**, 409–416.

Galewsky, J., Silver, E.A., Gallup, C.D., Edwards, R.L. and **Potts, D.C.** (1996) Foredeep tectonics and carbonate platform dynamics in the Huon Gulf, Papua New Guinea. *Geology*, **24**, 819–822.

Ginsburg, R.N. (1971) Landward movement of carbonate mud: new model for regressive cycles in carbonates. *AAPG Bull.*, **55**, 340.

Grotzinger, J. and **Royden, L.** (1990) Elastic strength of the Slave craton at 1.9 Ga and implications for the thermal evolution of the continents. *Nature*, **347**, 64–66.

Gurnis, M. (1992a) Long-term controls on eustatic and epeirogenic motions by mantle convection. *GSA Today*, 2, 141, 144–145, 156–157.

Gurnis, M. (1992b) Rapid continental subsidence following initiation and evolution of subduction. *Science*, **255**, 1556–1558.

Gustadt, A.M. (1958) Upper Ordovician stratigraphy in eastern interior region. *AAPG Bull.*, **42**, 513–547.

Hatfield, C.B. (1968) Stratigraphy and paleoecology of the Saluda Formation (Cincinnatian) in Indiana, Ohio, and Kentucky. *Geol. Soc. Am. Spec. Pap.*, **95**, 1–34.

Hauser, E.C. (1993) Grenville foreland thrust hidden beneath the Eastern U.S. Midcontinent. *Geology*, **21**, 61–64.

Haynes, J.T. (1994) The Ordovician Deicke and Millbrig K-Bentonite beds of the Cincinnati Arch and the southern Valley and Ridge Province. *Geol. Soc. Am. Spec. Pap.*, **290**, pp. 80.

Hohman, J.C. (1998) *Depositional history of the Upper Ordovician Trenton Limestone, Lexington Limestone, Maquoketa Shale and equivalent lithologic units in the Illinois Basin; an application of carbonate and mixed carbonate-siliciclastic sequence stratigraphy.* Unpublished Ph.D. Dissertation, University of Indiana, pp. 187.

Holland, S.M. (1993) Sequence stratigraphy of a carbonate-clastic ramp: the Cincinnatian Series (Upper Ordovician) in its type area: *Geol. Soc. Am. Bull.*, **105**, 306–322.

Holland, S.M. and **Patzkowsky, M.E.** (1996) Sequence stratigraphy and long-term paleoceanographic change in the Middle and Upper Ordovician of the eastern United States. In: *Paleozoic Sequence Stratigraphy: Views from the North American Craton* (Eds B.J. Witzke, G.A. Ludvigson and J.D. Day), Boulder, Colorado, *Geol. Soc. Am. Spec. Pap.*, **306**, 117–129.

Holland, S.M. and **Patzkowsky, M.E.** (1997) Distal orogenic effects on peripheral bulge sedimentation: Middle and Upper Ordovician of the Nashville Dome. *J. Sed. Res.*, **67**, 250–263.

Holland, S.M. and **Patzkowsky, M.E.** (1998) Sequence stratigraphy and relative sea-level history of the Middle and Upper Ordovician of the Nashville Dome, Tennessee. *J. Sed. Res.*, **68**, 684–699.

Horrell, M.A. (1981) *Stratigraphy and depositional environments of the Oregon Formation (Middle Ordovician) of central Kentucky.* Unpublished M.S. Thesis, University of Kentucky, Lexington, Kentucky, 121 pp.

Hrabar, S.V., Cressman, E.R. and **Potter, P.E.** (1971) Crossbedding of the Tanglewood Limestone Member of the Lexington Limestone (Ordovician) of the Blue Grass region of Kentucky. *Brigham Young Univ. Geol. Stud.*, **18**, 99–114.

Jacobi, R.D. (1981) Peripheral bulge – a causal mechanism for the Lower/Middle Ordovician disconformity along the western margin of the northern Appalachians. *Earth Planet. Sci. Lett.*, **56**, 245–251.

Jacobs, G.W. (1986) Mixed siliciclastic-carbonate storm deposits in the Upper Ordovician of central Kentucky. *Geol. Soc. Am. Program with Abstracts*, 18, 228.

Jewell, H.R. and **Ettensohn, F.R.** (2004) An ancient seismite response to Taconian far-field forces: The Cane Run Bed, Upper Ordovician (Trenton) Lexington Limestone, central Kentucky. *J. Geodynam.*, **37**, 487–511.

Jordan, T.E. and **P.B. Flemings** (1991) Large-scale stratigraphic architecture, eustatic variation, and unsteady tectonism: A theoretical evaluation. *J. Geophys. Res.*, **96**, 6681–6699.

Keith, B.D. (1988a) Regional facies of Upper Ordovician Series of eastern North America. In: *The Trenton Group (Upper Ordovician Series) of eastern North America* (Ed. B.D. Keith). *AAPG Stud. Geol.*, **29**, 1–16.

Keith, B.D. (1988b) The Trenton Group (Upper Ordovician Series) of eastern North America; deposition, diagenesis, and petroleum. In: *The Trenton Group (Upper Ordovician Series) of eastern North America* (Ed. B.D. Keith). *AAPG Stud. Geol.*, **29**, 267–276.

Kolata, D.R., Huff, W.D. and **Bergström, S.M.** (1996) Ordovician K-bentonites of eastern North America. *Geol. Soc. Am. Spec. Pub.* **313**, 84 pp.

Kolata, D.R., Huff, W.D. and **Bergström, S.M.** (1998) Nature and significance of unconformities associated with the Middle Ordovician Hagan K-bentonite complex in the North American midcontinent. *Geol. Soc. Am. Bull.*, **110**, 723–739.

Kolata, D.R., Huff, W.D. and **Bergström, S.M.** (2001) The Ordovician Sebree Trough: an oceanic passage to the midcontinent United States. *Geol. Soc. Am. Bull.*, **113**, 1067–1078.

Kreisa, R.D. (1981) Storm-generated sedimentary structures in subtidal marine facies with examples from the Middle and Upper Ordovician of southwestern Virginia. *J. Sed. Petrol.*, **51**, 823–848.

Kuhnhenn, G.L. and **Haney, D.C.** (1986) The Middle Ordovician High Bridge Group and Kentucky River Fault System in central Kentucky. In: *Centennial Field Guide, Southeastern Section of the Geol. Soc. Am.* (Ed. T.L. Neathery), Vol. 6, pp. 25–29.

Kuhnhenn, G.L., Grabowski, G.J., Jr. and **Dever, G.R., Jr.** (1981) Paleoenvironmental interpretation of the Middle Ordovician High Bridge Group in central Kentucky. In: *Stratigraphy, Sedimentology* (Ed. T.G. Roberts), *Geol. Soc. Am. Cincinnati Meeting Field Trip Guidebooks*, pp. 1–30.

Kulp, M.A. (1995) *Paleoenvironmental interpretation of the Brannon Member, Middle-Upper Ordovician Lexington Limestone, central Bluegrass region of Kentucky*. Unpublished M.S. thesis, University of Kentucky, Lexington, Kentucky, 222 pp.

Kulp, M.A. and **Ettensohn, F.R.** (1995) Paleoenvironmental interpretation of the Brannon Member: evidence of synsedimentary tectonism in the Middle Ordovician Lexington Limestone, central Kentucky. *Geol. Soc. Am. Abstracts with Programs, Southeastern Section*, **27**, 67.

Lash, G.G. (1988) Middle and Late Ordovician shelf activation and foredeep evolution, central Appalachian Orogen. In: *The Trenton Group (Upper Ordovician Series) of eastern North America* (Ed. B.D. Keith). *AAPG Stud Geol.*, **29**, 37–53.

Lavoie, D. (1995) A Late Ordovician high-energy temperate-water carbonate ramp, southern Quebec, Canada: implications for Late Ordovician oceanography. *Sedimentology*, **42**, 95–116.

Mackey, R.T. (1972) *Lithostratigraphy and depositional environment of the Perryville Limestone and related members of the Lower Lexington Limestone (Ordovician) of south-central Kentucky*. Unpublished Masters thesis, University of Kentucky, Lexington, Kentucky, 76 pp.

MacQuown, W.C., Jr., Kuhnhenn, G.L., Shepard, R.G. and **Gooding, P.G.** (1984) Fieldtrip 6, Lithostratigraphy and depositional environments of the Middle Ordovician limestones in central Kentucky. In: *Fieldtrip Guidebooks for the Southeastern/North-central Section Geological Society of America Meeting* (Eds N. Rast and H.B. Hay), University of Kentucky, Lexington, Kentucky, pp. 66–83.

McBride, E.F. (1962) Flysch and associated beds of the Martinsburg Formation (Ordovician), central Appalachians. *J. Sed. Petrol.*, **32**, 39–91.

McLaughlin, P.I. and **Brett, C.E.** (2004) Eustatic and tectonic control of marine seismites; examples from the Upper Ordovician of Kentucky, USA. *Sed. Geol.*, **168**, 165–192.

McLaughlin, P.I., Brett, C.E., Taha McLaughlin, S.L. and **Cornell, S.R.** (2004) High-resolution sequence stratigraphy of a mixed carbonate-siliciclastic, cratonic ramp (Upper Ordovician; Kentucky-Ohio, USA): insights into the relative influence of eustasy and tectonics through analysis of facies gradients. *Palaeogeogr. Palaeoclimatol. Palaeoecol.*, **210**, 267–294.

Mitrovica, J.X., Beaumont, C. and **Jarvis, G.T.** (1989) Tilting of continental interiors by the dynamical effects of subduction. *Tectonics*, **8**, 1079–1094.

Mussman, W.J. and **Read, J.F.** (1986) Sedimentology and development of a passive-to convergent-margin unconformity: Middle Ordovician Knox unconformity. *Geol. Soc. Am. Bull.*, **97**, 282–295.

Nádai, A. (1963) *Theory of Flow and Fracture of Solids*. McGraw Hill Book Co., New York, 411 pp.

Onasch, C.M. and **Kahle, C.F.** (1991) Recurrent tectonics in a cratonic setting: an example from northwestern Ohio. *Geol. Soc. Am. Bull.*, **103**, 1259–1269.

Onasch, C.M. and **Kahle, C.F.** (2002) Seismically induced soft-sediment deformation in some Silurian carbonates, eastern U.S. Midcontinent. In: *Ancient Seismites* (Eds F.R. Ettensohn, N. Rast and C.E. Brett). *Geol. Soc. Am. Spec. Pap.*, **359**, 165–176.

Patzkowsky, M.E. and **Holland, S.M.** (1993) Biotic response to a Middle Ordovician paleoceanographic event in eastern North America. *Geology*, **21**, 619–622.

Patzkowsky, M.E. and **Holland, S.M.** (1999) Biofacies replacement in a sequence stratigraphic framework: Middle and Upper Ordovician of the Nashville Dome, Tennessee, USA. *Palaios*, **14**, 301–323.

Pope, M.C. and **Read, J. F.** (1995) Late Ordovician Tanglewood Bank: a highly compartmentalized buildup reflecting tectonics and moderate amplitude eustasy on the Cincinnati Arch. *Geol. Soc. Am. Abstracts with Programs*, **27**, A450.

Pope, M.C. and **Read, J.F.** (1997a) High-frequency cyclicity of the Lexington Limestone (Middle Ordovician), a cool-water carbonate clastic ramp in an active foreland basin. In: *Cool-Water Carbonates* (Eds N.P. James and J.P. Clarke). *SEPM Spec. Publ.* **56**, 411–429.

Pope, M.C. and **Read, J.F.** (1997b) High-resolution surface and subsurface sequence stratigraphy of Middle to Late Ordovician (Late Mohawkian to Cincinnatian) foreland basin rocks, Kentucky and Virginia. *AAPG Bull.*, **81**, 1866–1893.

Pope, M.C. and **Read, J.F.** (1998) Ordovician metre-scale cycles: Implications for Ordovician climate and eustatic fluctuations in the central Appalachian Basin, USA. *Palaeoclimatol. Palaeogeogr. Palaeoecol.*, **138**, 27–42.

Pope, M.C., Read, J.F., Bambach, R.K. and **Hofmann, H.J.** (1997) Late Middle to Late Ordovician seismites of Kentucky, Southwestern Ohio and Virginia: Sedimentary recorders of Earthquakes in the Appalachian Basin. *Geol. Soc. Am. Bull.*, **109**, 489–503.

Pratt, B.R. and **James, N.P.** (1986) The tidal flat islands model of peritidal shallowingupward sequences: St. George Group, western Newfoundland. *Sedimentology*, **33**, 313–345.

Quinlan, G. M. and **Beaumont, C.** (1984) Appalachian thrusting, lithospheric flexure, and the Paleozoic stratigraphy of the eastern interior of North America. *Can. J. Earth Sci.*, **21**, 973–996.

Rast, N. and **Goodman, P.T.** (1995) Tectonic and sedimentary consequences of Late Proterozoic and Early and Mid-Paleozoic overthrusting in Kentucky and adjacent states. *Northeastern Geol. Environ. Sci.*, **17**, 2–9.

Read, J.F. (1980) Carbonate ramp-to-basin transitions and foreland basin evolution, Middle Ordovician, Virginia Appalachians. *AAPG Bull.*, **64**, 1575–1612.

Read, J.F. (1989) Controls on evolution of Cambrian–Ordovician passive margin, U.S. Appalachians. In: *Controls on Carbonate Platform and Basin Development* (Eds P.D. Crevello, J.L. Wilson, J.F. Sarg and J.F. Read). *SEPM Spec. Publ.*, **44**, 147–165.

Riding, R. (2004) Solenopora is a chaetetid sponge, not an alga. *Palaeontology*, **47**, 117–122.

Rodgers, J. (1971) The Taconic orogeny. *Geol. Soc. Am. Bull.*, **82**, 1141–1178.

Ruppel, S.C. and **Walker, K.R.** (1984) Petrology and depositional history of a Middle Ordovician carbonate platform; Chickamauga Group, northeastern Tennessee. *Geol. Soc. Am. Bull.*, **95**, 568–583.

Scharpf, C.D. (1990) *Stratigraphy and associated faunas of the Middle Ordovician (Rocklandian) Millbrig K-bentonite in central Kentucky*. Unpublished Masters thesis, University of Cincinnati, Cincinnati, Ohio, 280 pp.

Shanmugam, G. and **Lash, G.G.** (1982) Analogous tectonic evolution of the Ordovician foredeeps, southern and central Appalachians. *Geology*, **10**, 562–566.

Shanmugam, G. and **Walker, K.R.** (1980) Sedimentation, subsidence and evolution of a foredeep basin in the Middle Ordovician, southern Appalachians. *Am. J. Sci.*, **280**, 479–496.

Shanmugam, G. and **Walker, K.R.** (1982) Anatomy of the Middle Ordovician Sevier Shale Basin, eastern Tennessee. *Sedimentology*, **34**, 315–337.

Simmons, G.C. and **Oliver, W.A., Jr.** (1967) Otter Creek Coral Bed and its fauna, east-central Kentucky. *U.S. Geol. Surv. Bull.*, **1244-F**, 1–13.

Sinclair, H.D. and **Allen, P.A.** (1992) Vertical vs. horizontal motions in the Alpine orogenic wedge: stratigraphic response in the foreland basin. *Basin Res.*, **4**, 215–232.

Sinclair, H.D., Coakley, B.J., Allen, P.A. and **Watts, A.B.** (1991) Simulation of foreland basin stratigraphy using a diffusion model of mountain belt uplift and erosion: an example from the central Alps, Switzerland. *Tectonics*, **10**, 599–620.

Sloss, L.L. (1963) Sequences in the cratonic interior of North America. *Geol. Soc. Am. Bull.*, **74**, 93–113.

Sweet, W.C. (1984) Graphic correlation of upper Middle and Upper Ordovician rocks, North American Midcontinent Province, U.S.A. In: *Aspects of the Ordovician System: Palaeontological Contributions form the University of Oslo* (Ed. D.L. Bruton), *Universitetsforlaget*. Oslo, University of Oslo, pp. 23–35.

Tankard, A.J. (1986) On the depositional response to thrusting and lithospheric flexure: examples from the Appalachian and Rocky Mountain basins. In: *Foreland Basins* (Eds P.A. Allen and P. Homewood). *Int. Assoc. Sedimentol. Spec. Publ.* 8. Blackwell Publications, Oxford, pp. 369–392.

Turcotte, D.L. and **Schubert, G.** (2002) *Geodynamics*, 2nd edition. Cambridge University Press, Cambridge, 456 pp.

Van Schmus, W. R., Bickford, M.E., Anderson, J.L., Bender, E.E., Anderson, R.R., Bauer, P.W., Robertson, J.M., Bowring, S.A., Condie-K.C., Denison, R.E., Gilbert, M.C., Grambling, J.A., Mawer, C.K, Shearer, C.K., Hinze, W.J., Karlstrom, K.E., Kisvarsanyi, E.B., Lidiak, E.G., Reed, J.C. Jr., Sims, P.K., Tweto, O., Silver, L.T.,Treves, S.B., Williams, M.L. and **Wooden, J.L.** (1993) Transcontinental Proterozoic provinces. In: *Precambrian; Conterminous U.S.* (Eds J.C. Reed, Jr., M.E. Bickford, R.S. Houston, P.L. Link, D.W. Rankin, P.K. Sims and W.R. Van Schmus). *The Geology of North America*, C-2, *Geol. Soc. Am.*, Boulder, CO, US, pp. 171–334.

Walker, K.R. and **Laporte, L.F.** (1970) Congruent fossil communities from Ordovician and Devonian carbonates of New York. *J. Paleontol.*, **44**, 928–944.

Waschbusch, P.J. and **Royden, L.H.** (1992a) Spatial and temporal evolution of foredeep basins: lateral strength variations and inelastic yielding in continental lithosphere. *Basin Res.*, **4**, 179–196.

Waschbusch, P.J. and **Royden, L.H.** (1992b) Episodicity in foredeep basins. *Geology*, **20**, 915–918.

Weir, G.W., Peterson, W.L. and **Swadley, W.C.** (1984) Lithostratigraphy of Upper Ordovician strata exposed in Kentucky. *US Geol. Surv. Prof. Pap.* **1151-E**, 121 pp.

Weiss, M.P., Edwards, W.R., Norman, C.W. and **Sharp, E.R.** (1965) The American Upper Ordovician standard. VII. Stratigraphy and petrology of the Cynthiana and Eden Formations of the Ohio Valley. *Geol. Soc. Am. Spec. Pap.* **81**, 76 pp.

Wilson, C.W., Jr. (1949) Pre-Chattanooga stratigraphy in central Tennessee. *Tennessee Division of Geol., Bull.*, **56**, 1–407.

Wilson, C.W., Jr. (1962) Stratigraphy and geologic history of Middle Ordovician rocks of central Tennessee. *Geol. Soc. Am. Bull.*, **73**, 481–504.

Wilson, C.W., Jr. and **Stearns, R.G.** (1963) Quantitative analysis of Ordovician and younger structural development of Nashville Dome, Tennessee. *AAPG Bull.*, **47**, 823–832.

Wolcott, D.E., Cressman, E.R. and **Connor, J.J.** (1972) Trend-surface analysis of the thickness of the High Bridge Group (Middle Ordovician) of central Kentucky and its bearing on the nature of the post-Knox unconformity. *US Geol. Surv. Prof. Pap.* **800**, B25–B33.

Int. Assoc. Sedimentol. Spec. Publ. (2009) **41**, 277–295

Reinterpreting a Proterozoic enigma: *Conophyton–Jacutophyton* stromatolites of the Mesoproterozoic Atar Group, Mauritania

LINDA C. KAH*, JULIE K. BARTLEY† and ALICE F. STAGNER‡

**Department of Earth & Planetary Sciences, University of Tennessee, Knoxville, TN 37996, USA (E-mail: lckah@utk.edu)*

†Department of Geosciences, University of West Georgia, Carrollton, GA 30118, USA

‡School of Geology and Geophysics, University of Oklahoma, Norman, OK 73019, USA

ABSTRACT

The Mesoproterozoic Atar Group, Taoudeni Basin, Mauritania, preserves a spectacular diversity of stromatolite morphologies, including stromatolitic biostromes comprised of the conical form *Conophyton*, the enigmatic branching conical form *Jacutophyton*, and a variety of irregularly branching forms, including *Tilemsina* and *Baicalia*. Until now, the peculiar juxtaposition of high- and low-relief stromatolite morphologies has posed a conundrum for environmental interpretation of stromatolite forms, and has led to interpretations of strict biological control over stromatolite morphology. Careful assessment of superpositional relationships among stromatolite elements, however, suggests that the diversity of stromatolite morphologies in the Atar Group can be readily explained via parasequence-scale sea-level changes and the incomplete and laterally discontinuous filling of accommodation space. In the Atar Group, biostrome growth initiates during relative rises in sea level with the widespread, subtidal nucleation of *Conophyton*. Exposure of *Conophyton* to wave energy during falls in relative sea level result in disruption of stromatolitic laminae, generation of inter-stromatolitic debris, and development of both superimposed and laterally adjacent branching stromatolite forms. In this scenario, the enigmatic stromatolite form *Jacutophyton* represents stromatolite growth through a complete depositional parasequence, and the unusual juxtaposition of stromatolite forms reflects growth of different forms that is separated by time and depositional environment. With subsequent rises in sea level, nucleation of new *Conophyton* in subtidal regions, and continued growth of branching forms and aggradation of the depositional substrate in intertidal regions, results in continued modification of sea-floor topography. In the model presented here, stratigraphic time is partitioned both vertically and laterally during biostrome growth, resulting in a complex internal architecture that is not readily discerned in outcrop. In connecting hydrodynamic variables to stromatolite morphology, this model provides a more comprehensive understanding of Atar Group stromatolites in terms of basin geometry, relative sea level and carbonate production.

Keywords Proterozoic, stromatolites, *Jacutophyton*, *Conophyton*, Mauritania, Atar Group, biostrome architecture, parasequence development.

INTRODUCTION

Whereas stromatolite microfabrics are generally believed to reflect a combination of microbial community growth, decomposition and lithification (Golubic, 1976; Bertrand-Sarfati, 1983; Turner *et al.*, 1993; Knoll & Sergeev, 1995; Kah & Knoll, 1996; Knoll & Semikhatov, 1998; Grotzinger & Knoll, 1999; Lee & Golubic, 2000; Reid *et al.*, 2000), stromatolite morphology appears to be affected more strongly by physical depositional factors, particularly water depth, wave energy and sediment influx (Cloud & Semikhatov, 1969; Semikhatov *et al.*, 1979; Semikhatov & Raaben, 1994, 1996; Andres & Reid, 2006). As a result, stromatolitic laminae, which record both microscale and macroscale growth processes, are arguably the most fundamental aspect of stromatolite morphology (Walter, 1992). Each lamina represents the active growth surface of the mat at the time of deposition and therefore records the topography of the depositional surface (i.e. synoptic relief, Hofmann, 1969; Donaldson, 1976), the relative rates of stromatolite growth and sediment infilling (i.e. wall structure, Hofmann, 1969; Semikhatov *et al.*, 1979) and the influence of wave and/or current action on mat growth (i.e. stromatolite elongation, Hoffman, 1967; Logan *et al.*, 1974). Not only do laminae closely record the morphology of a microbial mat on the seafloor, but the geometry of successive laminae also produces a record of microbial mat growth over time that is reflected in stromatolite morphology. More recently, process-oriented models (Grotzinger & Rothman, 1996; Grotzinger & Knoll, 1999; Dupraz *et al.*, 2006) have emphasized the importance of laminae geometry by using variable interface parameters including surface roughness (mat growth and degradation), upward growth (mineral precipitation) and surface dampening (draping sedimentation) to simulate lamina accretion and stromatolite morphogenesis.

The relationship between stromatolite morphology and depositional setting has been examined in a number of key studies (Hoffman, 1974; Donaldson, 1976; Grey & Thorne, 1985; Bertrand-Sarfati & Moussine-Pouchkine, 1985). These studies, summarized by Hoffman (1976) and Grotzinger (1989), indicate that specific stromatolite morphologies represent distinct depositional settings and are well correlated with inferred palaeowater depth, water energy and sediment supply. Upper intertidal to supratidal environments are characterized by stromatolites with low synoptic relief, frequently associated with intraclastic debris. Intertidal to shallow subtidal (above fair weather wave-base) environments are characterized by relatively low-relief domal and irregular columnar stromatolites associated with intraclastic and/or fine-grained detrital material. Deeper subtidal (below fair weather wave-base) environments are characterized by high-relief columnar to conical stromatolite forms that lack evidence of interstromatolitic sediment deposition. Empirically, relationships between stromatolite morphology and inferred depositional parameters appear relatively robust, yet similar relationships observed in modern environments highlight potential variability resulting from local and regional variation in accommodation space, hydrodynamic regime and patterns of sediment deposition (Andres & Reid, 2006).

The Proterozoic Atar Group, exposed along the northern edge of the Taoudeni Basin in Mauritania, Mali and Algeria, preserves a spectacular diversity of stromatolite morphologies, including three distinct biostromes composed of the conical stromatolite form *Conophyton,* the enigmatic branching conical stromatolite *Jacutophyton* and a variety of irregularly branching forms, including *Tilemsina* and *Baicalia* (Bertrand-Sarfati, 1972; Bertrand-Sarfati & Moussine-Pouchkine, 1999). These bioherms, herein referred to as Atar Formation biostromal units R1–R3, show striking stratigraphic similarity in both western (Mauritania) and eastern (Mali-Algeria) sections of the Taoudeni Basin, and have been interpreted as regional reef tracts that played a critical role in basin development (Bertrand-Sarfati & Moussine-Pouchkine, 1992, 1999; Moussine-Pouchkine & Bertrand-Sarfati, 1997).

Two aspects of the Atar Formation biostromes, however, pose a conundrum for an environmental interpretation of stromatolite morphology. First, biostromal intervals do not exhibit the large-scale changes in geometry that are commonly observed in stromatolite reefs and interpreted as resulting from changes in sea level or accommodation space (Grotzinger, 1989; Narbonne & James, 1996; Turner *et al.*, 1997; Lemon, 2000). Second, peculiar juxtaposition of conical and branching stromatolite forms (Fig. 1), such as laterally adjacent *Conophyton* and *Jacutophyton* or *Conophyton* and *Tilemsina*, which typically occur in distinct depositional environments, has been used to conclude that microbial community structure, rather than depositional environment, may exert the primary control over stromatolite morphology (Bertrand-Sarfati & Moussine-Pouchkine, 1985).

Recent work in the Mesoproterozoic Dismal Lakes Group provides a framework for interpreting the enigmatic relationships observed in the Atar Group biostromes (Kah *et al.*, 2006). The Dismal Lakes reef consists of a thin (<40 m), laterally extensive (>2400 km^2) biostromal complex that contains both exceptionally high-relief (up to 15 m) conical stromatolites and lower relief domal and columnar forms. In the Dismal Lakes Group, relationships among stromatolites are best explained in terms of the sea-level history of the platform, wherein high-frequency (i.e. parasequence-scale) sea-level changes produce a complex internal reef architecture, which results from lateral juxtaposition of facies across non-horizontal timelines (Kah *et al.*, 2006). High-relief conical stromatolites initiated during major transgression and reflect transgressive and highstand stromatolite growth. Subsequent sea-level fall exposed the tops of these conical stromatolites, creating a sequence boundary that traced the rugged seafloor topography delineated by the decametre-scale synoptic relief of the stromatolites. During the sea-level cycles that followed, lower-relief domal and columnar stromatolites

Fig. 1. Enigmatic juxtaposition of stromatolite forms in the Atar Formation. (a) *Conophyton–Jacutophyton* in the R1 reef interval. (b) *Jacutophyton–Tilemsina* in the R2 reef interval.

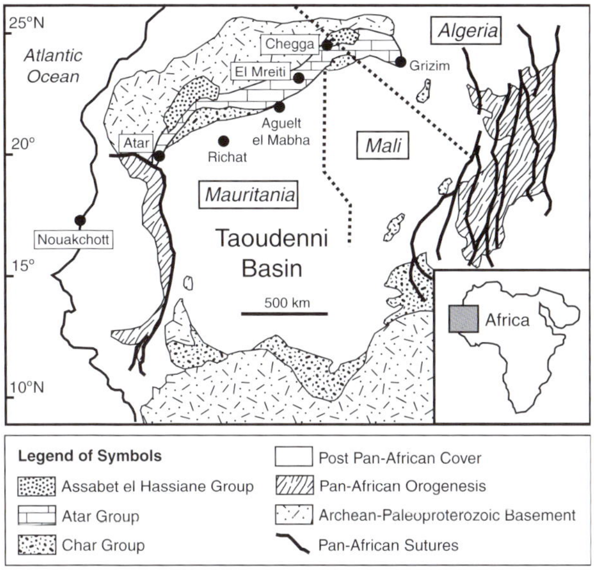

Fig. 2. Regional extent of the Atar Group. Atar Group strata are exposed in a continuous belt across the West African craton from Mauritania to western Algeria. Atar Group strata unconformably overlie predominantly siliciclastic strata of the Char Group, onlap Archaean–Palaeoproterozoic rocks of the Regubiat Shield and are overlain by siliciclastic strata of the Assabet el Hassiane Group. Variously deformed, stromatolitic strata in the Hoggar Uplift, Mali and Algeria, may represent depositional equivalents to the Atar Group (Bertrand-Sarfati & Trompette 1976; Bertrand-Sarfati *et al.*, 1987).

grew on this rugged surface, filling space between the high-relief cones. The resultant stromatolite biostrome exhibits lateral partitioning of time, with conical stromatolite growth dominating during relative sea-level rise and domal to columnar stromatolite growth occurring during times of lower accommodation space (Kah *et al.*, 2006).

In this paper, we provide a stratigraphic interpretation of the enigmatic Atar Formation *Conophyton–Jacutophyton* biostromes. Here, we (1) document superpositional relationships among stromatolitic elements in the Atar Formation biostromes; (2) interpret stromatolite growth in terms of relative sea-level change; and (3) explore stromatolite and sea-level relationships in terms of basin geometry and Proterozoic carbonate dynamics.

GEOLOGICAL SETTING AND AGE

The Taoudeni Basin, West Africa, preserves up to 1300 m of gently dipping (<1°), unmetamorphosed, and virtually undeformed Proterozoic strata that crop out in a sinuous belt spanning >1500 km from Mauritania to northern Mali and Western Algeria (Fig. 2). Strata unconformably overlie Archaean amphibolite and quartzo-feldspathic schists and Palaeoproterozoic granitic intrusions of the Reguibat Shield (Black & Fabre, 1983; Trompette & Carozzi, 1994) and are in turn unconformably overlain by Vendian (or Cambrian) and younger strata (Clauer & Deynoux, 1987; Bertrand-Sarfati *et al.*, 1995).

Proterozoic strata of the Taoudeni Basin are typified by well-exposed sections in the Adrar Mountains of Mauritania, where the succession is subdivided into three unconformity-bounded groups (Figs 2 and 3). Potentially correlative and similarly subdivided strata are also exposed in the Hoggar uplift of Algeria (Moussine-Pouchkine *et al.*, 1988; Bertrand-Sarfati *et al.*, 1991, 1996; Moussine-Pouchkine & Bertrand-Sarfati, 1997) and in several poorly studied exposures in the southern Taoudeni Basin (Bertrand-Sarfati & Moussine-Pouchkine, 1983; Bertrand-Sarfati *et al.*, 1987). At the base of the succession, coarse- to

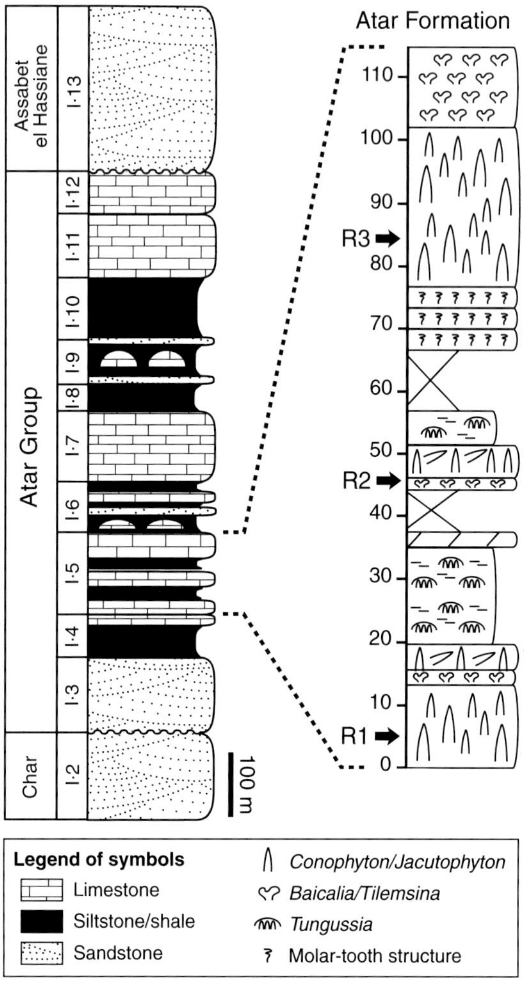

Fig. 3. Stratigraphy of the Atar Group. The ~750 m thick Atar Goup consists of ten formations (I-3, Foum Chor; I-4, Ksar Torchane; I-5, Atar; I-6, Oued Tariofet; I-7, Tawaz; I-8, Oued Terrarit; I-9, Aouinet Ould Bou Derbala; I-10, Aouleigate; I-11, Touiderguilt; I-12, Tifounke) that record deposition of fine-grained siliciclastic and carbonate sediment. The Atar Formation (I-5) is marked by deposition of three prominent *Conophyton–Jacutophyton* reefs (R1–R3). Each reef interval initiates at a marine transgression and represents transgressive to highstand systems tract deposition. Deposition of the lower two reef intervals terminates at an abrupt fall in relative sea level (marked by asterisks) and is followed by deposition of a predominantly siliciclastic lowstand systems tract. The uppermost reef interval is marked by a gradual fall in relative sea level, and is overlain by lowstand siliciclastic, carbonate, and evaporite deposition of the Oued Tariofet Formation (I-6).

fine-grained siliciclastic strata and subordinate carbonates of the Char Group are preserved as erosional remnants in cratonal depressions and represent initial mantling of the craton by fluvial, coastal aeolian and shallow-marine deposits (Benan & Deynoux, 1988). In contrast to the Char Group, fine-grained siliciclastic strata, stromatolite-bearing carbonate and minor evaporitic strata (Goodman & Kah, 2004) of the Atar Group represent craton-wide flooding and deposition in predominantly shallow marine environments (Bertrand-Sarfati & Moussine-Pouchkine, 1988). The overlying Assabet el Hassiane Group is composed of deep to shallow marine, fine-grained siliciclastic strata (Trompette & Carozzi, 1994; Moussine-Pouchkine & Bertrand-Sarfati, 1997).

The most lithologically variable of these successions, the ~750 m thick Atar Group, consists of ten formations that record alternating deposition of stromatolite-bearing carbonate and siliciclastic material (Fig. 3; Bertrand-Sarfati & Moussine-Pouchkine, 1988, 1999). The basal Foum Chor (unit I-3) Formation is markedly more coarse-grained than the remainder of the Atar Group, and consists primarily of fining-upward, coarse- to medium-grained fluvial to fluvial–marine sandstones that onlap both the Reguibat Shield and deeply incised strata of the Char Group. Remaining formations (Ksar Torchane, unit I-4; Atar, unit I-5; Oued Tariofet, unit I-6; Tawaz, unit I-7; Oued Terrarit, unit I-8; Aouinet Ould Bou Derbala, unit I-9; Aouleigate, unit I-10; Touiderguilt, unit I-11; and Tifounke, unit I-12) preserve regionally extensive depositional packages of siltstone and shale alternating with stromatolite-bearing carbonate and subordinate evaporites. Spectacular stromatolite biostromes of the Atar Group (R1–R3 of the Atar Formation, unit I-5) are the focus of this study. The presence of hummocky cross-stratification, wave and interference ripples, and lenses of flat-pebble conglomerates in siliciclastic strata suggests predominantly shallow subtidal to intertidal depositional environments. Subtidal to intertidal depositional environments are also inferred from stromatolite morphologies in the Atar Group (Bertrand-Sarfati, 1972; Bertrand-Sarfati & Moussine-Pouchkine, 1985), suggesting that the alternation of siliciclastic and carbonate strata may result primarily from regional climatic or tectonic activity and the episodic influx of siliciclastic material from source areas outside the Taoudeni basin (Bertrand-Sarfati & Moussine-Pouchkine, 1988; Moussine-Pouchkine & Bertrand-Sarfati, 1997).

The age of the Atar Group is poorly constrained. Rb–Sr geochronology (Clauer, 1976, 1981; Clauer *et al.*, 1982) performed on glauconite and illite in shaly intervals provide ages from 998 ± 32 Ma (Azougi Formation, unit I-2, underlying Char Group) to >694 Ma (Assabet el Hassiane Group, unit I-15), with most formations in the Atar

Group constrained by a single age (890 ± 35 Ma, Atar Formation, unit I-5; 874 ± 22 Ma, Oued Tarioufet Formation, unit I-6; 866 ± 67 Ma, Oued Terrarit Formation, unit I-8; 775 ± 52 Ma, Aouleigate Formation, unit I-10).

Although these Rb–Sr ages clearly represent diagenetic mineralization, the consistent decrease in ages through the stratigraphic column has been used to argue for early diagenetic stabilization of clay minerals and Neoproterozoic depositional ages for these strata. By contrast, recent C-isotope data from the Atar Group (Fairchild *et al.*, 1990; Teal & Kah, 2005) reveal moderately positive $\delta^{13}C$ values near +2‰ with several distinct negative excursions to nearly –2.5‰. The range of C-isotope values preserved in the Atar Group is inconsistent with the strongly positive values ($\delta^{13}C$ >+5‰) recorded in well-constrained isotopic compilations for the post-850 Ma Neoproterozoic (Kaufman & Knoll, 1995; Halverson *et al.*, 2005). Atar Group chemostratigraphy is also distinctly different from that of strata probably deposited in the earliest Neoproterozoic (Knoll *et al.*, 1995; Bartley *et al.*, 2001) and more closely reflects isotopic patterns preserved globally in mid to late Mesoproterozoic strata (Knoll *et al.*, 1995; Kah *et al.*, 1999a; Bartley *et al.*, 2001, 2007; Frank *et al.*, 2003;), suggesting that the Atar Group may be as old as ~1200 Ma.

STROMATOLITE BIOSTROMES OF THE ATAR FORMATION

Lithology and sequence stratigraphic framework

Like the rest of the Atar Group, the ~125 m thick Atar Formation (unit I-5) consists of alternating fine-grained siliciclastic strata and stromatolite-bearing carbonate (Fig. 3). Siliciclastic intervals are composed primarily of dark coloured shale containing laterally discontinuous interbeds of fine-grained sandstone and a variable carbonate component. Basal siliciclastic strata contain abundant thin (<1 m thick) interbeds of fine-grained, flat-laminated microbial carbonate and flat-clast breccia, and mark shallow-water lowstand deposition following regional flooding and deposition of shallow-marine carbonate strata of the Ksar Torchane Formation (unit I-4). Similarly, stratigraphically higher siliciclastic intervals represent lowstand deposition following craton-wide flooding and development of the Atar Formation *Conophyton–Jacutophyton* biostromes (Bertrand-Sarfati & Moussine-Pouchkine, 1988). Disappearance of flat-clast breccias and flat-laminated microbial carbonate interbeds in stratigraphically higher siliciclastic intervals and their replacement by small (<2 m thick) discontinuous bioherms of irregularly branching columnar stromatolites (*Tungussia confusa*; Bertrand-Sarfati & Moussine-Pouchkine, 1999), as well as rare intervals of 'molar-tooth' carbonate (Furniss *et al.*, 1998; James *et al.*, 1998; Pollock *et al.*, 2006) suggest shallow-marine deposition and overall greater accommodation space across the platform.

Carbonate-dominated intervals in the Atar Formation form three distinct stromatolite biostromes (R1–R3, 9–39 m thick; Fig. 3) that are the focus of this study. Biostromal intervals typically preserve a thin (<1 m thick) and discontinuous transgressive horizon at their base, composed of flat-clast breccia and low relief, divergent branching stromatolites (*Tilemsina divergens*; Bertrand-Sarfati & Moussine-Pouchkine, 1999). The main biostromes are composed of conical to branching conical stromatolites (*Conophyton ressoti*, *Conophyton jacqueti* and *Jacutophyton sahariensis*; Bertrand-Sarfati & Moussine-Pouchkine, 1999) that are interpreted to have developed as deep-water reef tracts (80–100 m; Bertrand-Sarfati & Moussine-Pouchkine, 1985) during highstand conditions. Termination of each biostrome appears to have resulted from loss of accommodation space. Termination of the R1 biostrome is marked by regional toppling of *Conophyton* (Fig. 4a), suggesting an abrupt decrease in accommodation space and impingement of storm wave base on the biostrome. Similarly, termination of the R2 biostrome is marked by a regional pavement of broken *Conophyton* and stromatolitic breccia (Fig. 4b), suggesting an abrupt decrease in accommodation space and prolonged exposure of the biostrome. By contrast, termination of the R3 biostrome is marked by a transition from dominantly conical stromatolites to a 10 m thick interval of irregularly branching forms (*Baicalia safia* and *Baicalia mauritanica*; Bertrand-Sarfati & Moussine-Pouchkine, 1999) that are overlain by black shale and 'molar-tooth' carbonate of the lower Oued Tarioufet Formation (unit I-6). Gradational contacts between stromatolite morphologies in the R3 biostrome suggest a gradual decrease in accommodation space driven by aggradational growth of the biostrome complex.

Combined, lithological and stratigraphic features indicate that the Atar Formation biostrome complexes developed during a series of

Fig. 4. Termination of stromatolite reef deposition. (a) Termination of the R1 reef interval is marked by regional toppling of *Conophyton*; oblique bedding plane view; Jacob staff marked in 10 cm intervals. (b) Termination of the R2 reef interval is marked by a regional pavement of broken *Conophyton* and stromatolitic breccia; bedding plane view.

thin (30–50 m thick) third-order stratigraphic sequences (Emery & Meyers, 1996; Miall, 1997). The base of each biostromal complex is marked by a transgressive surface and initiation of deep-water stromatolite growth. Stromatolite growth continued during transgressive and highstand phases and was terminated by a loss of accommodation space, either by a fall in sea level or continued aggradation of the biostrome complex. Finally, biostromes are overlain by late highstand to lowstand, shale-dominated deposits. Relatively thin sequences and abrupt transitions between siliciclastic and carbonate facies reflect a cratonal sedimentation regime (Bertrand-Sarfati & Moussine-Pouchkine, 1988), wherein minimal subsidence and the extremely low relief of the sedimentary substrate resulted in dramatic changes in depositional environment with even small changes in accommodation space.

Reef elements

Conical stromatolites

Conical stromatolites (*Conophyton* spp.) are the most common element of the Atar Formation biostromes (Fig. 5a), and a number of form-taxa have been recognized, including *Conophyton ressoti* and *Conophyton jacqueti* (Bertrand-Sarfati & Moussine-Pouchkine, 1985, 1999). Conical stromatolites have steeply dipping (75°–90°) wall-parallel laminae, narrow and nearly vertical axial zones and high synoptic relief. Synoptic relief of a stromatolite above the seafloor is delineated by the height of a single lamina, which represents the morphology of the seafloor at the time of lamina formation (Hofmann, 1969). Inheritance of antecedent topography in these stromatolites is high, and a single columnar structure often represents several distinct phases of stromatolite growth, with each cone growing directly atop the previous structure (Fig. 6a). Thus, a column may be up to 4 m tall; however, no single lamina exceeds 2 m in height, indicating that individual stromatolites stood no more than 2 m above the seafloor at any one time. Despite a generally uniform shape, differences in morphology are apparent. Cross-sectional shapes range from circular (*C. ressoti*) to elliptical (*C. jacqueti*; Fig. 6b; see discussion in Bertrand-Sarfati & Moussine-Pouchkine, 1999), and individual cones may display thickening of outer laminae, forming superficial protrusions along the outer surface of the cone (Fig. 6b). Cones also show variable modification after growth, including delamination (Fig. 6c) and erosional incision (Fig. 6d) of cone margins.

Individual conical stromatolites range in diameter from 10 to 50 cm and show interstromatolitic spacing of 5–70 cm. Interstromatolitic regions contain both platy breccia and fine-grained detrital carbonate, as well as several generations of precipitated carbonate cement (cf. Fig. 9; see descriptions below). The high synoptic relief of stromatolitic laminae suggests that conical stromatolites formed largely below wave base, in quiet-water environments, and that lithification occurred largely through *in situ* carbonate precipitation (cf. Bertrand-Sarfati & Moussine-Pouchkine, 1985; Kah *et al.*, 2006). Although high synoptic relief may have inhibited incorporation of sediment from the water column into the *Conophyton*,

Fig. 5. *Conophyton–Jacutphyton* association of the Atar Formation. (a) High synoptic relief and nearly vertical, parallel laminae characterize *Conophyton* within the Atar Formation reefs. *Conophyton ressoti* (shown here) bears a single central axial zone and occurs in R1 and R2 reef intervals; *Conophyton jacqueti* has an expanded axial zone, an elongate cross-section and occurs primarily in the R3 reef interval. (b) The enigmatic stromatolite *Jacutophyton sahariensis* is characterized by low synoptic relief branches that surround a conical central core. (c) Detailed cross-sectional view of *Jacutophyton* showing oblique growth of branching elements. (d) Oblique bedding plane view of *Jacutophyton* highlighting the petaloid nature of branching elements that encompass the central cone.

the continuous walled nature of stromatolite margins and general absence of erosion resulting from sediment entrained within this water column supports the paucity of detrital sediment at the time of cone growth. Rarely, successive laminae in the basal 3–5 cm of conical stromatolites record an abrupt onlapping of fine-grained interstromatolitic sediment, suggesting that at least small amounts of detrital sediment may have been deposited during *Conophyton* growth. Combined, these observations suggest that interstromatolitic debris represents sediment deposition that occurred predominantly after stromatolite growth.

Branching conical stromatolites

Along with *Conophyton*, branching-conical stromatolites (*Jacutophyton sahariensis*) form conspicuous elements of the Atar Formation biostromes. *Jacutophyton* consists of a central conical region, indistinguishable from *Conophyton*, that is surrounded by low synoptic relief, unwalled to moderately walled, columnar branches (Figs 5b, c and 7a). Branching elements are moderately to highly divergent from the central cone, extend up to 50 cm from the cone margin (Fig. 5b), and are subcylindrical to petaloid in form, forming elliptical to elongate cross-sections that encircle the central cone (Fig. 5d). With rare exception, branches initiate along a single laminae of the central cone (Fig. 7a). In *Jacutophyton*, branching occurs along the entire height of the central cone, although in the Atar Formation biostromes, branching elements variously attributable to *Baicalia* or *Tilemsina* (Fig. 7b) are occasionally concentrated in the uppermost portion of the central cone or on pavements of broken or toppled *Conophyton* (Fig. 7c).

Fig. 6. Morphological features of conical stromatolites. (a) Secondary growth of *Conophyton* without disruption of initial conical laminae. (b) Elliptical cross section of *Conophyton* showing irregular thickening of outermost laminae. (c) Delamination of partially lithified *Conophyton* laminae. (d) Eroded margins of *Conophyton* indicate exposure to wave action after lithification, but prior to deposition of interstromatolitic material. Jacob staff divisions are 10 cm.

Interstromatolitic regions of *Jacutophyton* consist exclusively of platy breccia and several generations of precipitated carbonate cement (Fig. 9; see descriptions below). The general absence of breccia between branching elements, and the presence of rare laminae that envelop numerous branches (Fig. 5b), however, suggest that interstromatolitic debris was deposited primarily after the growth of branches.

Branching columnar stromatolites

Low- to moderate-relief columnar stromatolites comprise several morphotypes of variably branching columnar stromatolites with synoptic relief of no more than a few centimetres (Fig. 8). Branches are typically moderately to highly divergent, producing stromatolites that widen upward. These forms have low to moderate inheritance and produce biostromes composed of numerous columns. *Tilemsina divergens* (Fig. 8a) and related forms are characterized by irregular unwalled margins of low synoptic relief (<2 cm) and variably divergent, asymmetric branching. Branching columns of moderate relief include the form-taxa *Tungussia confusa* (Fig. 8b), *Baicalia safia and Baicalia mauritanica* (Fig. 8c and d). These stromatolites have columns that widen upward and branch at irregular intervals. Synoptic relief varies from nearly flat (horizontal lamina) to ~8 cm. In the Atar Formation, stromatolites attributable to *Tungussia* occur only in the inter-reef shaly intervals, whereas *Tilemsina* and *Baicalia* both occur as reef elements.

Intercolumnar regions of branching stromatolites vary widely and can contain platy breccia, detrital microspar and several generations of precipitated carbonate cement (see descriptions below). *Tilemsina*, however, is associated exclusively with fine-grained platy breccia (Fig. 8a), and *Tungussia* is associated exclusively with detrital microsparitic carbonate (Fig. 8b). By contrast, intercolumnar regions of *Baicalia* are quite variable, often consisting of a generation

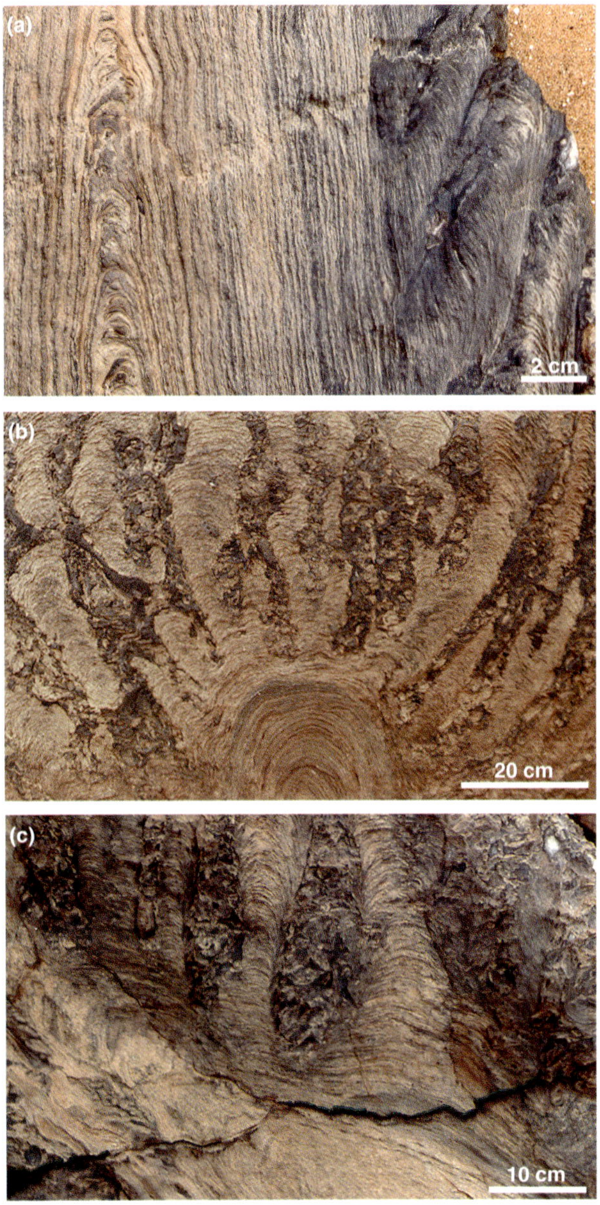

Fig. 7. Development of branching elements on *Conophyton*. (a) Detail of *Jacutophyton* sp. showing branch initiation along a single laminae horizon. (b) Divergent branching columns of *Tilemsina* sp. overgrowing modified top of *Conophyton* sp. (c) Upward growth of *Baicalia* sp. above toppled *Conophyton* sp.

of isopachous carbonate cement followed by deposition of detrital microsparitic carbonate (Fig. 8d).

Interstromatolitic detrital carbonate

Two distinct forms of interstromatolitic debris occur within the Atar Formation reefs: interstromatolitic breccia and carbonate microspar. Interstromatolitic breccia consists of variably convoluted, platy fragments (Fig. 9a–d) that show distinct lamination and petrographic fabric indistinguishable from that of intact conical stromatolites. Clasts range in size from 1 × 3 mm to 7 × 25 mm and show variable degrees of hydrodynamic sorting. Coarse-grained debris, which shows the widest range of grain sizes and poor to moderate sorting, occurs exclusively in association with *Conophyton* and *Jacutophyton* (Fig. 9a–d). Fine-grained debris shows the most hydrodynamic sorting and is found in association with all stromatolite types, although it is most prevalent in its association with *Tilemsina* (Fig. 8a). Carbonate microspar is locally abundant as an interstromatolitic element, forming discontinuous, massive to weakly laminated, undulatory beds a few millimetres to >50 cm thick. In addition to discontinuous beds, detrital microsparitic carbonate also occurs as later-stage, occasionally geopetal fill in voids formed by platy breccia (Fig. 9b and c).

The lateral distribution of interstromatolitic material within the Atar Formation reefs is extremely variable. Beds of detrital microsparitic carbonate greater than 30 cm thick can envelop several adjacent *Conophyton*, yet not extend laterally for more than a few metres. Single *Conophyton* columns can also be enveloped by detrital microsparitic carbonate at one side and coarse-grained platy breccia at the other. The vertical distribution of interstromatolitic facies is equally as variable, with the margin of a single 1.5 m tall *Conophyton* contacting up to five distinct layers of interstromatolitic fill.

Marine cement

Atar Formation stromatolite reefs contain locally abundant marine cement (up to 25% of interstromatolitic material). Marine cement forms multi-layered isopachous rims within voids defined by interstromatolitic debris (Fig. 9a–d) and, more rarely, along stromatolite edges and the depositional substrate. These first-generation cements are composed entirely of herringbone carbonate. Herringbone carbonate is an unusual carbonate morphology consisting of elongate crystals in which the *c*-axis rotates throughout growth, from parallel to perpendicular to crystal elongation (Sumner & Grotzinger, 1996a,b), which results in a characteristic sweeping extinction parallel to crystal elongation direction (Fig. 9e–g).

Fig. 8. Branching columnar stromatolites of the Atar Formation. (a) *Tilemsina divergens* is characterized by low synoptic relief, irregular margins and repeated outward-diverging branching. (b) Irregular branching and upward-broadening columns of *Tungussia confusa*. *Tungussia* occurs primarily as small biostromes in inter-reef shales. (c) Strongly clustered, irregular branching of *Baicalia* in the uppermost R3 reef interval. (d) Irregular branching columns of moderate synoptic relief are represented in the Atar Formation by *Baicalia safia,* which preserves irregular coniform laminae, and *Baicalia mauritania* (shown here), which preserves domal laminae in cross-section.

This unusual mode of crystal growth is most prevalent as a seafloor precipitate in the Archaean and as a void-filling cavity cement in younger successions (de Wet *et al.*, 1999), and has been attributed to both the presence of ferrous iron under conditions of regional anoxia (Sumner & Grotzinger, 1996b) and to the presence of locally elevated concentrations of dissolved inorganic carbon derived from remineralized organic matter (Tourre & Sumner, 2000). In the Mesoproterozoic, extensive precipitation of herringbone carbonate is generally restricted to basinal facies (Kah *et al.*, 1999b), with its presence in platform deposits closely associated with transgressive surfaces (Kah *et al.*, 2006). Later stage void-fill typically includes a single layer of isopachous bladed calcite, coarse blocky calcite spar and ferroan dolomite cement (Fig. 9a–d).

PARASEQUENCE DEVELOPMENT OF STROMATOLITE BIOSTROMES

Superpositional relationships between and among reef elements can be used to determine the relative timing of deposition (or growth) of elements, as well as the morphology of the sedimentary substrate. Combined, these relationships permit a detailed evaluation of changes in accommodation space during development of the Atar Formation biostromes. The cratonal deposition regime represented by the Atar Formation biostromes, and the Atar Group as a whole (Bertrand-Sarfati & Moussine-Pouchkine, 1988), suggests that changes in accommodation space were most likely driven by changes in sea level. In terms of stromatolite development, sea-level changes would result primarily in changes in ambient

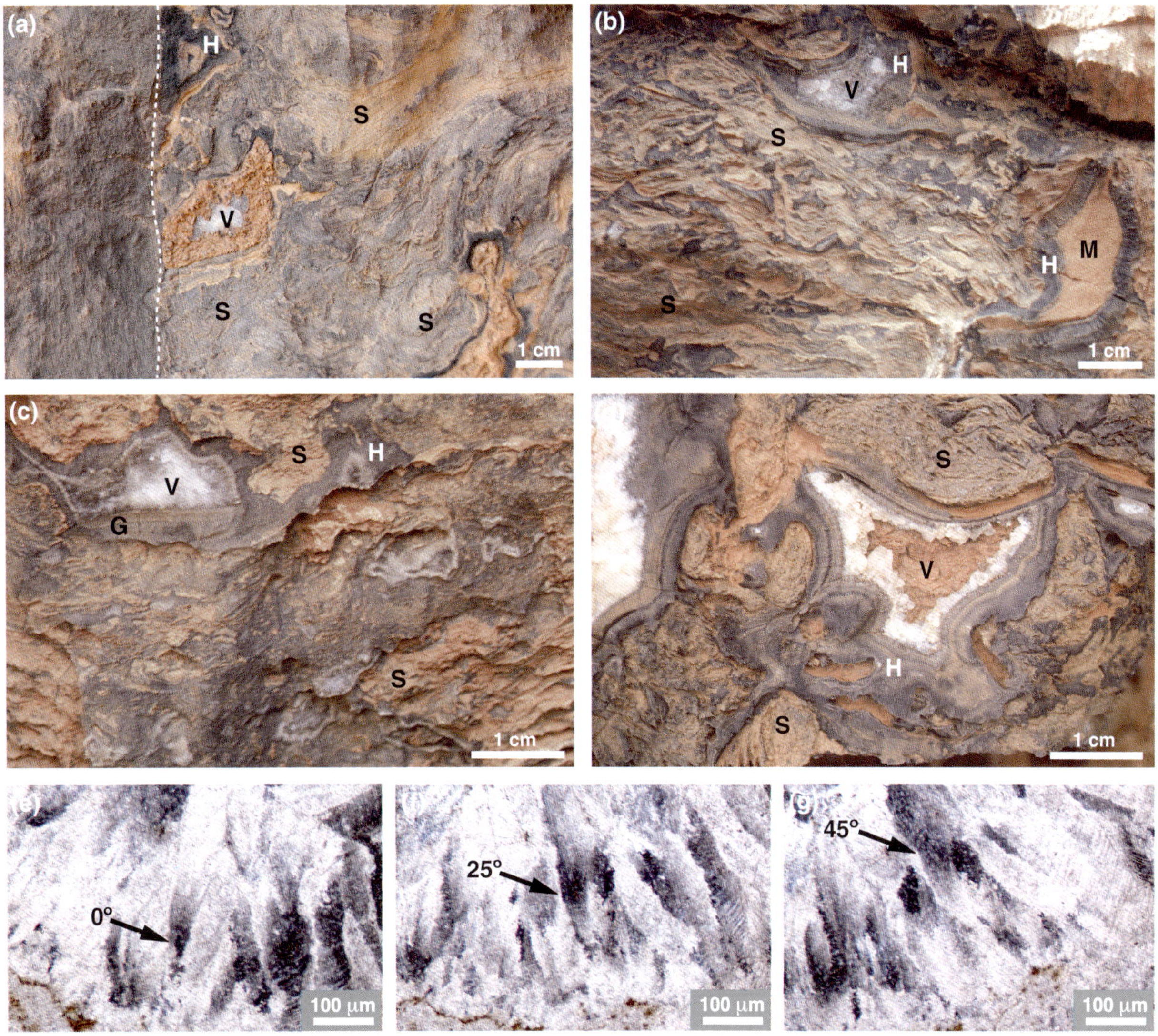

Fig. 9. Interstromatolitic material typical of *Conophyton–Jacutophyton* intervals. (a) Relationship between *Conophyton* (left of dotted line) and interstromatolitic material, including distorted stromatolitic debris (S), herringbone carbonate cement (H) and void-filling cement (V). (b) Void fills show several discrete generations of deposition. The first generation of cement always consists of marine herringbone carbonate cement; later generations include one or more of the following: microsparitic crystal silt (M), blocky white spar, or ferroan dolomite cement. (c) The presence of geopetal fill (G) implies that interstromatolitic void space remained open to fluid and sediment transport even after precipitation of marine herringbone cement. (d) Late-stage calcitic and dolomitic void fills imply significant porosity of interstromatolitic regions well after deposition and perhaps into burial environments. (e–g) The petrographic expression of herringbone carbonate, which shows sweeping extinction along the crystal axis, results from a rotation of the *c*-axis during crystal growth.

water energy and the position of storm and fair weather wave-base.

On the scale of a single biostromal package (interpreted to represent a third-order sequence), sea-level history is fairly straightforward. The base of each biostrome (R1–R3) is marked by the presence of a thin, discontinuous interval of branching columnar stromatolites, typically of the form genus *Tilemsina*, whose irregularly divergent branching, low synoptic relief, and unwalled column structure indicate that accretion of stromatolitic laminae kept pace with deposition of hydrodynamically well-sorted, fine-grained platy breccia. These thin, discontinuous, high-energy beds are interpreted as transgressive deposits. By contrast, the main body of the reef consists primarily of *Conophyton*, whose steep-sided walls, synoptic relief and large height-to-width ratio indicate deposition in a low-energy environment with little or no sedimentary influx. Such environments are typical of marine highstands, when accommodation space is at its greatest and backstepping of the shoreline inhibits sediment influx into the basin. Finally, each reef interval concludes with either a regional pavement of broken *Conophyton* (R1, R2) or the appearance of irregularly branching elements, each of which can be interpreted as reflecting a loss in accommodation space associated with either sea-level fall or aggradational stromatolitic growth (Bertrand-Sarfati & Moussine-Pouchkine, 1988).

Within each third-order package, however, unusual vertical and lateral juxtapositions of stromatolite forms suggest a much more complex developmental history. Detailed superpositional relationships between and among reef elements, however, suggest that the complex juxtaposition of stromatolite forms can be interpreted in terms of parasequence-scale (fourth to fifth order) changes in relative sea level. The most common and conspicuous elements of parasequences in the Atar Formation reef are *Conophyton*. These stromatolites represent dominantly subtidal (beneath storm wave-base; Donaldson, 1976; Bertrand-Sarfati & Moussine-Pouchkine, 1985; Kah *et al.*, 2006) stromatolite nucleation during marine transgression and upward growth throughout transgression and early highstand (Fig. 10a, V1–V6). Rare variation in *Conophyton* morphology, such as the presence of *Conophyton* with elliptical cross sections, irregular margins and occasional branch development (Fig. 6b; *Conophyton jacqueti*; Bertrand-Sarfati & Moussine-Pouchkine, 1999) may represent the presence of low-energy currents during *Conophyton* growth. The trajectory of parasequence formation, however, and the resultant complexity of individual parasequence (Fig. 10; see below) depends critically upon the position of *Conophyton* with respect to wave base, the magnitude of parasequence-scale sea-level changes, and the lithification history of the *Conophyton*.

If *Conophyton* tops are sufficiently beneath wave base, aggradational growth may continue even during falls in relative sea level, resulting in *Conophyton* with the greatest synoptic relief. Alternatively, upward growth may simply terminate and become a nucleus for later Conophyton development (Figs 10a–c, V1 and 6a). Parasequence complexity increases dramatically if, during falls in relative sea level, *Conophyton* tops are subjected to subaerial exposure or increased wave energy. High-energy storm events or prolonged subaerial exposure may topple cones, forming a substrate for subsequent microbial growth or terminating parasequence development (Figs 10b, V2 and 4a and b). Wave energy may also deeply erode fully lithified cones (Fig. 6d) or delaminate poorly lithified outer margins of the cones (Fig. 6c). In the former case, upon subsequent rise in sea level, stromatolite growth may resume, with stromatolite morphologies (conical vs. branching) reflecting the hydrodynamic conditions within the environment (Fig. 10b, V2). In the latter case, stromatolitic debris may be transported, hydrodynamically sorted and deposited in interstromatolitic regions (Fig. 10b, V3), with lowstand nucleation and growth of stromatolites such as *Tilemsina* in high-energy environments between adjacent *Conophyton* (Fig. 10b, V4). During subsequent rises in sea level, nucleation of new *Conophyton* may occur on either exposed cone tops or on interstromatolitic detritus.

Additional complexity of parasequences occurs when falls in relative sea level expose *Conophyton* to wave energy, but interstromatolitic detritus is generally absent. In this case, disruption of the outermost laminae of living *Conophyton* (i.e. those with unlithified outer margins) by wave energy would result in *Conophyton* encircled by partially delaminated microbial elements that, in turn, would form initial substrate for the petaloid branching elements characteristic of *Jacutophyton* (Fig. 10b, V5). Laterally adjacent non-living or more heavily lithified *Conophyton* may experience either no delamination, or more extreme erosion (Fig. 6d) during lowstand. The absence of platy breccia between petaloid branches suggests that wave energy may not be sufficient to create or transport significant breccia, although some breccia may be deposited during latest lowstand. In the Atar Formation biostromes, *Jacutophyton* parasequences show the most variability upon subsequent sea-level rise, with new *Conophyton* nucleating adjacent to *Jacutophyton*, initiating atop the central cone of *Jacutophyton* (Figs 10c, V5a and 11a), or developing directly on *Jacutophyton* branches (Figs 10c, V5b and 11b). In rare cases, subsequent growth of branching stromatolites (Fig. 10c, V5c) suggests sea-level rise insufficient to place microbial growth beneath wave-base.

Finally, if falls in relative sea level expose only the tip of *Conophyton* to wave energy, either because of the position of wave-base relative to stromatolites or the presence of abundant detrital

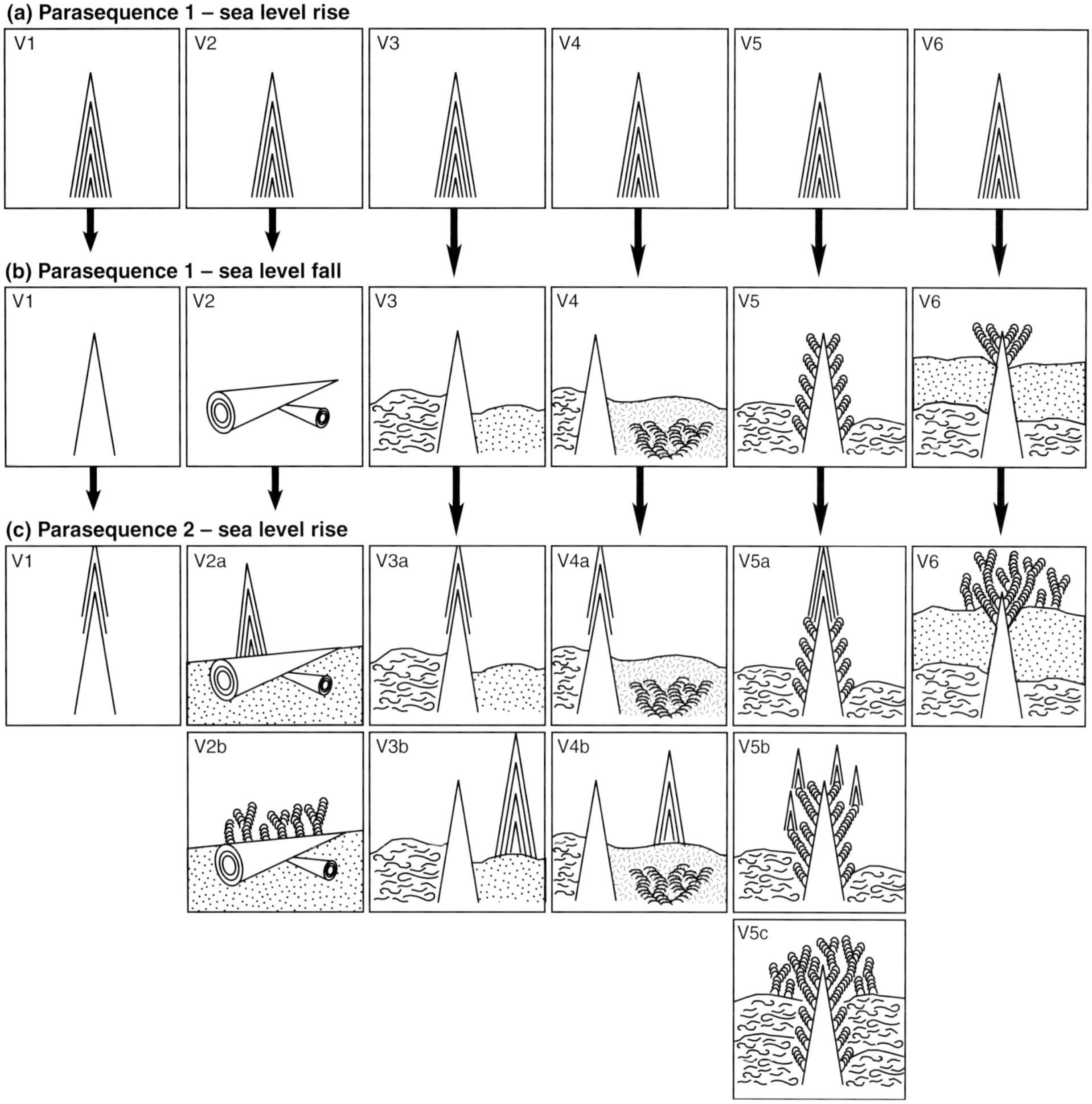

Fig. 10. Parasequence development of the Atar reefs. (a) Parasequences initiate upon sea level rise with the nucleation and growth of *Conophyton* (V1–V6). (b) During sea level fall, accretion of cones is halted (V1) and cones are subject to erosion or toppling (V2), partial burial by interstromatolitic material (stromatolitic debris or detrital microspar; V3), or development of lowstand phase stromatolites, including adjacent bioherms of Tilemsina (V4), concentric branches of *Jacutophyton sahariensis*. (V5), or restricted growth of irregular columnar branches (*Baicalia* or *Tilemsina*; V6). (c) Subsequent sea level rises, of sufficient magnitude to place the substrate beneath wave base, results in reinitiation of coniform growth atop previous upright or toppled *Conophyton* or *Jacutophyton* (V1–V5a), nucleation of new *Conophyton* (V3b–V4b), and development of *Conophyton* from branching elements of *Jacutophyton* (V5b). If, during subsequent sea level rises, the substrate remains within wave base, stromatolite growth is marked by development of irregular branching columns of *Tilemsina* and *Baicalia* (V6, V2b, V5c).

material, branching can develop at the tops of living cones (Fig. 10b, V6). During subsequent sea-level rise, branching can develop into new cones, or into biostromes of irregularly branching columns depending on the hydrodynamic state of the environment. In the Atar Formation, this scenario is observed in the formation of *Tilemsina* on cone tops in the R1 and R2 biostromes (Fig. 7b), in the appearance of branches on *Conophyton jacqueti* in the R3 biostrome

Fig. 11. Stacked parasequences in *Jacutophyton sahariensis*. (a) Branching elements of *Jacutophyton sahariensis* (PS1) are onlapped by conical laminae during initial development of a second-generation *Jacutophyton* (PS2) whose central cone nucleated directly atop the central cone of PS1. Jacob staff is marked in 10 cm intervals. (b) Several branching elements of *J. sahariensis* (PS1) develop into central cones and form the base of a second generation of *Jacutophyton* (PS2).

(cf. Bertrand-Sarfati & Moussine-Pouchkine, 1999) and in the gradational transition from *Conophyton* to *Baicalia Safia* and *Baicalia mauritanica* at the top of the R3 biostrome (Fig. 10c, V6).

The extraordinary variability of biostromal parasequences in the Atar Formation results, in part, from the vertically and laterally variable deposition of interstromatolitic material. Outcrop exposures record the small spatial scales (a few to a few tens of metres laterally, a few centimetres to tens of centimetres vertically) of variability, but do not permit detailed mapping of interstromatolitic debris. Laterally variable deposition probably results from the cratonal, regionally low-relief environments of the Atar Group (Bertrand-Sarfati & Moussine-Pouchkine, 1988), wherein variation in water energy (e.g. channelling of tidal surges, relief of the substrate), localized production of water-column carbonate (e.g. whitings) and restricted production of stromatolitic breccia (e.g. delamination of non-lithified cones) act together to produce a mosaic of interstromatolitic facies.

Platy breccia facies are episodically stabilized by the precipitation of herringbone carbonate cement. Rarely, herringbone cement is observed coating both stromatolites and the sedimentary substrate. In these cases, the thickness of the stromatolitic component of a single parasequence (i.e. the height of a *Conophyton*) is typically 100–150 cm, yet the thickness of the detrital component may vary from only a few centimetres to tens of centimetres in thickness. As a result, parasequence boundaries can show dramatic lateral variability. Furthermore, if the presence of herringbone carbonate indicates transport of basinal waters onto the shelf during marine transgression (Kah *et al.*, 2006), thick encrustations of herringbone carbonate within coarse-grained platy breccias (Fig. 9d) suggest that stabilization of the sedimentary substrate occurred over multiple parasequences. This scenario is supported by the presence of microsparitic detrital carbonate within herringbone-lined voids, which probably represents the winnowing of fine-grained detritus through a highly permeable substrate. If substrate stabilization is protracted through several sea-level cycles, it is possible that some parasequences within the Atar Formation reefs may have experienced significant post-depositional modification.

DISCUSSION

Idealized model for reef growth

Taken together, the development of Atar Formation biostromes on a low-relief, cratonal platform (Bertrand-Sarfati & Moussine-Pouchkine, 1988), the laterally variable thicknesses of parasequences and the possible post-depositional modification of parasequences make deciphering the precise architecture of the Atar reefs difficult. Our understanding of parasequence development, however, allows construction of an idealized model for biostrome development that is a function of (1) regional platform geometry, which controls the position of the substrate relative to wave base and the total accommodation space available for stromatolite growth, and (2) the magnitude of parasequence scale (fourth or fifth order) sea-level changes, which act to modify total accommodation space and the position of wave base.

At distal craton margins, the steeper gradient of the substrate and potentially greater rates of subsidence should result in available accommodation space that is large relative to parasequence-scale sea-level changes. In this scenario, upward growth of conical stromatolites would remain uninterrupted by small-scale changes in sea level and low water energy would prohibit growth of branching stromatolites and the production and deposition of interstromatolitic detritus, resulting in a stromatolite tract that is dominated by high relief *Conophyton* (cf. Dismal Lakes reef, Donaldson, 1976; Kah *et al.*, 2006). Branching stromatolites such as *Baicalia safia* and *Baicalia mauritanica* or the incipient branching of *Conophyton jacqueti* would form only with large magnitude falls in sea level or with the continued aggradation of stromatolite growth, which could potentially bring the biostrome top above wave base. *Jacutophyton*, which requires a significant fall in eustatic sea level and exposure of entire cones to wave energy, would not be expected to form.

By contrast, over much of the cratonal platform, intermediate water depths and low depositional relief should result in environments in which parasequence-scale changes in sea level would exert a greater influence over stromatolite form, resulting in greater variability of stromatolite morphology within biostromes. In these regions, total accommodation space available for stromatolite growth relative to the magnitude of parasequence-scale sea-level changes is critical in the distribution of different stromatolite morphologies. First, larger magnitude sea-level rises would bring more of the cratonal platform beneath wave base, resulting in widespread development of conical stromatolites. Similarly, smaller magnitude sea-level changes would result in restricted spatial development of conical stromatolites. Second, during eustatic falls in sea level, deeper-water portions of the platform would experience less wave energy (perhaps episodic wave energy if between fair weather and storm wave-base), resulting in biostromes dominated by *Conophyton* and *Jacutophyton*. Similarly, shallower-water portions of the shelf would experience wave energy sufficient to produce quantities of detrital material. In these regions, the biostrome would be dominated by *Jacutophyton* and the branching form *Tilemsina*, with sparse *Conophyton* reflecting conical forms that were well-lithified at the time of sea-level fall. Throughout these platformal regions, unusual juxtaposition of stromatolite forms results not from penecontemporaneous growth, as previously inferred (Bertrand-Sarfati & Moussine-Pouchkine 1985, 1999), but from laterally adjacent growth during distinct phases of sea-level cycles.

Shallowest regions of the craton should also contain a third, distinct stromatolite tract. In these regions, low overall accommodation space will result in deposition controlled mainly by parasequence-scale eustatic sea-level changes. Biostrome development should be absent except during brief periods of cratonal inundation and should be dominated by low-relief, irregularly branching stromatolite forms. In fact, high water energy might fully preclude development of *Conophyton*, with coniform growth represented only by the low-relief conical laminae of *Baicalia safia*. Low synoptic relief of stromatolites and

Bertrand-Sarfati, J. and **Moussine-Pouchkine, A.** (1992) Formation et comblement d'une depression intra-plateforme engendree par la croissance d'un biostrome stromatolitique, Proterozoique superieur, Sahara algerien. *CR Acad. Sci. Montrouge*, **315**, 837–843.

Bertrand-Sarfati, J. and **Moussine-Pouchkine, A.** (1999) Mauritanian microbial buildups: Meso-Neoproterozoic stromatolites and their environment; six days field trip on the Mauritanian Adrar. *Assoc. Sediment. Fr.*, **31**, 1–103.

Bertrand-Sarfati, J. and **Trompette, R.** (1976) Use of stromatolites for intrabasinal correlation; example from the late Proterozoic of the northwestern margin of the Taoudenni Basin. In: *Stromatolites* (Ed. M.R. Walter). Elsevier, Amsterdam, pp. 517–522.

Bertrand-Sarfati, J., **Moussine-Pouchkine, A.** and **Ait Kaci Ahmed, A.** (1996) Subdivisions stratigraphiques nouvelles dans la couverture néoprotérozoique au Nord-Est du basin de Taoudenni (Algérie). In: *Géodynamique du craton ouest africain central et oriental: héritage et évolution post-panafricains* (Eds L. Bitam and J. Fabre). *Serv. Geol. Algérie*, **1996**, 63–90.

Bertrand-Sarfati, J., **Moussine-Pouchkine, A.** and **Caby, R.** (1987) Les correlations du Protérozoique au Cabrien en Afrique de l'Ouest: nouvelle interprétation géodynamique. *Bull Soc. Géol. Fr.*, **3**, 855–865.

Bertrand-Sarfati, J., **Moussine-Pouchkine, A.**, **Affaton, P.**, **Trompette, R.** and **Bellion, Y.** (1991) Cover sequences of the West African Craton. In: *The West African Orogens and Circum-Atlantic Correlatives* (Eds R.D. Dallmeyer and J.P. Lecorche). Springer-Verlag, Berlin, pp. 65–82.

Bertrand-Sarfati, J., **Moussine-Pouchkine, A.**, **Amard, B.** and **Ait Kaci Ahmed, A.** (1995) First Ediacaran fauna found in western Africa and evidence for an Early Cambrian glaciation. *Geology*, **23**, 133–136.

Black, R. and **Fabre, J.** (1983) A brief outline of the geology of West Africa. In: *Afrique de l'Ouest: introduction geologique et termes stratigraphique* (Ed J. Fabre). Pergamon, Oxford, pp. 17–26.

Clauer, N. (1976) Géochimie isotopique du strontium des milieux sédimentaires: application a la geochronology de la couverture du craton ouest-africain. *Mem. Sci. Geol., Strasbourg*, **45**, 1–256.

Clauer, N. (1981) Rb–Sr and K–Ar dating of Precambrian clays and glauconites. *Precambrian Res.*, **15**, 331–352.

Clauer, N. and **Deynoux, M.** (1987) New information on the probable isotopic age of the late Proterozoic glaciation in West Africa. *Precambrian Res.*, **37**, 89–94.

Clauer, N., **Caby, R.**, **Jeanette, D.** and **Trompette, R.** (1982) Geochronology of sedimentary and metasedimentary Precambrian rocks of the West African craton. *Precambrian Res.*, **18**, 53–71.

Cloud, P.E. and **Semikhatov, M.A.** (1969) Proterozoic stromatolite zonation. *Am. J. Sci.*, **267**, 1017–1061.

De Wet, C.B., **Dickson, J.A.D.**, **Wood, R.A.**, **Gasworth, S.B.** and **Frey, H.M.** **(1999)** A new type of shelf margin deposit: rigid microbial sheets and unconsolidated grainstones riddled with meter-scale cavities. *Sed. Geol.*, **128**, 13–21.

Donaldson, J.A. (1976) Palaeoecology of conophyton and associated stromatolites in the Precambrian Dismal Lakes and Rae groups, Canada. In: *Stromatolites* (Ed. M.R. Walter). Elsevier, Amsterdam, pp. 523–534.

Dupraz, C. **Pattisina, R.** and **Verrecchia, E.P.** (2006) Translation of energy into morphology: simulation of stromatolite morphospace using a stochastic model. *Sed. Geol.* **185**, 185–203.

Emery, D. and **Myers, K.** (1996) *Sequence Stratigraphy*. Blackwell, Oxford, UK.

Fairchild, I.J., **Marshall, J.D.** and **Bertrand-Sarfati, J.** (1990) Stratigraphic shifts in carbon isotopes from Proterozoic stromatolitic carbonates (Mauritania): influences of primary mineralogy and diagenesis. *Am. J. Sci.*, **290-A**, 46–79.

Frank, T.D., **Kah, L.C.** and **Lyons, T.W.** (2003) Changes in organic matter production and accumulation as a mechanism for isotopic evolution in the Mesoproterozoic ocean. *Geol. Mag.*, **140**, 397–420.

Furniss, G., **Rittel, J.F.** and **Winston, D.** (1998) Gas bubble and expansion crack origin of 'molar-tooth' calcite structures in the Middle Proterozoic Belt Supergroup, western Montana. *J. Sed. Res.*, **68**, 104–114.

Golubic, S. (1976) Organisms that build stromatolites. In: *Stromatolites* (Ed. M.R. Walter). Elsevier, Amsterdam, pp. 113–126.

Goodman, E.E. and **Kah, L.C.** (2004) Trace sulfate concentrations as an indicator of depositional environment: examination of some possible calcitized evaporates from the Proterozoic Atar Group, Mauritania. *Geol. Soc. Am., Abstr. Prog.*, **36**, 78.

Grey, K. and **Thorne, A.M.** (1985) Biostratigraphic significance of stromatolites in upward shallowing sequences of the early Proterozoic Duck Creek Dolomite, Western Australia. *Precambrian Res.*, **29**, 183–206.

Grotzinger, J.P. (1989) Facies and evolution of Precambrian depositional systems: emergence of the modern platform archetype. In: *Controls on Carbonate Platform and Basin Development* (Eds P.D. Crevello, J.J. Wilson, J.F. Sarg and J.F. Read). *SEPM Spec. Pub.*, **44**, 79–106.

Grotzinger, J.P. and **Knoll, A.H.** (1999) Stromatolites in Precambrian carbonates: evolutionary mileposts or environmental dipsticks? *Ann. Rev. Earth Planet. Sci.*, **27**, 313–358.

Grotzinger, J.P. and **Rothman, D.** (1996) An abiotic model for stromatolite morphogenesis. *Nature*, **383**, 423–425.

Halverson, G.P., **Hoffman, P.F.**, **Schrag, D.P.**, **Maloof, A.C.** and **Rice, A.H.N.** (2005) Toward a Neoproterozoic composite carbon-isotope record. *Geol. Soc. Am. Bull.*, **117**, 1181–1207.

Hoffman, P.F. (1967) Algal stromatolites: use in stratigraphic correlation and paleocurrent determination. *Science*, **157**, 1043–1045.

Hoffman, P.F. (1974) Shallow and deep-water stromatolites in lower Proterozoic platform-tobasin facies change, Great Slave Lake, Canada. *AAPG Bull.*, **58**, 856–867.

Hoffman, P.F. (1976) Environmental diversity of Middle Precambrian stromatolites. In: *Stromatolites* (Ed. M.R. Walter). Elsevier, Amsterdam, pp. 599–611.

Hofmann, H.J. (1969) Attributes of stromatolites. *Geol. Surv. Can., Pap.*, **69–39**, 80.

James, N.P., **Narbonne, G.M.** and **Sherman, A.B.** (1998) Molar-tooth carbonates: shallow subtidal facies of the Mid-to Late Proterozoic. *J. Sed. Res.*, **68**, 716–722.

Kah, L.C. and **Knoll, A.H.** (1996) Microbenthic distribution in Proterozoic tidal flats: environmental and taphonomic considerations. *Geology*, **24**, 79–82.

Kah, L.C., Sherman, A.B., Narbonne, G.M., Kaufman, A.J. and **Knoll, A.H.** (1999a) $\delta^{13}C$ stratigraphy of the Proterozoic Bylot Supergroup, northern Baffin Island: Implications for regional lithostratigraphic correlations. *Can. J. Earth. Sci.*, **36,** 313–332.

Kah, L.C., Bartley, J.K., Frank, T.D. and **Lyons, T.W.** (1999b) Reef facies and possible chemosynthetic communities of the Sulky formation, ~1.3 Ga Dismal Lakes Group, NWT, Canada. *Geol. Assoc. Can./Mineral. Assoc. Can., Abstr.*, **24,** 60.

Kah, L.C., Bartley, J.K., Frank, T.D. and **Lyons, T.W.** (2006) Reconstructing sea-level change from the internal architecture of stromatolite reefs: an example from the Mesoproterozoic Sulky Formation, Dismal Lakes Group, arctic Canada. *Can. J. Earth Sci.*, **43,** 653–669.

Kaufman, A.J. and **Knoll, A.H.** (1995) Neoproterozoic variations in the C-isotopic composition of seawater: stratigraphic and biogeochemical implications. *Precambrian Res.*, **73,** 27–49.

Knoll, A.H. and **Semikhatov, M.A.** (1998) The genesis and time distribution of two distinctive Proterozoic stromatolite microstructures. *Palaios*, **13,** 408–422.

Knoll, A.H. and **Sergeev, V.N.** (1995) Taphonomic and evolutionary changes across the Mesoproterozoic–Neoproterozoic boundary. *Neues Jb. Miner. Geol. Paläontol.*, **195,** 289–302.

Knoll, A.H., Kaufman, A.J. and **Semikhatov, M.A.** (1995) The carbon isotopic composition of Proterozoic carbonates: Riphean successions from northwestern Siberia (Anabar Massif, Turukhansk Uplift). *Am. J. Sci.*, **295,** 823–850.

Lee, S.-J. and **Golubic, S.** (2000) Biological and mineral components of an ancient stromatolite: Gaoyuzhuang Formation, Mesoproterozoic of China. In: *Carbonate Sedimentation and Diagenesis in the Evolving Precambrian World* (Eds J.P. Grotzinger and N.P. James). *SEPM Spec. Pub.*, **67,** 91–102.

Lemon, N.M. (2000) A Neoproterozoic fringing stromatolite reef complex, Finders Ranges, South Australia. *Precambrian Res.*, **100,** 109–120.

Logan, B.W., Hoffman, P.F. and **Gebelein, C.D.** (1974) Algal mats, cryptalgal fabrics, and structures, Hamelin Pool, Western Australia. In: *Evolution and Diagenesis of Quaternary Carbonate Sequences, Shark Bay, Western Australia* (Eds B.W. Logan, J.F. Read and G.M. Hagan). *AAPG Mem.*, **69,** 140–194.

Pollock, M.D., Kah, L.C. and **Bartley, J.K.** (2006) Morphology of molar-tooth structures in Precambrian carbonates: influence of substrate rheology and implications for genesis. *J. Sed. Res.*, **76,** 310–323.

Miall, A.D. (1997) *The Geology of Stratigraphic Sequences.* Springer-Verlag, Berlin, Germany.

Moussine-Pouchkine, A. and **Bertrand-Sarfati, J.** (1997) Tectonosedimentary subdivisions in the Neoproterozoic to Early Cambrian cover of the Taoudenni Basin (Algeria-Mauritania-Mali). *J. Af. Earth Sci.*, **24,** 425–443.

Moussine-Pouchkine, A., Bertrand-Sarfati, J., Ball, E. and **Caby, R.** (1988) Les series sedimentaires et volcaniques anorogeniques proterozoiques impliquees dans la chaine pan africaine; la region de l'Adrar Ahnet (NW Hoggar, Algerie). *J. Af. Earth Sci.*, **7,** 57–75.

Narbonne, G.M. and **James, N.P.** (1996) Mesoproterozoic deep-water reefs from Borden Peninsula, Arctic Canada. *Sedimentology*, **43,** 827–848.

Reid, R.P., Visscher, P.T., Decho, A., Stolz, J.K., Bebout, B.M., Dupraz, C., MacIntyre, I.G., Paerl, H.W., Pinchney, J.L., Prufert-Bebout, L., Steppe, T.F. and **DesMarais, D.J.** (2000) The role of microbes in accretion, lamination and early lithification of modern marine stromatolites. *Nature* **406,** 989–992.

Semikhatov, M.A. and **Raaben, M.E.** (1994) Dynamics of the global diversity of Proterozoic stromatolites. Article 1: Northern Eurasia, China, and India. *Strat. Geol. Correl.*, **2,** 492–513.

Semikhatov, M.A. and **Raaben, M.E.** (1996) Dynamics of the global diversity of Proterozoic stromatolites. Article II: Africa, Australia, North America, and general synthesis. *Strat. Geol. Correl.*, **4,** 24–50.

Semikhatov, M.A., Gebelein, C.D., Cloud, P., Awramik, S.M. and **Benmore, W.C.** (1979) Stromatolite morphogenesis – progress and problems. *Can. J. Earth Sci.*, **19,** 992–1015.

Sumner, D.Y. and **Grotzinger, J.P.** (1996a) Herringbone calcite: petrography and environmental significance. *J. Sed. Res.*, **66,** 419–429.

Sumner, D.Y. and **Grotzinger, J.P.** (1996b) Were kinetics of Archean calcium carbonate precipitation related to oxygen concentration? *Geology*, **24,** 119–122.

Teal, D.J. and **Kah, L.C.** (2005) Using C-isotopes to constrain interbasinal stratigraphic correlations, Mesoproterozoic Atar Group, Mauritania. *Geol. Soc. Am., Abstr. Prog.*, **37,** 45.

Tourre, S.A. and **Sumner, D.Y.** (2000) Geochemistry of herringbone calcite from an ancient Egyptian quarry. *Geol. Soc. Am., Abstr. Prog.*, **32,** 215.

Trompette, R. and **Carozzi, A.V.** (1994) *Geology of Western Gondwana (2000–500 Ma).* A.A. Balkema, Rotterdam, 340 pp.

Turner, E.C., James, N.P. and **Narbonne, G.M.** (1997) Growth dynamics of Neoproterozoic calcimicrobial reefs, Mackenzie Mountains, northwest Canada. *J. Sed. Res.*, **67,** 437–450.

Turner, E.C., Narbonne, G.M. and **James, N.P.** (1993) Neoproterozoic reef microstructures from the Little Dal Group, northwestern Canada. *Geology*, **21,** 259–262.

Walter, M.R. (1992) Proterozoic Stromatolites. In: *The Proterozoic Biosphere: A Multidisciplinary Study* (Eds J.W. Schopf and C. Klein), Cambridge, New York, pp. 253–260.

Int. Assoc. Sedimentol. Spec. Publ. (2009) **41**, 297–304

Layering: what does it mean?

HAROLD R. WANLESS

Department of Geological Sciences, University of Miami, P.O. Box 249176, Coral Gables, FL 33124, USA (E-mail: hwanless@miami.edu)

ABSTRACT

'Layering. What does it mean?' Robert Ginsburg made this passing comment in the mid-1960s when the author was looking at a core sequence from Biscayne Bay, Florida. The core was a burrowed carbonate wackestone with a Hurricane Betsy storm layer at the top. It was a timely comment as the question has been asked many times since then. While a few important answers have emerged, there are still some serious challenges ahead. Understanding layering is more than just visually looking at the layers. It includes an appreciation of grain size, sorting and abrasion, influences of bioturbation and subsequent diagenesis. Together, these provide the critical tools to reconstruct palaeoenvironmental setting, define depositional processes and determine the spectrum of energy levels operating in the system.

Keywords Layering, texture, burrowing, diagenesis, Ginsburg.

PRIMARY STRATIFICATION

The question of layering was actively pursued by Robert Ginsburg, Laurie Hardie and others on the tidal flats on Andros Island in the Bahamas. These studies documented that laminated carbonates formed on the tidal flat supratidal channel levees because of winter storm flooding events. Layering thickened on the lower levee backslope approaching the pond, but grazing and infaunal bioturbation quickly destroyed deposited layers (Hardie, 1977). It was a revelation that not all laminated carbonates were deep basinal. This work caused a sudden trend for reinterpretations of laminated carbonates. Many laminated carbonates interpreted as deep basinal were now realized to be shallow tidal flat (and of course, some that were in fact deep basinal were misinterpreted to be tidal flat).

The specific form of the laminae produced (flat, crinkly and such) was the result of cyanobacterial mats growing on the surface (Fig. 1), and the distribution of the different mat types was a function of 'Exposure Index' (the inverse of the frequency of flooding; Hardie, 1977) or, as later documented on Caicos, the frequency of sedimentation events (Wanless *et al.*, 1988b).

Working on Caicos, and with the help of Hurricane Kate in 1985, Wanless *et al.* (1988b) demonstrated that some carbonate tidal flats tend to be built of centimetre-thick bedding not millimetre laminae. These are flats that are seldom influenced by winter-storm events but rely on hurricanes to build their sequence. Centimetre-thick bedding occurs on the inner marsh on Andros where the cyanobacterial mats were punctuated by thicker sediment layers from hurricanes, but not reached by winter storms. Based on field observations, channelled carbonate flats subjected to winter storms build laminated levees, and flats or portions of flats affected only by larger hurricane events produce centimetre-thick bedding.

GRAIN SIZE

This centimetre-thick supratidal bedding described above presented a problem because it was now recognized that hurricanes produce centimetre-thick stratification in both normally subtidal and normally supratidal environments. These can be differentiated by grain size. The sediment swept across a tidal flat by Hurricane Kate was that which could be carried in suspension. In contrast, the offshore storm layer contained a total mix of coarse to fine grain sizes moved around the offshore bottom by the storm (Wanless *et al.*, 1988b). Bagnold (1966), working on desert sands and in rivers in Africa, and McCave (1970, 1971), working on sands

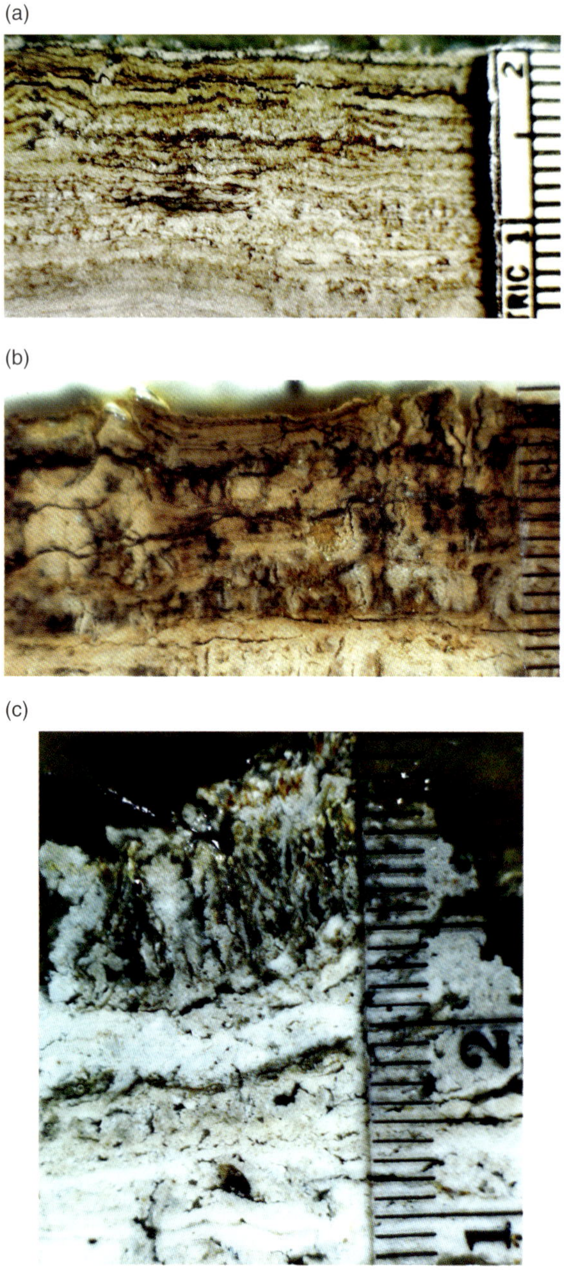

Fig. 1. With less frequent sedimentation events on carbonate tidal flats, the tufted cyanobacteria, *Scytonema* has an increasing influence on layering form. (a) Frequent sedimentation events permit only the beginning of tufts to form and the layering takes on a crinkly appearance. (b) With more time between sedimentation events tufts become more pronounced and disrupt layer continuity. (c) In areas removed from significant sedimentation events for years, a pronounced tufted cushion mat forms and adsorbs minor sedimentation events. Millimetre or thicker sediment layers (light) smother and kill the mat which will take 0.5–1 year to recolonize. The decayed tufted mat and its contained sediment form the thicker, darker layers in the sequence. Andros Island tidal flats, Bahamas.

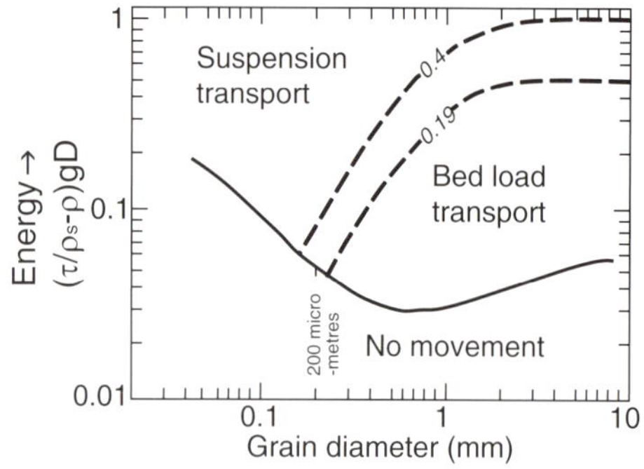

Fig. 2. Entrainment diagram for quartz sands. When there is sufficient wave or current energy to move grains finer than 200 μm, then there is also enough turbulence to transport them in suspension (τ = shear stress; ρ_s = density of grain; ρ = density of fluid; and D = grain diameter). Modified from McCave (1970).

at the bottom of the North Sea, showed that grains finer than 200 μm tend to move in suspension (Fig. 2). Specifically, when there is enough energy to move particles finer than about 200 μm, there is enough turbulence to put them in suspension. So even in hurricanes surges, except within about 100 m of the shore, the tidal flat layers are nearly entirely composed of less than 200 μm, suspension-transported sand. Risi *et al.* (1995) found that this was true for the 1–10 cm thick layer swept 10 km inland through the broad mangrove swamp of southwest Florida during Hurricane Andrew, a Category 5 storm in 1992.

Beaches display this principle well. Sand finer than 200 μm will not stay on a beach because whenever the sand moves (by water or wind), it will move in suspension and be carried from the beach. The minimum grain size of nearly all beaches is around 200 μm. In areas with higher persistent energy, the bedload boundary (what remains on the beach) will shift to the coarser grain size as indicated in Fig. 2. The rapid erosion of many beach renourishment projects is the result of ignoring this fundamental behaviour of sands.

Grain size explains a lot about the origin of stratification in the modern and ancient environment, whether one is looking at the centimetre-thick sand and shale beds in the Cambrian Bright Angel Shale in the Grand Canyon, water-density-driven canal sedimentation in Marco Island, Florida, or the sediments washing off to the flanks of Great Bahama Bank during storms. Rapid sediment dispersal of suspension-transported sands can, with sufficient sediment supply, totally overwhelm bedload transport and is thus a dominating influence on texture of event stratification on epeiric seas and broad carbonate platforms and their margins.

Grain-size analysis by sieving is not always a sufficient approach when analysing modern carbonates. Carbonate grains are commonly hollow, porous, and/or platy and their sieved grain size may not be a good measure of the grains' behaviour. Settling analysis, the settling of grains through a column of water, is a better measure of grain size. Ooids and abraded mollusc fragments are solid carbonate, equant, and have about the same density as quartz, so sieve analysis will give a good assessment. However, crinoid columnals, other echinodermal plates, foraminiferal tests, bryozoan fragments, the plates of the calcareous alga, *Halimeda,* and faecal pellet grain aggregates are very porous and will settle together with quartz grains half to quarter their size (carbonates – Braithwaithe, 1973; Wanless *et al.*, 1981; siliciclastics – Haven & Morales-Alamo, 1968). Sieving is, therefore, a poor measure of their behaviour.

Applying this to ancient limestones is challenging as the original porosity within grains is not always apparent. The first time a modern crinoid fragment is handled it is surprising how little the grain weighs and how porous it is. This contrasts strongly with the massive crinoidal limestones in which they are commonly preserved in the Palaeozoic.

THE POINT BAR–TEMPESTITE–TURBIDITE–TSUNAMIITE DILEMMA

Great strides have been made in understanding the meaning of layering in turbidites (Kuenen, 1948, 1950; Bouma, 1962; Walton, 1967; Enos, 1969), tempestites (Ager, 1974; Aigner, 1985), and fluvial point-bar sequences (Jackson, 1976). These processes – turbidity currents, storm transport and river flood transport – actually produce very similar products which are still easily confused and not rigorously characterized. Most researchers seem best able to define what these layers represent when they look at associated facies and fauna – the pelagic E unit of the turbidite, the shelf to lagoonal fauna and trace fossils of tempestites, and the terrestrial signature in floodplains capping meandering river sequences. In fact, copious literature is devoted to characterizing the details of these environments and deposits, but very little in defining criteria to distinguish them based on the physical sedimentary structures. It would first be helpful to dispose of the terms tempestite and turbidite as they essentially force the observer into one camp or another in the description phase. The tempestites in particular are probably still a lumping of stratification from several different types of storm transport processes, including local resuspension and redeposition, strong lateral cross-shelf transport induced by density of sediment-laden water (Aigner & Reineck, 1982), and deposition on subtidal storm deltas (Aigner, 1985). Siliciclastic and carbonate sedimentologists have characterized storm deposition quite differently, mostly reflecting different energy settings and different preserved components. Siliciclastic storm units are thought of as being made of hummocky and swaley stratification of sand (Harmes *et al.*, 1975; Dott & Bourgeois, 1982). In carbonates, undulating sand stratification is only a part of the sequence, and is complemented by basal skeletal shell concentrations and capping muds and post-event faunal processes and community remains (Aigner, 1985).

Tsunami deposits must now be added to this list. There is perhaps hope for differentiating tsunami deposits as they are commonly a series of 2–6 very closely spaced, but distinct events, although not all tsunami deposits show multiple units (Dawson & Shi, 2000). In addition, it is important to acknowledge that tsunami deposits can vary from thin layers of mud or sand that extend kilometres inland, or more localized giant boulders (i.e. Grand Cayman).

Reineck & Wunderlich (1968) helped to differentiate sequences built of repetitive sand and mud alternations by tides from a stacking by less frequent storm events. In sands, flaser bedding – repeated, commonly reversing, ripple stratification separated by thin, partly eroded mud drapes – represent alternations of frequent, repetitive transport and quiet water settle out, commonly tidal in origin. More widely temporally spaced event stratification commonly tends to be an alternation of sand layers interbedded with slightly thicker and more continuous mud layers, and the sand layers are characteristically more planar to hummocky stratified with a thin ripple top (Fig. 3). For skeletal gravels, storm deposition is generally characterized by a loose and chaotically packed, flat to undulating layer. Some of the skeletal material is deposited articulated and most exhibits little abrasion unless the deposit has been repeatedly

Kuenen, Ph.H. (1948) Turbidity currents of high density. *Eighteenth Sess. Intern. Geol. Cong.*, 1948.

Kuenen, Ph.H. (1950) *Marine Geology*. John Wiley & Sons, New York, 568 pp.

Logan, B.W. and **Semeniuk, V.** (1976) Dynamic metamorphism; processes and products in Devonian carbonate rocks, Canning Basin, Western Australia. *Geol. Soc. Austral. Spec. Pub.*, **16**, 138 pp.

McCave, I.N. (1970) Sand waves in the North Sea off the coast of Holland. *Mar. Geol.*, **10**, 199–255.

McCave, I.N. (1971) Wave effectiveness at the sea bed and its relationship to bed-forms and deposition of mud. *J. Sed. Petrol.*, **41**, 89–96.

Perlmutter, M. (1982) *The role and recognition of storm deposits in the subtidal sediments of the ten thousand islands, Southwest Florida*. Ph.D. dissertation, University of Miami, 230 pp.

Reineck, H.E. and **Wunderlich, F.** (1968) Classification and origin of flaser and lenticular bedding. *Sedimentology*, **11**, 99–104.

Risi, J.A. (1998) *Event sedimentation from hurricane andrew along the Southwest Florida Coast*. Ph.D. dissertation, University of Miami, 198 pp.

Risi, J.A., **Wanless, H.R.**, **Tedesco, L.P.** and **Gelsanliter, S.** (1995) Catastrophic sedimentation from Hurricane Andrew along the southwest Florida coast. *J. Coastal Res.*, **21**, 82–102.

Shepard, F.P. and **Moore, D.G.** (1960) Bays of central Texas coast. In: *Recent Sediments, Northwest Gulf of Mexico* (Eds F.P. Shepard, F.B. Phleger and T.H. van Andel), pp. 117–152, AAPG, Tulsa, Oklahoma.

Tedesco, L.P. and **Wanless, H.R.** (1989) Role of burrow excavation and infilling in creating the preserved depositional fabric of Pennsylvanian phylloid mounds of southeastern Kansas. In: *Sequence Stratigraphic Interpretations and Modeling of Cyclothems* (Eds W.L. Watney, J.A. French and E.K. Franseen), *Kansas Geol. Soc. **41**st Annual Field Trip*, pp. 179–192.

Vlaswinkel, B. and **Wanless, H.R.** (2009) Rapid recycling of organic-rich carbonates during transgression: a complex coastal system in southwest Florida. In: *Perspectives in Sedimentary Geology: A Tribute to the Career of Robert Nathan Ginsburg* (Eds P.K. Swart, G.P. Eberli and J. McKenzie). Blackwell, Oxford. *Spec. Publ. Int. Assoc. Sedimentol.*, **41**, 91–112.

Walton, E.K. (1967) The sequence of internal structures in turbidites. *Scot. J. Geol.*, **3**, 305–317.

Wanless, H.R. (1979) Limestone response to stress: solution and dolomitization. *J. Sed. Petrol.*, **49**, 437–462.

Wanless, H.R. (1981) Limestone response to stress: solution and dolomitization. In: *Diagenesis of Carbonate Rocks: Cement-Porosity Relationships* (Eds G.M. Friedman and S.A. Ali), *SEPM Reprint Series*, **10**, 251–276.

Wanless, H.R. (1982) Limestone response to stress: pressure solution and dolomitization-reply. *J. Sed. Petrol.*, **52**, 328–332.

Wanless, H.R. and **Tagett, M.G.** (1989) Origin and dynamic evolution of carbonate mudbanks in Florida Bay. *Bull. Mar. Sci.*, Symposium on Florida Bay, **44**, 454–489.

Wanless, H.R., **Burton, E.A.** and **Dravis, J.J.** (1981) Hydrodynamics of carbonate fecal pellets. *J. Sed. Petrol.*, **51**, 27–36.

Wanless, H.R., **Tedesco, L.P.** and **Tyrrell, K.M.** (1988a) Production of subtidal tubular and surficial tempestites by Hurricane Kate, Caicos Platform, British West Indies. *J. Sed. Petrol.* **58**, 739–750.

Wanless, H.R., **Tedesco, L.P.**, **Cottrell, D.J.**, **Tagett, M.G.** and **Warzeski, E.R.** (1995) Origin and growth of carbonate mud banks in south Florida. *Int. Assoc. Sedimentol. Spec. Publ.*, **23**, 439–473.

Wanless, H.R., **Tyrrell, K.M.**, **Tedesco, L.P.** and **Dravis, J.J.** (1988b) Tidal-flat sedimentation from Hurricane Kate, Caicos Platform, British West Indies. *J. Sed. Petrol.*, **58**, 724–738.

Int. Assoc. Sedimentol. Spec. Publ. (2009) **41**, 305–325

Falling-stage systems tract in tropical carbonate rocks

WOLFGANG SCHLAGER* and GEORG M.D. WARRLICH†

**Vrije Universiteit/Earth and Life Sciences, 1081HV Amsterdam, Netherlands*
(E-mail: wolfgang.schlager@falw.vu.nl)
†*Petroleum Development Oman, DSC 72, P.O. Box 81, Muscat 113, Sultanate of Oman*
(E-mail: georg.warrlich@pdo.co.om)

ABSTRACT

The standard model of sequence stratigraphy postulates that the falling limb of a sea-level cycle leaves no depositional record on the shelf and upper slope – highstand and lowstand systems tracts are separated by a hiatus and erosional unconformity. There are now many examples of sediment accumulations formed during relative sea-level fall, here referred to as 'falling-stage systems tract'. This study focuses on tropical carbonates. Numerical modelling is used to identify key parameters for the development of the falling-stage systems tract and the standard-model anatomy, and to determine their stability domains in this parameter space. Well-documented case studies are used to compare model results and field observations. Key parameters are the rates of sea-level fall, erosion and carbonate production with slope angle as a modifying factor. The falling-stage systems tract is favoured by low rates of fall, low rates of erosion, high rates of carbonate production, and low slope angle. The effects of subaerial and marine erosion are similar. The two variables are linearly correlated in a wide range of conditions and the sum of subaerial and marine erosion rates can be plotted on one axis after appropriate conversion. Numerical reconnaissance and field observation indicate that the geologically probable space for the falling-stage systems tract is large and the phenomenon should be common in tropical carbonates.

Keywords Falling-stage systems tract, sequence stratigraphy, tropical carbonates, stratigraphic modelling.

INTRODUCTION

The concept of sequence stratigraphy as introduced in the open literature includes a consistent, detailed model of the depositional anatomy of sequences, here referred to as the 'standard model' (Mitchum *et al.*, 1977; Vail *et al.*, 1977; Vail, 1987; Van Wagoner *et al.*, 1987, 1988; Jervey, 1988 Posamentier & Vail, 1988; Posamentier *et al.*, 1988; Sarg, 1988). The standard model of sequence stratigraphy (STM) assumes that the lowstand systems tract is separated from the preceding highstand tract by an erosional unconformity that extends continuously from the highstand shoreline to the lowstand minimum of sea level (Fig. 1). A depositional record of sea-level fall exists only on the lower slope and basin floor in the form of turbidite fans and wedges of slumps and debris flows.

Subsequent field studies produced many examples where the falling sea level also left a record in the form of downstepping or downward-migrating shelf and upper slope deposits (Hunt & Tucker, 1992, 1993; Nummedal *et al.*, 1992; Posamentier & Allen, 1999; with more case studies in Hunt & Gawthorpe, 2000). Reviews by Plint & Nummedal (2000) and Posamentier & Morris (2000) concluded

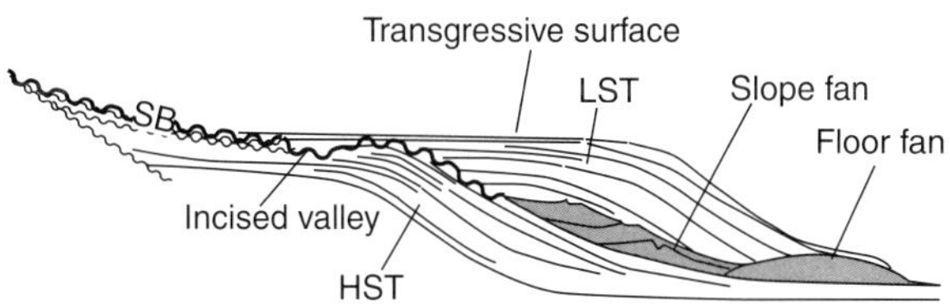

Fig. 1. Standard model of sequence stratigraphy in the vicinity of the sequence boundary (SB). Highstand systems tract (HST) beneath the SB is dissected by an incised valley; elsewhere erosion of HST is minor as indicated by the absence of truncated timelines. During sea-level fall sediment accumulated only on lower slope and basin floor (grey shading). On upper slope, no falling-stage systems tract is present and lowstand systems tract (LST) abuts against clinoforms of preceding highstand tract. After Vail (1987), modified.

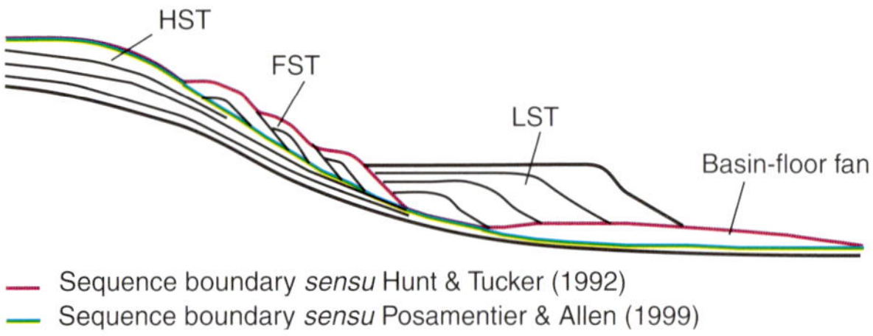

Fig. 2. Model of forced-regressive wedge systems tract (here called falling-stage systems tract, FST) according to Hunt & Tucker (1992). The FST consists of (1) downstepping pairs of topsets and foresets and (2) slope and basin-floor fans already included in the standard model; both units represent sediment bodies formed during sea-level fall. In colour: two common ways of positioning the sequence boundary if FST exists.

that some form of 'falling-stage systems tract' (FST) is rather common in the sequence record even though the systems-tract terminology and the position of the sequence boundary remain controversial (Fig. 2).

This study examines the situation with respect to carbonate rocks, in particular the deposits of the tropical, or T factory *sensu* Schlager (2003, 2005). The key parameters controlling the presence or absence of the FST were identified via basic principles of sedimentation and numerical modelling. Well-documented case studies were used as 'ground truth' for the validity of the numerical models. Data from numerical models and case studies were combined to establish the parameter space for the development of FST and the STM respectively. Finally, it was estimated what portion of the stability domains of FST and STM is geologically probable, i.e. supported by geological observation.

METHODS

CARBONATE-3D (C3D)

All modelling was done with CARBONATE-3D (C3D), a finite difference, forward-modelling program that simulates the evolution of carbonate platforms and mixed carbonate siliciclastic systems up to millions of years with time steps of tens to thousands of years. The program is written in IDL (Information Data Language). The basic mathematical structure and its justification in view of sedimentological principles is given in Warrlich *et al.* (2002). Below follows a brief summary of the program structure. The most relevant input settings for specific runs are given in the captions or the accompanying text.

The C3D program simulates the major processes operating in carbonate systems after input of time and spatial scales, initial three-dimensional surface, and sea-level history. In the program, accommodation can be controlled by sea-level changes, differential subsidence and compaction during the model run. Carbonate sediments are simulated to be produced in platform margin, platform interior and pelagic environments. Each of these carbonate factories has production rates that vary with water depth and with respect to a restriction effect related to distance from the platform edge. Platform-margin production increases to a maximum toward the margin, platform-interior production increases with distance inshore from the margin and pelagic sedimentation increases basinward away from the margin (for details see Warrlich *et al.*, 2002). Simulated strata may be eroded at a user-defined rate to give coarse and fine reworked sediment. If the shear stress is above a critical value specified for a particular grain size, sediment becomes entrained and transported. Transport is governed by a shear stress vector field that is a combination of the shear stress due to currents and waves as well as slope. As only the shear stress due to slope is grain-size dependent, coarse-grained sediments tend to follow slope and fine sediments follow currents and waves (Warrlich *et al.*, 2002). Thus the program simulates bed-load and suspended-load transport respectively. Deposition occurs in areas where shear stresses are below the critical value. The C3D program calculates sedimentary geometries as time surfaces and simulates facies. The facies assigned to a cell is either that of the process that forms the largest fraction of the accumulated sediment within that cell or the facies defined by the user as a fixed combination of grain types from a particular production belt or from erosion (see 'digital facies' of Boylan *et al.* (2002) for details).

The set-up used to investigate the FST was a topographic surface 0.6–5 km long and 0.1–0.5 km wide, consisting of flat top and adjacent slope, with initial sea level covering the shallow part; one carbonate production function with production rates of 100–6000 $\mu m\ yr^{-1}$, decreasing with depth as described in Warrlich *et al.* (2002). Information on sea-level curves, production rates and erosion rates are indicated with the specific runs.

RESULTS

Numerical modelling

The program C3D was used to explore under which conditions a falling-stage systems tract would develop and when the configuration of the standard model would be realized. 'Standard-model configuration' implies that sea-level fall created a continuous surface of erosion or non-deposition and left no sediment accumulations in the area swept by the falling sea level. The depositional record of the time of sea-level fall consists only of deeper-water sediments on the lower slope and basin floor, i.e. the area below the sea-level minimum. This section presents the most important parameters controlling FST development and summarizes their effects as observed in about 300 modelling runs.

Rate of sea-level fall

Arguing from fundamental principles of erosion and deposition, one can already predict that the development of a FST is more likely if sea level falls slowly whereas rapid falls will favour the standard model. The numerical models clearly confirm this view and provide quantitative estimates on the rates of fall required to create FST or STM anatomy respectively (Fig. 3). The runs were made for a tropical carbonate platform able to build a wave-resistant margin and maintain steep slopes. Both subaerial and shallow-marine erosion are set to zero in Fig. 3. Under these conditions, a sinusoidal sea-level cycle produces a large falling-stage systems tract. Increasing asymmetry of the sea-level cycle and increasing rates of fall reduce the size of the FST and finally lead to the standard model. It should be noted that the rate required to completely suppress the FST is also a function of the spatial and temporal resolution used as input for the modelling runs.

Rate of erosion

The C3D program allows one to separately model the effects of subaerial and marine erosion. The effect of increasing subaerial erosion is shown in Fig. 4. Much like changing rates of sea-level fall, varying rates of erosion shift the system from the FST mode to the mode of the standard model and vice versa. In the present context, the effects of shallow-marine erosion (by currents and waves, or by boring or grinding organisms) are similar to those of subaerial erosion. Increasing shallow-marine erosion again reduces the size of the FST and eventually creates the anatomy of the standard model (Fig. 5).

The overall effect of marine erosion on FST development in C3D models is similar to subaerial erosion but there are important differences with regard to the rates required to suppress the FST (expressed as surface lowering in $\mu m\,yr^{-1}$). The two erosive processes also create somewhat different morphology on the erosion surface. Subaerial carbonate erosion is assumed to be uniform over the exposed area and to proceed by dissolution, thus all eroded material leaves the model space. Marine erosion varies with water depth and much eroded material is redeposited within the model. Marine erosion is less effective in destroying the FST because its maximum rates are only realized in a narrow depth zone and most eroded material is redeposited further downslope and thus needs to be reworked repeatedly during sea-level falls. In this study, the 'FST destruction efficiency' of marine and subaerial erosion was found to be linearly correlated in a wide range of conditions (Fig. 6). This allows the substitution of marine for subaerial erosion and vice versa in modelling runs and plot them on one axis in parameter space.

In summary, the modelling runs clearly indicate that changes in rates of sea-level fall and rates of erosion have similar effects on depositional anatomy. In both instances, increasing rates shift the depositional system in the direction of the STM, decreasing rates shift it in the direction of the FST.

Carbonate production

Carbonate production is a third important control on FST development. Production supplies sediment for the construction of the FST and therefore directly counteracts erosion. Figure 7 illustrates the effect of increasing production while the rates of erosion and sea-level fall remain unchanged. Both highstand and lowstand tract increase in volume as production increases. The FST appears when production rate is so high that erosion is unable to remove all the sediment produced during sea-level fall.

Slope angle

The FST sediments are deposited on slopes and this may cause resolution problems if slopes

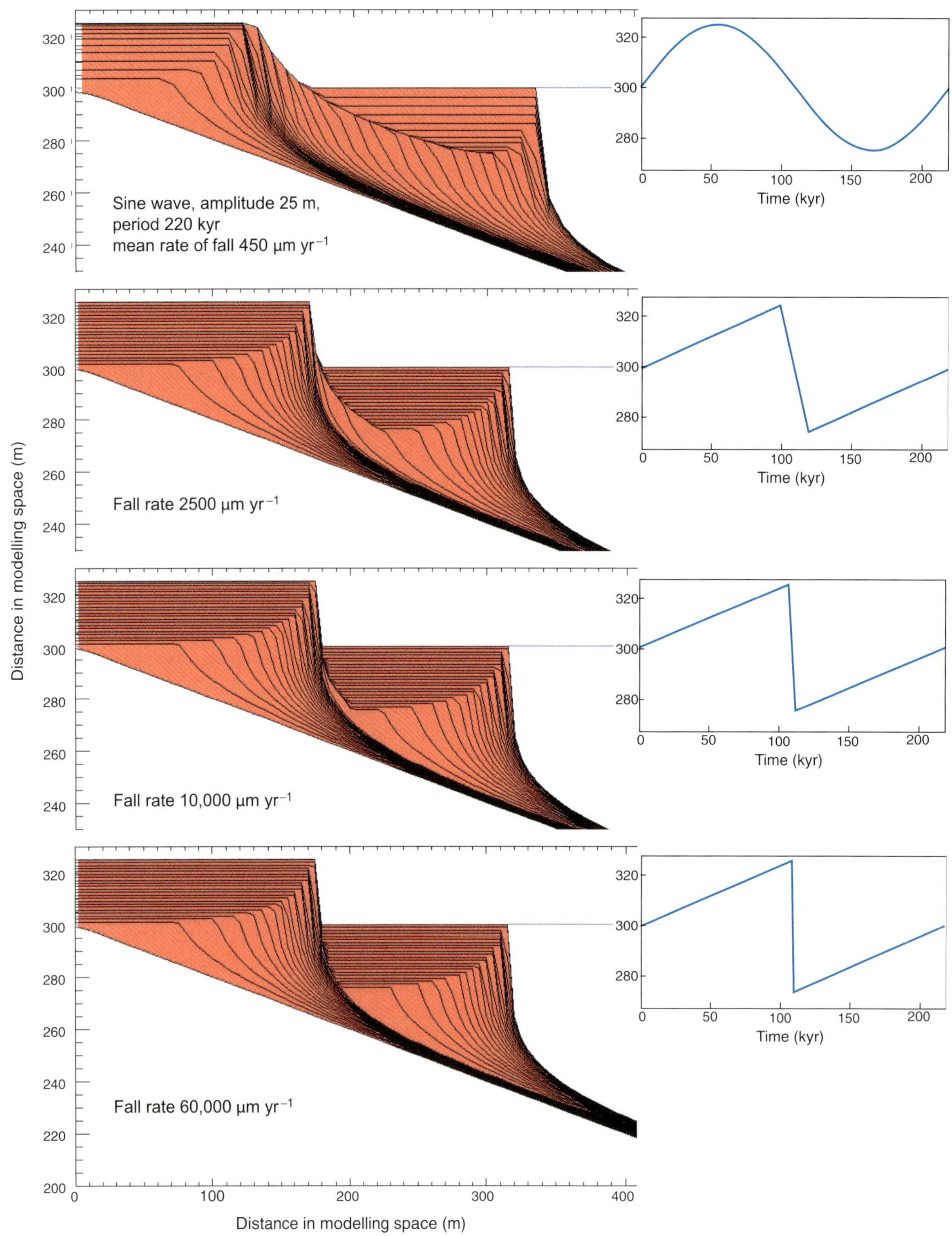

Fig. 3. Rates of sea-level fall and FST development. Carbonate accumulation (red) on the left, forcing sea-level curve (blue line) on the right. Top panel: sine wave with mean rate of fall 450 μm yr^{-1} (maximum rate 700 μm yr^{-1}) produces a large FST that records the sea-level wave in all important aspects. Lower panels: with increasing rates of fall, FST shrinks and falls below resolution at rate of fall of 60,000 μm yr^{-1}. (In all runs of this figure: carbonate production 5000 μm yr^{-1}, no erosion, run time 220 kyr, reef-rich platform-margin facies capable of maintaining slopes of over 42°). Numbers on axes of red figures indicate distances (metres) in modelling space, thin lines are time lines, spaced at 5 kyr.

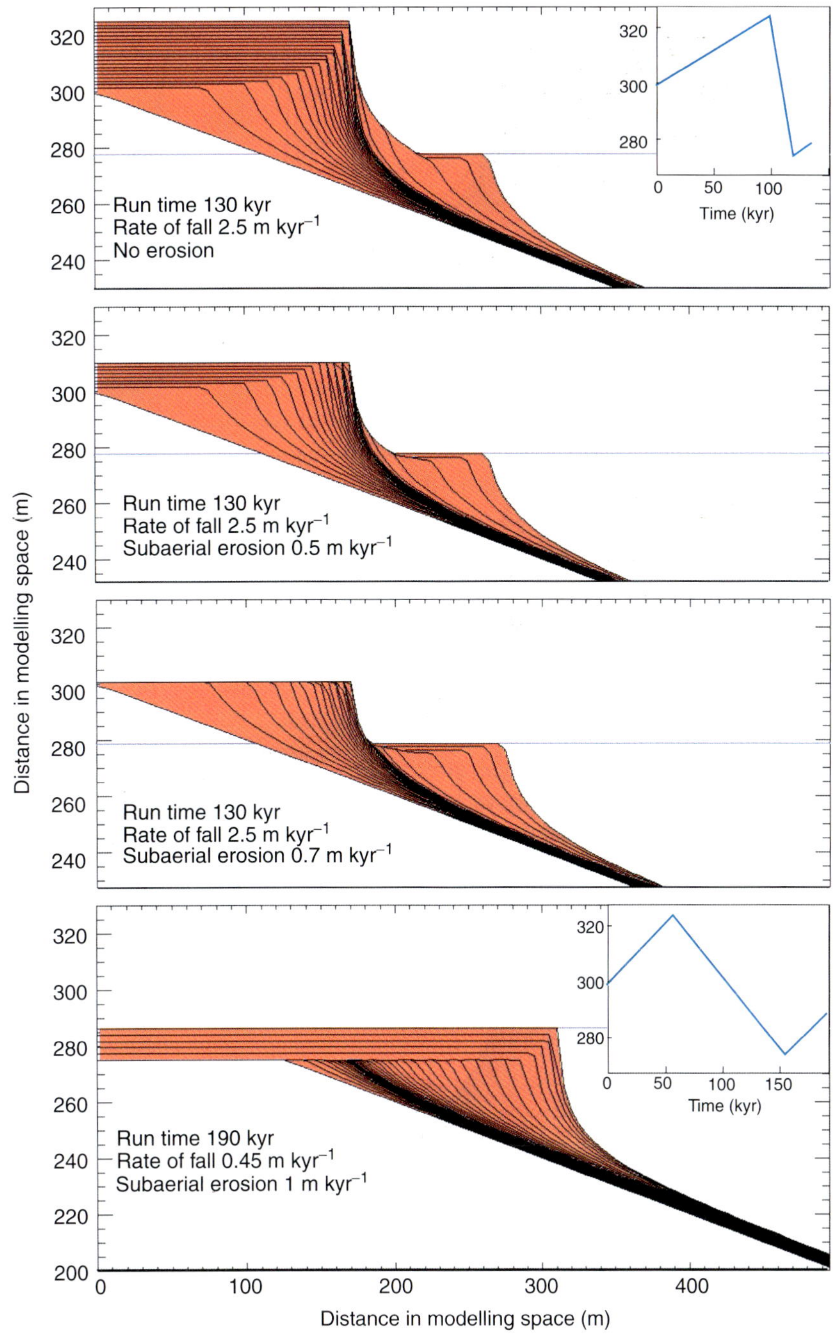

Fig. 4. Subaerial erosion and FST development. The FST disappears at an erosion rate of 700 μm yr^{-1}. In bottom panel, platform is eroded down to lowstand level as rate of erosion significantly exceeds rate of sea-level fall. Time lines and numbering on axes as in Fig. 3.

become too steep. This is because the lateral separation of the time lines used to define the FST in a computer output is a function of the cosine of the slope and thus decreases with increasing slope angle while the thickness of the printed line remains nearly constant. This problem was alleviated by conducting the FST experiments in a relatively narrow range of intermediate slope angles. The vast majority of FST presence/absence tests were done on slopes of 15–25°, only in exceptional circumstances slopes up to 40° were considered.

Parameter space

The modelling results on FST development can best be summarized by viewing the stability domains of the FST and STM anatomies in a space determined by the principal controlling parameters. The basic controls are the rates

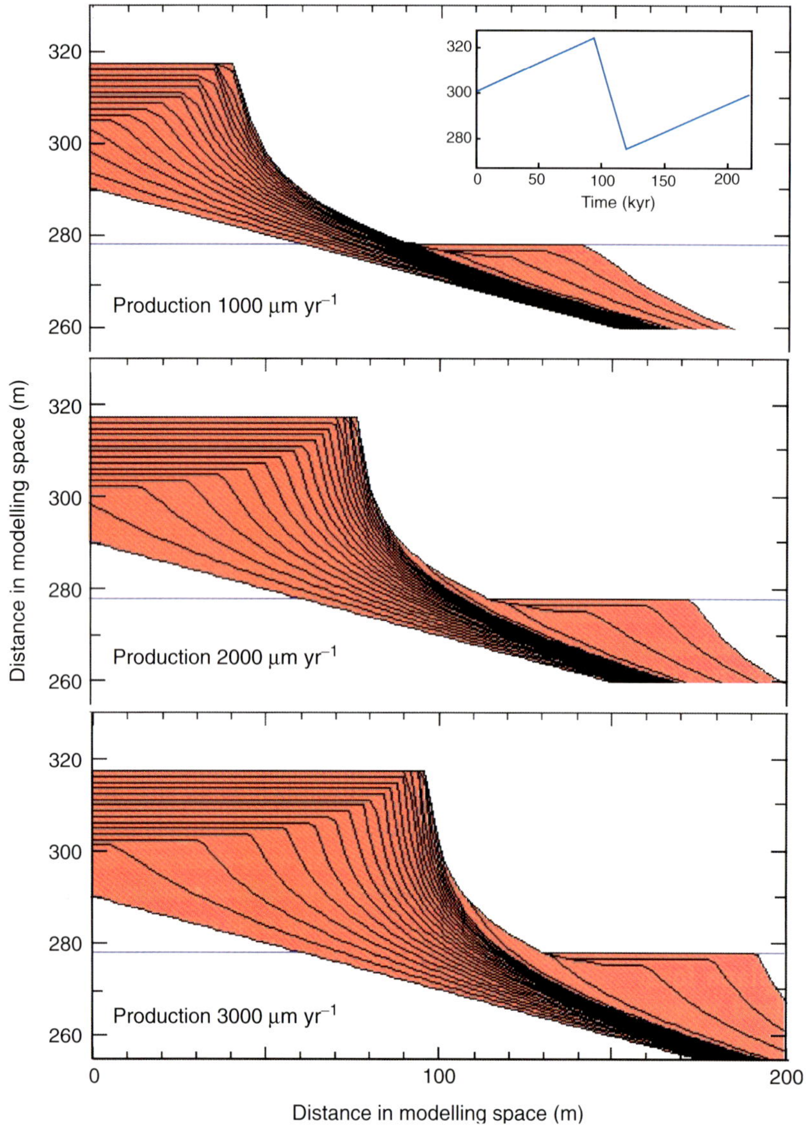

Fig. 7. Carbonate production and FST development. From top to bottom, production increases from 1000 to 3000 $\mu m\, yr^{-1}$, other parameters remain unchanged (rate of fall 3000 $\mu m\, yr^{-1}$, erosion 275 $\mu m\, yr^{-1}$). Top panel shows STM, middle and bottom panel FST. Note that appearance of FST goes hand in hand with volume increase of all systems tracts. Time lines and numbering on axes as in Fig. 3.

part of the sediment cover of the Arabian craton. It was deposited during the Aptian stage of the Early Cretaceous in an epeiric sea that was differentiated into carbonate platforms and shallow basins with argillaceous sediments. Sedimentological models of Shu'aiba anatomy have evolved considerably during the past two decades. Now there is broad agreement that the main phase of platform building, the mid-Shu'aiba of Boote & Mou (2003), was followed by a late phase of basinward progradation and downstepping of platform facies, the upper Shu'aiba (Figs 13 and 14). Most or all of the upper Shu'aiba can be interpreted as a falling-stage systems tract deposited after the mid-Shu'aiba highstand tract. The C3D modelling with these assumptions also produces the best match with observation (Fig. 15).

Other case studies

The Early Cretaceous Habshan Formation of Oman shows platform–basin transitions with strongly prograding clinoforms, repeatedly including FST geometries. Droste & Van Steenwinkel (2004) and H. Droste (personal communication) present a detailed description with sufficient constraints on ages and range of fall to include the example here.

In the Maldives, Belopolsky & Droxler (2004) established a very detailed seismic stratigraphy that is calibrated by core borings of the Ocean Drilling Program (ODP) and exploration wells. In the Miocene, several FSTs (called 'forced regressive wedges' by the authors) were observed. Ranges of sea-level fall were gleaned from seismic

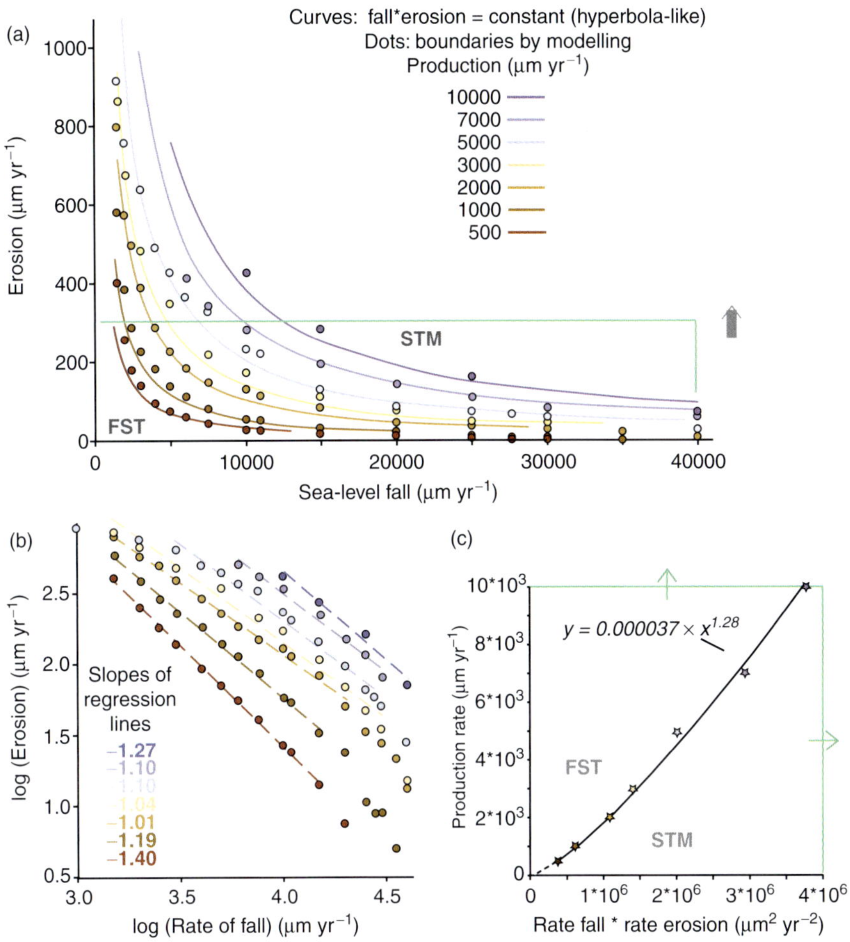

Fig. 8. Three-dimensional parameter space illustrating stability domains of FST and STM. (a) Plane defined by rate of sea-level fall and rate of erosion. Coloured dots: FST/STM boundaries determined by modelling; colour indicates rate of carbonate production (see code at upper right). Boundaries of FST and STM domains resemble hyperbolas centred on the origin. Coloured lines are best-fitting exact hyperbolas of type $x \times y$ = constant. Green box: outer limit of geologically probable space. (b) Bi-logarithmic plot of sea-level/erosion plane of (a). Linear trends of FST–STM boundaries confirm their nature as power functions approaching hyperbolas; however, slopes of regression lines vary and differ from 1; lines also bend near the ends, probably because of limited spatial and temporal resolution of the models. (c) Cross plot of rate of production versus product of rates of sea-level fall and erosion; coloured stars may be viewed as intersection points with boundary hyperbolas in (a). The FST–STM boundary in this plane is parabola-like but the exponent is smaller than 2. Green arrows indicate that geologically probable space in this plane extends beyond limits of graph.

geometry. Ages of the relevant seismic reflectors are sufficiently well constrained to estimate rates of fall.

The Miocene of Mallorca in the Mediterranean is well known from detailed studies of sea cliffs and numerous water wells (Pomar, 1991, 2000). The rocks are well dated by biostratigraphy. These dates and the close match between observation and forward modelling (Bosence *et al.*, 1994) support the orbital interpretation of the platform cycles advanced by Pomar (1991). The orbital interpretation of cycles and the clearly observed ranges of sea-level fall in outcrop provide narrow constraints on the rates of sea-level fall.

Case studies in FST parameter space

Figure 16 shows the case studies in the plane of erosion versus sea-level fall of the parameter space established by modelling (Fig. 8a). The rates of erosion and sea-level fall with their error bars were derived as follows.

Rates of subaerial carbonate erosion were taken from the data on extant systems in White (1984): warm temperate settings 60–125 µm yr^{-1} (assumed to apply to Nijar and Mallorca); wet tropical settings 90–200 µm yr^{-1} (assumed to apply to Maldives, Bu Hasa, Safah and Habshan). The rates in White (1984) were measured over time spans of months to years. In the case studies, the time spans for forming the FST are in the range of 10^3–10^5 yr. Gardner *et al.* (1987) have argued that erosion rates, like sedimentation rates, decrease with increasing length of the time window and presented a quantitative estimate of this scaling trend based on eight orders of magnitude in time. In this study, the scaling of erosion rates is accepted as a possibility but not as a certainty. This increases the uncertainty of the rate estimates and the range of erosion rates in Fig. 16 reflects this situation. The upper limit of the range bars is drawn by assuming that the rate scaling does not apply and the maximum rates in White (1984) remain valid guesses also for the long time

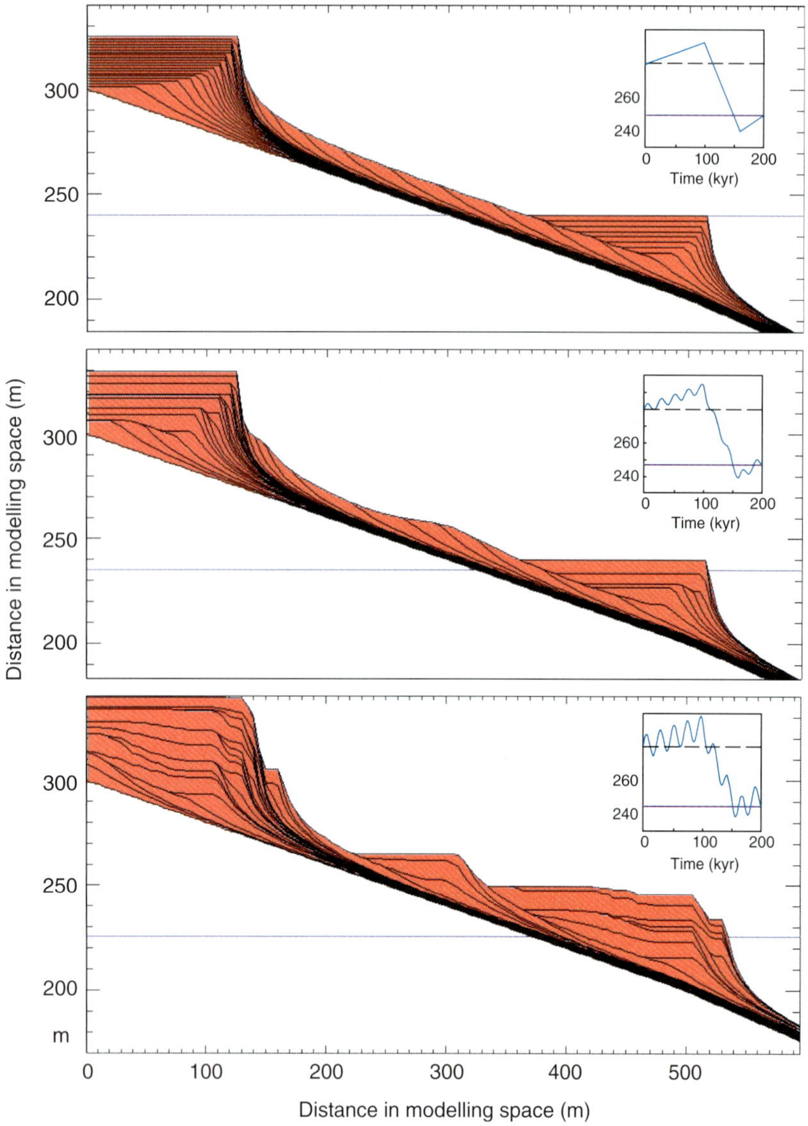

Fig. 9. Modelling runs showing superposition of short sea-level oscillations on long falling trend. Top panel: no short oscillations; extended FST forms on the slope. Middle panel: sine waves of 5 m range, 23 kyr period slightly modulate the surface of the FST. Bottom panel: sine wave of 15 m range, 23 kyr period subdivides the FST into a series of downstepping platforms. (Other settings: production rate 2000 μm yr^{-1}, no erosion.)

intervals of the case studies. The minimum rates were derived by assuming that rate scaling does apply and the minima in White (1984) had to be further reduced using the scaling formula of Gardner *et al.* (1987).

Marine erosion is very poorly constrained on a formation scale. Rates sustained for 50 kyr and longer were assumed to vary between 75 and 750 μm yr^{-1}, which translates into 10–100 μm yr^{-1} subaerial equivalents using the regression formula of Fig. 6. It seems unlikely that marine erosion rates of 200 μm yr^{-1} (subaerial equivalent) derived from the maximum rates of notch erosion in the Holocene can be sustained for time intervals represented by the case studies in question. No rate scaling was applied to the minimum because the effect on the total erosion rate was considered negligible.

Rates of sea-level fall were gleaned directly from sequence geometry and age dates of each case study (arranged in stratigraphic order).

(1) *Habshan* (after Droste & Van Steenwinkel, 2004; H. Droste, pesonal communication): the Early Cretaceous Habshan Formation of Oman consists of six cycles with forced-regression (= FST) deposits. The range of fall is 60–70 m, duration of cycles about 1–3 Myr. Assuming that the FST represents 50% of the cycle duration yields rates of fall of 40–140 μm yr^{-1}.

(2) *Shu'aiba of Bu Hasa* (after Fitchen, 1997): the FST of the Upper Shu'aiba Formation occupies the time between the sequence boundary 109.5 and 108.7 Ma, i.e. 0.8 Myr. The range of fall is

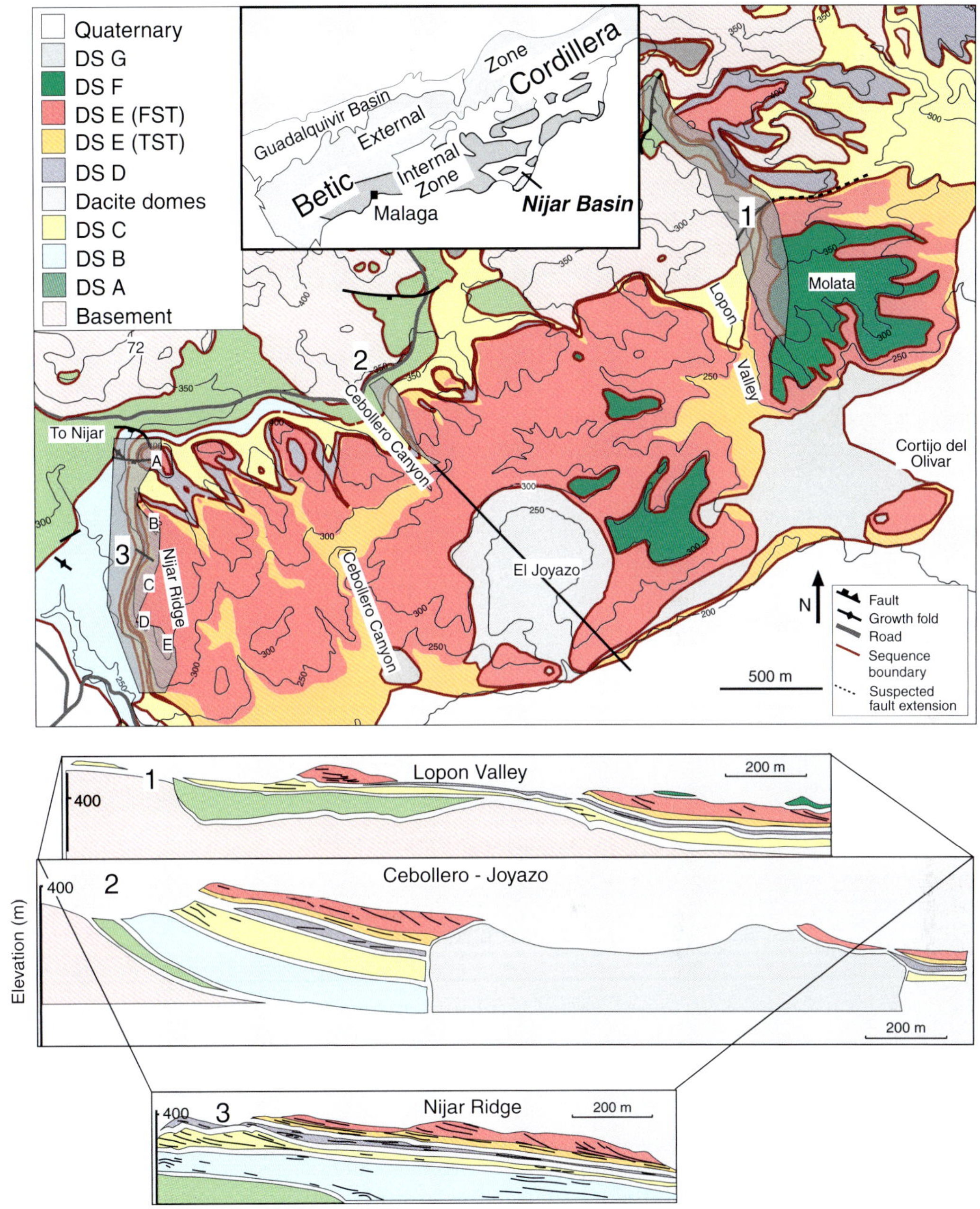

Fig. 10. Map and sections of Nijar Basin, showing hardrock units and depositional sequences (DS, A–E) of the sedimentary basin fill. TST, transgressive systems tract; FSST, falling-stage systems tract. Contours in metres. (After Warrlich *et al.*, 2005, modified.).

estimated as 51–64 m (based on Fitchen, 1997, fig. 32). The resulting range of rates of fall is about 64–80 μm yr^{-1}.

(3) *Shu'aiba of Safah* (after Boote & Mou, 2003, fig. 29; Van Buchem *et al.*, 2002, fig. 15): the FST of Upper Shu'aiba is estimated to occupy 0.57–1.28 Myr of time; the range of fall is 11.5–30 m. This translates into rates of fall of 9–53 μm yr.

(4) *Maldives* (after Belopolsky & Droxler, 2004): Miocene unit M2 is estimated to represent 0.9–3.4 Myr of time; based on thickness proportions, the FST is assumed to represent 25% of this time, i.e. 0.22–0.85 Myr. Range of fall is 50 ms two-way travel time; estimated interval velocity at this level is 2000 m s^{-1}, which yields a range of fall of 50 m and rates of fall of 60–230 μm yr^{-1}.

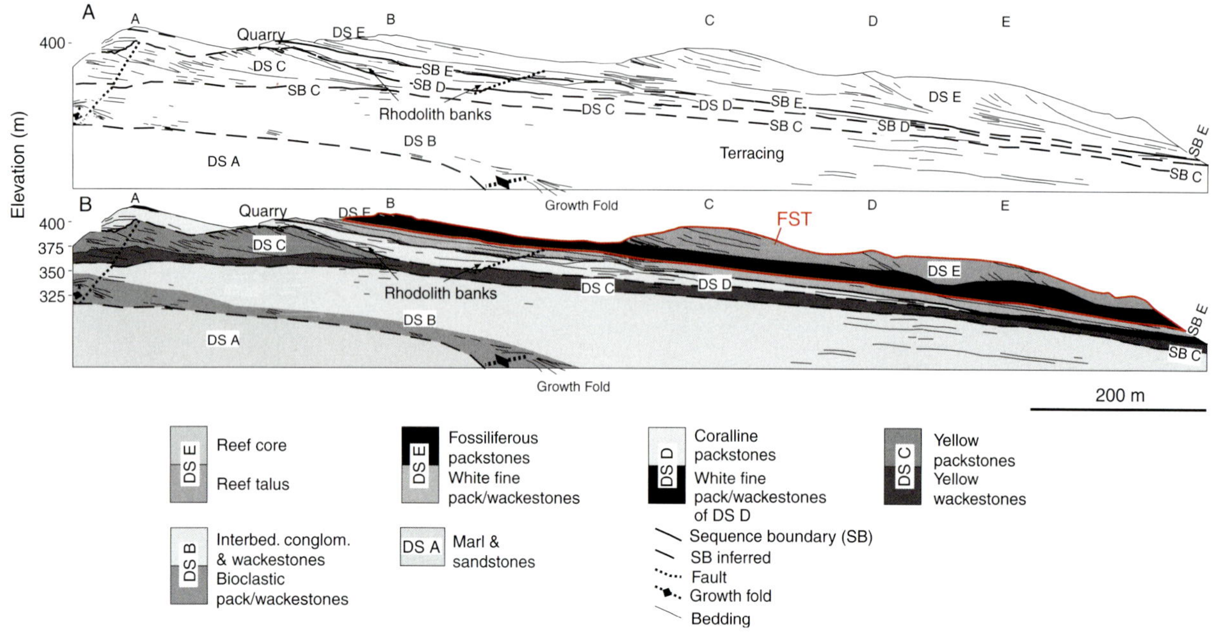

Fig. 11. Nijar Ridge – bedding, lithology and depositional sequences. Number 400 in top left corner of both profiles indicates elevation in metres. Letters A–E denote hills used as topographic reference points. After Warrlich *et al.* (2005), modified.

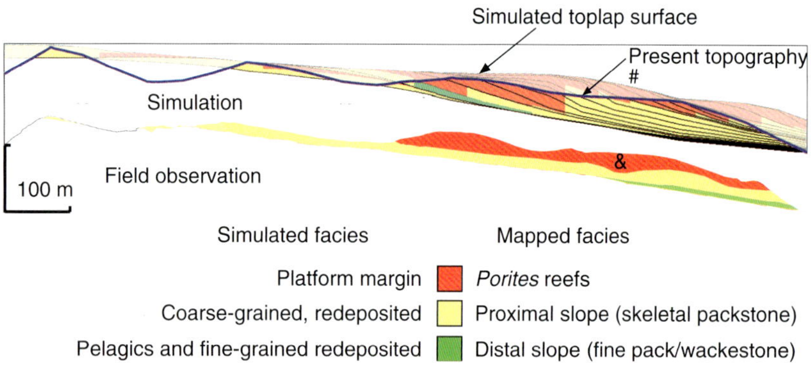

Fig. 12. Nijar Ridge – field observations and computer simulation of FST. After Warrlich (2001). Important modelling settings: platform margin production 4500 μm yr^{-1}, marine erosion 1500 μm yr^{-1}, mean rate of sea-level fall 74 μm yr^{-1}.

(5) *Mallorca* (after Pomar, 1991, 2000): a hierarchy of prograding sigmoids and sigmoid bundles was recognized and tentatively correlated with orbital cycles. Plotted are rates for co-sets of sigmoids, interpreted as expression of the 100 kyr eccentricity cycle. Range of sea-level fall in the co-sets is estimated as 60–70 m based on Pomar (2000). Assuming the FST represents half of a 100-kyr cycle, the range of rates of fall is 1200–1400 μm yr^{-1}.

(6) *Nijar* (after Warrlich *et al.*, 2005): the FST of depositional sequence E is well constrained by field data and a good match with modelling runs. Duration of the FST interval is estimated as 0.266 Myr and the range of fall as 168–177 m. This yields rates of fall of about 630–665 μm yr^{-1}.

Figure 16 shows that all FST case studies considered actually plot in the stability domain of the FST predicted by modelling, even if one assumes that their production rates were only 500 μm yr^{-1}.

DISCUSSION

The modelling runs provide a reasonably clear picture about the conditions required for developing a FST or the continuous unconformity of the standard model during sea-level fall. The key

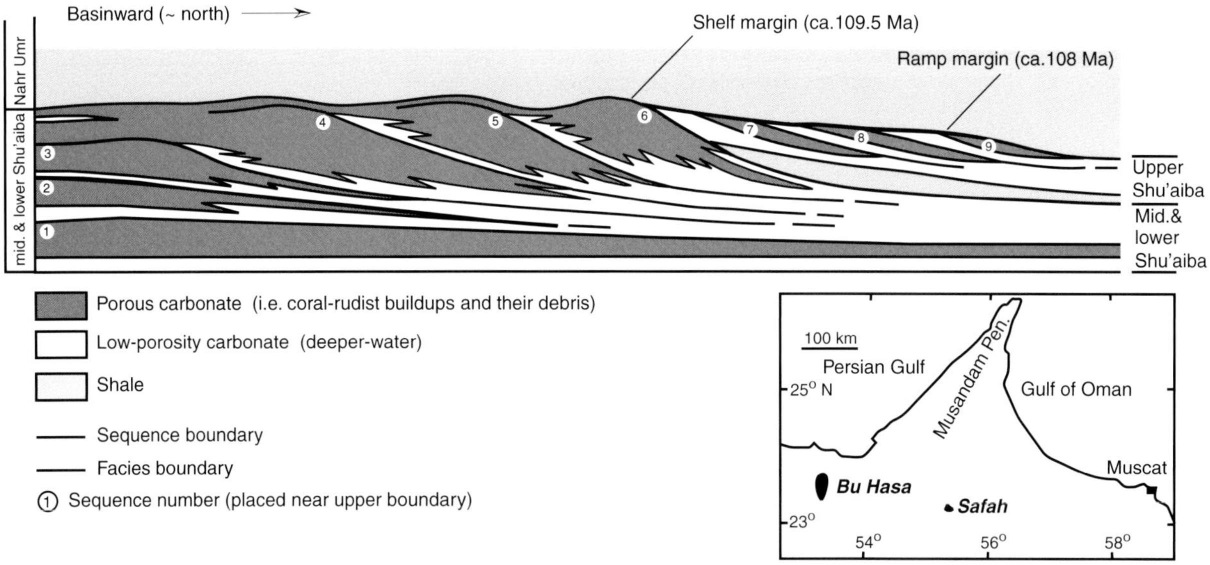

Fig. 13. Schematic section of Shu'aiba Formation (Aptian), Bu Hasa Field, Abu Dhabi. After Fitchen (1997), modified. Tropical carbonate platform (algae, foraminifera, rudist-coral biostromes) progrades into a shallow cratonic basin where argillaceous sediments accumulate. During early progradation (units 1–6), slopes steepen and each sequence has recognizable topset beds – the units constitute a highstand systems tract. During late stage of progradation (units 7–9), slopes flatten, sequences step down and lack topset beds – the units form a falling-stage systems tract (called 'late highstand tract' by Fitchen, 1997). Rate of fall estimated as 15–20 μm yr, based on age dates and thicknesses in Fitchen (1997, Fig. 32).

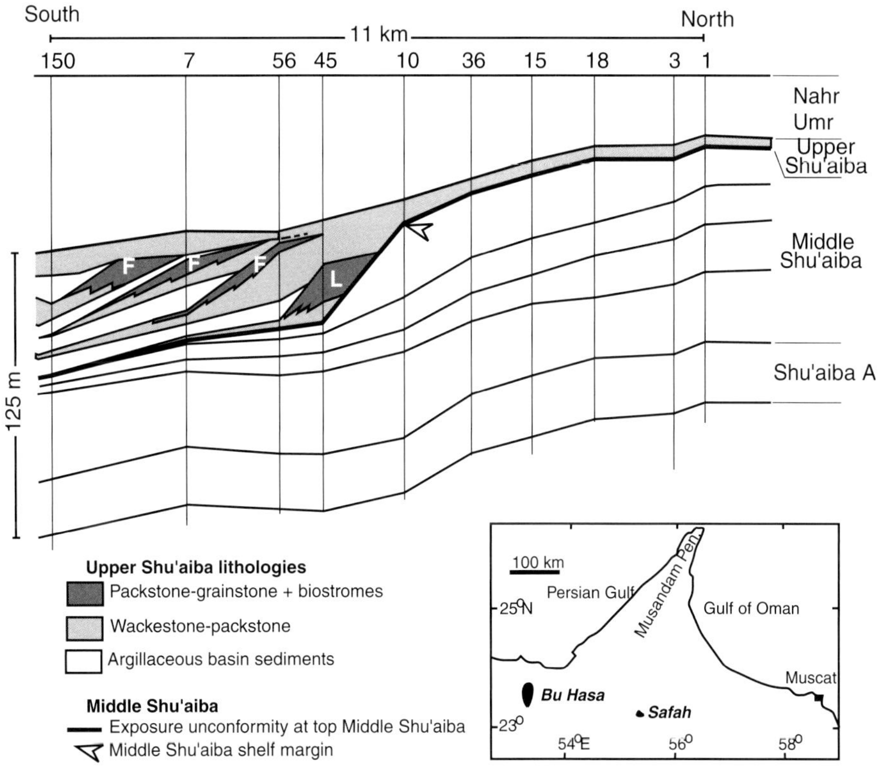

Fig. 14. Cross-section of Shu'aiba Formation of Safah Field, Oman, based on correlation of wells (well numbers shown above section). After Boote & Mou, (2003), modified. Section is flattened on an Eocene datum to remove Miocene deformation. Depositional evolution is very similar to Bu Hasa Field (Fig. 13). Lower Shu'aiba lacks significant depositional relief; middle-Shu'aiba deposits evolve from ramp to rimmed platform (with rudists, and coral biostromes). After significant sea-level fall, upper Shu'aiba deposition starts with a lowstand systems tract (L) that abuts against the mid-Shu'aiba slope; subsequently, the systems aggrades. The final stage of deposition consists of three downstepping limestone tongues (F) that form an FST.

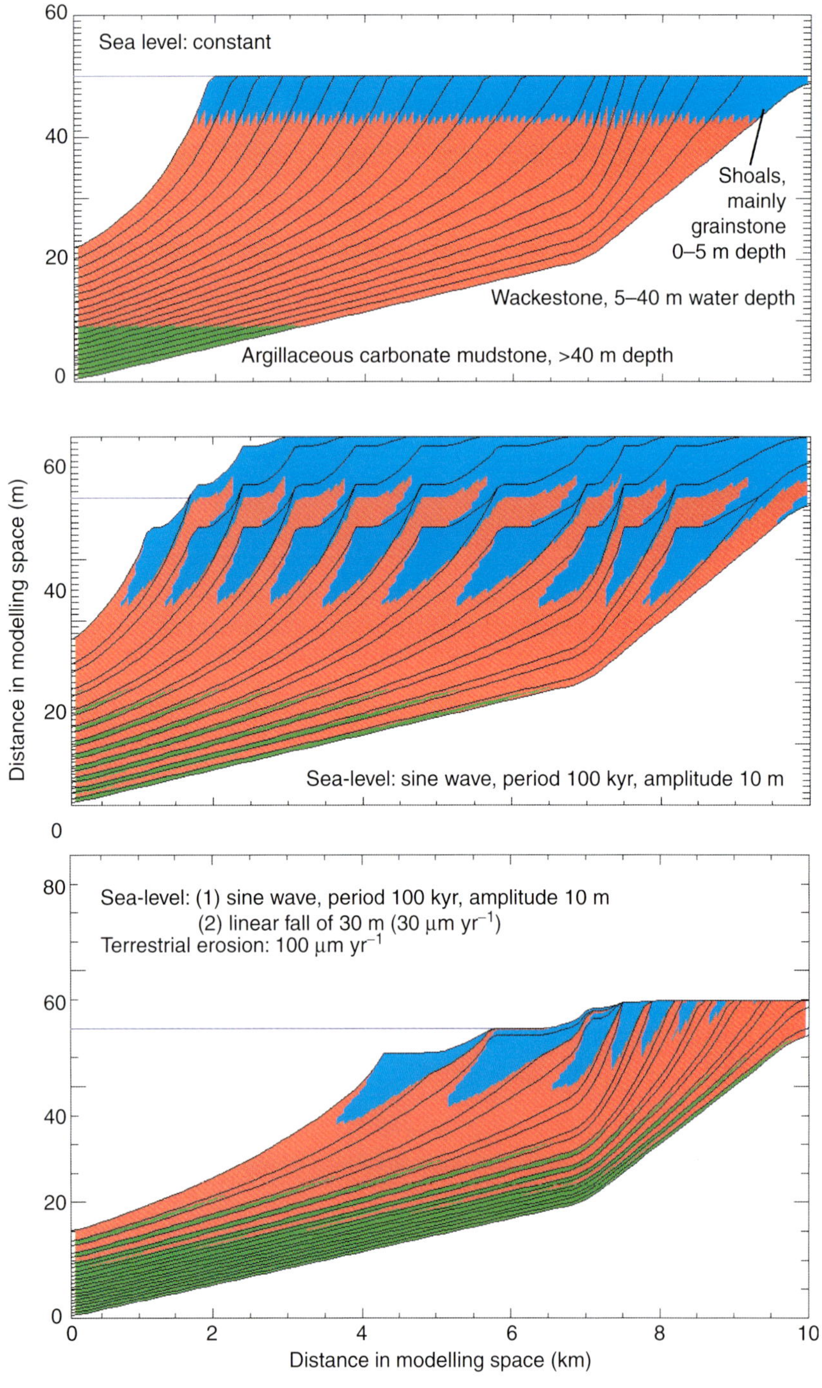

Fig. 15. The C3D simulations of upper Shu'aiba depositional anatomy indicate that the scenario of an erosionally truncated FST best matches observations in both Safah and Bu Hasa fields.

question to be pursued in this section is: how do the results from modelling compare with geological reality? In other words: which part of the parameter space is geologically probable and supported by observations? The situation is evaluated separately for each of the principal controls.

Rates of sea-level fall

It is well established that the 'power' (or variance) of sea-level fluctuations changes with the frequency of the wave patterns. This implies that rates and amplitudes of sea-level change are not independent of the length of the time interval

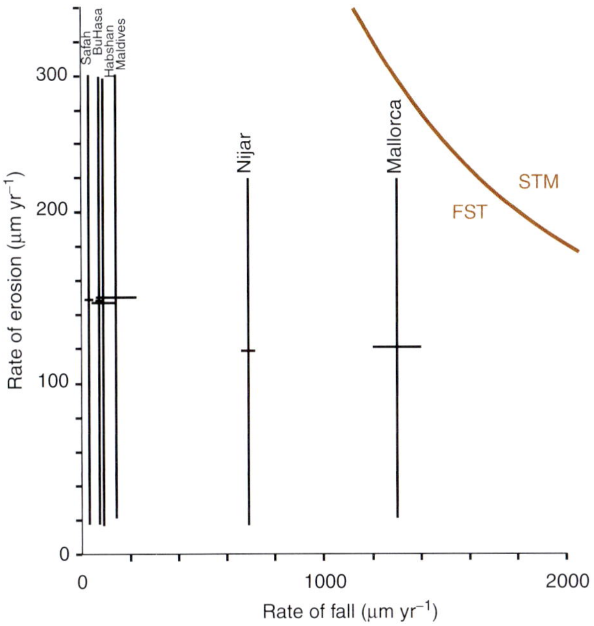

Fig. 16. The FST case studies plotted with estimated rates of sea-level fall and erosion in the fall-erosion plane of the parameter space of Fig. 8. Brown line in upper right is apex of brown hyperbola in Fig. 8a, separating stability domain of FST and STM at production rates of 500 μm yr^{-1}. See text for data sources and justification of error bars.

of observation (Harrison, 2002). Considering the sea-level curves of the modelling runs, the most critical observation interval for the present discussion is 0.5–500 kyr. This constraint makes the Pleistocene Series the most important source of data because sufficiently accurate time control for sea-level movements of the more distant past is very rare. For the Pleistocene (and Holocene), there exists a well-established relationship between oxygen isotope ratios in calcareous shells, ice volume and eustatic sea level. Sea-level curves gleaned from oxygen-isotope data repeatedly show rapid falls in the range of 20,000–40,000 μm yr^{-1} (Labeyrie *et al.*, 1987; Shackleton, 1987). These rates are in the range required to suppress the FST and create the STM in tropical carbonates without much additional contribution from erosion. However, it should be kept in mind that the extreme Pleistocene rates remain somewhat speculative. U–Th age data generally provide insufficient resolution and data coverage to confirm or reject the rapid sea-level falls derived from oxygen isotopes. In summary, it seems highly probable that rates required to suppress the FST by rapid sea-level fall occurred in the Pleistocene. By inference, it seems likely that the same holds for other times of large ice bodies, such as the Permo-Carboniferous and parts of the Late Proterozoic.

Rates of erosion

Discussion of erosion rates should separately address marine and terrestrial environments because the processes and their controls are vastly different.

Subaerial erosion of tropical carbonates

Tropical carbonates have a high content of metastable aragonite and magnesian calcite. These minerals dissolve rapidly in rain water and re-precipitate as calcite cement. Thus tropical carbonate sediments commonly lithify upon exposure. This lithification plus organic binding and frame-building in the depositional environment reduce the rates of mechanical erosion. Denudation rates of large terrestrial surfaces of carbonate rocks strongly depend on dissolution. Most data and theoretical models were contributed by studies of extant karst systems. There is fairly good agreement between rates deduced from field observations and those predicted from theoretical models based on reaction chemistry (White, 1984; Dreybrodt, 1988). Denudation rates of karst surfaces in temperate and cold climates are in the range of 60–125 μm yr^{-1}, tropical rates 90–200 μm yr^{-1}. Purdy & Winterer (2001) postulate theoretical rates of up to 650 μm yr^{-1} assuming strong undersaturation of the waters and no damping effect of evapo-transpiration; both assumptions seem rather unlikely. Erosion rates in cave conduits may exceed 1000 μm yr^{-1}, i.e. an order of magnitude higher than rates of surface denudation. The higher rates in conduits lead to efficient subsurface drainage, thus reducing surface denudation and often saving the depositional surface from wholesale destruction. The Miami Oolite is a good example. The formation was deposited as a carbonate sand shoal during the last interglacial, about 120 ka. Despite more than 100 kyr of karst erosion, the oolite surface still shows the depositional pattern of sand bars and intervening channels and the relief of this pattern still is in the range of modern analogues in the Bahamas (Halley & Evans, 1983; Schlager, 2005). In sequence stratigraphic terms one may conclude that the depositional surface of the highstand systems tract (HST) of the Miami Oolite essentially survived

The maximum rates of sea-level fall are taken from the studies on the Pleistocene (Labeyrie *et al.*, 1987; Shackleton, 1987) as discussed above. The upper limit of subaerial erosion is set equal to the maximum denudation rates of extant tropical karst terrains, i.e. 200 μm yr^{-1} (White, 1984). The maximum rate of marine erosion, arguably the least constrained quantity of the plot, is assumed to be 100 μm yr^{-1} (see discussion above). This yields a maximum total erosion rate of 300 μm yr^{-1} in geologically probable space. The lower limits of the rates of sea-level fall and erosion are assumed to be indistinguishable from zero at the resolution of the diagram.

The third important parameter, carbonate production, is shown on the vertical axis of Fig. 8c. The upper limit of geologically probably space is set at a production rate of 40,000 μm yr^{-1}. However, this high rate is only probable for time intervals of a few thousand years and shorter (see discussion above). Only the low-rate part of the production space has been explored by modelling and the limit of the geologically probable production space is not shown in Fig. 8c.

Finally, it must be emphasized that what is 'geologically probable' strongly depends on the state of knowledge at a given point in time. At present it seems very difficult to conceive of natural laws that would preclude rates of sea-level fall, carbonate production or erosion to proceed faster than observed up to now. In this regard, numerical experiments remain heuristically valuable even where they lie outside the parameter space currently supported by observation.

Tropical carbonates versus siliciclastics

The tropical carbonate factory studied here differs in many aspects from siliciclastic systems that provided the principal data base for the standard model. These differences need to be considered when applying the results of this study to siliciclastics. The notion that FST development is fundamentally determined by the rate of sea-level fall and the rate of erosion should apply to siliciclastics, too. Consequently, the distribution of FST and STM anatomy of siliciclastics may be described in the same parameter space (with external sediment supply replacing carbonate production).

Major differences can be expected with regard to erosion and local reworking. (1) Mechanical erosion in siliciclastic settings can be expected to exceed those of tropical carbonates because the clastics lack the stabilizing effect of organic frame-building and submarine lithification. (2) Dissolution is a dominant factor in erosion of tropical carbonates and this means that the erosion products are highly mobile and likely to disappear (as ions) from the space considered here. In siliciclastic systems, most eroded material can be expected to be redeposited in falling stage or lowstand systems tracts.

Siliciclastic case studies indicate high and highly variable rates of erosion in siliciclastic systems during sea-level falls. In particular, complete erosion of delta topsets and severe truncation of the adjacent prodelta foresets are commonly observed (Anderson *et al.*, 2004; Roberts *et al.*, 2004). Severe erosion may also occur during the ravinement stage of transgression. Erosion rates deduced from the carefully compiled geohistory plots of Fillon *et al.* (2004) yield rates of surface lowering of up to 1100 μm yr^{-1} for the Quaternary of the northeastern Gulf of Mexico. The high rate of erosion may suppress the development of a falling-stage systems tract and favour the formation of a continuous surface of erosion that connects highstand and lowstand tracts. However, the intense erosion that creates the erosive sequence boundary also severely truncates the underlying highstand tract – a significant difference to the standard model of sequence stratigraphy (Fig. 1) where erosional truncation of the underlying highstand tract is minor.

Terminology

Since the first reports on the existence of significant sediment accumulations formed during sea-level fall, several proposals on terminology have been made. Hunt & Tucker (1992) called them 'forced-regressive wedge systems tract' or 'forced regressive systems tract' (Hunt & Tucker, 1995) to emphasize that these sediment bodies were clear evidence of forced regression in the sense of Plint (1991) and Posamentier *et al.* (1992). Nummedal *et al.* (1992) coined the term 'falling sea level systems tract'. Nummedal *et al.* (1995) and Plint & Nummedal (2000) simplified this to 'falling-stage systems tract'.

The position of the sequence boundary in the presence of the FST is hotly debated. Should it be put at the base of the FST, i.e. at the level where sea-level fall commenced (Posamentier & Morris, 2000), or at the top of the FST (Nummedal *et al.* 1995)? In a detailed review, Plint & Nummedal (2000) succinctly summarize the pros and cons of the two options; they ultimately chose the second option because the unconformity on top of the FST can be expected to be more pronounced and easier to trace than the surface at the base of the FST. In this study, the first option is favoured for reasons given below.

The differentiation into transgressive and highstand systems tracts is a function of the balance between the rate of accommodation creation and the rate of sediment supply (Jervey, 1988; Schlager, 1993). Constructing a sea-level curve from these patterns requires independent information on the variations of sediment supply or data on the variations in the rate of aggradation. Lowstand and falling-stage systems tracts, on the other hand, offer unambiguous evidence of a fall of relative sea level independent of the rate of sediment supply. The surface at the base of the FST provides the timing of the onset of this fall; timing of the termination of fall as well as the range of fall can be gleaned from the lowstand systems tract. Thus, FST and the LST of the standard model have very similar functions in sea-level reconstructions and this similarity is a strong argument in favour of drawing the sequence boundary between HST and FST (Posamentier & Allen, 1999; Posamentier & Morris, 2000) and considering the FST as a subdivision or a close relative of the lowstand tract.

CONCLUSIONS

1 Numerical modelling linked to field observations indicates that the falling limb of a sea-level cycle commonly generates a sediment accumulation in tropical carbonates – the falling-stage systems tract (FST). This sequence anatomy differs sharply from the standard model (STM) of sequence stratigraphy where highstand and lowstand systems tract are separated by an erosional unconformity.

2 Whether tropical carbonates develop FST or STM anatomy depends on the balance of the rates of sea-level fall, erosion and carbonate production. Slope angle is a fourth, modulating control parameter.

3 The FST is favoured by slow sea-level fall, slow erosion and high carbonate production as well as by small slope angles.

4 Numerical modelling allows one to define the stability fields of FST and STM in a parameter space of sea-level, production and erosion. The FST domain fills about half the geologically probable part of this parameter space, i.e. the part supported by geological observation. Falling-stage systems tracts and the continuous unconformity of the standard model should, therefore, both be common in the sequence record of tropical carbonates.

5 This study found no support for two commonly expressed generalizations: (1) The FST should always be present, albeit difficult to recognize. (2) The anatomy of the standard model can serve as a prototype for all sequences.

ACKNOWLEDGEMENTS

We thank David A. Waltham and Bart Wassing for advice on modelling matters and Henk Droste for data on the Cretaceous of Oman. C.G. St. C. Kendall and J.A. Simo provided important reviews.

REFERENCES

Anderson, J., Rodriguez, A., Abdulah, K., Fillon, R.H., Banfield, L., McKeown, H. and **Wellner, J.** (2004) Late Quaternary stratigraphic evolution of the northern Gulf of Mexico margin: a synthesis. In: *Late Quaternary Stratigraphic Evolution of the Northern Gulf of Mexico Margin* (Eds J.B. Anderson and R.H. Fillon), *SEPM Spec. Publ.*, **79**, 1–24.

Belopolsky, A.V. and **Droxler, A.W.** (2004) Seismic expressions and interpretation of carbonate sequences: the Maldives platform, equatorial Indian Ocean. *AAPG Stud. Geol.*, **49**, 1–46.

Boote, D.R.D. and **Mou, D.** (2003) Safah field, Oman: retrospective of a new-concept exploration play, 1980 to 2000. *GeoArabia*, **8**, 367–430.

Bosence, D.W.J., Pomar, L., Waltham, D.A. and **Lankester, T.H.G.** (1994) Computer modeling a Miocene carbonate platform, Mallorca, Spain. *Am. Assoc. Petrol. Geol. Bull.*, **78**, 247–266.

Boylan, A.L., Waltham, D.A., Bosence, D.W.J., Badenas, B. and **Aurell, M.** (2002) Digital rocks: linking forward modelling to carbonate facies. *Basin Res.*, **14**, 401–415.

Dabrio, C.J., Esteban, M. and **Martin, J.M.** (1981) The coral reef of Nijar, Messinian (uppermost Miocene), Almeria province, SE Spain. *J. Sedim. Petrol.*, **51**, 521–539.

Dreybrodt, W. (1988) *Processes in Karst Systems: Physics, Chemistry, and Geology*. Springer, Berlin, 290 pp.

Droste, H. and **Van Steenwinkel, M.** (2004) Stratal geometries and patterns of platform carbonates: the Cretaceous of Oman. In: *Seismic Imaging of Carbonate Reservoirs and Systems* (Eds G.P. Eberli, J.L. Masaferro, J.F. Sarg). *AAPG Mem.*, **81**, 185–206.

Ebren, P. (1996) *Impact des variations rapides du niveau marin sur le developpement des atolls au Quaternaire: Mururoa (Polynesie Francaise).* Unpublished Ph.D. thesis, University Provence, Marseille, 309 pp.

Fillon, R.H., Kohl, B. and **Roberts, H.H.** (2004) Late-Quaternary deposition and paleobathymetry at the shelf-slope transition, ancestral Mobile River delta complex, northeastern Gulf of Mexico. In: *Late Quaternary Stratigraphic Evolution of the Northern Gulf of Mexico Margin* (Eds J.B. Anderson and R.H. Fillon), *SEPM Spec. Publ.*, **79**, 111–141.

Fitchen, W.M. (1997) Carbonate sequence stratigraphy and its application to hydrocarbon exploration and reservoir development. In: *Carbonate Seismology* (Eds I. Palaz and K.J. Marfurt) *Geophys. Developments*, **6**, 121–178, Society Exploration Geophysicists, Tulsa.

Franseen, E.K. and **Mankiewicz, C.** (1991) Depositional sequences and correlation of middle (?) to late Miocene carbonate complexes, Las Negras and Nijar areas, southeastern Spain. *Sedimentology*, **38**, 871–898.

Franseen, E.K., Goldstein, R.H. and **Whitesell, T.E.** (1993) Sequence stratigraphy of Miocene carbonate complexes, Las Negras area, southeastern Spain: Implications for quantification of changes in relative sea level. In: *Carbonate Sequence Stratigraphy* (Eds R.G. Loucks and J.F. Sarg), *AAPG Mem.*, **57**, 409–434.

Gardner, T.W., Jorgensen, D.W., Shuman, C. and **Lemieux, C.R.** (1987) Geomorphic and tectonic process rates: effects of measured time interval. *Geology,* **15**, 259–261.

Grammer, G.M., Ginsburg, R.N. and **Harris, P.M.** (1993) Timing of deposition, diagenesis, and failure of steep carbonate slopes in response to a high-amplitude/high-frequency fluctuation in sea level, Tongue of the Ocean, Bahamas. In: *Carbonate Sequence Stratigraphy* (Eds R.G. Loucks and J.F. Sarg), *AAPG Memoir*, **57**, 107–132.

Halley, R.B. and **Evans, C.C.** (1983) *The Miami limestone. A guide to selected outcrops and their interpretation (with a discussion of diagenesis in the formation).* Miami Geological Society, Miami, 67 pp.

Harrison, C.G.A. (2002) Power spectrum of sea level change over 15 decades of frequency. *Geochem. Geophys. Geosystems*, **3**, 1–17.

Hopley, D. (1982) *The Geomorphology of the Great Barrier Reef: Quaternary Development of Coral Reefs.* Wiley, New York, 453 pp.

Hunt, D. and **Gawthorpe, R.L.** (Eds) (2000) Sedimentary responses to forced regressions. *Geol. Soc. Sp. Publ.*, 172. Geological Society, London, 383 pp.

Hunt, D. and **Tucker, M.E.** (1992) Stranded parasequences and the forced regressive wedge systems tract: Deposition during base-level fall. *Sedimentary Geol.*, **81**, 1–9.

Hunt, D. and **Tucker, M.E.** (1993) Sequence stratigraphy of carbonate shelves with an example from the mid-Cretaceous (Urgonian) of southeast France. In: *Sequence Stratigraphy and Facies Associations* (Eds H.W. Posamentier, C.P. Summerhayes, B.U. Haq and G.P. Allen). *IAS Spec. Publ.*, **18**, 307–341, Blackwell, London.

Hunt, D. and **Tucker, M.E.** (1995) Reply to discussion. *Sedimentary Geol.*, **95**, 147–160.

James, N.P. (1997) The cool-water carbonate depositional realm. In: *Cool-water Carbonates* (Eds N.P. James and J.A.D. Clarke), *SEPM Spec. Publ.*, **56**, 1–20.

James, N.P. and **Macintyre, I.G.** (1985) Carbonate depositional environments, modern and ancient. Part 1 – Reefs: Zonation, depositional facies, diagenesis. *Colorado School of Mines Quarterly*, **80**, 70 pp.

James, N.P., Bone, Y., Collins, L.B. and **Kyser, T.K.** (2001) Surficial sediments of the Great Australian Bight: facies dynamics and oceanography on a vast cool-water carbonate shelf. *J. Sedim. Res.*, **71**, 549–567.

Jervey, M.T. (1988) Quantitative geological modeling of siliciclastic rock sequences and their seismic expression. In: *Sea-level Changes: An Integrated Approach* (Eds C.K. Wilgus, B.S. Hastings, C.G.St.C. Kendall, H.W. Posamentier, C.A. Ross and J.C. Van Wagoner), *SEPM Spec. Publ.*, **42**, 47–69.

Kenter, J.A.M. (1990) Carbonate platform flanks: Slope angle and sediment fabric. *Sedimentology*, **37**, 777–794.

Labeyrie, L.D., Duplessy, J.C. and **Blanc, P.L.** (1987) Variations in mode of formation and temperature of oceanic deep waters over the past 125,000 years. *Nature*, **327**, 477–481.

Mankiewicz, C. (1996) The middle to upper Miocene carbonate complex of Nijar, Almeria province, southeastern Spain. In: *Models for Carbonate Stratigraphy from Miocene Reef Complexes of Mediterranean Regions* (Eds E.K. Franseen, M. Esteban, W.C. Ward and J.M. Rouchy), *SEPM Concepts Sedimentol. Paleontol.*, **5**, 141–157.

Mitchum, R.M., Vail, P.R. and **Thompson, S.** (1977) Seismic stratigraphy and global changes of sea level, part 2: the depositional sequence as a basic unit for stratigraphic analysis. In: *Seismic Stratigraphy – Applications to Hydrocarbon Exploration* (Ed. C.E. Payton), *AAPG Mem.*, **26**, 53–62.

Neumann, A.C. and **Hearty, P.J.** (1996) Rapid sea-level changes at the close of the last interglacial (substage 5e) recorded in Bahamian island geology. *Geology*, **24**, 775–778.

Nummedal, D., Gupta, S., Plint, A.G. and **Cole, R.D.** (1995) The falling stage systems tract: Definition, character and expression in several examples from the Cretaceous from the U.S. Western Interior. In: *Sedimentary Responses to Forced Regressions* (Eds D. Hunt, R.L.Gawthorpe and M. Dogherty). *Geol. Soc. London,* Abstract Volume, pp. 45–48.

Nummedal, D., Riley, G.W., Cole, R.D. and **Trevena, A.S.** (1992) The falling sea level systems tract in ramp settings (abstract). In: *Mesozoic of the Western Interior.* Society Economic Paleontologists and Mineralogists, Theme Meeting, Fort Collins, Colorado, pp. 50.

Plint, A.G. (1991) High frequency relative sea level oscillations in Upper Cretaceous shelf clastics of the Alberta foreland basin: possible evidence of a glacio-eustatic control? In: *Sedimentation, Tectonics and Eustasy* (Ed. D.I.M. MacDonald), *IAS Spec. Publ.*, **12**, 409–428.

Plint, A.G. and **Nummedal, D.** (2000) The falling stage systems tract: recognition and importance in sequence stratigraphic analysis. In: *Sedimentary Responses to Forced Regressions* (Eds D. Hunt and R.L. Gawthorpe), *Geol. Soc. London Spec. Publ.*, **172**, 1–17.

Plotnick, J.E. (1986) A fractal model for the distribution of stratigraphic hiatuses. *J. Geol.*, **94**, 885–890.

Pomar, L. (1991) Reef geometries, erosion surfaces and high-frequency sea-level changes, upper Miocene reef complex, Mallorca, Spain. *Sedimentology*, **38**, 243–270.

Pomar, L. (2000) *Course E209 Excursion Guide*. Dept. Ciencias de la Terra, Univ. Balearic Islands, Palma de Mallorca, 81 pp.

Posamentier, H.W. and **Allen, G.P.** (1999) Siliciclastic sequence stratigraphy – concepts and applications. *SEPM Concepts Sedimentol. Paleontol.*, **7**, 209 pp.

Posamentier, H. and **Morris, W.R.** (2000) Aspects of the stratal architecture of forced regressive deposits. In: *Sedimentary Responses to Forced Regressions* (Eds D. Hunt and R.L. Gawthorpe), *Geol. Soc. Spec. Publ.*, **172**, 19–46.

Posamentier, H.W. and **Vail, P.R.** (1988) Eustatic controls on clastic deposition II – sequence and systems tract models. In: *Sea-level Changes: an Integrated Approach* (Eds C.K. Wilgus, B.S. Hastings, C.G.St.C. Kendall, H.W. Posamentier, C.A. Ross and J.C. Van Wagoner), *SEPM Spec. Publ.*, **42**, 125–154.

Posamentier, H.W., Allen, G.P., James, D.P. and **Tesson, M.** (1992) Forced regressions in a sequence stratigraphic framework: Concepts, examples, and exploration significance. *Am. Assoc. Petrol. Geol. Bull.*, **76**, 1687–1709.

Posamentier, H.W., Jervey, M.T. and **Vail, P.R.** (1988) Eustatic controls on clastic deposition I – conceptual framework. In: *Sea-level Changes: an Integrated Approach* (Eds C.K. Wilgus, B.S. Hastings, C.G.St.C. Kendall, H.W. Posamentier, C.A. Ross and J.C. Van Wagoner), *SEPM Spec. Publ.*, **42**, 109–124.

Purdy, E.G. and **Winterer, E.L.** (2001) Origin of atoll lagoons. *Geol. Soc. Am. Bull.*, **113**, 837–854.

Roberts, H.H., Fillon, R.H., Kohl, B., Robalin, J.M. and **Sydow, J.C.** (2004) Depositional architecture of the Lagniappe delta: Sediment characteristics, timing of depositional events, and temporal relationship with adjacent shelf-edge deltas (Eds J.B. Anderson and R.H. Fillon). *SEPM Spec. Publ.*, **79**, 143–188.

Sadler, P.M. (1981) Sediment accumulation rates and the completeness of stratigraphic sections. *J. Geol.*, **89**, 569–584.

Sarg, J.F. (1988) Carbonate sequence stratigraphy. In: *Sea-Level Changes: An Integrated Approach* (Eds C.K. Wilgus, B.S. Hastings, C.G.St.C. Kendall, H.W. Posamentier, C.A. Ross and J.C. Van Wagoner), *SEPM Spec. Publ.*, **42**, 155–182.

Schlager, W. (1993) Accommodation and supply – a dual control on stratigraphic sequences. *Sed. Geol.*, **86**, 111–136.

Schlager, W. (1999) Scaling of sedimentation rates and drowning of reefs and carbonate platforms. *Geology*, **27**, 183–186.

Schlager, W. (2003) Benthic carbonate factories of the Phanerozoic. *Int. J. Earth Sci.*, **92**, 445–464.

Schlager, W. (2005) Carbonate sedimentology and sequence stratigraphy. *SEPM Concepts Sedimentol. Paleontol.*, 8, 209 pp.

Shackleton, N.J. (1987) Oxygen isotopes, ice volume and sea level. *Quaternary Sci. Rev.*, **6**, 183–190.

Spencer, T. (1992) Bioerosion and biogeomorphology. In: *Plant–Animal Interactions in the Marine Benthos* (Eds D.M. John, S.J. Hawkins and J.H. Price), Systematic Association Special Volume, **46**, pp. 493–509.Clarendon Press, Oxford.

Vail, P.R. (1987) Seismic stratigraphic interpretation using sequence stratigraphy. Part 1: Seismic stratigraphy interpretation procedure. In: *Atlas of Seismic Stratigraphy* (Ed. A.W. Bally) *AAPG Stud. Geol.*, **27–1**, 1–10.

Vail, P.R., Mitchum, R.M. and **Thompson, S.** (1977) Seismic stratigraphy and global changes of sea level, part 3: relative changes of sea level from coastal onlap. In: *Seismic Stratigraphy – Applications to Hydrocarbon Exploration* (Ed. C.E. Payton). *AAPG Mem.*, **26**, 63–81.

Van Buchem, F., Pittet, B. and **Hillgärtner, H.** (2002) High-resolution sequence stratigraphic architecture of Barremian/Aptian carbonate systems in northern Oman and the United Arab Emirates (Kharaib and Shu'aiba Formations). *GeoArabia*, **7**, 461–500.

Van Wagoner, J.C., Mitchum, R.M., Posamentier, H.W. and **Vail, P.R.** (1987) Seismic stratigraphy interpretation using sequence stratigraphy. Part 2: Key definitions of sequence stratigraphy. In: *Atlas of Seismic Stratigraphy* (Ed. A.W. Bally), *AAPG Stud. Geol.*, **27–1**, 11–13.

Van Wagoner, J.C., Posamentier, H.W., Mitchum, R.M., Vail, P.R., Sarg, J.F., Loutit, T.S. and **Hardenbol, J.** (1988) An overview of the fundamentals of sequence stratigraphy and key definitions. In: *Sea-level Changes: an Integrated Approach* (Eds C.K. Wilgus, B.S. Hastings, C.G.St.C. Kendall, H.W. Posamentier, C.A. Ross and J.C. Van Wagoner), *SEPM Spec. Publ.*, **42**, 39–45.

Warrlich, G.M.D. (2001) *Computer modelling of carbonate platform evolution.* Ph.D. thesis, Royal Holloway University of London, London, 268 pp.

Warrlich, G.M.D., Waltham, D.A. and **Bosence, D.W.J.** (2002) Quantifying the sequence stratigraphy and drowning mechanisms of atolls using a new 3-D forward stratigraphic modelling program (CARBONATE 3D). *Basin Res.*, **14**, 379–400.

Warrlich, G.M.D., Waltham, D.A. and **Bosence, D.W.J.** (2005) 3D and 4D controls on carbonate depositional systems: Sedimentologic and sequence-stratigraphic analysis of an attached carbonate platform and atoll (Miocene, Nijar Basin, SE Spain). *Sedimentology*, **52**, 363–389.

Warrlich, G.M.D., Bosence, D.W.J., Waltham, D.A., Wood, C., Boylan, A. and **Badenas, B.** (2008) 3-D stratigraphic forward modeling for analysis and prediction of carbonate platform stratigraphies in exploration and production. *Marine Petrol. Geol.*, **25**, 35–58.

White, W.B. (1984) Rate processes: Chemical kinetics and karst landform developments. In: *Groundwater as a Geomorphic Agent* (Ed. R.G. LaFleur), pp. 227–248. Allen Unwin, Boston.

Int. Assoc. Sedimentol. Spec. Publ. (2009) **41**, 327–336

Early load-induced fracturing in a prograding carbonate margin

DONALD F. MCNEILL and GREGOR P. EBERLI

Comparative Sedimentology Laboratory, Marine Geology and Geophysics, University of Miami, 4600 Rickenbacker Causeway, Miami, FL 33149, USA (E-mail: dmcneill@rsmas.miami.edu)

ABSTRACT

The description of early non-tectonic fracturing of carbonates is rare because (1) fractures in young carbonates occur in areas not easily accessible, and (2) burial and diagenesis often precludes the determination of the exact timing of (early) fracturing in older rocks. The Bahamas Drilling Project core Clino recovered the upper part of a young prograding platform margin containing a set of fractures interpreted to be non-tectonic; formed by rapid loading of a clinothem package. Three intervals with open and partially infilled fractures were recovered in Clino, and those fractures are closely related to the lithological variation in the slope sediments. Fracturing occurred specifically within coarse, well-cemented skeletal deposits. These coarser deposits are generally margin-derived sediments within the background sediment. The overlying and underlying finer-grained sediments are significantly less fractured. The fractures occur within a 45-m-thick interval immediately below a surface of non-deposition or erosion with a hiatus of >0.5 Myr. A proposed model suggests that early cementation of the skeletal sands, followed by rapid margin progradation, provided sufficient load to generate brittle fracturing (shear and extensional) in lithified strata. Laboratory deformation experiments, magnetic fracture-plane orientation and the timing of load emplacement are consistent with load-induced fracturing. Fracture planes preferentially dip westward which is parallel to the depositional slope. Similar early fracture porosity can act as high permeability conduits for diagenetic fluids and hydrocarbons in platform margin and slope strata.

Keywords Fracture, carbonate, load-induced, clinothem, Bahamas, porosity.

INTRODUCTION

Early fracturing of lithified carbonates is a rarely reported phenomenon. Observation of fractures in young carbonates is difficult because such features generally occur in relatively inaccessible near-surface settings. In older rocks, burial and diagenetic history may preclude the identification and the exact timing of fracturing, making it difficult to determine that fracturing occurred *early*. The Bahamas Drilling Project core Clino recovered, for the first time, a complete section of a young (Neogene) prograding platform margin along the western margin of Great Bahama Bank (Fig. 1). This core recovered a set of fractures in the clinothem package, and they are described here. These fractures are natural and clearly not drilling induced (Kulander *et al.*, 1990). Our interpretation of the fractures benefits greatly from existing studies. A detailed lithological (Kenter *et al.*, 2001), stratigraphic (Eberli *et al.*, 2001; McNeill *et al.*, 2001) and diagenetic (Melim *et al.*, 2001; Swart *et al.*, 2001) framework has been produced for core Clino. This extensive dataset allows a comprehensive evaluation of the conditions that contributed to early rock fracturing. The objectives of this paper are to use the fracture inclination, fracture orientation, fracture density and constraints on fracture timing to illustrate a mechanism of early fracturing. The hypothesis is that early diagenesis and rapid progradation of the platform margin to invoke fracturing. The mechanism of early fracturing in the model has implications for preferential fluid exchange and might be a contributing factor in the collapse of platform margins (Hine *et al.*, 1992).

Stratigraphic setting of the fracture interval

Three intervals of open and partially infilled fractures were recovered in core Clino within mid-Pliocene seismic sequences *e* and *f*, just below their respective seismic sequence boundaries (SSB) D and E (Fig. 2) (see Eberli *et al.* (2001) for a description of the seismic units and sequences). In these two sequences, core Clino penetrated the relatively steep (~4°) part of the prograding clinoform (Eberli & Ginsburg, 1987). The two main fracture intervals (upper and middle) are within sequence *e*, and a third, thin fracture interval (lower) is at the top of sequence *f* (Fig. 2). The SSB-D and a disconformity just above it, represent an interval of non-deposition and/or erosion with a duration of about 0.5–0.8 Myr. Above SSB-D, a ~275-m-thick package of sediment associated with prograding sequence *d* was deposited in a period of between 0.1 and 0.2 Myr. The three fractured intervals are closely tied to the lithological variation in the slope sediments. The majority of fractures (upper and middle interval) occur specifically within coarse, well-cemented skeletal deposits. These skeletal deposits, termed 'interruptions' by Kenter *et al.* (2001), represent margin-derived sediments deposited within (or interrupting) the background platform-derived sediment. The sediment 'interruption' beds are interpreted by Kenter *et al.* (2001) as transported to the slope during times when most of the platform was subaerially exposed. The overlying and underlying finer-grained hemipelagic or 'background' sediments that are shed to slope during times of platform flooding and high sedimentation rates are generally less indurated and contain less fractures. Seismic sequence *e* represents sediments deposited during a major backstep of the platform margin (to the east) and a period of platform aggradation (McNeill, 2005). The slightly coarser grain-size probably promoted more rapid cementation relative to the finer-grained and less permeable hemipelagic sediments (Westphal & Munnecke, 1997; Kenter *et al.*, 2001). This preferential early cementation is also supported by rock magnetic studies on the fractured and non-fractured interval, where (very) early cementation appears to have preserved chains of magnetite crystal from magnetotactic bacteria (McNeill, 1997).

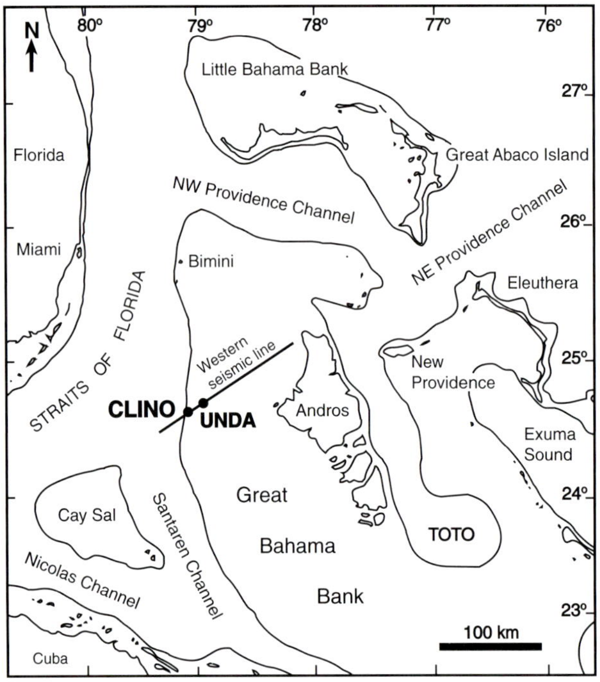

Fig. 1. Map showing the location of core Clino on the western margin of Great Bahama Bank. The location of the seismic line on which the line drawing (Fig. 2) is based is also shown. TOTO stands for Tongue of the Ocean.

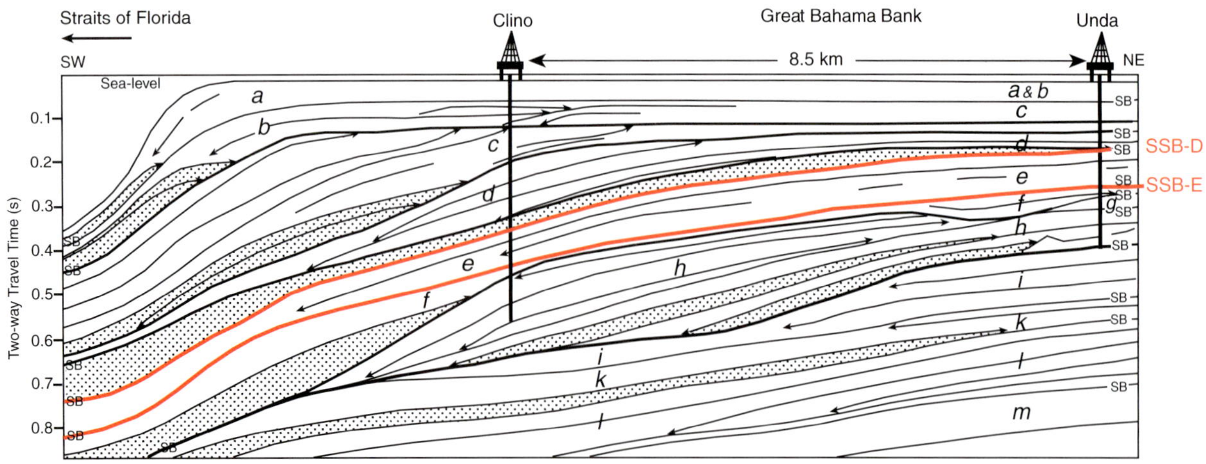

Fig. 2. Line drawing of the seismic sequences penetrated by core Clino (from Eberli *et al.*, 2001). Clino fractures occur in sequence *e* (bounded by red lines). Fractures in sequence *e* occur primarily in beds of coarse skeletal sediment redeposited downslope from a shallow-water source area east of core Unda. The shaded areas indicate onlapping packages.

METHODS

Core data and fracture characterization

The high core recovery (>90%) in the middle and lower parts of core Clino provide a nearly continuous record of early diagenetic features such as the fractures reported here. One-third of the whole core (6.3 cm diameter) was cut and separated for archive purposes. The remaining two-thirds of the core was used to measure and to collect samples. The depth values reported here are consistent with those in Ginsburg (2001), where the depths are in metres below mud pit (subtract 7.3 m to obtain depth below mean sea level). The depth interval from about 375 to 450 m contains distinct fractures that were first noted on site during drilling. The fractures in the slabbed core were characterized by measuring fracture plane inclination, fracture length, fracture aperture, and noting if the fracture was open or filled. The vertical extent of the fracture was also recorded and was used to provide a rough estimate of total fracture length per metre of core. This calculation is used to approximate fracture density.

Fracture orientation

The dip azimuth of the fracture plane was calculated using magnetic declination. This procedure consisted of measuring the fracture dip orientation relative to an arbitrary red line marked on the core (used to confirm the up direction). The offset angle of the red line was then measured relative to a palaeomagnetic plug drilled from that piece of core. The declination direction measured from the plug (see McNeill *et al.*, 2001) was then used to reorient the core piece (red line), and calculate the true dip azimuth of the fracture plane. This reorientation method is fairly robust given the strength and stability of the magnetic signal in this section of the core (McNeill 1997, McNeill *et al.*, 2001). This method does, however, have a certain amount of inherent error (such as magnetometer error, least-squares calculation of declination, preservation of true declination), and the error range is *estimated* at ±5°. All of the orientation data are from pieces of core that contain either a magnetic plug sample or are from adjacent pieces that could be fitted together with a plug-oriented core piece.

Sr-isotope ratio on fracture fill

A single strontium-isotope ratio on the mineralized fill material (celestite) was determined by Prof. Swart and provided for this study. This data point followed the methodology described by Swart *et al.* (2001) for the strontium-isotope stratigraphy of core Clino material. The celestite value was normalized to a NIST-987 standard of 0.710248 and the age determined by the look-up table of McArthur *et al.* (2001), which is calibrated to the Berggren *et al.* (1995) timescale.

Uniaxial deformation experiments

To obtain an estimate of the force needed to fracture rock of similar composition to that which hosts the fractures, two uniaxial compression tests were conducted. Two whole-core samples from the fractured zone (440 and 447 m) were subjected to compression tests at ETH-Zurich. The effective pressure of brittle failure was determined with standard methods of comparing the confining pressure minus the pore pressure (water at 10 MPa). The nature of the fracture and fault produced was also recorded.

RESULTS

Fracture data set

The fractures in core Clino were grouped into three classes: through-going shear, axial split, or cataclastic shear (following the usage of Paterson, 1978). The through-going shear type was often a simple, single fracture or fault with relatively little effect on the adjacent host sediment (Fig. 3a and b). The axial fractures are those oriented vertically or near vertically (>70°) along the axis of the core (Fig. 3c). This type often had multiple fractures that radiated outward from a slightly larger central fracture, although it was difficult to trace these subfractures in a 6.3 cm wide core. The cataclastic shear fracture type showed a zone, 10–20 cm long, that contained abundant fractures of differing orientations, lengths and aperture sizes (Fig. 3d and e). The cataclastic zones sometimes had a central through-going fracture surrounded by subfractures, or terminated abruptly on a bed of differing composition and integrity.

The inclination of the fractures ($n = 151$) showed a skewed distribution, with most fractures biased to higher angles between 30° and vertical (Fig. 4). The mean inclination was ~57°, with a large standard deviation (~24°). The fracture inclination with depth in each of the intervals shows a

Fig. 3. Photographs of fracture core. (a, b) through-going shear fracture at 441 m depth; (c) axial fracture at 379 m, note central fracture with small subfractures radiating outward; (d) cataclastic shear fracture with a central through-going shear, 431 m depth; (e) cataclastic fracture at a bed contact at 437 m; celestite infill of larger fracture openings, this photograph illustrates how minor differences in the degree of cementation probably affect fracture potential; (f) fracture surface with large crystals of celestite at 401 m depth.

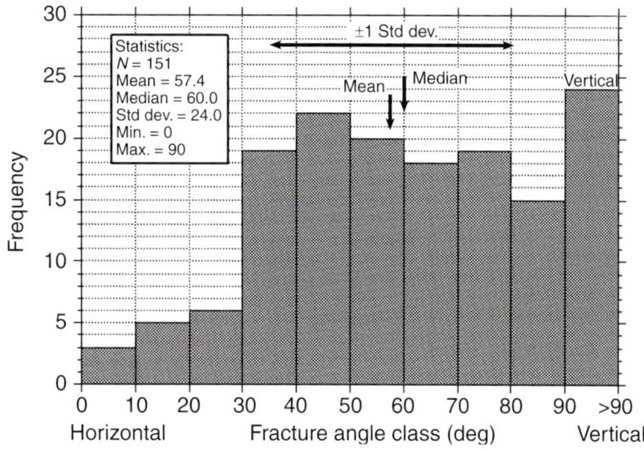

Fig. 4. Histogram of inclination angles for the whole fracture interval in core Clino. The domination by intermediate to vertical fractures suggests compressional and perhaps some extensional forces are influencing the fracture formation.

broad scatter of values (Fig. 5). It is interesting to note that the two intervening hemipelagic intervals between the three fracture intervals do contain several fractures, and that these fractures are very steep to vertical. Unlike the fracture inclination with depth, the cumulative fracture length (total fracture length in centimetres per 3 m interval) does show a generally higher fracture density near the tops of the three intervals and a trend of decrease downward (poorly defined) (Fig. 6).

Host rock lithology, log signature and fracture fill

The two main fracture intervals are hosted by the well-cemented skeletal grainstones and packstone units, termed 'interruptions' and derived from the shallow platform to the east. The differential cementation between the fracture-host sediment and the hemipelagic sediments is well illustrated in the downhole log signature (Fig. 7). The cemented skeletal intervals tend to show a higher sonic velocity relative to the background hemipelagic intervals. Similarly, the caliper log across the fractured intervals shows borehole diameters very close to the size of the diamond core bit, whereas the hemipelagic units above, within and below the cemented skeletal intervals show larger diameters due to less cementation (Fig. 7). Within the fracture intervals, both the sonic and caliper logs show high-frequency variations, a signature that probably reflects the variability of fracture-derived porosity and density.

The more intensely fractured rock in the middle and upper intervals sometimes contain secondary mineralization in the form of celestite (Melim *et al.*, 2001) ($SrSO_4$ also known as celestine, Fig. 3f). The infill of fractures by celestite is highly variable with respect to depth and aperture size. In some cases, celestite has nearly filled the fracture (see Fig. 3e), while in some of the larger fractures elongate individual celestite crystal are found lining the fracture surface (as in Fig. 3f). The celestite infill clearly demonstrates the natural origin of these fractures. The timing of the celestite infill can be only poorly constrained. A celestite crystal at ~386 m produced a Sr-isotope ratio with an age range of 1.34–2.15 Ma. The uncertainty is high for this age as the time of crystallization depends on the two possible origins of the strontium: either from dissolution of the aragonite or from seawater entering the platform. Most likely, it is some

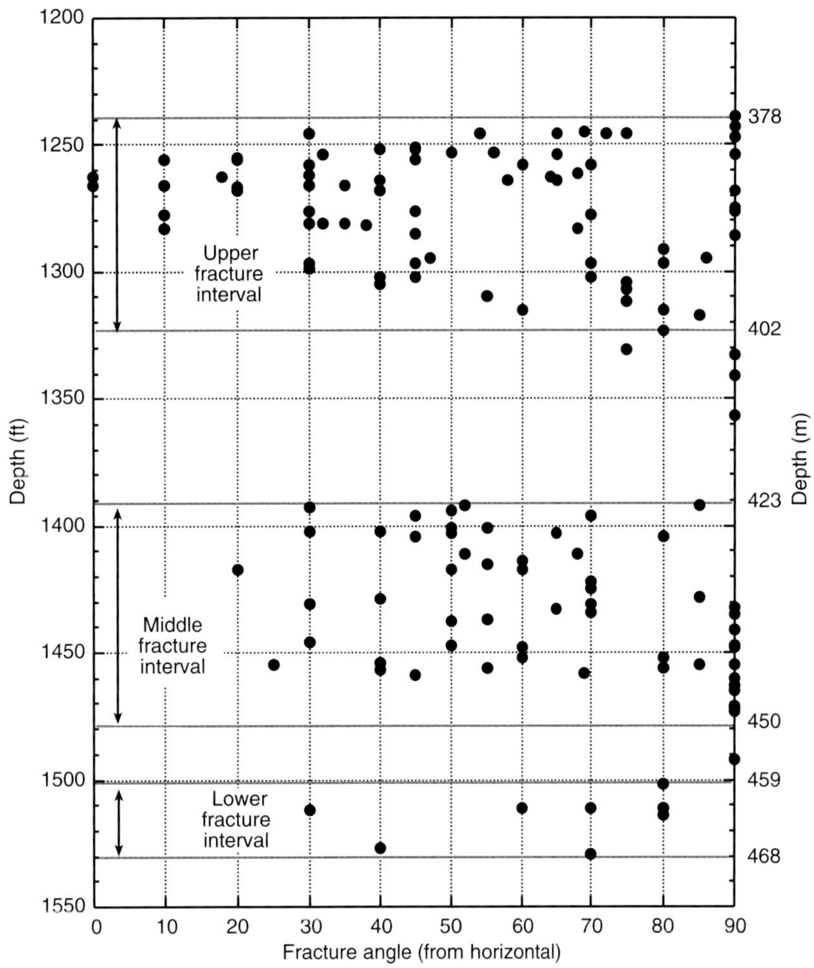

Fig. 5. Fracture plane angle versus depth. The skeletal 'interruptions' exhibit a wide scatter in inclination, but most are in excess of 30°. Note that relatively few fractures occur in the hemipelagic intervals between the fracture intervals.

combination of the two, as Melim *et al.* (2001) suggest a temporal overlap of celestite precipitation and aragonite neomorphism. They also argue that abundant celestite precipitation would require the import of sulphate by active fluid circulation. Even with the temporal uncertainty, it is interesting to note that the celestite age is coincident with the end of a major progradational pulse in the latest Pliocene (see Discussion below).

Fracture orientation

The dip azimuth of 19 through-going shear fractures, determined using magnetic declination, shows a bi-directional, conjugate distribution (Fig. 8). Two general groupings are approximately 180° apart, one westward and one eastward. These two orientations are roughly equivalent to seaward (west) and bankward (east) of the bank margin (oriented just east of north). The majority of the fracture planes dip to the west (14 of 19) in the direction of the depositional slope and basin.

DISCUSSION OF FRACTURE ORIGIN

Origin of the fractures: a loading model

The origin of the fractures is clearly not related to drilling (as celestite partly infills many fractures), and there is no evidence of local tectonic forces that could generate such localized fractures. Several key features and lines of evidence (shear/cataclastic fracture type, steep angle, conjugate fracture sets orientated perpendicular to the margin, and downward cumulative fracture length) suggest that fracture formation is related to vertical compression from sediment loading. The proposed model for fracture formation involves several factors: the integration of the coarse, permeable skeletal sand 'interruptions', a prolonged period of non-deposition or erosion associated with a sequence boundary, and the subsequent rapid deposition of overlying sediment related to clinothem progradation (Fig. 9). The fractures occur within a 45-m-thick interval immediately

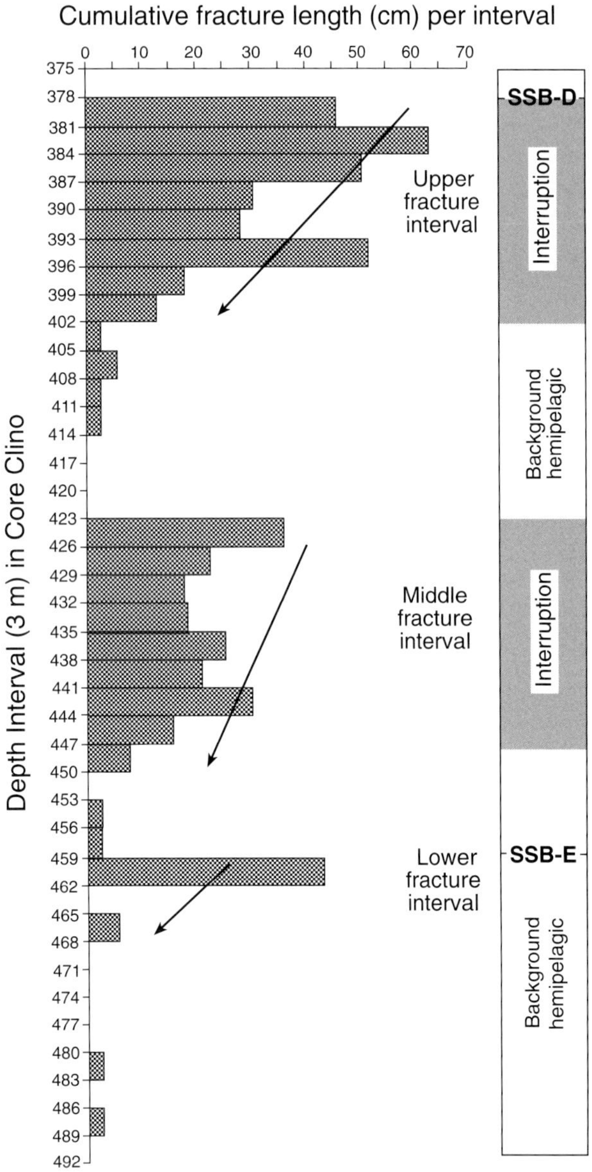

Fig. 6. Cumulative fracture length (total fracture length in centimetres) per 3 m interval. A slight trend of decreased fracture density with depth can be observed in these histogram plots. This trend may be a result of greater cementation at the sequence boundaries that cap the upper and lower interval, or simply related to the higher permeability in the skeletal 'interruptions'.

below a non-depositional surface (SSB-D). This period of non-deposition (3.5–3.0 Ma) may have promoted cementation within the skeletal sediments of sequence *e* and at the top of sequence *f*. Malone *et al.* (2001) dated similar Quaternary 'lowstand' deposits and corroborated the interpretation that the better cemented coarser-grained slope deposits are indeed deposited during times of lowered sea level and reduced sedimentation rates. The circulation of seawater through the permeable skeletal 'interruptions' likely promoted lithification relative to the less permeable and finer-grained hemipelagic 'background' sediments. The onset of platform progradation in the late Pliocene rapidly loaded the partially cemented 'interruption' intervals and promoted shear and extensional fracturing (see model summarized in Fig. 10). Experimental rock mechanics has shown that compressional loading can produce conjugate shear fractures with inclinations ranging from 40° to 70°, as well as near-vertical extensional (axial splitting) fractures. The inclinations of the fracture planes in core Clino are generally consistent with the features generated from laboratory experiments.

Laboratory support for loading and early fracturing

Uniaxial deformation experiments of the two Clino 'interruption' samples (at a depth of 440 and 447 m) showed brittle behaviour at laboratory strain rates of ~10 MPa (effective pressure). The resultant brittle failure produced a single, steeply inclined, through-going fault (D. Olgaard, personal communication). This brittle fracturing is consistent with the current overburden pressure of a 425 m rock column (at 2.1 g cm^{-3}), within today's depth range of the fractures. If fracturing occurred after the progradation load (275 m column) was emplaced (~1.5 Ma) it is reasonable to assume that the 'interruption' horizon was at an earlier stage in its lithification history and thus susceptible to early load-induced fracture.

Ancient analogues for loading and early fracturing

Documentation of early fracturing on carbonate platform margins is rare. One example from a prograding margin is the Permian, Capitan reef of west Texas and southeast New Mexico where marine cement and fossils filling fractures indicate early fracturing (Saller, 1996; Kosa & Hunt, 2006). The Capitan reef also shows a conjugate fracture set oriented perpendicular to the shelf margin. The fractures parallel to the Capitan reef front are generally steeper than Clino, and are concentrated in reefal facies. Differential compaction and basinward tilting (extension) are invoked to explain the steep axial fractures (Saller, 1996; Harwood & Kendall, 1999; Hunt *et al.*, 2003; Stanton & Pray, 2004). Although a different setting, this Permian

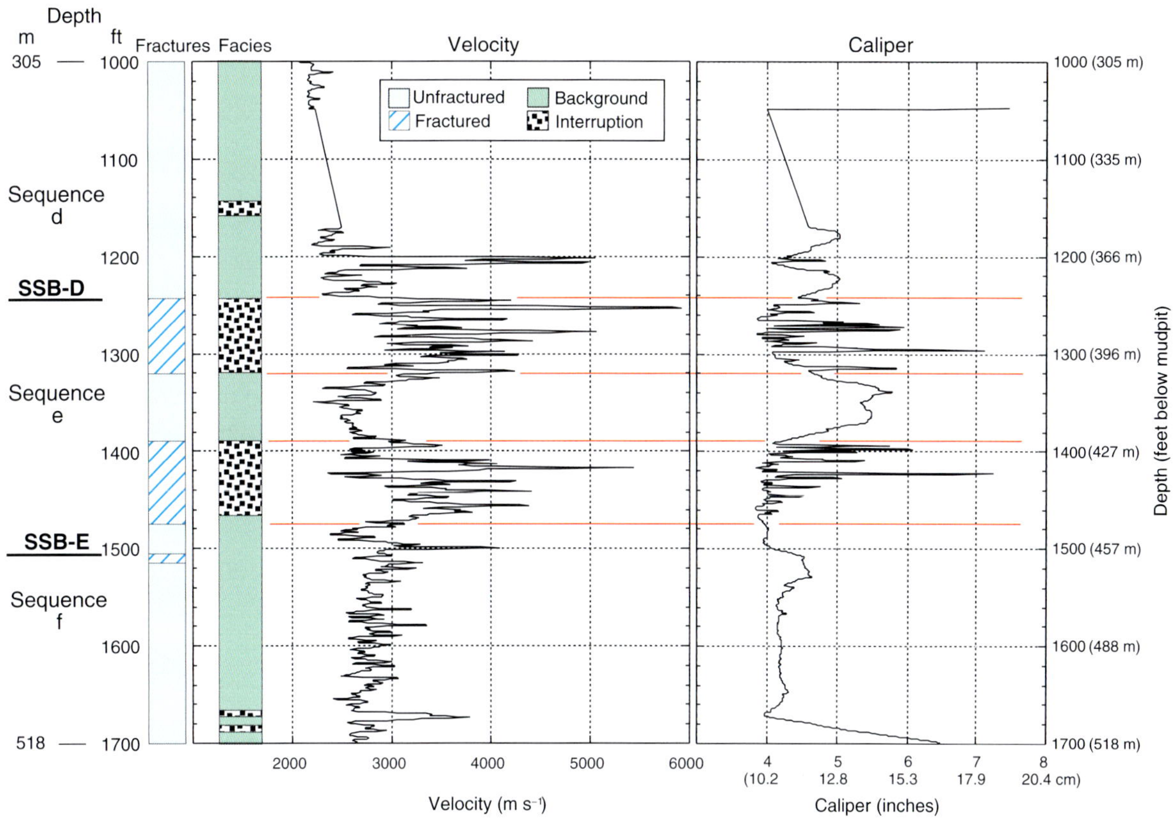

Fig. 7. Sonic velocity and caliper logs over the fracture interval. The fractures show greater velocity than the adjacent hemipelagic sediments, a result of greater cementation. Likewise, the caliper log reflects a smaller diameter borehole in the skeletal interruptions relative to the less cemented hemipelagic intervals.

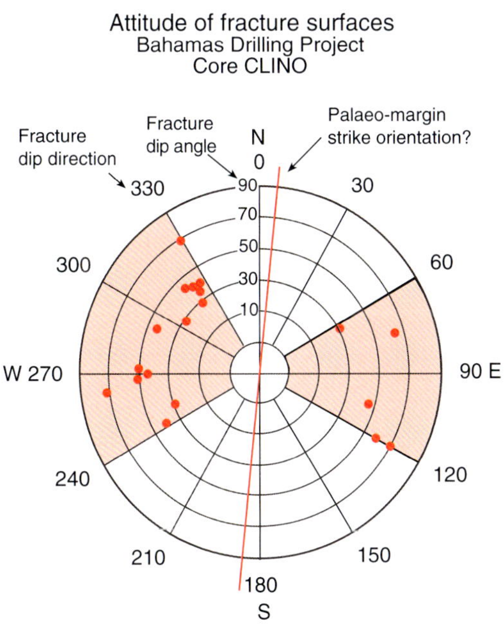

Fig. 8. Attitude of the fracture surfaces (dip angle from the horizontal) in Clino as determined by palaeomagnetic orientation. The conjugate fractures reflect an orientation perpendicular to the platform margin, with the majority of the fractures dipping seaward.

example suggests that early cementation (in this case the reef margin) can fracture relatively early under load-related influences. In the case of Clino, the affect of down-to-the-basin compaction is hard to assess (in core), but perhaps the preferential fracture dip direction to the basin (west) (Fig. 8) might be explained by this influence.

CONCLUSIONS

Studies of fracturing within prograding carbonates are relatively rare. In one example, Saller (1996) and others describe fractures within the reefal facies of the Capitan that have dip directions normal to the depositional strike of the platform margin. The Capitan fractures range between 75° and 85° and were interpreted to have formed because of extension during down-to-the-basin differential compaction. The Bahamas fractures may have also experienced some of the same differential compaction, but fracturing was probably driven largely by the loading effects of the overlying, rapidly emplaced prograded

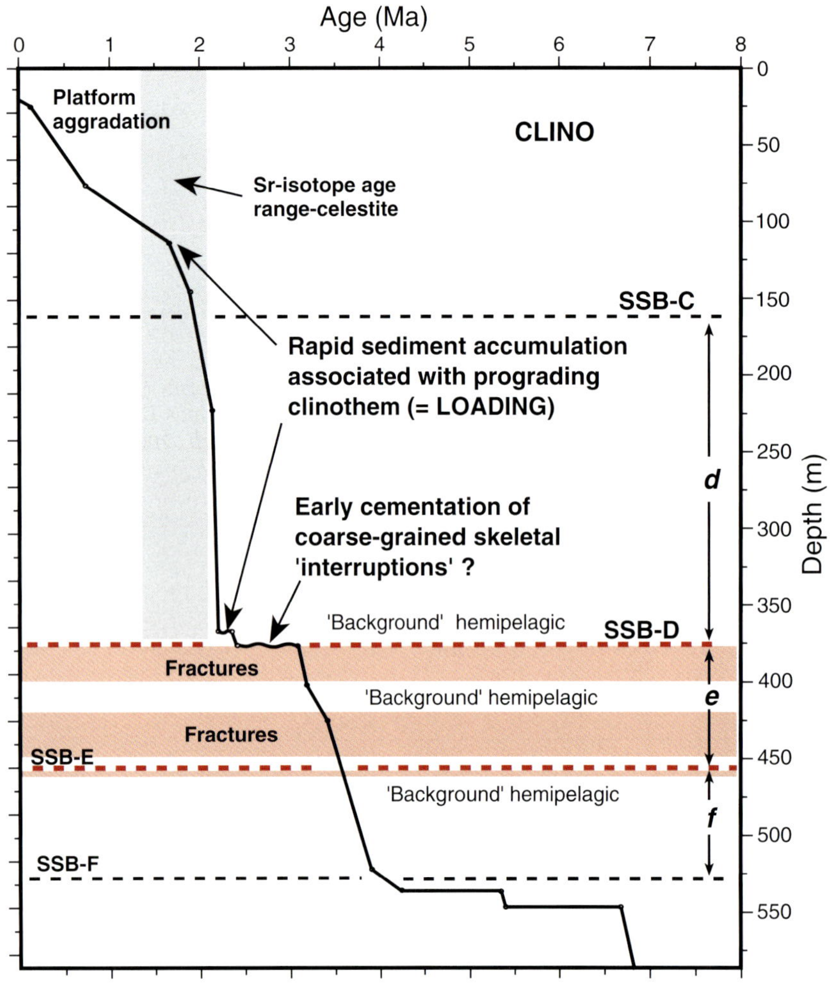

Fig. 9. Age–depth plot for core Clino from McNeill *et al.* (2001). The upper, middle and lower fracture intervals are shown in middle Pliocene age host sediments. Progradation loading above SSB-D in the latest Pliocene rapidly added over 300 m of sediment on top of the skeletal 'interruptions' to produce compression fractures.

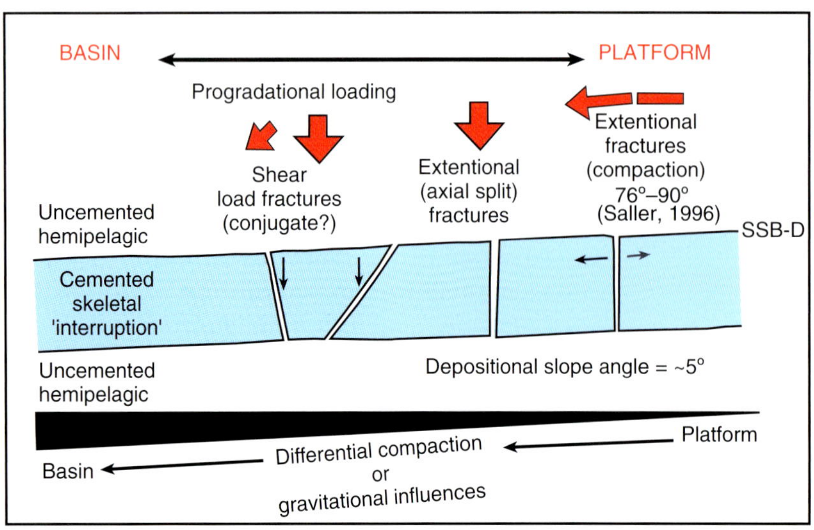

Fig. 10. A cartoon summarizing the general forces and fracture types at the progradational margin of western Great Bahama Bank. The key to early fracture is the preferential early cementation of the skeletal 'interruption' beds.

sediment that accumulated on the seaward (basin) side of the Clino location. It is proposed that this fracturing formed during a period of tectonic quiescence on Great Bahama Bank and is thus essentially non-tectonic in origin.

The Clino fracture data support a non-tectonic mechanism for the formation of macroscopic fractures. The combination of early diagenesis (cementation) and the subsequent load emplacement by the prograding platform margin have produced very early fracturing in the burial history of the platform. These fractures (at least for their early burial history) have distinct geophysical log signatures, and have largely remained opened and largely unfilled (except for some celestite infill). The original porosity, or later re-opened fracture porosity, may act as major fluid-flow pathways within the platform margin slope. Those pathways may be conduits for the upward movement of burial fluids and focus diagenetic alteration associated with those burial fluids. In the deeper subsurface, associated fracture permeability may enhance vertical and lateral fluid flow.

ACKNOWLEDGEMENTS

We first and foremost want to thank Robert Ginsburg for his longstanding drive and leadership to make the Bahamas Drilling Project a reality. The US National Science Foundation supported this project, both the drilling and subsequent analyses (OCE 8917295 and OCE 9104294). We also thank the Industrial Associates of the Comparative Sedimentology Laboratory for providing funds for core collection and (continued) analysis. The Donors to the American Chemical Society-Petroleum Research Fund (25818-AC8) supported DFM. The work of Jeroen Kenter, Peter Swart and Leslie Melim provided the framework critical to interpreting the Clino fractures, and we thank them for their input. We are grateful to David Olgaard for conducting the uniaxial deformation experiments on Clino core samples while at ETH-Zurich. Prof. Harry Elderfield's laboratory at University of Cambridge kindly provided the Sr data on the celestite in collaboration with Peter Swart (University of Miami). We thank Al Hine for suggesting a possible role of early fracture on the collapse of platform margins. Art Saller and Taury Smith provided thoughtful review and constructive comments on this manuscript.

REFERENCES

Berggren, W.A., Kent, D.V., Swisher, C.C. III and **Aubry, M.-P.** (1995) A revised Cenozoic geochronology and chronostratigraphy. In: *Geochronology Times Scales and Global Stratigraphic Correlation* (Eds W.A. Berggren, D.V. Kent, and M.-P., Aubry). *SEPM Spec. Publ.*, **54**, 129–212.

Eberli, G.P. and **Ginsburg, R.N.** (1987) Segmentation and coalescence of platforms, Tertiary, NW Great Bahama Bank. *Geology*, **15**, 75–79.

Eberli, G.P., Anselmetti, F., Kenter, J.A.M., McNeill, D.F., Ginsburg, R.N., Swart, P.K. and **Melim, L.A.** (2001) Facies, diagenesis, and timing of prograding sequences on western Great Bahama Bank. In: *Subsurface Geology of a Prograding Carbonate Platform Margin, Great Bahama Bank: Results of the Bahamas Drilling Project* (Ed. R.N. Ginsburg). *SEPM Spec. Pub.*, **70**, 241–265.

Ginsburg, R.N. (2001) *Subsurface Geology of a Prograding Carbonate Platform Margin, Great Bahama Bank: Results of the Bahamas Drilling Project. SEPM Spec. Pub.*, **70**, 271 pp.

Harwood, G.M. and **Kendall, A.C.** (1999) Reef margin collapse, gully formation and filling within the Permian Reef; Carlsbad Caverns, New Mexico, USA. *Sedimentology*, **46**, 443–461.

Hine, A.C., Locker, S.D., Tedesco, L.P., Mullins, H.T., Hallock, P., Belknap, D.F., Gonzales, J.L., Neumann, A.C. and **Snyder, S.W.** (1992) Megabreccia shedding from modern, low-relief carbonate platforms, Nicaraguan Rise. *Geol. Soc. Am. Bull.*, **104**, 928–943.

Hunt, D.W., Fitchen, W.M. and **Kosa, E.** (2003) Syndepositional deformation of the Permian Capitan reef carbonate platform, Guadalupe Mountains, New Mexico, USA, *Sedimentary Geol.*, **154**, 89–126.

Kenter, J.A.M., Ginsburg, R.N. and **Troelstra, S.R.** (2001) Sea-level-driven sedimentation patterns on the slope and margin. In: *Subsurface Geology of a Prograding Carbonate Platform Margin, Great Bahama Bank: Results of the Bahamas Drilling Project* (Ed. R.N. Ginsburg). *SEPM Spec. Pub.*, **70**, 61–100.

Kosa, E. and **Hunt, D.W.** (2006) Heterogeneity in fill and properties of karst-modified syndepositional faults and fractures; Upper Permian Capitan Platform, New Mexico, U.S.A. *J. Sedimentary Res.*, **76**, 131–151.

Kulander, B.R., Dean, S.L. and **Ward, B.J., Jr.** (1990) Fractured core analysis: interpretation, logging, and use of natural and induced fractures in core. *AAPG Methods in Exploration Series* 8, Tulsa, Oklahoma, 88 pp.

Malone, M.J., Slowey, N.C. and **Henderson, G.M.** (2001) Early diagenesis of shallow-water periplatform carbonate sediments, leeward margin, Great Bahama bank (Ocean Drilling Program Leg 166). *Geol. Soc. Am. Bull.*, **113**, 881–894.

McArthur, J.M., Howarth, R.J. and **Bailey, T.R.** (2001) Strontium isotope stratigraphy: LOWESS Version 3: Best fit to the marine Sr-isotope curve for 0–509 Ma and accompanying look-up table for deriving numerical ages. *J. Geol.*, **109**, 155–170.

McNeill, D.F. (1997) Facies and early diagenetic influence on the depositional magnetization of carbonates. *Geology*, **25**, 799–802.

McNeill, D.F. (2005) Accumulation rates from well-dated late Neogene carbonate platforms and margins. *Sedimentary Geol.*, **175**, 73–87.

McNeill, D.F., Eberli, G.P., Lidz, B.H., Swart, P.K. and **Kenter, J.A.M.** (2001) Chronostratigraphy of a prograded carbonate platform margin: a record of dynamic slope sedimentation, western Great Bahama Bank. In: *Subsurface Geology of a Prograding Carbonate Platform Margin, Great Bahama Bank: Results of the Bahamas Drilling Project* (Ed. R.N. Ginsburg). *SEPM Spec. Pub.*, **70**, 241–265.

Melim, L.K., Swart, P.K. and **Maliva, R.G.** (2001) Meteoric and marine-burial diagenesis in the subsurface of Great Bahama Bank. In: *Subsurface Geology of a Prograding Carbonate Platform Margin, Great Bahama Bank: Results of the Bahamas Drilling Project* (Ed. R.N. Ginsburg). *SEPM Spec. Pub.* **70**, 137–161.

Paterson, M.S. (1978) *Experimental Rock Deformation – The Brittle Field*. Springer-Verlag, Berlin, 254 pp.

Saller, A.H. (1996) Differential compaction and basinward tilting of the prograding Capitan reef complex, Permian, west Texas and southeast New Mexico, USA. *Sedimentary Geol.*, **101**, 21–30.

Stanton, R.J. and **Pray, L.C.** (2004) Skeletal-carbonate neptunian dikes of the Capitan Reef; Permian, Guadalupe Mountains, Texas, U.S.A. *J. Sedimentary Res.*, **74**, 805–816.

Swart, P.K., Elderfield, H. and **Beets, K.** (2001) The $^{87}Sr/^{86}Sr$ ratios in carbonates, phosphorites, and fluids collected during the Bahamas Drilling Project cores Clino and Unda: Implications for dating and diagenesis. In: *Subsurface Geology of a Prograding Carbonate Platform Margin, Great Bahama Bank: Results of the Bahamas Drilling Project* (Ed. R.N. Ginsburg). *SEPM Spec. Pub.* **70**, 175–185.

Westphal, H. and **Munnecke, A.** (1997) Mechanical compaction versus early cementation in fine-grained limestones: differentiation by the preservation of organic microfossils. *Sedimentary Geol.*, **112**, 33–42.

Int. Assoc. Sedimentol. Spec. Publ. (2009) **41**, 337–360

Markov models for linking environments and facies in space and time (recent Arabian Gulf, Miocene Paratethys)

BERNHARD M. RIEGL and SAMUEL J. PURKIS

National Coral Reef Institute, Nova Southeastern University Oceanographic Center, 8000 N. Ocean Drive, Dania, FL 33004, USA (E-mail: rieglb@nova.edu)

ABSTRACT

If, as comparative sedimentology maintains, knowledge of the Holocene can sometimes be helpful to explain the past (and vice versa), common quantitative denominators might exist between recent and fossil systems. It may also be possible to describe dynamics and find linkages between space and time with a unique set of quantitative tools. To explore such conceptual links, spatial facies patterns mapped using satellite imagery were compared with temporal patterns in analogous ancient outcropping facies using Markov chains and graphs. Landsat and Ikonos satellite imagery was used to map benthic facies in a nearshore carbonate ramp (Ras Hasyan) and offshore platform system (Murrawah, Al Gharbi) in recent deposits of the Arabian Gulf (United Arab Emirates), and results were compared with the Fenk Quarry outcrop in Burgenland, Austria, a carbonate ramp of the Miocene (Badenian) Paratethys. Facies adjacencies (i.e. Moore neighbourhood of colour-coded image pixels of satellite image or outcrop map) were expressed by transition probability matrices, which showed that horizontal (spatial) facies sequences and vertical (temporal) outcrop sequences had the Markov property (knowledge of tth state defines likelihoods of $t + 1$st state) and that equivalent facies were comparable in frequency. The transition probability matrices are expressed as weighted digraphs and calculated fixed probability vectors which encapsulate information on both the spatial and temporal components (size of and time spent in each facies). Models of temporal functioning were obtained by modifying matrices (digraphs) of spatial adjacency to matrices (digraphs) of temporal adjacency by using the same vertices (facies) but adjusting transitions without changing paths. With this combined spatio-temporal model, changes in facies composition in falling and rising sea-level scenarios are investigated by adjusting transition likelihoods preferentially into shallower (falling sea level) or deeper (rising sea level) facies. The model can also be used as a numerical analogue to a Ginsburg-type autocyclic model. The fixed probability vector was used as a proxy for final facies distribution. Using Markov chains it is possible to use vertical outcrop data to evaluate the relative contribution of each facies in any time-slice which can aid, for example, in estimation of reservoir sizes and to gain insight into temporal functioning as derived from spatial pattern.

Keywords Arabian Gulf, Leitha Limestone, Paratethys, Markov chain, comparative sedimentology, Walther's Law, landscape pattern, remote sensing, carbonate sedimentation.

INTRODUCTION

The actualistic principle and comparative sedimentology see the present as a key to the past (Ginsburg, 1974). Additionally, a liberal interpretation of Walther's Law (Middleton, 1973) might suggest that in some special cases the past may even hold a key to the present or future, since knowledge of the conformable strata in outcrop, if time-transgressive, might allow some reconstruction of the lateral facies mosaic in any time-slice, be it in the past or the present (Doveton, 1994; Parks *et al.*, 2000). All this suggests that a quantitative expression of temporal information from core or outcrop into spatial information on a map of a landscape, and vice versa, would be interesting and in the tradition of comparative sedimentology (Ginsburg, 1974, 1975).

Invoking the present for reconstruction of the past obviously requires some understanding of its functioning. Geology and biology merge in shallow, tropical carbonate systems since ecology and successional stages of biotic communities influence the development of facies and vice versa (Ginsburg, 1957; Toomey, 1981; Schlager, 2005). While the environment (climate, tectonism, eustasy, etc.) controls lateral dimensions and vertical successional pattern (James, 1984; Wilgus *et al.*, 1989; Tucker & Wright, 1990; Wilkinson *et al.*, 1997, 2003; Lehrmann & Goldhammer, 1999; Rankey, 2004; Schlager, 2005), ecology often defines facies at a finer scale (Flügel, 2004).

In numerous previous studies, Markov chains have shown that conditional probabilistic relationships (knowledge of an event/facies will define probability of a following event/facies) exist among vertical facies successions (Krumbein, 1967; Krumbein & Greybull, 1967; Gingerich, 1969; Doveton, 1971, 1994; Miall, 1973; Carle & Fogg, 1997; Wilkinson *et al.*, 1997, 2003; Lehrmann & Rankey, 1999; Parks *et al.*, 2000; Davis, 2002). Stages of ecological community and landscape successions have Markovian properties (Pielou, 1969; Horn, 1975; Usher, 1979; Turner *et al.*, 2001; Urban & Wallin, 2002), which can be expressed using graphs (Roberts, 1976; Cantwell & Forman, 1993; Bunn *et al.*, 2000; Urban & Keitt, 2001; Urban, 2005). Markov chains and graphs may therefore emerge as useful quantitative tools for investigations into potential linkages between lateral and temporal, ecological and sedimentological dynamics in the tradition of comparative sedimentology (Ginsburg, 1974).

There clearly exist relations between sedimentological pattern and ecological pattern (Ginsburg, 1957). Landscape patterns result from ecological and sedimentological processes that are intricately linked and that feed back into each other (Ginsburg & Shinn, 1994). The lateral aspects of the present landscape can be efficiently described using remote sensing (Harris & Kowalik, 1994; Rankey & Morgan, 2002; Rankey, 2004; Purkis & Riegl, 2005; Purkis *et al.*, 2005; Hetzinger *et al.*, 2006) and vertical patterns can be obtained from outcrop or core. In this study, we use satellite images to describe recent facies patterns in the Arabian Gulf, one of the classic carbonate ramp settings (Kirkham, 1998). These facies (coral reefs, grain shoals, hardgrounds, etc.) express quantifiable neighbourhood patterns (Purkis *et al.*, 2005) and good understanding exists on ecological (Riegl, 2002; Purkis & Riegl, 2005) and sedimentological successions (Shinn, 1969; Uchupi *et al.*, 1996; Alsharhan & Kendall, 2003).

The horizontal Holocene facies pattern in the Arabian Gulf will be compared with vertical outcrop data from the Miocene (Badenian) Paratethys in the Austrian Leitha Limestone, previously interpreted as a fossil equivalent of today's Arabian Gulf (Riegl & Piller, 2000). The presence of comparable facies frequencies and conditional transition probabilities (i.e. the Markov property) in both systems would suggest a quantitative link between the two time-slices and between vertical and lateral succession of facies – since from Walther's Law it may be deduced that one should exist (Doveton, 1994; Parks *et al.*, 2000; Elfeki & Dekking, 2001, 2005). Using discrete mathematical tools the expression of a quantitative link of temporal and spatial dynamics will be attempted in order to explore whether the relationship among facies in a landscape can be reduced to a simple graph or matrix encapsulating a model of the system's functioning that can then be used to predict changes in the spatial characteristics of facies through time, for example in changing environments. The model will be applied to investigate possible facies change due to small-amplitude sea-level change.

REGIONAL SETTING

The Holocene sediments of the United Arab Emirates (UAE) include shallow-water marine carbonates in one of the world's classic, modern carbonate systems (Kirkham, 1998). Our study area was defined by the gently sloping bathymetry of the Arabian homocline and highs produced primarily by salt diapirism that form banks with well-defined lateral facies successions (Purser & Seibold, 1973). Coastal and near-shore sedimentology was shaped by rapidly changing sea-level in the Pleistocene and Holocene (Lambeck, 1996; Uchupi *et al.*, 1996; Kirkham, 1998) and, in the shorter term, by wind and waves. Daily afternoon breezes and especially the prevailing (though largely seasonal) Shamal winds influence the local wave-induced currents, which in turn are a major factor in developing the sedimentology of the coast and the offshore banks (Kirkham, 1998; Sheppard *et al.*, 1992). The northwest trending coastline of the UAE and the northern edges of the banks are strongly influenced by the Shamal, due to their oblique position to this north wind and the lack of shelter from offshore barriers. The shallow sea floor throughout the study area lies above wave-base typical for Shamal conditions – likely to extend to a depth of at least 20 m (Purser & Evans, 1973).

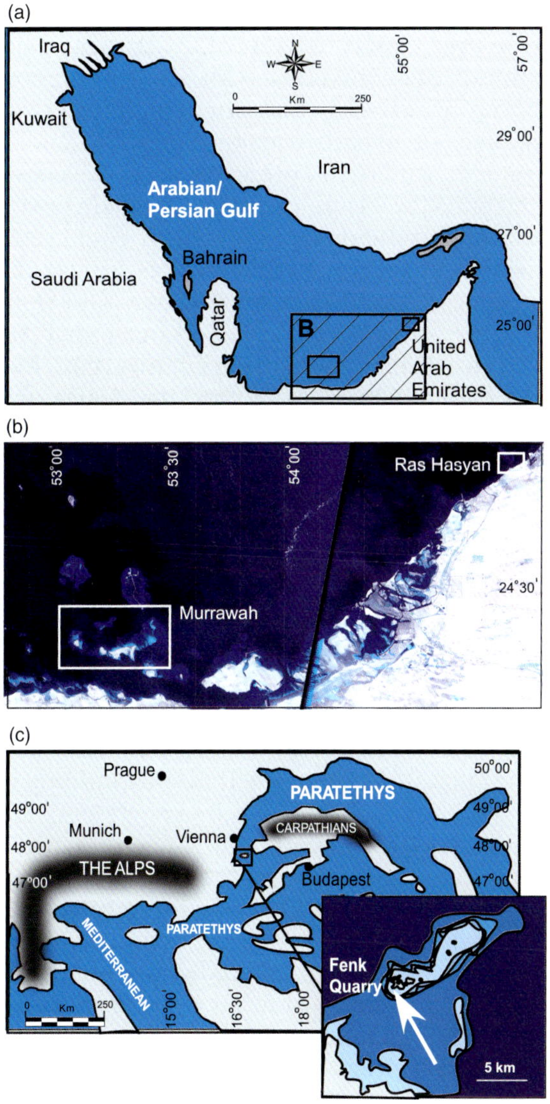

Fig. 1. (a) Study area in the Arabian Gulf. The landscape evaluated was a mixture of sandy and hardground facies with biostromal coral frameworks in a typical carbonate ramp setting. (b) Landsat Image showing location of study areas. (c) Palaeogeographical setting of the Leitha Limestone in the Vienna basin. An island arc on Triassic basement spanned the Alpine and Carpathian piedmont that formed the coastline around the flooded Vienna basin, which was a peripheral basin of the Paratethys epicontinental sea. On this arc of islands and shoals, Leitha Limestone was deposited. Around islands, corals and shallow-water carbonate facies are found. Contour lines at 20 m depth interval.

Two systems were studied in detail. A fine-scale study covered only a few square-kilometres that were investigated with Ikonos satellite imagery. The details of the imagery and of image processing are given in Purkis *et al.* (2005) and Purkis (2005). This system is near Ras Hasyan, in the UAE (Fig. 1) and extended shore-parallel for ~7 km and ~1.5 km offshore, a region with a typical depth of 8 m beneath lowest astronomical tide (mean slope angle 0.5°). Previous studies (Riegl, 1999, 2002; Riegl *et al.*, 2001) had identified coral carpets (= biostromes), areas covered by unconsolidated carbonate sand, macro-algae and seagrass, underlain in wide areas by hardgrounds consisting of early diagenetically cemented calcarenites (Shinn, 1969; Evans *et al.*, 1973). Six coral assemblages of variable live cover were known from within this area.

(A) Large, widely spaced *Porites lutea* and other *Porites* mixed with several other massive species; widely distributed on hardgrounds.
(B) Coherent patches of tabular colonies of *Acropora clathrata* and *A. downingi* covering 40–90% of the substratum.
(B1) Edges of (B) with less space cover (<25%) but still *Acropora* dominated.
(C) Clusters of faviids (most notably *Platygyra lamellina*, *P. daedalea*, *Cyphastrea serailia*, *Favia* spp.) either widely spaced or densely packed.
(D) Widely spaced *Siderastrea savignyana* colonies on sandy hardgrounds.
(E) Patches of densely spaced (80% coral cover) columnar *Porites harrisoni* intermingled with faviids (*Favia* spp., *Platygyra* spp.).

As a second study area, a major island/shoal complex associated with the Great Pearl Bank (Purser, 1973) was investigated using Landsat imagery at the Murrawah and Al Gharbi shoals complex, which includes shoals around the islands of Murrawah, Al Heel, Fiyya and Al Gharbi. The shoals are roughly circular or oblong, with a fringe of reefal sediments, followed by grainstones and a largely muddy interior (Purser, 1973; Schlager, 2005).

A system comparable to the present-day Arabian Gulf existed in the Miocene Paratethys and is partly represented by the Austrian Leitha Limestone. This unit, of the Miocene Badenian central Paratethys stage (= Langhian to Lower Serravalian Mediterranean stage = Foraminifera Biozones M5 to M7 of Berggren *et al.*, 1985), represents a shallow subtidal ramp sloping into a peripheral basin of the subtropical Paratethys (Fig. 1) (Piller & Vavra, 1991; Piller *et al.*, 1996). The term "Leitha Limestone" is used here in a purely descriptive manner since a redefinition in Papp *et al.* (1978) was not in accordance with the stratigraphic codes (Steininger & Piller, 1999).

Its greatest thickness is about 50 m and Dullo (1983) recognized 10 microfacies types (bioclastic algal debris facies, bioclastic rhodolite debris, bioclastic algal mollusc, bioclastic, bafflestone, foraminiferal algal mollusc, pavement, foraminiferal rhodolite, foraminiferal algal debris, foraminiferal), which are broadly comparable to modern Arabian Gulf facies (Riegl & Piller, 2000). These microfacies types were also applied to similar rocks in the Styrian basin by Friebe (1990, 1991), who grouped the Leitha Limestone together with other shallow-water sediments into the 'Weissenegg Formation'. However, the term Leitha Limestone is more widely recognized. A characteristic element is coralline algae that occur as rhodolite facies or calcarenites consisting mainly of branch fragments. Coral buildups of limited size occurred only locally as patch reefs (Piller & Kleemann, 1991) and biostromes (Riegl & Piller, 2000). The best developed coral buildups are present at the southern tip of the Leitha Mountains where the limestones reach their greatest spatial extent and thickness (about 50 m). During the Badenian, the tips of the Leitha mountains extended above sea level as a chain of islands (Fig. 1c). Due to the islands' position and absence of continuous terrigenous influx, coral growth was relatively prolific. The position in a marginal basin at high latitude suggests a similar temperature regime to today's Arabian Gulf (Riegl & Piller, 2000) with repeated death of corals due to temperature extremes.

METHODS

Image basis for landscape analysis

The basis of all optical classifications and maps were spaceborne images (Ikonos and Landsat ETM7 11-bit multispectral satellite image, 4 m and 30 m pixel-size respectively). Details about image processing, ground-truthing and the mapping approach are given in Purkis (2005) and Purkis *et al.* (2005). Full bathymetric datasets for depth correction of spectral reflectance values of the Ikonos were obtained from acoustic surveys and were tidally corrected against data obtained from *in situ* loggers deployed during field work. The bathymetry data were then used for calibration of the Landsat image. Image classification was conducted using an unsupervised *k*-means clustering routine and the multivariate normal probability driven classifier described by Purkis & Pasterkamp (2004) after masking of pixels that were outside the known depth-resolution of the images (~8 m). Extensive ground-truthing illustrated that the classified Ikonos map had an overall accuracy of 81%, and the classified Landsat map had an overall accuracy of 72%. Many errors occurred in distinguishing between algae and seagrass, which could not be satisfactorily resolved. Some errors in classification existed between coral bioherms/biostromes and algae, since dead coral areas (many had been killed in thermal anomalies of 1996–1998; Riegl, 1999; Purkis & Riegl, 2005) had dense algal overgrowth. Accuracy differences between the images were largely due to different spatial resolution (pixel size) and the size of the studied areas. Also bathymetry was extracted from the Landsat image using a modification of the algorithm by Stumpf *et al.* (2003).

At the Fenk Quarry in Burgenland, Austria, two complete top-to-bottom and several incomplete sections in the Badenian Leitha Limestone were measured at a centimetre-scale to obtain a good understanding of present facies and then, to the highest possible resolution, the entire outcrop was mapped since Lehrmann & Rankey (1999) stress the importance of incorporating two-dimensional information into Markov chain analysis. The measured sections were superimposed on a photograph of the outcrop and facies units were followed laterally for the production of polygons showing the outlines of facies over the entire outcrop. This outcrop was chosen because Riegl & Piller (2000) previously demonstrated the equivalence of the Leitha Limestone to the modern Arabian Gulf by qualitative facies analysis.

Ecological background data

Data regarding the biology of the system include series of 155 line transects of 10 m and 50 m taken in 1995, 1999, 2000, 2002 and 2005. One hundred and eighty point observations for ground-truthing maps and satellite imagery, noting the species composition and taphonomic status of 1 × 1 m or 2 × 2 m sample areas, were taken in 1995, 1996, 1998, 2000, 2002, 2003, 2005. Taphonomic status and breakdown sequence were evaluated by sampling corals and skeletons in 1995, 1996, 1999, 2000 and 2005. Since these data are already published elsewhere (Riegl, 1999, 2002; Purkis & Riegl, 2005; Purkis *et al.*, 2005), they are not shown in detail but used here as background information for the calibration of models.

Markov chains

Markov models are mathematical models of probabilistic processes, which generate random sequences of outcomes to certain probabilities. Detailed treatments can be found in Kemeny & Snell (1960), Roberts (1976), Iosifescu (1980) and examples applied to sedimentology in Krumbein & Dacey (1969), Miall (1973), Doveton (1971, 1994), Carle & Fogg (1997), Davis (2002) and many others. The basic premise of a Markov process is that if the outcomes of all the first *t–n* events of a series of events are known, then the probabilities of outcomes in the *t*th experiments are also known. The *t*th step transition probability is given by $p_{ij}(t) = \mathbf{Pr}[\mathbf{f}_t = \mathbf{s}_j \mid \mathbf{f}_{t-1} = \mathbf{s}_i]$, where p_{ij} is the transition probability (**Pr**) of an event *i* to *j* in which the outcome function *f* takes a value s_j at a time *t* that depends only on the directly preceding outcome function *f* at time $t-1$ having value $\mathbf{s}_i$. If the set of all possible outcomes is finite, it is a finite stochastic process. Markov chains describe Markov processes if (Kemeny & Snell, 1960):

(1) There is a finite set of outcomes depending only on the outcomes before the *t*-th set, i.e. $\mathbf{Pr}[\mathbf{f}_t = \mathbf{s}_j \mid (\mathbf{f}_{t-1} = \mathbf{s}_i)\wedge\mathbf{p}] = \mathbf{Pr}[\mathbf{f}_t = \mathbf{s}_j \mid \mathbf{f}_{t-1} = \mathbf{s}_i]$. This condition is called the Markov property.
(2) The probability $p_{t+1}(O)$ that outcome *O* will occur at trial *t* is known if we know what outcome occurred on trial $t-1$, i.e. $p_{ij}(t) = \mathbf{Pr}[\mathbf{f}_t = \mathbf{s}_j \mid \mathbf{f}_{t-1} = \mathbf{s}_i]$.
(3) The dependence of $p_{t+1}(O)$ on the previous outcome is independent of *t*, i.e. it is the same for trial 2 as for 1000.

There are several types of Markov chains (Kemeny & Snell, 1960; Roberts, 1976).

(1) Absorbing: where all states (the facies in our studied system are each a 'state' in the Markov chain) move towards one final state, into which they are finally absorbed and cannot leave it anymore. The long-term behaviour of such a chain is dependent on its starting state. Absorbing chains consist of transient states (which the chain will only use a few times, then they will decline to zero) and absorbing states (which will eventually 'absorb' the transient states, i.e. the chain remains exclusively within the absorbing states). They can be reduced to a canonical form from which can be derived their fundamental matrix. The basic questions to ask such a chain are: (1) the probability of entering an absorbing state, given that the chain did not start there; (2) the expected number of times that the chain will be in any non-absorbing state before absorption; (3) the expected number of steps until the chain enters an absorbing state (Roberts, 1976).
(2) Regular: if all states can be reached from each other in the same number of steps (i.e. the chain's period $d = 1$), which also means that P^N has no zero entries. Regular chains have a fixed-point probability vector with all positive entries. The basic questions to ask such a chain are (1) the probability to be in state u_j after *t* steps if the chain starts in state u_i; (2) the expected number of steps before returning to a particular state.
(3) Ergodic: if the underlying digraph (see below) is strongly connected, i.e. every state in the chain can be reached by any other state but the chain's period *d* is other than 1. Many of the theorems developed for absorbing and regular chains are, with modification, applicable to ergodic chains. For example, every Markov chain with an er et has a fixed point probability vector with zeros in the transient states but positive entries in the ergodic set and it also has a fundamental matrix. This allows similar questions to be asked regarding an absorbing chain and a regular chain.

Embedded Markov chains (Krumbein & Dacey, 1969; Davis, 2002) are special models that do not allow self-to-self transitions, i.e. have zero entries in the dominant diagonal. Markov processes can be one-dimensional or act in higher dimensions and can be conditioned on boundary conditions or not (Elfeki & Dekking, 2001, 2005). Furthermore, single-step or multistep processes can be evaluated. In this paper, emphasis is on unconditioned, one-dimensional processes with single-step transitions analogous to the expression of ecological dynamics (Horn, 1975; Logofet & Lesnaya, 2000) since the paper is interested in capturing the successions within the landscape in the most simple and intuitive way. Quantification of the overall contribution of facies to the composition of the landscape and the spatial relationships among facies is attempted. More complex Markov models can be applied to use the information gained by our approach to predict facies outside the known outcrop (Parks *et al.*, 2000; Elfeki & Dekking, 2001, 2005).

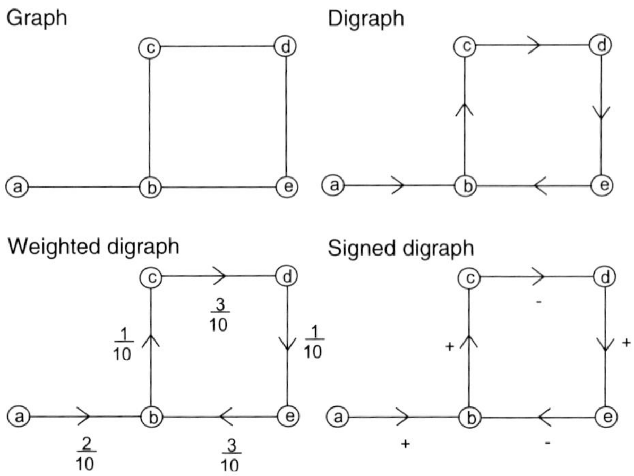

Fig. 2. Examples of graphs defined by the sets of $p = 5$ vertices {a,b,c,d,e} and $q = 5$ edges {ab,bc,cd,de,eb}. The digraphs identify an ergodic set {b,c,d,e}. An ergodic set is a combination of vertices from which, once entered, there is no escape since no paths lead out of it. Therefore, once the ergodic set is entered, the transient states outside this set will never again be visited.

Graph theory

Graph theory, like finite Markov chains, is also firmly founded in discrete mathematics and has, since its invention by Euler in 1736 to settle the famous Königsberg Bridge problem, been used widely in the natural and engineering sciences (Roberts, 1976). It is in wide use in population (Caswell, 1982, 2001; Wardle, 1998) and landscape ecology (Cantwell & Forman, 1993; Urban & Keitt, 2001; Urban, 2005). Graphs allow a convenient visualization of any type of problem that has a spatial or temporal component that can be visualized as states being connected by transitions. Due to this property, graph theory is also closely connected to matrix theory, and all graphs can be conveniently expressed as matrices. Each facies is expressed in a graph (Fig. 2) as a circle, called a 'vertex' or 'node'. The fact that facies neighbour each other, or transit into each other, can be shown on the graph by a line, or 'edge' that connects vertices. These connections can be given a direction in a digraph (short for directed graph), be assigned weights (such as in the presented case here transitional probabilities) in a weighted digraph, or be simply assigned positive or negative signs in a signed digraph (Fig. 2). Each graph has an adjacency matrix (indicating the number of paths between the vertices), a reachability matrix (indicating the linkage between vertices, i.e. whether any paths exist between vertices) and a distance matrix (indicating the distances between vertices, i.e. how many paths over how many vertices lead from any one vertex to another) (Roberts, 1976). This paper is concerned only with weighted adjacency matrices, which is the transition frequency/probability matrix of our Markov model.

Processing steps to obtain neighbourhood statistics

Maps derived from the satellite images and the outcrop photograph (gridded to a pixel-resolution of 10 × 10 cm) were evaluated by counting the Moore neighbourhood (all eight pixels adjoining any given pixel) of every pixel. Pixels were colour coded according to the image classification and therefore each pixel class corresponded to a facies. Only at the edges of the outcrop or satellite image, or next to masked areas, could pixels neighbour 'nothing' (white or black on the image), so all 'colour-to-white/black' adjacencies were excluded from further analysis, since they were only relevant to the specific outcrop/satellite image situation, but cannot be generalized. Throughout this paper, facies remain encoded by pixel-classes or pixel-colours and the terms can be interchanged.

From the raw counts of neighbourhood frequency, the transition frequency matrix that was then transferred into a transition probability matrix was populated. In order to decide an adequate gridding size for the outcrop photograph, it was gridded to several sizes, the transition probabilities were evaluated by neighbour counting, the transition probability matrices were made and their fixed probability vectors compared. Differences in fixed probability vector meant that facies had been lost due to too coarse gridding.

ANALYSES AND RESULTS

Distribution of facies in the recent Arabian Gulf and comparison with Miocene Leitha Limestone

The classified Ikonos imagery of the Jebel Ali carbonate ramp draped over the digital elevation model (Fig. 3) shows three shore-parallel zones. In the first 500 m from shore occurs a zone without coral growth, instead dominated by sand, seagrass and algae. A middle zone from 500 to 1000 m offshore consists mainly of sand and hardground. A deep zone is characterized by coral growth of variable density, interspersed by dense algae.

Eight bottom types (facies) with a patchy distribution could be discriminated (Fig. 3). Landscape components clearly fell into two biologically and sedimentologically functional groups. The first group, found only on unconsolidated sediment, consisted of sand, seagrass and shallow macroalgae. The second group, found only on consolidated sediment, consisted of hardground, sparse corals, dense corals, dead corals and algae. The ecological data suggested that four coral assemblages were resolved on the image:

- large but sparse live corals (mostly *Porites* and faviids) on hardground (magenta in Fig. 3);
- sparse, mostly dead *Acropora* mixed with live faviids and *Porites*. (light blue in Fig. 3);
- dense and interlocking dead *Acropora* (dark purple in Fig. 3);
- sparse, small, live *Siderastrea savignyana* on sandy hardgrounds. (orange in Fig. 3 and not differentiated from bare hardgrounds).

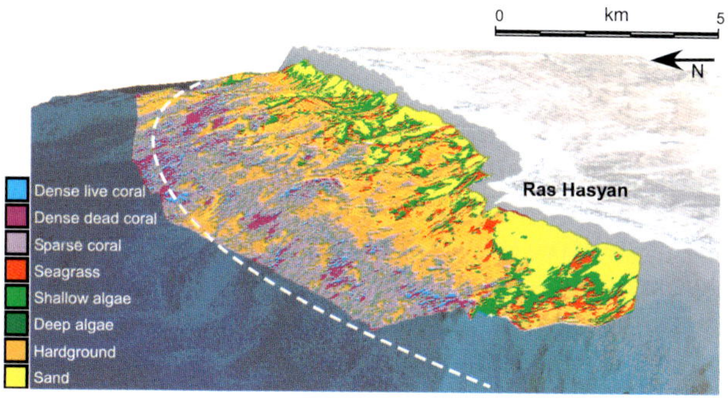

Fig. 3. The classified landscape in the coral biostromal system and its associated landscapes in the Arabian Gulf (Jebel Ali). The white line shows the depth cut-off (8 m), deeper pixels did not carry enough information for classification and were discarded.

These coral assemblages form part of a successional cycle, driven by temperature-related coral death (Riegl, 1999, 2002; Purkis & Riegl, 2005). Coral assemblages show a halo-like structure with patches of dense dead coral surrounded by sparse live coral and then by hardground with only occasional corals (Fig. 3). The repetitive and most likely cyclic nature of coral succession strongly suggests a temporal Markovian property of the system. Several positive temperature excursions killed wide areas of mostly *Acropora*. Once killed, the corals were soon settled by *Chama aspera* bivalves, and then by red algae *Lithophyton kotschyanum* (Fig. 4). A variety of green and brown algae also settled on the skeletons, the

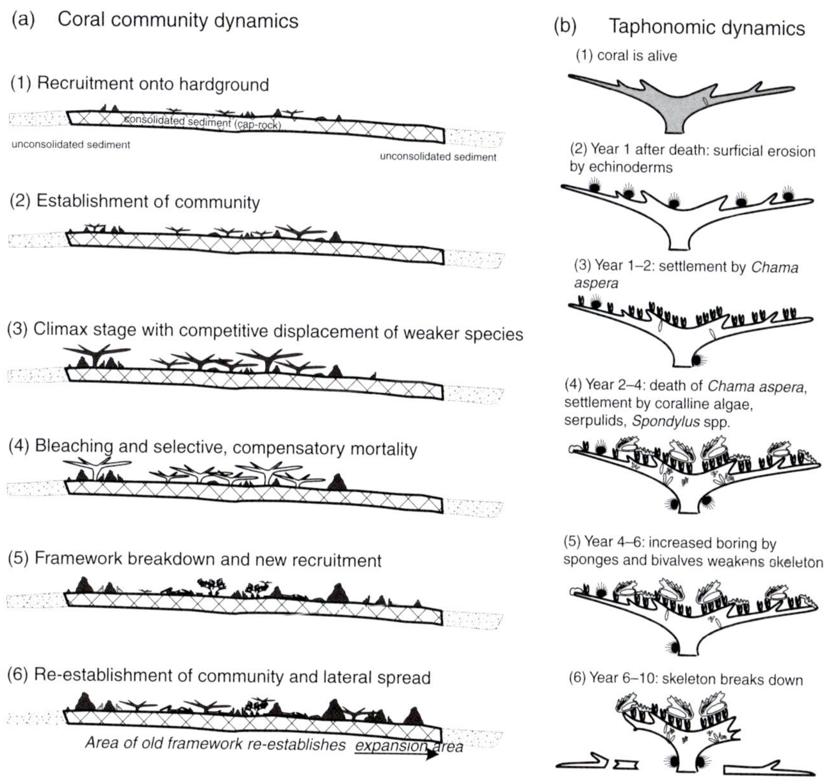

Fig. 4. Ecological (a) and taphonomic (b) dynamics within the coral communities that make up the patches of coral biostrome in the Jebel Ali landscape. Modified from Purkis & Riegl (2005) where discussed in detail.

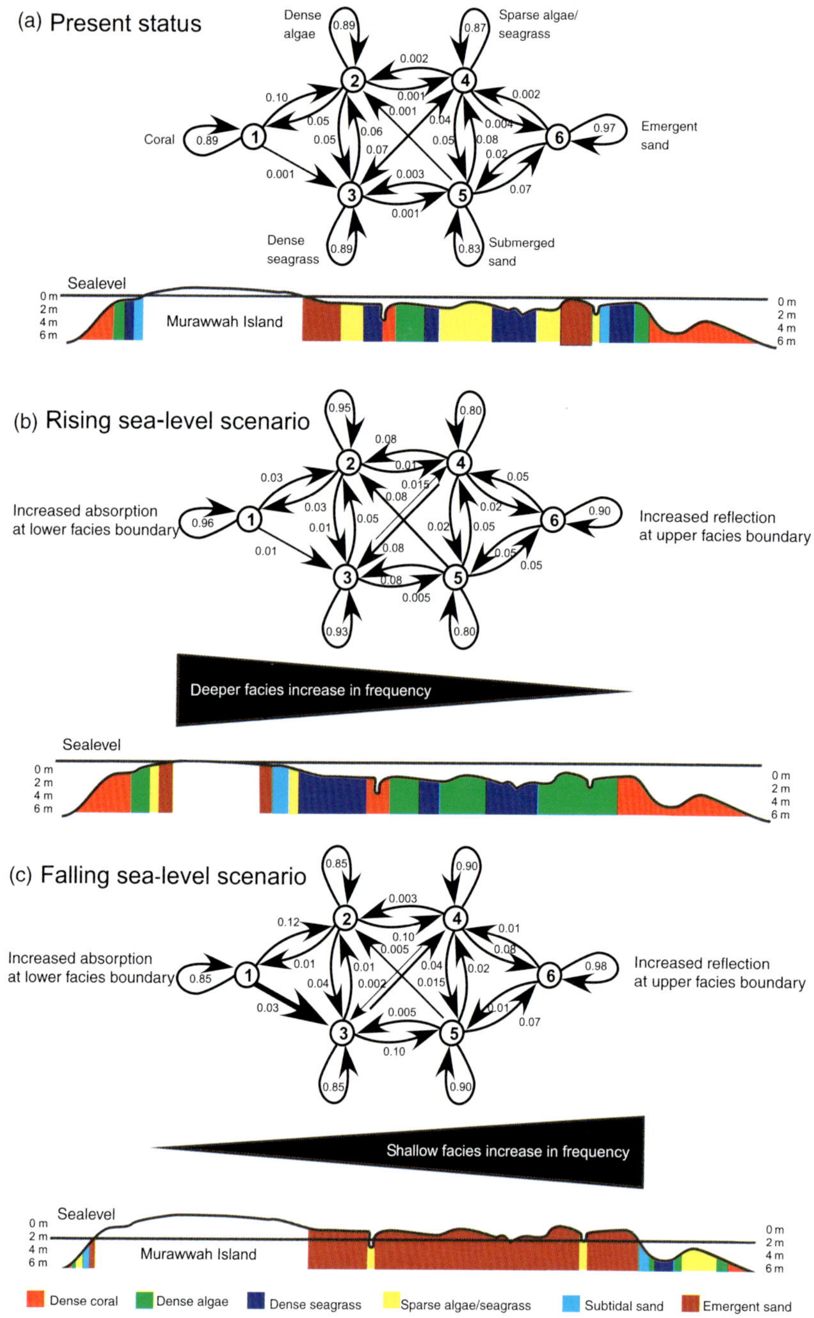

Fig. 14. If we accept that the spatial digraph is also a proxy for temporal transitions, we can predict changes in facies abundance in reaction to a changing environment by modifying the transition probabilities between vertices. (a) The digraph of the present situation from which the changes in sea-level were derived. Facies distribution along a transect shown in Fig. 12c is illustrated. The thickness of arrows in (b) and (c) shows the preferred direction of facies transition and the wedge underneath the digraphs thickens in the direction of preferential facies change. (b and c). Digraphs of transition changes and anticipated changed facies distribution according to the models of sea-level change. Facies frequency in the transect follow the FPVs in Fig. 15.

(the islands are not that big) and the gently sloping edges of the platform will provide much suitable new habitat for algae and corals without much habitat loss due to drowning. Emergent sand would be flooded and colonized by algae and seagrass. The newly flooded areas of formerly dry land (the islands) would become shallow, at least partly emergent sand. In the falling sea-level scenario, emergent and shallow subtidal sands would greatly expand at the cost of the deeper facies, which would move into yet deeper areas (masked pixels) and maybe decline less dramatically than illustrated in Fig. 15. However, since the deeper, masked areas were not included in the analysis, a precise numeric evaluation was not possible. Within the 0–10 m depth zone the decline was adequately represented (Fig. 15).

DISCUSSION

A computationally and conceptually simple way of harnessing spatial patterns in a landscape to extrapolate information for modifying spatial into

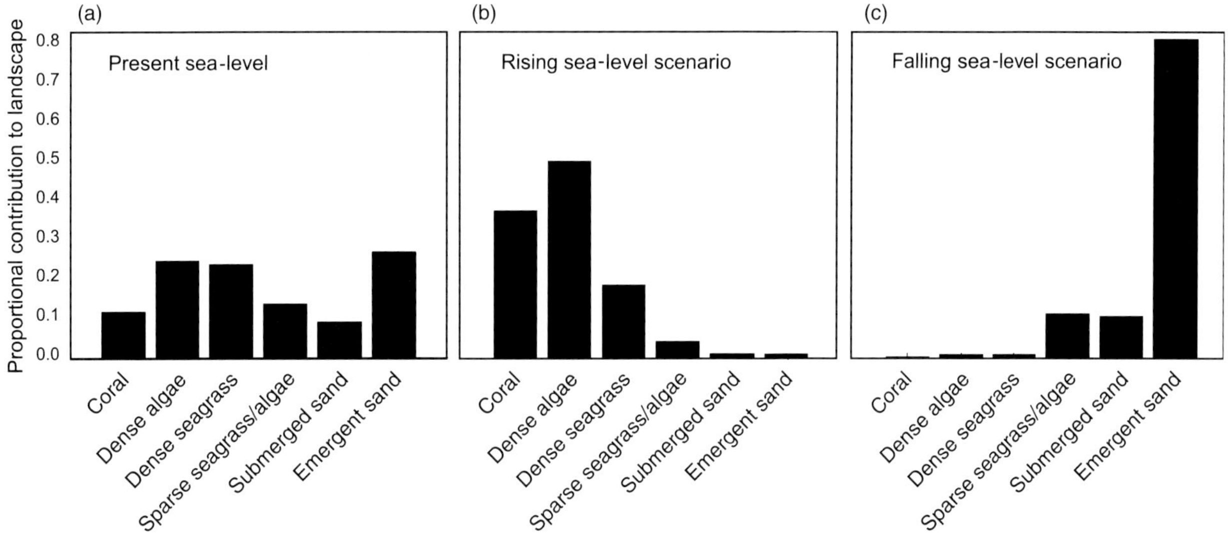

Fig. 15. The fixed probability vectors obtained from the transition probability matrices underlying the digraphs in Fig. 14. The *y* axis shows proportional space cover by the facies, which is the expected distribution of facies in the rising/falling sea-level scenarios. The FPV does not *per se* give information about the time needed to achieve this. The amplitude of the sea-level excursion is defined in the transition probabilities. In our model, sea-level rise and fall is in the range of 1.5–2 m.

temporal transitions has been explored above. Markov chains and weighted digraphs provided the opportunity to forecast changes in facies by manipulating transition likelihoods obtained from lateral patterns in a living landscape. This allows translating knowledge of facies transitions in core or outcrop (= time) for the reconstruction of ancient landscapes (= space) and vice versa (Doveton, 1994; Parks *et al.*, 2000; Elfeki & Dekking, 2001, 2005) and the frequency of facies that can be expected in any time slice. Thus, the estimation of the lateral dimensions of reservoir rocks can be improved, for example. The present study also builds on and expands the concepts developed in Lehrmann & Rankey (1999), who showed the value of incorporating knowledge of lateral facies extension in outcrop to obtain improved understanding of cyclicity.

The sedimentary systems investigated were characterized by facies of fairly uniform thickness for which we assumed comparable sedimentation rates, thus no rate-adjustments were made for translations from space into time and vice versa. Differential sedimentation rates and their realization in Markov chains are discussed by Parks *et al.* (2000) who scaled transitions by a constant relative to sedimentation rate (Schwarzacher, 1969). Bioherms, for example, may accrete faster vertically rather than laterally, and faster than surrounding sediments. Such a situation would lead to a stronger self-transition (loop) in the vertical than the lateral. Additional to the method of Parks *et al.* (2000), the use of embedded Markov chains would also be useful since these ignore the self-transitions that encode within-facies deposition rate. However, such an analysis would limit itself to evaluating adjacencies and lose any information contained in the loops. Coral frameworks investigated in this paper were mostly biostromal (i.e. coral carpets *sensu* Riegl & Piller, 2000), thus not accreting faster to much greater thickness than surrounding sediments and no corrections by loop-multiplication were deemed necessary.

The modern and the Miocene situations explored in this study have in common that both were evaluated only in a small piece of the entire system – whatever was imaged in the Arabian Gulf and the relatively small Miocene outcrop. The Ras Hasyan area was described from a 7 × 1.5 km plane area and the Leitha Limestone from an outcrop 20 m high and 60 m long. This difference in scale and the obvious incompleteness with regards to the entire landscape need not matter since first the relative frequency of facies is of prime concern and second the living Arabian Gulf landscape has fractal properties (Purkis *et al.*, 2005). The correspondence of the FPVs of Miocene and living Arabian Gulf system may suggest that, since the Miocene system was very similar (Riegl & Piller, 2000), it may also have been fractal. Therefore, the different sizes of study areas may indeed be of no importance.

Different outcrop or map scales simply will detect patches of different sizes but in comparative proportion. Thus, in the illustrated case here, it is believed that both systems were similarly patterned in space and therefore the fixed-point vector as the bridge between space and time would hold at all scales and between scales. This would remove the need to compare landscapes and successions of comparable sizes. In this study, facies were at best several metres thick and a small outcrop of few hundred square-metres could be plausibly interpreted as accommodating a comparable number of facies (bearing in mind that the assignment of facies always remains somewhat subjective) as the moderately large modern study area of about 70 km^2. Difficulties may arise when estimating facies of vastly varying thickness, as would be expected to be caused by variable deposition rates. While the assumption of the fractal nature of the landscape may still hold, care would be needed to assure sampling at appropriate scales – which would have to be evaluated using spatial statistics.

Transition weightings between facies depend strongly on environmental factors and a complex interplay exists between ecology and sedimentology. When deriving the temporal facies models from the spatial adjacencies, the ecology of facies-determining organisms (coral, algae, seagrass) guided the choice of which transitions were possible. In particular, the dynamics of the corals, as producers of bioherms, biostromes or rubble in the recent Arabian Gulf and the Miocene Leitha Limestone, strongly influenced the sedimentology by producing well-defined biofacies that were directly equivalent to sedimentary facies. Algae and seagrass interacted with the sedimentary environment in a more passive way, presumably primarily as binders and bafflers. However, simply their presence and cover of unconsolidated sediments would to a certain extent protect the latter from erosion, thus giving them sedimentological relevance. Their clear ecological preferences and distribution on the banks made them useful indicators for facies on the satellite images and proxies in the models. Dense algae clearly was found primarily on the grainstone-dominated platform edges and as secondary cover on dead coral frameworks, while dense and sparse seagrass was found primarily on the finer-grained bank interior (Purser, 1973; Schlager, 2005).

An obvious key environmental factor with the potential of shaping facies distribution is the depth of the water column through which light, thermal and wave energy transit to the benthic biota and sediments. The importance of small-scale sea-level oscillations and their influence on the distribution of facies in shallow water is a matter of active debate (Lehrmann & Goldhammer, 1999; Rankey, 2004; Burgess, 2006). Directional changes in the facies transition likelihoods were introduced to simulate sea-level, and via the fixed probability vector the expected distribution of facies in a new environment was obtained. Such thought-experiments have the potential of increasing our understanding of both lateral as well as vertical facies succession, however, they are only valid if the facies can indeed realistically transit into each other. In this study area, a relatively mild depth gradient was observed (Fig. 15d) with the potential of facies moving relatively unhindered up and down the bank and expanding or shrinking at each other's expense in response to a small change in sea-level. Other than eustatic, such changes could be caused by gentle subsidence and subsequent filling of accommodation space. If seen from this perspective, the model could be used as a quantitative analogue to Ginsburg-type autocyclic sedimentation (Ginsburg, 1971; Hardie & Shinn, 1986; Schlager, 2005).

As can be expected in an offshore bank setting, shallow and nearshore facies become more and more limited in area in the rising sea-level scenario, and in the falling sea-level scenario the shallow infratidal and the deeper facies became restricted. This disadvantage of deeper facies was caused by the model not considering seafloor >8 m. Interestingly, this depth is the boundary between typical high light/shallow and low light/deeper biota and sedimentary environments in the recent Arabian Gulf. Yet by masking, an artificial lower boundary was created that need not exist. Facies occurring in deeper water could simply move downslope and therefore not require as markedly increased reflection (translated into decrease of facies extent) at the lower boundary as in the model. However, since the bathymetric information was derived from the same satellite image as the information about facies, it was limited to the same depth and speculation was restricted to what lay beyond. In particular, slope geometry will have to be considered when parameterizing such models, since it will define how facies can migrate up or down while tracking sea-level. Any cycles generated from such a Markov model would be necessarily non-random and while it would be a good tool to model landscapes from cores and vice versa, it would not be an adequate forward modelling tool for questions of order and noise in sequences (Wilkinson *et al.*, 1997, 2003; Lehrmann & Goldhammer, 1999; Rankey, 2004; Burgess, 2006).

When using the derived temporal models to predict changes in facies, as for the Landsat scene in rising/falling sea-level scenarios presented here, the fixed probability vector gives information about the contribution of facies in the new environment but not about the time needed to achieve this. It is possible to evaluate how many matrix time-steps are needed (i.e. to what power the TPM needs to be raised) until a distribution of facies close to that of the FPV is obtained. Using an absorbing model, it has been shown that it is possible to calculate time until absorption of the transient facies (Fig. 11). However, the timescale is always one of matrix multiplication steps and careful calibration of a matrix step's meaning in years is required. Such information could be obtained in outcrop by dating the length of time required for a facies transition predicted from the spatially derived model. Here, no cores through the living Arabian Gulf environment were available, nor were reliable dating from the Miocene outcrop and therefore time information was lacking. Clearly more work on this aspect is required.

The depth of sea-level fall and height of sea-level rise need to be encoded in the transition probabilities. Here, assumptions of sea-level rise and fall were small, in the range of 1.5–2 m. The biological, and to some degree the sedimentary environments encountered on the banks studied are relatively sensitive depth indicators and occurred within relatively specific depth ranges (Figs 3 and 12). Therefore, they could expand and restrict at the cost of each other and no new facies had to be introduced. Models such as presented here are limited to situations without dramatic changes in facies arrangement and composition and require at least a certain facies–depth relationship, which is not always unequivocal (Rankey, 2004).

The application of Markov chains to sedimentary systems or to ecological succession is not new (Krumbein, 1967; Pielou, 1969; Schwarzacher, 1969; Horn, 1975; Usher, 1979), and its conceptual link to Walther's Law for the forward modelling of sedimentary landscapes has also been noted before (Doveton, 1994; Parks *et al.*, 2000; Elfeki & Dekking, 2005). What is new in this analysis is the addition of graphs for the easy visualization of functional pathways within the system and derivation of a spatial model from the temporal – or vice versa. This allows ecology and sedimentology to be linked, since heed can be paid to ecological dynamics that may need to be introduced into the derived models of facies dynamics via modified transition probabilities. This is an important new step to the existing techniques used for forward modelling with Markov chains – the visualization of spatial relationships and the derived temporal functioning in graphs. Rather than adapting the transition matrix via weightings by scaling factors (Schwarzacher, 1969; Parks *et al.*, 2000) the graph-evaluation step allows a quite realistic derivation of a temporal matrix (or spatial matrix, depending on starting point) from the spatial matrix. The fixed point vector is the link between the spatial and temporal models and any modification must result in statistically similar FPVs. This essentially makes the fixed point vector a quantitative expression of Walther's Law and the unifying link between (matrix – and thus its encoded facies) space and time.

ACKNOWLEDGEMENTS

We thank R.N. Ginsburg for being an inspiration and an enthusiastic as well as patient teacher. He planted more ideas in our minds than we ourselves are aware of. He taught us how to think, ask questions, and keep an open, yet critical, mind. His contribution to our science is unrivalled. And by making us ask the 'so what?' question after every major thought or posit, he made us go and think further and better than we would have otherwise. Work on this paper was sponsored among others by the Austrian Science Foundation (FWF), Austrian Geological Survey, the Dolphin Energy/WWF Coral Reef Project, the Ocean Drilling Project and NOAA grant NA16OA1443 to NCRI at NSU. E. Rankey contributed many useful thoughts, discussions and carefully edited the manuscript. W. E. Piller helped with measured sections and evaluating the Fenk outcrop. We thank M. Chandler, G. Rae, R. al Mubarak, Th. Abdessalaam, A. al Cibahy and F. Launay for unwavering support during work in the Gulf. This is NCRI publication number 83 and a contribution of the NCRI Monitoring Network.

REFERENCES

Alsharhan, A.S. and **Kendall C.G.St.C** (2003) Holocene coastal carbonates and evaporites of the southern Arabian Gulf and their ancient analogues. *Earth Sci. Rev.* **61**, 191–243.

Berggren, W.A., Kent, D.V., Flynn, J.J. and **Van Couvering, J.A.** (1985) Cenozoic geochronology. *Geol. Soc. Am. Bull.*, **96**, 1407–1418.

Bunn, A.G., Urban, D.L. and **Keitt, T.H.** (2000) Landscape connectivity: a conservation application of graph theory. *J. Environ. Manage.*, **59**, 265–278.

Burgess, P.M. (2006) The signal and the noise: forward modelling of allocyclic and autocyclic processes influencing peritidal carbonate stacking patterns. *J. Sed. Res.*, **76**, 962–977.

Cantwell, M.D. and **Forman, R.T.T.** (1993) Landscape graphs: ecological modeling with graph theory to detect configurations common to diverse landscapes. *Landscape Ecol.*, **8**, 239–255.

Carle, F.S. and **Fogg, G.E.** (1997) Modeling spatial variability with one and multidimensial continuous-lag Markov chains. *Math. Geol.*, **29**, 891–918.

Caswell, H. (1982) Stable population structure and reproductive value for populations with complex life cycles. *Ecology*, **63**, 1223–1231.

Caswell, H. (2001) *Matrix Population Models*. Sinauer Associates, Sunderland, 722 pp.

Cox, D.R. & Miller, H.D. (1965) *The Theory of Stochastic Processes*. John Wiley and Sons, New York, 422 pp.

Davis, J.C. (2002) *Statistics and Data Analysis in Geology*, 3rd edn. John Wiley and Sons, New York, 637 pp.

Doveton, J.H. (1971) An application of Markov chain analysis to the Ayreshire coal measures succession. *Scott. J. Geol.*, **7**, 11–27.

Doveton, J.H. (1994) Theory and application of vertical variability measures from Markov chain analysis. In: *Computer Applications in Geology 3* (Eds J.M. Yarus and A.L. Chamberlain). *Am. Assoc. Petrol. Geol.*, 55–64.

Dullo, W.C. (1983) Fossildiagenese im miozänen Leitha-Kalk der Paratethys von Österreich: Ein Beispiel für Faunenverschiebung durch Diageneseunterschiede. *Facies*, **8**, 112.

Elfeki, A. and **Dekking, M.** (2001) A Markov chain model for subsurface characterization: theory and applications. *Math Geol*, **33**(5), 569–589.

Elfeki, A. and **Dekking, M.** (2005). Modelling subsurface heterogeneity by coupled Markov chains: directional dependency, Walther's law and entropy. *Geotechn. Geol. Eng.*, **23**, 721–756.

Evans, G., Murray, J.W., Biggs, H.E.J., Bate, R. and **Bush, P.R.** (1973) The oceanography, ecology, sedimentology and geomorphology of parts of the Trucial Coast barrier island complex, Persian Gulf. *The Persian Gulf. Holocene Carbonate Sedimentation and Diagenesis in a Shallow Epicontinental Sea* (Ed. B.H. Purser). Springer, Berlin, Heidelberg, New York, pp. 233–278.

Flügel, E. (2004) *Microfacies of Carbonate Rock*. Springer, Berlin, 976 pp.

Friebe, J. (1990) Lithostratigraphische Neugliederung und Sedimentologie der Ablagerungen des Badeniens (Miozän) um die Mittelsteirische Schwelle (Steirisches Becken, Österreich) *Jahrb. Geol. BA*, **133**, 223–257.

Friebe, J.G. (1991) Carbonate sedimentation within a siliciclastic environment: the Leithakalk of the Weißenegg Formation (Middle Miocene, Styrian Basin, Austria). *ZB Geol. Paläont.*, **1**, 1671–1687.

Gingerich, P.D. (1969) Markov analysis of cyclic alluvial sediments. *J. Sed. Petrol.*, **39**, 330–332.

Ginsburg, R.N. (1957) Early diagenesis and lithification of shallow-water carbonate sediments in South Florida. In: *Regional Aspects of Carbonate Deposition. SEPM Spec. Publ.*, **5**, 80–100.

Ginsburg, R.N. (1971) Landward movement of carbonate mud: new model for regressive cycles in carbonates (abs.). *AAPG Bull.*, **55**, 340.

Ginsburg, R.N. (1974) Introduction to comparative sedimentology of carbonates. *AAPG Bull.*, **58**(5), 781–786.

Ginsburg, R.N. (Ed.) (1975) *Tidal Deposits. A Casebook of Recent Examples and Fossil Counterparts*. Springer, Berlin, 428 pp.

Ginsburg, R.N. and **Shinn, E.** (1994) Preferential distribution of reefs in the Florida Reef Tract: The past is the key to the present. In: *Global Aspects of Coral Reefs: Health, Hazards and History* (Ed. R.N. Ginsburg). University of Miami Rosenstiel School of Marine and Atmospheric Science, Miami, pp. 21–26.

Hardie, L.A. and **Shinn, E.A.** (1986) Carbonate depositional environments, modern and ancient. Part 3 – tidal flats. *Colorado School of Mines Quart.*, **80**, 74.

Harris, P.M. and **Kowalik, W.S.** (1994) Satellite images of carbonate depositional settings: examples of reservoir- and exploration-scale geological facies variation. *AAPG Methods Exploration Ser.*, **11**, 147.

Hetzinger, J., Halfar, J., Riegl, B. and **Godinez-Orta, L.** (2006) Sedimentology and acoustic mapping of modern rhodolith beds on a non-tropical carbonate shelf (Gulf of California, Mexico). *J. Sedim. Res.*, **76**, 670–682.

Horn, H.S. (1975) Markovian properties in forest succession. In: *Ecology and Evolution of Communities* (Eds M.L. Cody and J.M. Diamond). Harvard University Press, Cambridge, MA, pp. 196–211.

Hubbell, S.P. (2001) *The Unified Neutral Theory of Biodiversity and Biogeography. Princeton Monogr. Population Biol.* **32**, 375.

Iosifescu, M. (1980) *Finite Markov Processes and their Applications*. John Wiley and Sons, New York.

James, N.P. (1984) Shallowing-upward sequences in carbonates. In: *Facies Models* (Ed. R.G. Walker). Geoscience Canada Reprint Series, Kitchener, Ontario, pp. 213–228.

Kemeny, J.G. and **Snell, J.L.** (1960) *Finite Markov Chains*. Springer Verlag, Berlin, 224 pp.

Kirkham, A. (1998) A Quaternary proximal foreland ramp and its continental fringe, Arabian Gulf, UAE. In: *Carbonate Ramps* (Eds V.P. Wright, T.P. Burchette). *Geol. Soc. (Lond.) Spec. Pub.*, **149**, 15–42.

Krumbein, W.C. (1967). FORTRAN IV computer program for Markov chain experiments on geology. *Computer Contributions 13, Kansas Geological Survey, Lawrence, Kansas*, 38 pp.

Krumbein, W.C. and **Dacey, M.F.** (1969) Markov chains and embedded Markov chains in geology. *Int. Assoc. Math. Geol.*, **1**, 79–96.

Krumbein, W.C. and **Greybull, F.A.** (1967) *An Introduction to Statistical Models in Geology*. McGraw-Hill, New York, 475 pp.

Lambeck, K. (1996) Shoreline reconstructions for the Arabian Gulf since the last glacial maximum. *Earth Planet Sci. Lett.*, **142**, 43–57.

Lehrmann, D.J. and **Goldhammer, R.K.** (1999) Secular variation in parasequence and facies stacking patterns of platform carbonates: a guide to application of stacking pattern analysis in strata of diverse ages and settings. In: *Advances in Carbonate Sequence Stratigraphy: Applications to Reservoirs, Outcrops, and Models* (Eds P.M. Harris, A.H. Saller & J.A. Simo). *SEPM Spec. Pub.*, **63**, 51–62.

Lehrmann, D.J. and **Rankey, E.C.** (1999) Do meter-scale cycles exist? A statistical evaluation from vertical (1-D) and lateral (2-D) patterns in shallow-marine carbonates-siliciclastics of the 'fall in' strata of the Capitan Reef, Seven Rivers Formation, Slaughter Canyon, New Mexico. In: *Geologic Framework of the Capitan Reef* (Eds A.H. Saller, P.M. Harris, B.L. Kirkland & S.J. Mazzullo). *SEPM Spec. Pub.*, **65**, 51–62.

Logofet, D.O. and **Lesnaya, E.V.** (2000) The mathematics of Markov models: what Markov chains can really predict in forest successions. *Ecol. Model.*, **126**, 285–298.

Miall, A.D. (1973) Markov chain analysis applied to ancient alluvial plain successions. *Sedimentology*, **20**, 347–364.

Middleton, G.V. (1973). Johannes Walther's Law of the Correlation of Facies. *Geol. Soc. Am. Bull.*, **84**, 979–988.

Papp, A., Cicha, I., Senes, J. and **Steininger, F.** (1978) *M4 – Badenien (Moravien, Wielicien, Kosovien). Chronostratigraphie und Neostratotypen. Miozän der Zentralen Paratethys.* Slowakische Akademie der Wissenschaften, Bratislava, 594 pp.

Parks, K.P., Bentley, L.R. and **Crowe, A.S.** (2000) Capturing geological realism in stochastic simulations of rock systems with Markov statistics and simulated annealing. *J. Sed. Res.*, **70**, 803–813.

Pielou, E.C. (1969) *An Introduction to Mathematical Ecology.* John Wiley and Sons, New York, 245 pp.

Piller, W.E. and **Kleemann, K.** (1991) Middle Miocene Reefs and related facies in eastern Austria. (1) Vienna Basin. *VI International Symposium on Fossil Cnidaria including Archaeocyatha and Porifera, Excursion Guidebook, Excursion B4*, pp. 1–28.

Piller, W.E. and **Vavra, N.** (1991) Das Tertiär im Wiener und Eisenstädter Becken. In: *Exkursionen im Tertiär Österreichs, Molassezone – Waschbergzone – Korneuburger Becken – Wiener Becken – Eisenstädter Becken* (Eds R. Roetzl and N. Nagel). Österreichische Paläontologische Gesellschaft, 161–216.

Piller, W.E., Decker, K. and **Haas, M.** (1996) *Sedimentologie und Beckendynamik des Wiener Beckens.* Sediment 96, 11. Sedimentologentreffen, Exkursion guide, Geologische Bundesanstalt, Wien, 41 pp.

Purkis, S.J. (2005) A 'reef-up' approach to classifying coral habitats from Ikonos imagery. *IEEE Trans. Geosci. Remote Sens.*, **43**, 1375–1390.

Purkis, S.J. and **Pasterkamp, R.** (2004) Integrating *in situ* reef-top reflectance spectra with Landsat TM imagery to aid shallow-tropical benthic habitat mapping. *Coral Reefs*, **23**(1), 5–25.

Purkis, S.J. and **Riegl, B.** (2005) Spatial and temporal dynamics of Arabian Gulf coral assemblages quantified from remote-sensing and *in situ* monitoring data (Jebel Ali, Dubai, U.A.E.). *Mar. Ecol. Progr. Ser.*, **287**, 99–113.

Purkis, S.J., Riegl, B. and **Andréfouët, S.** (2005) Remote sensing of geomorphology and facies patterns on a modern carbonate ramp (Arabian Gulf U.A.E.). *J. Sed. Res.*, **75**, 861–876.

Purser, B.H. (1973) Sedimentation around bathymetric highs in the southern Persian Gulf. *The Persian Gulf. Holocene Carbonate Sedimentation and Diagenesis in a Shallow Epicontinental Sea* (Ed. B.H. Purser). Springer, Berlin, Heidelberg, New York, pp. 157–179.

Purser, B.H. and **Evans, G.** (1973) Regional sedimentation along the Trucial Coast, SE Persian Gulf. *The Persian Gulf. Holocene Carbonate Sedimentation and Diagenesis in a Shallow Epicontinental Sea* (Ed. B.H. Purser). Springer, Berlin, Heidelberg, New York, pp. 211–233.

Purser, B.H. and **Seibold, E.** (1973) The principal environmental factors influencing Holocene sedimentation and diagenesis. *The Persian Gulf. Holocene Carbonate Sedimentation and Diagenesis in a Shallow Epicontinental Sea* (Ed. B.H. Purser). Springer, Berlin, Heidelberg, New York, pp. 234–269.

Rankey, E.C. (2004) On the interpretation of shallow water carbonate facies and habitats: how much does water depth matter? *J. Sed. Res.*, **74**, 2–6.

Rankey, E.C. and **Morgan, J.** (2002) Quantified rates of geomorphic change on a modern carbonate tidal flat, Bahamas. *Geology*, **30**, 583–586.

Rasser, M.W. and **Riegl, B.** (2002) Holocene coral reef rubble and its binding agents. *Coral Reefs*, **21**, 57–72.

Riegl, B. (1999) Coral communities in a non-reef setting in the southern Arabian Gulf (Dubai, UAE): fauna and community structure in response to recurrent mass mortality. *Coral Reefs,* **18**, 63–73.

Riegl, B. (2002) Effects of the 1996 and 1998 SST anomalies on corals, coral diseases and fish in the Arabian Gulf (Dubai, UAE). *Mar. Biol.*, **140**, 29–40.

Riegl, B. and **Piller, W.E.** (2000) Biostromal coral facies – a Miocene example from the Leitha Limestone (Austria) and its actualistic interpretation. *Palaios*, **15**, 399–413.

Riegl, B., Korrubel, J.L. and **Martin, C.** (2001) Mapping and monitoring of coral communities and their spatial patterns using a surface-based video method from a vessel. *Bull. Marine Sci.*, **69**, 869–880.

Roberts, F.S. (1976) *Discrete Mathematical Models.* Prentice-Hall, Englewood Cliffs, NJ, 559 pp.

Schlager, W. (2005) Carbonate sedimentology and sequence stratigraphy. *Concepts Sedimentol. Paleontol.*, **8**, 200.

Schwarzacher, W. (1969) The use of Markov chains in the study of sedimentary cycles. *Math. Geol.*, **1**, 17–39.

Sheppard, C.R.C., Price, A.R.G. and **Roberts, C.M.** (1992) *Marine Ecology of the Arabian Region: Patterns and Processes in Extreme Tropical Environments.* Academic Press, London, 257 pp.

Shinn, E.A. (1969) Submarine lithification of Holocene carbonate sediments in the Persian Gulf. *Sedimentology*, **12**, 109–144.

Steininger, F.F. and **Piller, W.E.** (1999) Empfehlungen (Richtlinien) zur Handhabung der stratigraphischen Nomenklatur. *Courier Forschungsinstitut Senckenberg*, **209**, 19.

Stumpf, R.P., Holderied, K. and **Sinclair, M.** (2003) Determination of water depth with high-resolution satellite imagery over variable bottom types. *Limnology and Oceanography*, **48**, 547–556.

Toomey, D.F. (1981) *European Fossil Reef Models. SEPM Special Publication*, **30**, 450 pp.

Tucker, M.E. and **Wright, V.P.** (1990) *Carbonate Sedimentology*. Blackwell, Oxford, 482 pp.

Turner, M.G., Gardner, R.H. and **O'Neill, R.V.** (2001) *Landscape Ecology in Theory and Practice. Pattern and Process*. Springer, New York, 400 pp.

Uchupi, E., Swift, S.A. and Ross, D.A. (1996) Gas venting and late quarternary sedimentation in the Persian (Arabian) Gulf. *Mar. Geol.*, **129**, 237–269.

Urban, D.L. (2005) Modelling ecological processes across scales. *Ecology*, **86**, 1996–2006.

Urban, D.L. and **Keitt, T.** (2001) Landscape connectivity: a graph-theoretic perspective. *Ecology*, **85**, 1205–1218.

Urban, D.L. and **Wallin, D.O.** (2002) Introduction to Markov models. In: *Learning Landscape Ecology. A Practical Guide to Concepts and Techniques* (Eds E. Gergel and M.G. Turner). Springer, New York, pp. 35–48.

Usher, M.B. (1979) Markovian approaches to ecological succession. *J. Anim. Ecol.*, **48**, 413–426.

Wardle, G.M. (1998) A graph theory approach to demographic loop analysis. *Ecology*, **79**, 2539–2549.

Wilgus, C.K., Hastins, Posamentier, H., Van Wagoner, J., Ross, C.A. and **Kendall, C.G.St.C.** (1989) Sea-level changes: an integrated approach. *SEPM Special Publication*, **42**, 407.

Wilkinson, B.H., Drummond, C.N., Rothman, E.D. and **Diedrich, N.W.** (1997) Stratal order in peritidal carbonate sequences. *J. Sed. Res.*, **67**, 1068–1082.

Wilkinson, B.H., Merrill, G.K. and **Kivett, S.J.** (2003) Stratal order in Pennsylvanian cyclothems. *Geol. Soc. Am. Bull.*, **115**, 1068–1087.

Int. Assoc. Sedimentol. Spec. Publ. (2009) **41**, 361–372

Evaluating validity and reliability in high-resolution stratigraphic analysis

CARL N. DRUMMOND and LAILAH A. MARLOW

Department of Geosciences and Office of Research and External Support, Indiana University–Purdue University Fort Wayne, Fort Wayne, IN 46805-1499 260 481 5750, USA (E-mail: drummond@ipfw.edu)

ABSTRACT

The analysis of stratigraphic sections has always been conducted through an integration of quantitative and qualitative observations. Recently, new techniques have been put forward as mechanisms for extracting high-resolution quantitative data from outcrops for the purpose of assessing controls on accommodation-space change. Evaluation of the validity and reliability of these analytical techniques provides an important framework for future implementation. First, the discrete facies rank technique is employed by stratigraphers as a mechanism for the construction of a proxy sea-level curve. Exclusion of highly diagnostic but thin or infrequently occurring facies, as driven by sampling interval size, will potentially have an impact on the results of analysis of stratigraphic organization. The probability of inclusion of a facies in a dataset of equally spaced lithological observations is controlled by the thicknesses of the occurrences of the lithology as well as its recurrence frequency in the section. Second, the grey-scale analysis of outcrop photographs is used as a quantitative proxy for lithological variation. Evaluation of this technique indicates that it is highly dependent upon the geometric relationships between the outcrop and the focal plane of the camera. Geometric foreshortening can result in quantitative distortion of the grey-scale series, and thus has impacts on interpretations drawn from the analysis of the data series. Additionally, scale-dependent averaging of grey-scale values across a row of pixels can result in quantifiable distortion in the spectral characteristics of the data series. While both the discrete facies rank technique and grey-scale analysis of outcrop photographs can provide useful mechanisms for the collection of uniformly spaced stratigraphic data, great care must be taken in how these techniques are used and how results from the analysis of resultant data are interpreted.

Keywords Cyclicity, Milankovitch, photographic interpretation, Fourier analysis.

INTRODUCTION

The analysis of stratigraphic successions has typically been achieved through application of an array of quantitative and qualitative techniques. Bed thickness, grain size, dip direction of cross-bedding and many other types of stratigraphic and sedimentological observations are intrinsically quantitative in their character. Measurements are made within the limitations of accuracy and precision and the resultant data are evaluated using a wide range of numerical techniques. Conversely, selection of the number and type of lithostratigraphic elements observed in a succession and their palaeoenvironmental relationships are subjective and qualitative measures of the nature of the stratigraphic record. Taken together, qualitative and quantitative analyses have been utilized in a variety of ways to evaluate spatio-temporal variation in depths, durations and modes of deposition. Stratigraphic analysis has likewise been complicated by the several orders of magnitude difference between the scales and resolutions of outcrop and subsurface observations. Subsurface methods return continuous records of geophysical parameters that are readily evaluated by a wide range of statistical methods. Conversely, field-based observations are commonly more discrete and more qualitative in their character. Recently, several techniques for converting some of the complex qualitative aspects of outcrop observation into a continuous data series suitable for spectral analysis have been advanced. It is the purpose of this paper to describe and evaluate two analytical techniques that have received use in the quantitative stratigraphic literature, with specific attention to assessing the validity and reliability of analytical results derived from them.

DISCRETE FACIES RANK TECHNIQUE

One of the central objectives of stratigraphy is the analysis of lithologically complex stratigraphic successions. The selection of an appropriate and representative suite of lithofacies, the parsing of the section according to occurrences of these lithofacies, and the ranking of the lithofacies by depth or some other palaeoenvironmentally significant characteristic requires a comprehensive synthesis of all the observable sedimentological and stratigraphic characteristics possessed by the succession. Of the many and varied depositional environments recognized in sedimentary rocks, shallow-water and peritidal carbonate settings present perhaps the greatest degree of textural and compositional variability and complexity. From the Proterozoic to the modern, sediments deposited in temperate and tropical carbonate environments display an astonishing variety of sedimentary features. In some cases, individual facies can be tied directly to specific depth ranges or palaeoenvironments. For instance, the framestone textures of reef environments are readily recognizable, as are the desiccation features of supratidal laminites. Difficulty arises, however, in distinguishing and ranking the depth or positional relationships between the large number of apparently substitutable subtidal facies that form across the range of depths affected by wave action. Such lagoonal, shallow-shelf or ramp sediments make up significant percentages of ancient carbonate rocks and therefore have been the subject of extensive study and debate.

Ranking of sedimentary facies by depth is a common and well-established practice in stratigraphic analysis (Fischer, 1964; Olsen, 1986; Bond *et al.*, 1991). Qualitative and quantitative analysis of patterns of depth-dependent facies transitions have likewise been conducted in a wide range of carbonate settings and the analysis and interpretation of carbonate facies cyclicity has been the subject of debate for several decades (Schwarzacher, 1975, 1993). More recently, however, workers have begun to approach the quantitative analysis of depth-ranked successions in a new way. By converting a standard stratigraphic column into a lithofacies series, where the depth-ranked value of the succession is established at a constant thickness interval Δt, the resultant series of stratigraphic data is readily evaluated for spectral characteristics (Fig. 1). This technique has, for example, been used by Preto *et al.* (2001) to argue for the presence of Milankovitch forcing of eustatic sea-level recorded by shallow carbonate sediments of the middle Triassic Latemar platform. In that example, the classic bi-lithological cycles of the Latemar platform were further subdivided into four depth-ranked sub-lithologies. The 160-m Cimon del Latemar section was parsed into 472 occurrences of four sublithofacies with an average thickness of 34 cm. This measured stratigraphic column was then sampled at 0.5 cm intervals to create a ranked lithofacies series of 32,000 equally spaced secondary observations. The resultant series was then taken as a proxy sea-level curve for the duration of the Cimon del Latemar deposition. After tuning the results of a spectral analysis of the lithofacies series to a precessional duration of 21.7 kyr, the Latemar carbonates are then interpreted to have formed under conditions of periodic low-amplitude (2–3 m) sea-level oscillations driven at Milankovitch orbital frequencies.

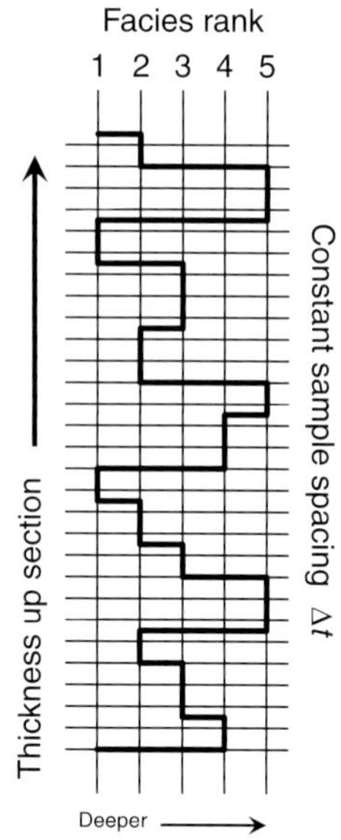

Fig. 1. Graphical example of the discrete facies rank technique where a stratigraphic column is parsed into five distinct facies types, ranked by depth or some other palaeoenvironmentally significant parameter, and then sampled at a constant spacing Δt. This technique allows for the creation of a continuous proxy sea-level curve derived from standard stratigraphc data.

While this approach has yielded compelling results in the case of the Latemar succession, there remain several important questions to be considered regarding the validity and reliability of the

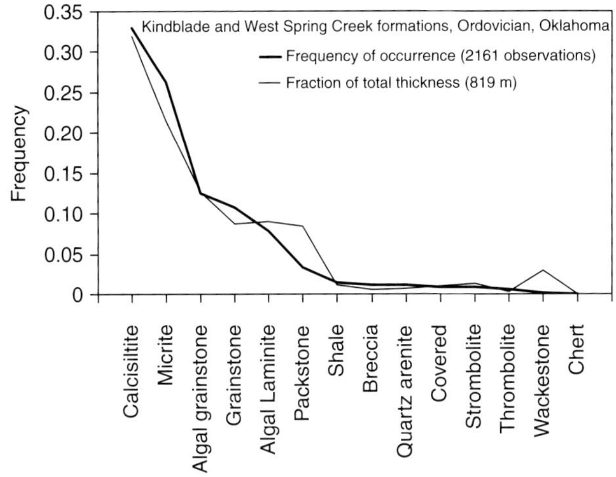

Fig. 2. Distribution of relative abundance of the 14 facies types identified by Wilkinson *et al.* (1997) in the Lower Ordovician Kindblade and West Spring Creek Formations along Interstate 35 in the Arbuckle mountains north of Ardmore, Oklahoma, USA. The fraction of total thickness was determined using standard section measuring techniques by summing the thicknesses of all occurrences of a given facies. The frequency of occurrence was calculated as the number of times that facies was identified when the section was measured a second time using a fixed interval observation technique.

discrete facies rank technique and its results. The first and most obvious question relates to the selection of a vertical sampling interval Δt. In a study of the recurrence characteristics of peritidal carbonate facies in the Lower Ordovician Kindblade and West Spring Creek Formations, Wilkinson *et al.* (1997) carried out detailed section measuring along Interstate 35 in the Arbuckle mountains north of Ardmore, Oklahoma. In this study, 2161 lithologically distinct beds were identified in 819 m of continuous section by standard section measuring procedures. Fourteen lithostratigraphic units were recognized in this succession. Subsequently, the outcrop was remeasured in the field utilizing a discrete sampling methodology where lithological composition was determined by direct observation every 0.38 m up-section throughout the succession. This sampling interval was derived from the average thickness of the 2161 beds measured using standard section measuring techniques. These two techniques produce similar, albeit not identical, results (Fig. 2). Using the discrete sampling method, micrite and grainstone facies are slightly overrepresented relative to their percentage of the total stratigraphic column thickness, while packstone and wackestone facies are underrepresented relative to their total thickness. It is expected, when utilizing the fixed-interval sampling technique that lithofacies exhibiting average thicknesses at or less than the average for the entire succession to be under sampled. Indeed this is found to be the case for both wackestone (0.39 m) and packstone (0.24 m) facies. Interestingly, however, the grainstone facies is also thinner than the outcrop average (0.22 m) but is actually overrepresented relative to its total thickness. Two factors lead to this observation: first, grainstones make up over 20% of the total number of observed beds but only 9% of the total thickness and second, grainstones are texturally distinct on a weathered outcrop and are therefore less likely to be misidentified. Thus, the probability of inclusion in a dataset composed of equally spaced samples of lithology is controlled by both the thicknesses of the occurrences of the lithology as well as its recurrence frequency in the section.

In the case of the Ordovician study (Wilkinson *et al.,* 1997), the choice of the sampling interval size was driven by an observed stratigraphic characteristic, average bed thickness. Additionally, because the technique used was the direct resampling of the outcrop, there were limitations associated with the collection of this primary lithological data. That is, resampling at shorter stratigraphic intervals, while possible, was impracticable given the vast thickness exposed at the Ardmore outcrop. Without question, this technique would not be considered ideal for quantitative analysis of spatio-temporal characteristics due to the relatively broad spacing between successive samples (0.38 m) but it does provide an interesting example of how interval sampling of an outcrop can result in a different understanding of the stratigraphy than does a traditional boundary measuring technique. Conversely, the analysis of the Latemar section (Preto *et al.*, 2001) utilized a very small (0.5 cm) secondary resampling interval for the collection of a proxy data series.

Within the Cimon del Latmar section, 161 of the 472 recognized lithological units are vadose altered dolostones and caliche soils, a facies that is equivalent to the vadose diagenetic cycle caps of Goldhammer *et al.* (1987, 1990). Representing intervals of platform exposure, this facies provides the most reliable measure of relative accommodation space of the four subfacies recognized by Preto *et al.* (2001). The stratigraphic distribution of Facies 1 throughout the Cimon del Latemar section is found to be fairly uniform, as might be expected for a facies that makes up over

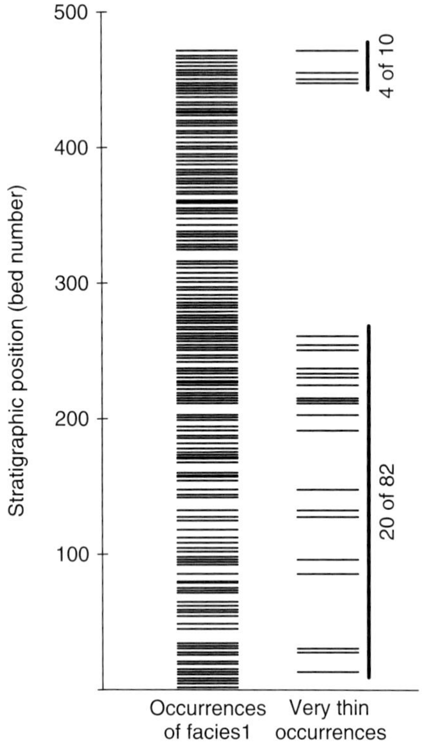

Fig. 3. Stratigraphic distribution of occurrences of the vadose altered dolostone and caliche subfacies (Facies 1) in the Cimon del Latemar section (Preto *et al.*, 2001). The left column illustrates the positions of occurrences of the facies, while the right column describes the position of those occurrences that are thinner than or equal to the scale of the secondary resample spacing (0.5 cm). A total of 24 of the 162 occurrences of Facies 1 are at or below the spacing interval.

a third of the observed units (Fig. 3). Interestingly, however, despite the very small scale resample spacing of 0.5 cm, nearly 15% of the 162 occurrences of the facies fall at or below the resampling size. Additionally, the stratigraphic distribution of these very thin caps is not uniform throughout the section. Rather, 20 of the first 82 and 4 of the last 10 occurrences are at or below the resolution of the discrete sampling technique (Fig. 3). Clearly, the observation that very thin occurrences of Facies 1 are not uniformly distributed throughout the section is of interest when considering potential controls of variation in long-term rates of accommodation space creation across the Triassic platform.

As the facies most representative of sea-level lowstands, Facies 1 is critical to any reconstruction of the temporal record of accommodation space change across the Latemar platform. The fact that occurrences of this facies could be skipped by the resampling technique raises two important questions about the discrete sampling teachnique. First, and most obviously, is the choice of the resampling interval. Unlike the case of the Ardmore section where there were practical physical limitations to making discrete interval observations throughout an outcrop, when creating a secondary resampled dataset, the size of the resample interval need only be dependent on computational efficiency. Ideally, resampling should be done at or below the Nyquist rate, 0.05 cm in the case of the Cimon del Latemar section. Conversely, if computational limitations would make resampling at very high resolution problematic, an alternative solution is to adjust the offset, or phase of the resampling interval. In the example given above it would be possible to shift the resampling from 0.5, 1.0, 1.5 cm ... to 0.4, 0.9, 1.4 cm.

By shifting the resampling in this way, it becomes possible to select an offset that maximizes inclusion of thin beds in the dataset. Likewise, each resampled dataset could be analysed using any of a variety of statistical techniques to evaluate sensitivity to the chosen resampling interval. Obviously, if comparable results are obtained at all offsets, the impact of skipping a few beds would be minimal.

Unquestionably, selection of an appropriate sampling interval is critical to any quantitative analysis of a section sampled using the discrete facies rank technique. What impact has exclusion of some of the thinnest occurrences of Facies 1 had on interpretations of sea-level change over the Latemar platform? It is reasonable to suppose that at least some aspects of the power and line spectrogram would be impacted by changes in the size of the resample interval. While as many as 24 occurrences of Facies 1 could be skipped by the technique used by Preto *et al.* (2001), in fact only eight were with no appreciable change in the resulting power spectrum when reanalysed at the Nyquist rate (L. Hinnov, personal communication, 2007).

While the exclusion of some of the occurrences of supratidal caliche soils from the numerical analysis of Preto *et al.* (2001) apparently did not adversely effect their interpretations, the exclusion of any occurrence of a palaeoenvironmentally significant subfacies such as the vadose cap and caliche soils of the Latemar has the potential to affect quantitative analysis. Facies 1 can be formed only when sea-level drops below the platform margin, making every occurrence, irrespective of thickness, significant. That is, it is conceivable that sea-level dropped below the platform margin and failed to produce a supratidal record, but the facies cannot be formed without subareal exposure. Therefore, the observed number of occurrences of Facies 1 represents the minimum number of intervals of platform exposure. Given that 162 occurrences of Facies 1 were recognized in the Cimon del Latemar section, and the duration of accumulation for that succession has been interpreted to range from ~0.6 Myr (Zühke *et al.*, 2003) to 3.1 Myr (Preto *et al.*, 2001) the most elementary calculation would indicate that the Latemar platform experienced subareal exposure at an average recurrence interval ranging from 3700 years to some 19,100 years, values that are midway between the short and long precessional frequencies calculated by Preto *et al.* (2001). Exclusion from this calculation of the eight occurrences not included in the Preto *et al.* (2001) analysis lengthens the recurrence estimate by 5%.

Establishing meaningful thickness-duration relationships for this Facies 1 is extremely difficult (Drummond, 2002) because of its importance in the interpretation of platform exposure and submergence, and because vadose diagenesis and caliche formation are not constrained to follow the law of superposition due to the fact that diagenetic alteration and soil formation occurs in a penetrative way as an alteration of existing sediments. Thus, the spatio-temporal occurrence of this facies is much more significant than the thickness of individual occurrences. Accounting for potential computational biases resulting from exclusion of very thin but palaeoenvironmentally significant facies is a critical topic for future work. This could be especially critical in peritidal ramps exhibiting more lithological complexity than the bilithological cycles found at Cimon del Latemer.

GREY-SCALE ANALYSIS OF OUTCROP PHOTOGRAPHS

Recently stratigraphic analyses have been conducted using digital outcrop photographs as a data source. In this technique, a field photograph of an outcrop is sampled along a continuous transect to create a series of grey-scale intensity value measurements. Since the data are generated from a pixilated image, the result is a uniformly sampled continuous series of data. Such a data series could, in the ideal, present a very useful foundation for quantitative analysis.

As field geologists, stratigraphers and sedimentologists are well-trained in the techniques of observation and interpretation of outcrops. One of the most important aspects of field geology is the ability to observe and understand the geological 'signal' that is often overprinted by various forms of 'noise' that can impede high-quality interpretations (Fichter, 1987). Specifically, the effects of weathering, erosion, and cover by scree or vegetation can adversely affect the quality of observation and interpretation in the field. These problems are compounded when the dataset is reduced from an integrated sedimentological and stratigraphic analysis to a set of pixellated grey-scale data drawn from across an image of an outcrop.

The grey-scale technique assumes that there is a direct relationship between lithological composition and grey-scale value. As discussed by Schwarzacher (2005), the shade of grey of any pixel in a photograph is the result of a complex set of variables that include lithology, weathering and cover. Additionally, there are less obvious factors that influence grey-scale intensity that include sharpness of contrast due to ambient light conditions, angle of the sun and depth of shadows, as well as the effect of any ground or surface waters on the outcrop. Taken together, these factors make interpretation of a stratigraphic section by grey-scale analysis an extremely complex and potentially unreliable process.

In a study of the Upper Triassic Dachstein Limestone of Italy, Cozzi *et al.* (2005) conducted a spectral analysis of 830 grey-scale measurements from a field photograph. The three segments of the Picco di Carnizza outcrop included in the study cover approximately 270 m of section and are interpreted to be composed of 112 Lofer-type cycles with an average thickness of 2.41 m. Given the scale of the image and the resolution of the scan, each pixel is calculated to represent 0.135 m of section and thus, on average, a cycle is composed of 18 pixels or approximately 0.225 mm on the photograph (Fig. 4). Thus, this technique creates over a 10,000 fold reduction in stratigraphic resolution, prior to the spectral analysis of the data. Variations in the grey-scale values of this record are interpreted by Cozzi *et al.* (2005) to be the product of differential weathering of the cycles, wherein lighter shades are taken to be resistant ledges of subtidal limestone and the darker shades are recessive weathered and vegetated portions of the outcrop.

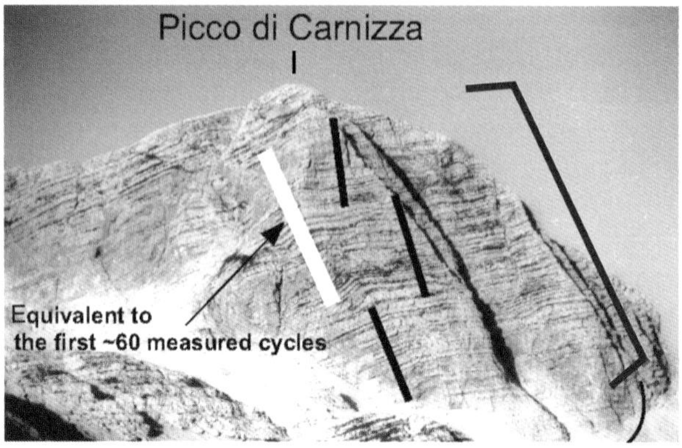

Fig. 4. Field photograph of the Picco di Carnizza section illustrating the lines of grey-scale analysis used by Cozzi *et al.* (2005, Geological Society of America data repository item 2005157). The three lines of section (dark, sub-parallel lines) in this image were used as the basis for the spectral analysis of Milankovitch periodicities in the succession. Sampled at 2000 dpi, the grey-scale data series derived from this image is the product of over a 10,000-fold reduction in stratigraphic resolution.

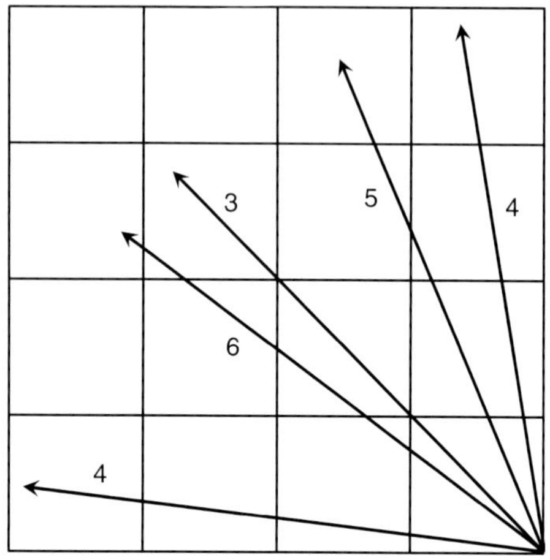

Fig. 5. Graphical representation of the significance of differences in the angle between a rectilinear pixel grid of a scanned image and the apparent up-section direction recognized on a field photograph. Five traces of equal length are positioned on the pixel grid where the number of cells intersected by those traces ranges from a minimum of three to a maximum of six. Such dependence upon angular position can have a major control on the interpretation of spatio-temporal characteristics interpreted from a grey-scale series.

ANGLE OF GREY-SCALE ANALYSIS

The analysis of grey-scale data taken from outcrop photographs is open to several criticisms. First, the scale of field photographs (e.g. Fig. 4) can result in an astonishingly large reduction in the resolution of data available for analysis. Second, variation in grey-scale value could be interpreted to be driven by differential weathering taken to be facies-specific in its character. Yet, establishing a correlation between grey-scale intensity and directly observed lithological composition is difficult. That is, without a set of baseline data that ties composition directly to grey-scale intensity through field checking, the validity of any analysis drawn from the data is at best suspect.

The somewhat obvious drawbacks discussed above mask other flaws in the grey-scale approach. An essential requirement of the spectral analysis technique is that there be a clear understanding of the spatio-temporal spacing of the data in the series. Figure 4, however, illustrates several problems. First, the composite section used in the spectral analysis was created from three separate lines taken at different angles across the image. Apparently, this was done in an attempt to keep the line of analysis perpendicular to strike as it appears on different faces of the outcrop. The result, however, is that the sampling across the rectilinear grid of pixels results in different amounts of vertical section per pixel (Fig. 5). As illustrated by the five arrows of equal length, the angle between the section and the grid controls how many pixels are represented in the section, and thus the average thickness represented by each pixel. In this simple example, the number of intersected pixels ranges from a maximum of six to a minimum of three. In the actual scan of the Picco di Carnizza, the differences in angle between the three traces are not as large as illustrated in Fig. 5, but the integrated effect over hundreds of pixels is significant even when dealing with small angular differences. As such, differences in the pixel–thickness relationship between outcrop segments introduce artificial trending to the dataset. The angle of section problem, while real, is perhaps minor in the Picco di Carnizza example. There remains, however, a second more significant problem associated with pixel scaling in the analysis.

ANGLE OF THE OUTCROP

The use of field photography as a data source for quantitative analysis presents a number of complex problems, the most obvious and most vexing is the complex geometry of the outcrop surface. Natural outcrops such as the Picco di Carnizza or artificially created outcrops such as road cuts are subjected to the effects of

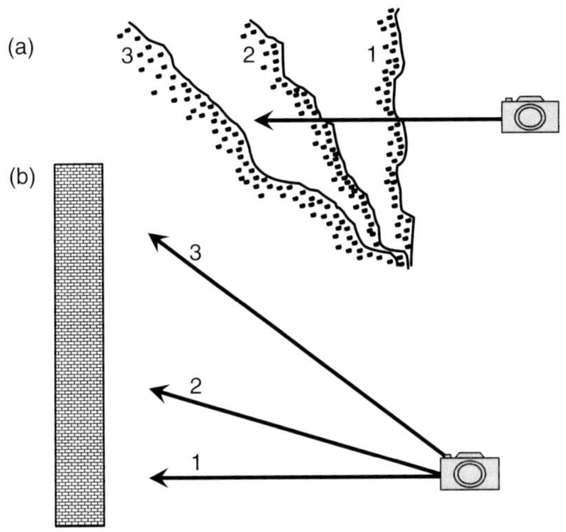

Fig. 6. Cartoon illustrating the modelling of outcrop step-back by changing the angle between a vertical brick face and the focal plane of the camera.

weathering and erosion. As a result, an outcrop surface nearly always steps back at an angle from the vertical. This recession of the face of the rocks away from an observer results in a foreshortening of rocks lower in the section relative to those in stratigraphically higher positions. In order to evaluate the effect of outcrop angle on grey-scale analysis of field photographs, a simple experiment was conducted. Modelling of different outcrop surface geometries was accomplished by photographs taken of a brick wall with the focal plane of the camera tilted at various angles relative to the vertical brick surface (Fig. 6).

With the focal plane held parallel to the brick surface, the degree of vertical distortion of the image is minimized. As the angle is changed, and the camera tilted up, the upper portion of the image becomes compressed relative to the lower portion thus modelling the effect produced by the geometry of back-stepping outcrop surface photographed with a vertical focal plane (Fig. 6). By changing the angle between the brick surface and the camera's focal plane from 0° to 20° to 30°, it is possible to evaluate the role of outcrop surface geometry on the grey-scale technique. Photographs of the brick surface were scanned, cropped to a half-brick width, and then sampled following the process described by Cozzi *et al.* (2005) using the image analysis shareware Scion Image.

In all three trials, the grey-scale series clearly shows the presence of brick-mortar couplets (Fig. 7a, d and g). Using the position of local minima in the grey-scale values of the data series allows for the thickness of each couplet to be calculated (average actual couplet thickness 6.8 cm, 5.5 cm brick, 1.3 cm mortar). When the photograph was taken with the focal plane perpendicular to the wall, the couplets were found to have an average thickness of 31 pixels, resulting in a resolution of approximately 2.2 mm of wall length per pixel. Importantly, because the focal plane is parallel to the wall surface, there is no trend in the thickness data (Fig. 7b). Fourier power spectrum calculated using standard techniques (Davis, 1986) from grey-scale data derived from the parallel focal plane photograph displays a prominent single peak at 31 pixels. Thus, when the 'outcrop' is vertical and the focal plane is placed in a parallel orientation, the positional data contained within the grey-scale series is reliable. Unfortunately, this is not the case when there is an angle between the focal plane and the surface.

When the camera's focal plane is turned 20° to the wall surface, as illustrated in Fig. 6, the foreshortening that is produced results in a strong linear trend of decreasing apparent brick–mortar couplet thickness up-section, as measured in pixels (Fig. 7e). Interestingly, while the cyclic pattern of the brick-mortar couplets is still evident in the grey-scale series, the trending in the data produces a significant degradation in the spectral power of the couplet (Fig. 7f). The single strong peak found in the parallel data is broken into a number of smaller peaks of much lower spectral power positioned from 33 to 26 pixels, the range of apparent couplet thicknesses found in the series (Fig. 7e).

If the angle between the focal plane and the surface is increased to 30°, an angle of step-back from the vertical fairly common to natural outcrops such as the Picco di Carnizza, the validity of interpretations drawn from grey-scale analysis becomes even more questionable. With increasing angle, the slope of the thickness trend increases, resulting in a further degradation of the quality of the spectral power of the couplet (Fig. 7i). Now, the known peak at 31 pixels is further reduced in power, with additional peaks down to a thickness of 20 pixels present. Obviously, in this simple example it would be possible to detrend the data so as to return a result similar to the parallel example (Fig. 7a–c). In a natural setting, however, establishing the correct stretching factor for detrending would be extremely difficult.

Unquestionably, the analysis of grey-scale renderings of field photographs can be considered valid only when great care is taken in the selection of the geometry of the photographic angle relative

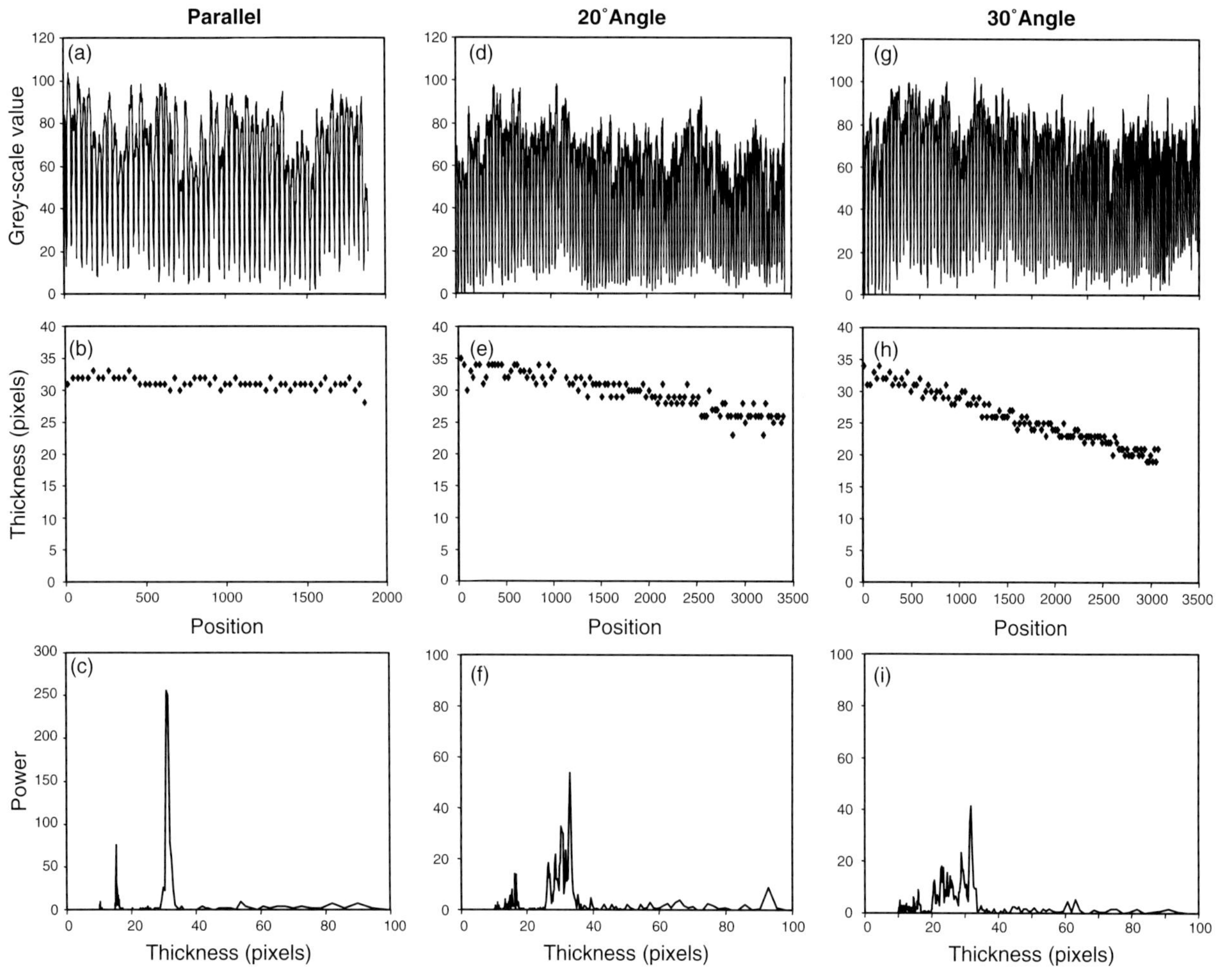

Fig. 7. Results of the outcrop step-back experiment. When the camera's focal plane is held parallel to the wall surface, the grey-scale series defining the brick–mortar couplets shows no spatial trend (a and b) and results in a strong periodic signal at a thickness of 31 pixels (c). As the angle between the camera and the wall is increased to 20° and 30°, foreshortening of the lower portion of the photograph relative to the upper part results in compression of the thicknesses of the brick–mortar couplets (d, e, g, h). Additionally, as the section becomes compressed, there is a marked degradation of the periodic signal (f and i). Given the simplicity and constancy of the brick–mortar couplets in this model, and the degree of signal degradation observed, foreshortening of an outcrop driven by step-back can have a major impact on any quantitative analysis of grey-scale data series.

to the stratigraphic section. As illustrated in the previous discussions, variation in the strike geometry across the outcrop as well as the degree of outcrop step-back can have an impact on the reliability of the positional relationships within the grey-scale data and, as such, can affect the validity of any conclusions drawn from a quantitative analysis of grey-scale data. In fact, the spectrogram derived from the Picco di Carnizza section (Cozzi *et al.*, 2005, Fig. A) shows a frequency shift across the section. This shift in frequency is at least as likely to be driven by such a photographic artefact as it is to reflect any real long-term changes in the rate of accommodation space creation during deposition.

THE ROLE OF LATERAL VARIABILITY IN GREY-SCALE DATA

One of the great strengths of traditional field observation is the way in which an experienced geologist is able to draw data from along a bed's lateral exposure. Given the complexity of the stratal expression of peritidal carbonate successions, can the grey-scale technique be adapted to integrate lateral stratigraphic content and if so, what impact does such integration have on the analysis of the resultant data series? Interestingly, the image analysis shareware Scion Image will average grey-scale values across a row of pixels. As such, if the geometry of the image is carefully

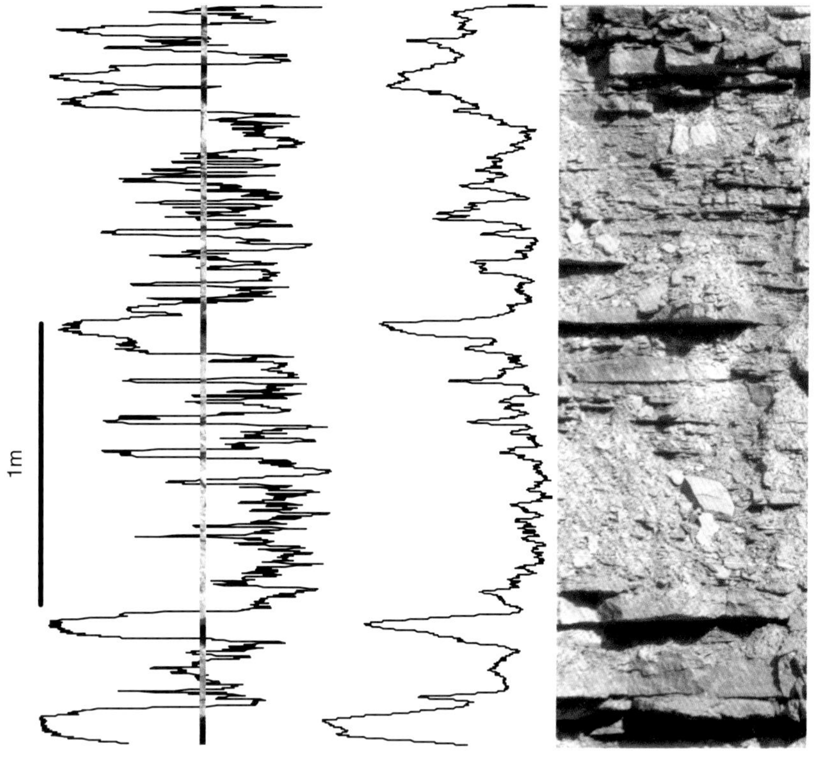

Fig. 8. Cropped photographs and grey-scale traces of a portion of an outcrop of the Kope Formation, Ordovician, northern Kentucky, USA. On the right is a 1-m-wide scan of the outcrop. To the left of the field photograph is the grey-scale trace averaged over each horizontal row of pixels. Farthest left is a composite of a 2-cm-wide scan through the centre of the 1-m-wide photograph and the grey-scale trace derived from that image. Note the significant degree of smoothing in the 1 m grey-scale relative to the 2 cm trace.

selected such that horizontal rows of pixels are parallel to bedding, then it is possible to evaluate the grey-scale character of the outcrop at different width scales.

In order to evaluate the effects of lateral averaging along beds, high-resolution field photographs of the Ordovician Kope and Bellview formations of northern Kentucky were analysed (Figs 8 and 9). Limestones of the Kope and Bellview formations were deposited on a shallow storm-dominated mixed siliciclastic-carbonate ramp. These rocks are part of the well-known Cincinnatian Series and have been studied extensively by a number of workers (Tobin & Pryor, 1981; Jennette & Pryor, 1993; Holland *et al.*, 1997). The Kope and Bellview formations are well-suited for photographic analysis because outcrops along the Alexandria–Ashland highway present a series of horizontally bedded exposures that are readily measured and photographed at high resolution with minimal outcrop step-back. Additionally, the interbedded limestone-shale stratigraphy presents perhaps a less complex lithological setting than other peritidal or subtidal carbonate sequences (Wilkinson *et al.*, 1997; Cozzi *et al.*, 2005) and, as such, is better suited to grey-scale analysis.

The Kope Formation consists of a series of fossil-rich beds ranging from pure grainstones to packstones and wackestones interbedded with fossiliferous shale. Drummond & Sheets (2001) identified the presence of stratigraphic organization within the Kope composite reference section of Holland *et al.* (1997) in the form of stratigraphic clustering of grainstone and packstone beds. As illustrated in Fig. 8, the Kope Formation is relatively shale-rich with prominent ledge-forming grainstone beds and small local scree cones. A field photograph of the Kope Formation was cropped to two sizes; in the first example, the width of the field of view is approximately 1 m while the second consists of approximately 2 cm taken from the centre of the 1 m image. Figure 8 illustrates 2.51 m of the stratigraphic section with a pixel scale of 0.71 mm. Using the shareware Scion Image, these cropped images were subjected to grey-scale analysis. Due to the averaging of pixel values across the wider row, the 1 m wide image produces a smoother grey-scale trace than does the 2 cm wide image. Both of the traces, however, display a great deal of similarity in their gross geometry (Fig. 8). Specific variations between the shape of the traces is due to a combination of lateral lithological variation and increased inclusion of stratigraphic 'noise' in the form of differential weathering, scree cover and vegetation when wider images are used.

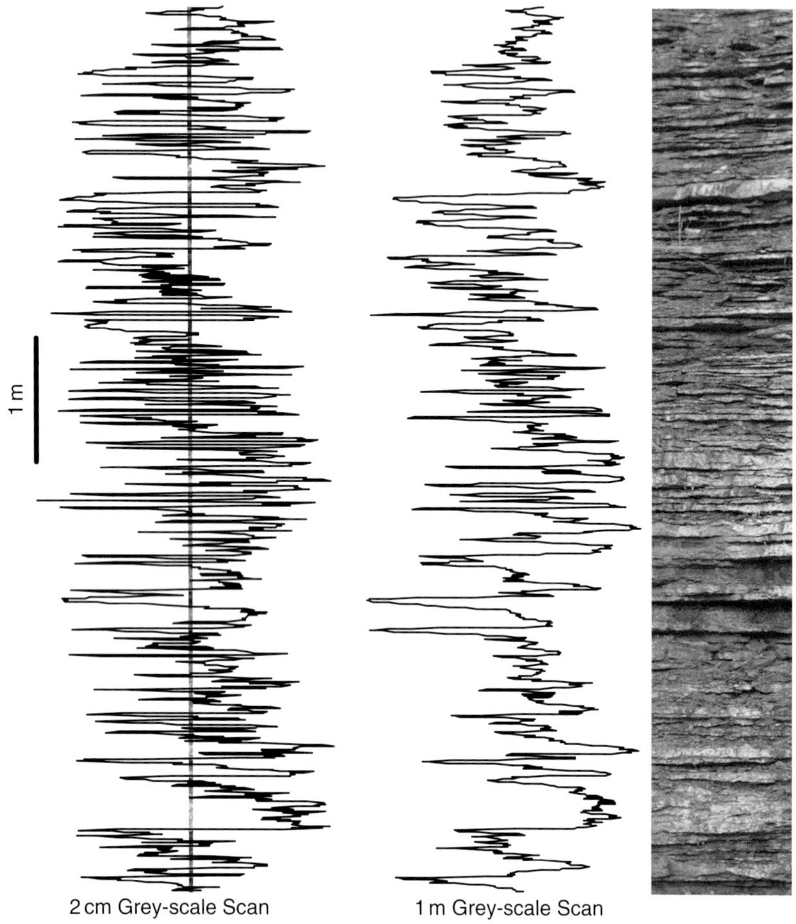

Fig. 9. Cropped photographs and grey-scale traces of a portion of an outcrop of the Bellview Formation, Ordovician, northern Kentucky, USA. On the right is a 1-m-wide scan of the outcrop. To the left of the field photograph is the grey-scale trace averaged over each horizontal row of pixels. Furthest left is a composite of a 2-cm-wide scan through the centre of the 1-m-wide photograph and the grey-scale trace derived from that image. Note that the thinly bedded grainstones of the Bellview Formation produce a trace with much more high-frequency and high-amplitude variation than exhibited by the shale-rich Kope Formation (Fig. 8).

When evaluating the stratal characteristics of the Bellview Formation several observations are readily apparent. First, the Bellview consists of a succession of thinly bedded grainstones and packstones with significantly less interbedded shale than found in the Kope. Second, the ambient light conditions at the time the photograph was taken and the lack of shale results in thinner and less pronounced shadows than in the case of the Kope. Following the procedure discussed above, two scans of the Bellview exposure were subjected to grey-scale analysis. The larger vertical extent of the Bellview outcrop coupled with the thin bedded nature of the grainstones results in a very complex grey-scale trace (Fig. 9). This illustration displays 4.79 m of section with a pixel scale of 3.14 mm. When considering the traces from the 1 m and 2 cm scans, the higher frequency variation in the grey-scale values makes visual comparison between the two somewhat difficult. However, gross trends and sharp breaks in contrast are recognizable in both datasets.

In order to expand the analysis of these grey-scale datasets, Fourier power spectra were calculated for each of the four scans using the standard techniques of Davis (1986). All four examples describe a log-log linear decrease in spectral power with increased harmonic number (Fig. 10). While the overall patterns of these spectra are similar, there are some striking differences. In the Kope example, there is a significant reduction in the slope of the log power-log harmonic number relationship wherein the slope changes from −2.05 to −1.47. This reduction in slope is a reflection of the greater degree of high-frequency variation in the 2-cm-wide scan. That is, there is more spectral power over higher frequencies than is observed in the smoother, low-frequency dominated, 1-m-wide scan (Fig. 8).

The grey-scale values derived from the photograph of the Bellview Formation (Fig. 9) display significantly greater degrees of high-frequency variation than do those taken from the Kope Formation photographs (Fig. 8). This difference is evident not only by visual inspection of the

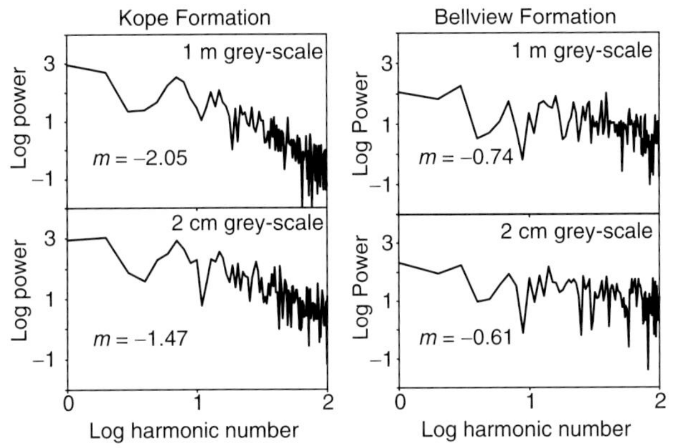

Fig. 10. Power spectrograms of the four grey-scale traces illustrated in Figs 8 and 9. For the 1-m-wide scan of the Kope Formation, the grey-scale pattern exhibits a Brownian-like shape, with log power–log harmonic number slope of −2.05. The 2-cm-wide scan of the Kope Formation, however, displays greater high-frequency variation and thus a shallower slope of −1.47. Conversely, the Bellview Formation grey-scale traces exhibit patterns much closer to a white noise spectrum (−0.74 and −0.61 log–log slopes). These traces are dominated by high-frequency variations associated with the thinly bedded grainstones typical of the Bellview Formation. Also note there is significantly less impact on the slope of the log–log relationship when a narrower scan of the Bellview Formation is measured.

traces, but also is clearly expressed in the power spectral analyses of the series (Fig. 10). In the case of the Bellview Formation, the slopes of the log power versus log harmonic number relationships are much shallower than the Kope Formation. Likewise, the difference in slope of the power–frequency relationship between the 1 m and 2 cm scans is much less in the case of the Bellview (−0.74 to −0.61) than in the Kope (−2.05 to −1.47) (Fig. 10).

How, if at all, can the spectral analysis of these grey-scale traces be used to evaluate the stratigraphic character of the two outcrops? In a wide variety of natural systems, the spectral power of a data series is found to scale inversely with frequency. The relationship $P(f) = 1/fb$, where $P(f)$ is the spectral power, f is the frequency, and b is slope of the power–frequency relationship (e.g. Fig. 10), is the mathematical description of these types of systems. Two well-known signal types, white noise $b = 0$ and random walks or Brownian noise $b = 2$, define commonly occurring natural end-member cases of the $1/f$ noise range. Lateral averaging of pixel grey-scale values tends to smooth the signal and thus increase the slope of the power–frequency relationship. Clearly, the two outcrops considered in this study display significantly different spectral characteristics. The Kope Formation is dominated by low-frequency shale–limestone alternations (closer to Brownian noise) while the Bellview Formation is dominated by high-frequency shale–limestone alternations (closer to white noise). Thus, when used in this way, the grey-scale technique does hold some promise as a tool for recognizing and quantifying bed geometries.

FUTURE DIRECTIONS

Recently, workers have begun to apply advanced imaging techniques to the analysis of sedimentary outcrops by combining traditional field observations conducted by standard section measuring techniques with positional data derived from GPS measurements and three-dimensional positional observations (Adams *et al.*, 2004; Bellian *et al.*, 2005; Redfern *et al.*, 2007). These approaches allow for the creation of highly precise digital outcrop models that can then be the subject of high-resolution quantitative analysis. While still being refined, these techniques provide an exciting approach to creating digital datasets suitable for quantitative stratigraphic analysis.

CONCLUSIONS

What conclusions can be drawn from this evaluation of quantitative techniques of outcrop analysis?

1 Choice of scale of sample spacing can affect the proxy sea-level curve generated from the discrete facies sampling technique. Likewise exclusion of highly diagnostic but thin or infrequently occurring facies is of significant concern when forming interpretations drawn from the proxy sea-level curve. Spatio-temporal variation in the recognition of thin facies can affect the reliability of time-series results drawn from incomplete records of relative accommodation space change.
2 Geometric orientation of outcrop bedding relative to the rectilinear structure of a grey-scale pixilated image can control the absolute stratigraphic thickness represented by each pixel. If the apparent strike of the beds changes across the photograph, then the angular relationship between up-section direction and the pixel geometry can result in artificial trending in the grey-scale data.
3 The angle of outcrop step-back can cause foreshortening of lower portions of the outcrop

and result in pronounced spatio-temporal compression of the upper portions of the outcrop, thus introducing another type of artificial trending in the dataset.

4 The analysis of grey-scale data can, in some cases, be used as tool for the interpretation of bedding geometry allowing for recognition of differences in bedding style within and among outcrops.

ACKNOWLEDGEMENTS

On this occasion of celebration of Professor Robert Ginsburg and his contributions to the fields of carbonate sedimentology and stratigraphy, it is only fitting that we acknowledge the vast impact his work has had on the research community's ongoing dialogue regarding the characteristics and origins of shallow-water carbonate deposition. Bob's consistent demand that we ask the hard questions, not be satisfied with easy solutions, and that we base our conclusions on sound observations and legitimate quantitative analysis has served as a model and guide for carbonate research for over four decades. Personally, and as part of the larger community of carbonate geologists, we offer a sincere 'thank you' to Bob. The manuscript was greatly improved by the comments, suggestions, and criticisms of Linda Hinnov and the editors of the volume.

REFERENCES

Adams, E.W., **Schroder, S.**, **Grotzinger, J.P.** and **McCormick, D.S.** (2004) Digital reconstruction and stratigraphic evolution of a microbial-dominated, isolated carbonate platform (Terminal Proterozoic, Nama Group, Namibia. *J. Sedimentary Res.*, **74**, 479–497.

Bellian, J.A., **Kerans, C.** and **Jennette, D.C.** (2005) Digital outcrop models: applications of terrestrial scanning LIDAR technology in stratigraphic modeling. *J. Sedimentary Res.*, **75**, 166–176.

Bond, G.C., **Kominz, M.A.** and **Bevan, J.** (1991) Evidence for orbital forcing of Middle Cambrian peritidal cycles: Wah Wah Range, south-central Utah. In: *Sedimentary Modeling: Computer Simulations and Methods for Improved Parameter Definition* (Eds E.K. Franseen, L. Watney, E. Franseen, C. Kendall, and W. Ross), pp. 293–318. *Kansas Geol. Survey Bull.*, **233**.

Cozzi, A., **Hinnov, L.A.** and **Hardie, L.A.** (2005) Orbitally forced Lofer cycles in the Dachstein Limestone of the Julian Alps (northeastern Italy). *Geology*, **33**, 789–792.

Davis, J.C. (1986) *Statistics and Data Analysis in Geology*, 2nd edition. John Wiley & Sons, New York, 646 pp.

Drummond, C.N. (2002) Superposition and soil formation – A stratigraphic paradox. *J. Geosci. Educ.*, **50**, 232.

Drummond, C.N. and **Sheets, H.** (2001) Taphonomic reworking and stratal organization of tempestite deposition: Ordovician Kope Formation, northern Kentucky, U.S.A. *J. Sedimentary Res.*, **71**, 621–627.

Fischer, A.G. (1964) The Loffer cyclothems of the Alpine Triassic. *Geol. Surv. Kansas Bull.* **169**, 107–149.

Fichter, L.S. (1987) Design and use of the stratigraphic strip log. *J. Geol. Educ.*, **35**, 71–80.

Goldhammer, R.K., **Dunn, P.A.** and **Hardie, L.A.** (1987) High frequency glacio-eustatic sea level oscillations with Milankovitch characteristics recorded in Middle Triassic platform carbonates in northern Italy. *Am. J. Sci.*, **287**, 853–892.

Goldhammer, R.K., **Dunn, P.A.** and **Hardie, L.A.** (1990) Depositional cycles, composite sea-level changes, cycle stacking patterns, and the hierarchy of stratigraphic forcing: examples from Alpine Triassic platform carbonates. *Geol. Soc. Am. Bull.*, **102**, 535–562.

Holland, S.M., **Miller, A.I.**, **Dattilo, B.F.**, **Meyer, D.L.** and **Diekmyere, S.L.** (1997) Cycle anatomy and variability in the storm-dominated type Cincinnatian (Upper Ordovician): coming to grips with cycle delineation and genesis. *J. Geol.*, **105**, 135–152.

Jennette, D.C. and **Pryor, W.A.** (1993) Cycle alternation of proximal and distal storm facies: Kope and Fairview formations (Upper Ordovician), Ohio and Kentucky. *J. Sedimentary Petrol.*, **63**, 183–203.

Olsen, P.E. (1986) A 40-million year lake record of early Mesozoic orbital climatic forcing. *Science*, **234**, 842–848.

Preto, N., **Hinnov, L.A.**, **Hardie, L.A.** and **De Zanche, V.** (2001) Middle Triassic orbital signature recorded in the shallow-marine Latemar carbonate buildup (Dolomites, Italy). *Geology*, **29**, 1123–1126.

Redfern, J., **Hodgetts, D.** and **Fabuel-Perez, I.** (2007) Digital analysis brings renaissance for petroleum geology outcrop studies in North Africa. *First Break*, **25**, 81–87.

Schwarzacher, W. (1975) *Sedimentation Models and Quantitative Stratigrpahy*. Elsevier, Amsterdam, 382 pp.

Schwarzacher, W. (1993) *Cyclostratigraphy and the Milankovitch Theory*. Elsevier, Amsterdam, 225 pp.

Schwarzacher, W. (2005) The stratification and cyclicity of the Dachstein Limestone in Lofer, Leogang and Steinernes Meer (Northern Calcareous Alps, Austria). *Sedimentary Geol.*, **181**, 93–106.

Tobin, R.C. and **Pryor, W.A.** (1981) Sedimentological interpretation of an Upper Ordovician carbonate-shale vertical sequence in northern Kentucky. In: *Geological Society of America* Cincinnati '81 *Field Trip Guidebooks*, Vol. 1: *Stratigraphy and Sedimentology* (Ed. T.G. Roberts), pp. 1–10. American Geological Institute, Falls Church, Virginia.

Wilkinson, B.H., **Drummond, C.N.**, **Diedrich, N.W.** and **Rothman, E.D.** (1997) Biological mediation of stochastic peritidal carbonate accumulation. *Geology*, **25**, 847–850.

Zühlke, R., **Bechstädt, T. and Mundil, R.** (2003) Sub-Milankovitch and Milankovitch forcing on a model Mesozoic carbonate platform – the Latemar (Middle Triassic, Italy). *Terra Nova* **15**, 69–80.

Index

Note: Page numbers in *italics* refer to figures, those in **bold** refer to tables